INSECT EVOLUTIONARY ECOLOGY

Proceedings of the Royal Entomological Society's 22nd Symposium

INSECT EVOLUTIONARY ECOLOGY

Proceedings of the Royal Entomological Society's 22nd Symposium

Edited by

M.D.E. Fellowes

School of Animal and Microbial Sciences
University of Reading
Reading, UK

G.J. Holloway

School of Animal and Microbial Sciences
University of Reading
Reading, UK

J. Rolff

Department of Animal and Plant Sciences
University of Sheffield
Sheffield, UK

CABI Publishing

CABI Publishing is a division of CAB International

CABI Publishing
CAB International
Wallingford
Oxon OX10 8DE
UK

Tel: +44 (0)1491 832111
Fax: +44 (0)1491 833508
E-mail: cabi@cabi.org
Website: www.cabi-publishing.org

CABI Publishing
875 Massachusetts Avenue
7th Floor
Cambridge, MA 02139
USA

Tel: +1 617 395 4056
Fax: +1 617 354 6875
E-mail: cabi-nao@cabi.org

A catalogue record for this book is available from the British Library, London, UK.

Library of Congress Cataloging-in-Publication Data
Royal Entomological Society of London. Symposium (22nd : 2003 : University of Reading, UK)
   Insect evolutionary ecology : proceedings of the Royal Entomological Society's 22nd Symposium / edited by M. Fellowes, G. Holloway, J. Rolff.
      p. cm.
   "Proceedings of the Royal Entomological Society 22nd Symposium held at the University of Reading, UK."
   Includes bibliographical references and index.
   ISBN 0-85199-812-7 (alk. paper)
   1. Insects--Evolution--Congresses. 2. Insects--Ecology--Congresses.
I. Fellowes, M.D.E. (Mark) II. Holloway, G.J. (Graham) III. Rolff, J. (Jens) IV. Title.

ISBN 0 85199 812 7

Typeset in Souvenir by Columns Design Ltd, Reading.
Printed and bound in the UK by Biddles Ltd, King's Lynn.

Contents

Contributors

J.L. Bella, *Departamento de Biología, Facultad de Ciencias, Universidad Autónoma de Madrid, E-28049, Madrid, Spain*

J.J. Boomsma, *Institute of Biology, Department of Population Biology, University of Copenhagen, Universitetsparken 15, DK-2100 Copenhagen, Denmark*

A.F.G. Bourke, *Institute of Zoology, Zoological Society of London, Regent's Park, London NW1 4RY, UK*

R.K. Butlin, *Department of Animal and Plant Sciences, University of Sheffield, Western Bank, Sheffield S10 2TN, UK*

R. Calsbeek, *Center for Tropical Research, 1609 Hershey Hall, University of California, Los Angeles, CA 90035, USA*

J.M. Cook, *Department of Biological Sciences, Imperial College London, Silwood Park Campus, Ascot, Berks SL5 7PY, UK*

S. Cotton, *The Galton Laboratory, Department of Biology, University College London, 4 Stephenson Way, London NW1 2HE, UK*

O. Duron, *Institut des Sciences de l'Évolution (UMR 5554), Laboratoire Génétique et Environnement, Cc 065, Université Montpellier II, 34095 Montpellier cedex 05, France*

C. Dytham, *Department of Biology, University of York, York YO10 5YW, UK*

G.W. Elmes, *Centre for Ecology and Hydrology (NERC), CEH Dorset, Winfrith Technology Centre, Winfrith Newburgh, Dorchester, Dorset DT2 8ZD, UK*

I. Emelianov, *Plant and Invertebrate Ecology Division, Rothamsted Research, Harpenden, Herts, AL5 2JQ, UK*

P. Fort, *Centre de Recherche en Biochimie des Macromolécules (FRE2593), CNRS, 1919 route de Mende, 34293 Montpellier cedex 05, France*

R.S. Fritz, *Department of Biology, Vassar College, Poughkeepsie, NY 12604, USA*

F. Gilbert, *School of Biology, Nottingham University, Nottingham NG7 2RD, UK*

K. Gotthard, *Department of Zoology, Stockholm University, SE-10691 Stockholm, Sweden*

J.K. Hill, *Department of Biology, University of York, York YO10 5YW, UK*

C.G. Hochwender, *Department of Biology, University of Evansville, Evansville, IN 47722, USA*

C.L. Hughes, *Department of Biology, University of York, York YO10 5YW, UK*

W.O.H. Hughes, *Institute of Biology, Department of Population Biology, University of Copenhagen, Universitetsparken 15, DK-2100 Copenhagen, Denmark. Present address: School of Biological Science, A12, University of Sydney, Sydney, NSW 2006, Australia*

G.D.D. Hurst, *Department of Biology, University College London, 4 Stephenson Way, London NW1 2HE, UK*

C.D. Jiggins, *Institute of Evolutionary Biology, School of Biological Sciences, University of Edinburgh, West Mains Road, Edinburgh EH9 3JT, UK*

F. Johansson, *Department of Ecology and Environmental Science, Umeå University, 90187 Umeå, Sweden*

T.M. King, *Department of Zoology, University of Otago, PO Box 56, Dunedin, New Zealand*

R. Knell, *School of Biological Sciences, Queen Mary, University of London, Mile End Road, London E1 4NS, UK*

P. Labbe, *Institut des Sciences de l'Évolution (UMR 5554), Laboratoire Génétique et Environnement, Cc 065, Université Montpellier II, 34095 Montpellier cedex 05, France*

M.E.N. Majerus, *Department of Genetics, Downing Street, Cambridge CB2 3EH, UK*

J. Mallet, *The Galton Laboratory, Department of Biology, University College London, 4 Stephenson Way, London NW1 2HE, UK*

G.H. Nygren, *Department of Zoology, Stockholm University, SE-10691 Stockholm, Sweden*

S. Nylin, *Department of Zoology, Stockholm University, SE-10691 Stockholm, Sweden*

N. Pasteur, *Institut des Sciences de l'Évolution (UMR 5554), Laboratoire Génétique et Environnement, Cc 065, Université Montpellier II, 34095 Montpellier cedex 05, France*

A. Pomiankowski, *The Galton Laboratory, Department of Biology, University College London, 4 Stephenson Way, London NW1 2HE, UK*

M. Raymond, *Institut des Sciences de l'Évolution (UMR 5554), Laboratoire Génétique et Environnement, Cc 065, Université Montpellier II, 34095 Montpellier cedex 05, France*

P. Schmid-Hempel, *ETH Zurich, Ecology and Evolution, ETH-Zentrum, NW, CH-8092 Zurich, Switzerland*

K. Schönrogge, *Centre for Ecology and Hydrology (NERC), CEH Dorset, Winfrith Technology Centre, Winfrith Newburgh, Dorchester, Dorset DT2 8ZD, UK*

D.M. Shuker, *School of Biological Sciences, University of Edinburgh, Ashworth Laboratories, King's Buildings, Edinburgh EH9 3JT, UK*

R. Stoks, *Laboratory of Aquatic Ecology, University of Leuven, Ch. De Bériostraat 32, 3000 Leuven, Belgium*

J.A. Thomas, *Centre for Ecology and Hydrology (NERC), CEH Dorset, Winfrith Technology Centre, Winfrith Newburgh, Dorchester, Dorset DT2 8ZD, UK*

J.N. Thompson, *Department of Ecology and Evolutionary Biology, University of California, Santa Cruz, CA 95064, USA*

K.M. Webberley, *Department of Biology, University College London, 4 Stephenson Way, London NW1 2HE, UK and School of Biological Sciences, Queen Mary, University of London, Mile End Road, London E1 4NS, UK*

N. Wedell, *School of Biosciences, University of Exeter in Cornwall, Penryn, UK*

M. Weill, *Institut des Sciences de l'Évolution (UMR 5554), Laboratoire Génétique et Environnement, Cc 065, Université Montpellier II, 34095 Montpellier cedex 05, France*

K. Wilson, *Department of Biological Sciences, Lancaster University, Lancaster LA1 4YQ, UK*

Introduction

The 21st Symposium of the Royal Entomological Society

Mark D.E. Fellowes, Graham J. Holloway and Jens Rolff

Evolutionary ecology is a hybrid science. At its heart lies a desire to understand the ecology of organisms from an evolutionary perspective, so evolutionary ecologists ask questions about how organisms adapt to their environment, and the consequences of adaptive change for interactions among and within species. Such an approach has brought together a wide range of workers, from pest managers interested in the evolution of pesticide resistance, to behavioural ecologists interested in sexually selected behaviours, and all shades of ecologist in between. The evolutionary perspective is powerful, simply because it provides a coherent framework for hypothesis testing. Evolutionary ecology has been reinvigorated by the integration of molecular genetic approaches with traditional mathematical and empirical studies. This revitalization of evolutionary ecology has resulted in the development of rapidly growing sub-disciplines such as ecological immunology, and the rejuvenation of important areas such as the study of phenotypic plasticity. This flourishing has resulted in an increase in publications as evidenced by the *ISI Web of Science* citations for 'evolutionary ecology' (1981–1985: 14; 1986–1990: 38; 1991–1995: 114; 1996–2000: 191; 2001–mid-2004: 219). These numbers, of course, underestimate the true numbers of publications in the field, but they do illustrate how the science is growing rapidly.

In this volume, we present a series of empirical case studies in evolutionary ecology using insects as model systems. These case studies vary in scale from questions of intraspecific sexual selection through simple interspecific interactions, building up to consider the ecological consequences of the adaptive genetic variation. In many ways, the work contained in this book reflects some of the tremendous advantages provided by insects as model systems. Insects have provided some of the most important insights into evolutionary ecology. In part, this is simply because of their tractability. Insects typically reproduce rapidly, are relatively easy to work with in the laboratory, and are amenable to manipulative field studies. As a result, there is perhaps no other taxon which has provided so many insights into why the biological world around us is the way it is.

We have divided the chapters into three groups, reflecting the primary scales that the studies focus on. We start with intraspecific interactions, covering sexual selection in stalk-eyed flies and sperm competition in butterflies, before moving on to consider the evolution of phenotypic plasticity and how genes and behaviour are linked in social insects. These first chapters illustrate the great within-species variation that exists, and how this variation can have implications for our understanding of insect population ecology.

The second section considers two-species interactions. We start with two chapters on mutualistic interactions between figs and fig wasps, and ants and fungi. Mutualisms rarely receive the attention from ecologists that they deserve, yet they can play a key role in determining the distribution of keystone species in natural communities. This is followed by a series of chapters on antagonistic interactions between hosts and parasites, plants and herbivores, and predators and prey. The evolutionary ecology of such systems is fascinating, illustrating how adaptive change can alter patterns of interaction in the short term, through the evolution of resistance, or in the long term through the evolution of mimicry patterns. This section also contains what is likely to be the most controversial chapter, a polemical defence of the peppered moth as a symbol of Darwinian evolution. At the symposium, this talk raised the loudest round of applause for any speaker, and also the most heated debates as to the potential pitfalls of debating evolutionary science with those from a Christian-faith-based perspective. Whether one approves of this rallying cry in defence of evolution or not, it is important that we realize that these debates matter outside of our academic cocoons. No matter what our perspective on the debate, it is one that we can no longer assume belongs only to those on the west of the Atlantic; in the UK the recent tacit support provided by Whitehall for schools which teach creationism means that this is an issue that has to be considered by all biologists, not just those with an evolutionary point of view.

The third section focuses on larger-scale studies, either geographically or in terms of the numbers of species interacting in the study. We start with studies of local adaptation, and how variation in local adaptation can lead to a geographic mosaic in patterns of interactions among species. This is followed by chapters on speciation and the consequences of species boundaries at hybrid zones. Finally, we end with a chapter that considers how species respond to anthropogenic change, through climate change. If we needed any riposte to the challenge squared up to by Mike Majerus, then these chapters thoroughly illustrate evolution at work.

We would like to finish by taking this opportunity to thank all of those at the Royal Entomological Society who helped make this meeting successful, especially Bill Blakemore (the Society's registrar), Amanda Callaghan (local organizer for the national meeting), the referees and, most importantly, the speakers and co-authors, whose work keeps pushing back the boundaries of knowledge.

1

Genetics, Relatedness and Social Behaviour in Insect Societies

ANDREW F.G. BOURKE

Institute of Zoology, Zoological Society of London, London, UK

1. Introduction

Kin selection theory occupies a dominant position in the study of social evolution. Although its meaning and potential uses were slow to be appreciated following its publication (as inclusive fitness theory) by W.D. Hamilton in the early 1960s (Hamilton, 1963, 1964), since the early to mid-1970s, following the publication of work by Wilson (1971), Hamilton (1972), Alexander (1974), West-Eberhard (1975) and especially Trivers (1974) and Trivers and Hare (1976), it has formed the basic theoretical tool with which investigators have attempted to unpick the complexities of social behaviour. The essence of kin selection is that the evolution of social interactions between conspecific individuals, whether cooperative or competitive, should be influenced by their genetic relatedness. This is because, from the viewpoint of genes for social behaviour, relatedness measures the value of other individuals as a route to future generations. This insight is formalized as Hamilton's rule, which specifies the conditions for the spread of a gene for social behaviour as a function of the relatedness of the interactants and the numbers of offspring gained or lost by them (benefits and costs) as a consequence of the behaviour (Hamilton, 1963, 1964). It is worth re-emphasizing the very fundamental nature of kin selection theory. Two conditions are needed. First, there must be genetic variation for social behaviour. Second, interacting conspecifics must express social behaviour non-randomly with respect to relatedness. If these two conditions are met, then kin selection is the relevant theory to apply to achieve an understanding of how natural selection will affect the genes in question and hence how the evolution of the relevant form of social behaviour will proceed. In short, kin selection is the genetic theory of natural selection (Fisher, 1930) logically extended to cover social phenomena (e.g. Grafen, 1985; Frank, 1998). This means that demonstrating the efficacy of kin selection in one context helps strengthen the case for the validity of the kin selection approach in all potential contexts, because the basis of the underlying theory is the same. This is the point missed in a recent critique of the application of kin selection theory to social insects (Alonso and Schuck-Paim, 2002). The fundamental nature of kin selection theory also implies that it is potentially universally applicable. This is why the theory has, rightly, been applied far beyond its traditional province to help explain such phenomena as the evolution of multicellularity (Maynard Smith and Szathmáry, 1995; Michod and Roze, 2001) and the evolution of social behaviour in microorganisms (Crespi, 2001). In sum, kin selection is the basis of any general theory of the evolution of conflict and cooperation in all organisms at all levels in the biological hierarchy (Maynard Smith and Szathmáry, 1995; Keller, 1999; Queller, 2000; Michod and Roze, 2001).

Studies of social insects, and in particular of the social Hymenoptera, have been at the forefront of the application and testing of kin selection theory. Overall, such studies have been notably successful in verifying the predictions of the theory and so helping us understand the basis of social evolution in insects (Bourke and Franks, 1995; Crozier and Pamilo, 1996; Choe and Crespi, 1997) and, by extension, in other organisms. They have also been successful in the sense that they have led to the discovery of new and important social phenomena (e.g. worker policing of other workers' eggs in honeybees: Ratnieks and Visscher, 1989; selective male-killing by worker ants: Sundström *et al.*, 1996). However, given that 40 years have passed since kin selection theory was first proposed, and that in this period scores of studies on social insects of its predictions and explanatory power have been published, a student new to the field, and even researchers who are growing old in it, might legitimately ask themselves, 'What remains to be done?'. In fact, the field remains dynamic and active, and this is for several reasons. One is that rapid progress in molecular biology holds out the prospect of achieving an integrated understanding of social evolution at the molecular-genetic and behavioural levels. Another is that new contexts in which kin selection theory can be applied in social insects, and hence new *a priori* predictions derived from the theory, continue to be identified. Finally, there are sufficient cases where the theory's predictions are not fulfilled to create a need to conduct multiple tests across several taxa and for the theory's predictions then to be assessed using a rigorous, comparative approach. For these reasons, this seems an appropriate moment both to look backwards and consider the status of the field to date, and to look forwards and speculate as to how it might develop. The present essay is an attempt at these tasks. I first review the evidence for kin selection in the social Hymenoptera (strictly, the eusocial Hymenoptera, i.e. those having a reproductive division of labour between queens and workers) and I then consider the directions that the field might take in future.

2. Evidence for Kin Selection in Social Insects

The chief predictions of kin selection theory concern effects of relatedness on social behaviour. Of course the theory also makes predictions about individual- and group-level benefits and costs, but because these quantities have generally been hard to measure in practice, most tests of the theory have focused on expected effects of relatedness. Many other factors are likely to influence social evolution, especially at the proximate level. However, unlike relatedness, none forms the basis of so general a theory. The following review is not comprehensive and does not employ rigorous comparative methods controlling for effects of phylogeny in among-species comparisons. Instead, I concentrate on recent studies of sex allocation, male parentage and kin-selected caste conflict in the social Hymenoptera that have added to or improved our knowledge regarding the influence of relatedness on social behaviour. In the process, I highlight current gaps in our knowledge and anomalies and inconsistencies in the data. I do not consider the effects of relatedness on the origin of eusociality. The reason is that, although this issue has long been discussed, there are still relatively few data on facultatively social Hymenoptera that might prove informative in this context (Bourke, 1997b). Neither do I further discuss the evolution of reproductive skew (partitioning of reproduction among multiple-breeder groups), which is an important and currently extremely active area in the field (Keller and Reeve, 1994; Johnstone, 2000). Again, this is because empirical tests in social insects of the predictions of skew models with respect to relatedness are still relatively few (e.g. Field *et al.*, 1998; Reeve *et al.*, 2000; Rüppell *et al.*, 2002; Seppä *et al.*, 2002; Sumner *et al.*, 2002; Hannonen and Sundström, 2003a). In addition, tests of skew models in social insects have been comprehensively reviewed by Reeve and Keller (2001). Recent reviews of the topics and themes considered in the present chapter include those of Chapuisat and Keller (1999), Keller and Chapuisat (1999), Keller and Reeve (1999), Ratnieks *et al.* (2001), Sundström and Boomsma (2001) and Mehdiabadi *et al.* (2003).

2.1 Sex allocation

Kin selection theory predicts patterns of sex allocation (relative investment in the sexes) at both the population and colony levels in the social Hymenoptera. At the population level, assuming random mating, the sex investment ratio (expressed as the female:male investment ratio) is predicted to equal the relatedness asymmetry of the party controlling sex allocation (with relatedness asymmetry defined as the relatedness to females divided by relatedness to males) (Trivers and Hare, 1976; Boomsma and Grafen, 1990, 1991). In this context, 'females' refers to new queens produced by the colony and 'males' refers to new males produced by the colony; workers, all being female, are excluded because they are (generally) sterile. This statement of kin selection theory as regards sex allocation is essentially Fisher's (1930) sex ratio theory reformulated to incorporate the concept of a party controlling sex allocation that need not be the parent. An implication is that there may be kin-selected conflict over sex allocation, because the relatedness asymmetries of different potential controlling parties (e.g. queens and workers) need not be equal (Trivers and Hare, 1976). For example, in a random-mating population of social Hymenoptera with one singly-mated queen per colony and control of sex allocation by sterile workers, the population sex investment ratio is predicted to equal the well-known value of 3:1 females:males, since the workers' relatedness asymmetry is 0.75/0.25. But if the queen controls sex allocation, the population sex investment ratio should equal 1:1, which is the queen's relatedness asymmetry (0.5/0.5). The workers' relatedness values stem from the haplodiploid genetic system found in the Hymenoptera, in which females derive from fertilized eggs and are diploid and males derive from unfertilized eggs and are haploid, with the result that full sisters are related by 0.75 due to their sharing the same paternal genes, and sisters are related to brothers by 0.25 because males lack paternal genes (Hamilton, 1964).

At the colony level, kin selection theory predicts sex ratio variation if workers control sex allocation and if their relatedness asymmetry varies across colonies within populations; for example, if some colonies are headed by singly-mated queens and others by multiply-mated queens (Boomsma and Grafen, 1990, 1991). Specifically, the theory then predicts that colonies with relatively high relatedness asymmetries should concentrate on female production, because their workers are comparatively more closely related to females, and colonies with relatively low relatedness asymmetries should concentrate on male production, because their workers are comparatively more closely related to males (i.e. Boomsma and Grafen's (1990, 1991) split sex ratio theory). If queens control sex allocation, the theory makes no specific predictions because the factors that cause workers' relatedness asymmetry to vary across colonies generally have no effect on queens' relatedness asymmetry (Boomsma and Grafen, 1990, 1991).

2.1.1 Population-level sex ratio variation

Analyses of sex ratio evolution in relation to kin selection have tended to concentrate on social Hymenoptera that reproduce entirely or at least partly by the release of winged sexuals (e.g. many monogynous species, i.e. those having one queen per colony), in contrast to species that reproduce entirely or mainly by colony division (e.g. honeybees, *Apis* spp.). This is because in the latter set of species competition between related queens to head daughter colonies (local resource competition) is predicted to increase male bias in the population sex investment ratio, but to a degree that is hard to quantify (Pamilo, 1991; Bourke and Franks, 1995). It is worth noting, however, that the general Fisherian and Trivers–Hare approach has been notably successful in predicting patterns of population-level sex ratio variation in social Hymenoptera across a wide variety of social and mating systems (Bourke and Franks, 1995; Crozier and Pamilo, 1996).

In monogynous ants in which relatedness asymmetry has been measured or strongly inferred using genetic markers, observed sex ratios are significantly female-biased (mean fraction of investment in females = 0.63; Table 1.1). However, although the quantitative fit to the predicted values of the sex investment ratio is frequently remarkably close, on average observed sex investment ratios are significantly lower than sex investment ratios predicted on the basis of population-wide relatedness asymmetries (0.63 versus 0.72; paired t-test, $t = 3.91$, d.f. $= 15$, $P < 0.01$; Table 1.1). The reasons for this pattern, which was also detected by Boomsma (1989) and Pamilo (1990), although these studies did not compare observed sex investment ratios with expected values inferred from genetic data, are not fully resolved. They could occur: (i) because in populations in a 'poor' habitat, workers cannot fully express their kin-selected interests due to a proximate influence of a lack of resources on female investment (e.g. Nonacs, 1986); or (ii) because workers are not fully in control of sex allocation, such that population sex investment ratios represent a compromise between their optimal values and those of queens (e.g. Passera *et al.*, 2001; Reuter and Keller, 2001; Mehdiabadi *et al.*, 2003). However, no theory apart from kin selection theory explains the significant female bias in the population sex investment ratios of monogynous ants. In particular, random mating in all the species in Table 1.1 rules out local mate competition (competition between related males for mates) as a general source of female bias (Alexander and Sherman, 1977).

Sex investment ratios in slave-making ants have also been used to support kin selection predictions. In these monogynous social parasites, all the brood is reared by 'slave' workers of other species. Trivers and Hare (1976) therefore proposed that the slave-maker queen should control sex allocation because any mechanisms adopted by her to achieve her optimal sex ratio would be met with indifference by the unrelated slaves and, contrary to the case in non-parasitic ants, could not be practically opposed by her workers. The resulting prediction of unbiased (1:1) population sex investment ratios is met in two well-studied species (*Epimyrma ravouxi* and *Harpagoxenus sublaevis*: Bourke and Franks, 1995). However, in three other species (*H. canadensis*, *Protomognathus americanus* and *Leptothorax duloticus*), Herbers and Stuart (1998) found six of 11 population sex investment ratios to be significantly different from 1:1, although the mean value across all 11 populations did not differ significantly from 1:1 (mean [95% confidence limits] = 0.51 [0.39–0.63] as the fraction of investment in females). In *P. americanus*, Herbers and Stuart (1998) also found a positive association across sites and years between slave-maker sex ratios and those of their free-living host. This suggested that, contrary to Trivers and Hare (1976), slave workers influence slave-maker sex allocation. Herbers and Stuart (1998) concluded that there is ongoing conflict over sex allocation between the slave-maker queen, the slave-maker workers, and the slave workers. In a later study, Foitzik and Herbers (2001) found that *P. americanus* workers are unusually reproductive, producing an estimated two-thirds of all males. This means that the optimal sex ratios of the slave-maker queen and her workers tend to converge, leading to a reduction in the expected degree of sex-ratio conflict between them (Foitzik and Herbers, 2001). Overall, these phenomena suggest that, at least in some species of slave-making ants, the expression of kin-selected conflict over sex allocation is likely to be complicated and hence that population sex investment ratios in these species do not provide the tidy test of kin selection theory that Trivers and Hare (1976) proposed.

In bumblebees, male-biased or unbiased (1:1) population sex investment ratios formerly appeared to contradict kin selection theory. However, unbiased population sex investment ratios in this group are likely to be a consequence of lack of worker control of sex allocation (Bourke, 1997a; Brown *et al.*, 2003;

Table 1.1. Tests of Trivers and Hare's (1976) kin selection theory for the population sex investment ratio in monogynous ants. Adapted from Bourke (1997b), with additional cases. Population sex investment ratio is expressed as proportion of investment in females. Expected sex investment ratios (assuming worker control) were calculated as measured or inferred workers' relatedness asymmetry (expressed as fractions). Lack of inbreeding inferred from Hardy–Weinberg equilibrium of genetic markers, or zero queen–mate relatedness, or both. In a study of *L. tuberum* (Pearson *et al.*, 1995, 1997), observed and expected population sex investment ratios differed significantly in four of seven site-years, but the actual values were not tabulated.

Species	Population sex investment ratio		Inbreeding present?	Number of colonies	Reference
	Expected	Observed			
Colobopsis nipponicus	0.75	0.75	No	–	Hasegawa (1994)
Formica truncorum	0.63	0.65	No	22	Sundström (1994)
Lasius niger					Van der Have *et al.* (1988)
Population 1	0.72	0.68	No	125	
Population 2	0.68	0.65	No	26	
Population 3	0.59	0.36	No	50	
Lasius niger					Fjerdingstad *et al.* (2002)
Population 1, Year 1	0.73	0.56	No	28	
Population 1, Year 2	0.72	0.54	No	34	
Population 2	0.73	0.66	No	52	
Leptothorax nylanderi					Foitzik *et al.* (1997), Foitzik and Heinze (2000)
Year 1	0.75	0.62	No	174	
Year 2	0.75	0.79	No data	179	
Year 3	0.75	0.59	No data	334	
Myrmica punctiventris					Banschbach and Herbers (1996)
Year 1	0.75	0.76	No	84	
Year 2	0.75	0.74	No	40	
Pheidole desertorum	0.74	0.50	No	348	Helms (1999)
Pheidole pallidula	0.74	0.64	No	22	Fournier *et al.* (2002, 2003)
Solenopsis invicta	0.75	0.61	No	50	Vargo (1996)
Mean (95% confidence limits)	0.72 (0.70–0.75)	0.63 (0.57–0.69)			

Duchateau *et al.*, 2004), in which case the theory predicts 1:1 sex investment ratios (Trivers and Hare, 1976). Queens appear to have power over sex allocation in bumblebees because essential reproductive decisions in these annual social insects are made before many of the workers have eclosed (Müller *et al.*, 1992). Male-biased sex investment ratios in bumblebees are apparently a consequence of a protandrous mating system (one in which adult males are produced earlier than females) (Bulmer, 1983; Bourke, 1997a; Beekman and Van Stratum, 1998), which represents a violation of the assumptions of the original kin-selection model.

In wasps, population sex investment ratios have rarely been estimated, either because it is hard to divide females unambiguously into young queens (gynes) and workers, or because measuring sexual output over the season in a sufficient sample of colonies is difficult, or both. It seems likely that, in many wasps, complexities in the mating system could also complicate the interpretation of population sex investment ratios (Strassmann and Hughes, 1986; Tsuchida *et al.*, 2003).

2.1.2 Colony-level sex ratio variation

Split sex ratio theory represents a powerful test of kin selection because it predicts within-population patterns of sex ratio variation. Such a test controls for differences other than those involving kin structure, which potentially confound comparisons between population sex ratios of different populations or species (Chapuisat and Keller, 1999). The theory predicts that, if workers control sex allocation and their relatedness asymmetry varies between colonies, sex ratios should be split and specifically should be relatively female-biased in colonies with high relatedness asymmetry and relatively male-biased in colonies with low relatedness asymmetry. Another, weak prediction of the theory is that, if workers' relatedness asymmetry does not vary, sex ratios should not be split. This is a weak prediction because Boomsma and Grafen's (1990, 1991) theory does not rule out other causes of split sex ratios in social Hymenoptera, and indeed such other causes undoubtedly exist (see below). In a set of cases in which the main prediction can be tested, the theory accounts for split sex ratios in 19 of 25 cases (76%: Table 1.2). This comparison is conservative because independent evidence suggests that in some of the negative cases sex allocation is under the control of queens not workers, so violating an assumption of the theory (e.g. population sex investment ratios at the queen optima in *B. hypnorum* (Paxton *et al.*, 2001) and in the *Lasius niger* populations studied by Fjerdingstad *et al.* (2002)). It also needs noting that, where workers' relatedness asymmetry is generated by variations in queen number, the predictions of Boomsma and Grafen's (1990, 1991) theory coincide with those stemming from the idea that monogynous colonies should invest relatively more in females to compensate for male-biased sex allocation by polygynous (multiple-queen) colonies due to local resource competition (Boomsma, 1993; Nonacs, 1993; Bourke and Franks, 1995). This argument does not of course apply where relatedness asymmetry varies due to variation in queen mating frequency alone (Sundström, 1994; Sundström *et al.*, 1996), or where, for example, relatedness asymmetry varies independently of queen number in polygynous populations (Evans, 1995; Heinze et al., 2001). Experimental manipulations of relatedness asymmetry that generate the predicted sex ratio shift also provide particularly strong support for the theory (Mueller, 1991). Support for split sex ratio theory suggests that workers are able to assess the type of colony to which they belong (with relatively high or low relatedness asymmetry). Evidence exists that workers can achieve this by assessing heritable variation in the cuticular hydrocarbon profiles of nestmate workers (Boomsma et al., 2003). Mechanisms that workers then use to bias the sex ratio include the selective killing of males (Sundström et al., 1996) or the biasing, during larval development, of the final caste (development as queen or worker) of females (Hammond et al., 2002).

From the several cases where there are split sex ratios but no variation in workers' relatedness asymmetry (Table 1.2), it is clear that split sex ratios can occur for reasons unconnected with colony kin structure. In some of these cases, there is evidence that queens control sex allocation and use strategies to do so that themselves generate split sex ratios, e.g. *Pheidole desertorum* (Helms, 1999), *Solenopsis invicta* (Passera et al., 2001) and *Bombus terrestris* (Bourke and Ratnieks, 2001) (Table 1.2). For example, in the ant *P. desertorum*, queens appear unusual in that they can apparently lay worker-biased female eggs, most probably via hormonal effects (Helms, 1999). Half of colonies were found to be male specialists, half were found to be female specialists, the population sex investment ratio was at the queen optimum of 1:1, and colonies did not differ in either productivity or the workers' relatedness asymmetry. Helms (1999) proposed that queens achieve control of sex allocation by, in half the colonies, laying only worker-biased eggs and haploid eggs, so forcing the sterile workers to raise males. This means that, in the other half of the colonies, the optimum sex ratio for both workers and the queen is all-females (Pamilo, 1982). The outcome is a queen-preferred population sex investment ratio of 1:1 and a split sex ratio in the absence of any variation in workers' (or queens') relatedness asymmetry. In short, the ESS (evolutionarily stable strategy) for a queen heading a colony is effectively to toss a coin and decide to lay, with a chance of 0.5, either worker-biased female eggs and haploid eggs or non-worker-biased female eggs. Other factors altogether that may contribute to sex ratio splitting include variation in colony productivity coupled with a degree of local mate competition within populations (constant male hypothesis: Frank, 1987; Hasegawa and Yamaguchi, 1995), local resource enhancement (related females cooperate to enhance group productivity: e.g. Cronin and Schwarz, 1997) and, in some polygynous species, the need to replace queens to maintain levels of polygyny (queen replenishment hypothesis: Brown and Keller, 2000, 2002; Brown et al., 2002).

2.2 Male parentage

Workers of many species of social Hymenoptera are capable of laying unfertilized haploid eggs that develop into males (Bourke, 1988b; Choe, 1988). Successful worker reproduction may nonetheless fail to occur, either because

Table 1.2. Summary of tests of Boomsma and Grafen's (1990, 1991) split sex ratio theory (predicting split sex ratios as a function of workers' relatedness asymmetry, RA). Adapted from Queller and Strassmann's (1998) review, with additional cases.

	Sex ratio patterns in species/population fit theory	Sex ratio patterns in species/population do not fit theory
Workers' RA varies	*i.e. sex ratio is split in direction predicted (high-RA colonies produce mainly females, low-RA colonies produce mainly males)* Formica exsecta (ant, monogynous population) (Sundström et al., 1996; Sundström and Ratnieks, 1998) F. podzolica (ant) (Deslippe and Savolainen, 1995) F. truncorum (ant) (Sundström, 1994) Leptothorax acervorum (ant, Reichswald population) (Heinze et al., 2001) L. acervorum (ant, Santon population) (Chan et al., 1999; Hammond et al., 2002) L. longispinosus (ant) (Herbers, 1984, 1990) Myrmica ruginodis (ant) (Walin and Seppä, 2001) M. sulcinodis (ant) (Elmes, 1987) M. tahoensis (ant) (Evans, 1995) Rhytidoponera chalybaea (ant) (Ward, 1983) Rhytidoponera confusa (ant) (Ward, 1983) Augochlorella striata (bee) (Mueller, 1991; Mueller et al., 1994) Halictus rubicundus (bee) (Yanega, 1988; Boomsma, 1991) Lasioglossum laevissimum (bee) (Packer and Owen, 1994) Brachygastra mellifica (wasp) (Hastings et al., 1998) Parachartergus colobopterus (wasp) (Queller et al., 1993b) Polybia occidentalis (wasp) (Queller et al., 1993b) P. emaciata (wasp) (Queller et al., 1993b) Protopolybia exigua (wasp) (Queller et al., 1993b)	*i.e. sex ratio is split in direction opposite to that predicted, or sex ratio variation is uncorrelated with workers' RA, or sex ratio is not split* Formica exsecta (ant, polygynous population) (Brown and Keller, 2000) F. sanguinea (ant) (Pamilo and Seppä, 1994) Lasius niger (ant) (Fjerdingstad et al., 2002) Proformica longiseta (ant) (Fernández-Escudero et al., 2002) Bombus hypnorum (bee) (Paxton et al., 2001) Xylocopa sulcatipes (bee) (Stark, 1992)
Total	19	6
Workers' RA does not vary	*i.e. a split sex ratio is absent* Colobopsis nipponicus (ant) (Hasegawa, 1994) L. acervorum (ant, Roydon population) (Chan et al., 1999)	*i.e. a split sex ratio is present* Leptothorax nylanderi (ant) (Foitzik and Heinze, 2000; Foitzik et al., 2003) Pheidole desertorum (ant) (Helms, 1999) P. pallidula (ant) (Aron et al., 1999; Fournier et al., 2003) Solenopsis invicta (ant, monogynous population) (Passera et al., 2001) B. terrestris (bee) (Duchateau and Velthuis, 1988; Duchateau et al., 2004)
Total	2	5

workers refrain from laying male eggs (self-restraint) or because worker-laid male eggs are destroyed by other workers (worker policing) or by queens (queen policing) (Cole, 1986; Ratnieks, 1988). Under conditions of monogyny and single queen mating, kin selection theory predicts that, other things being equal, workers gain greater inclusive fitness from rearing sons (relatedness, $r = 0.5$) or the sons of other workers (nephews, $r = 0.375$) rather than the queen's male offspring (brothers, $r = 0.25$), whereas the queen gains greater inclusive fitness from the rearing of sons ($r = 0.5$) rather than workers' male offspring (grandsons, $r = 0.25$) (Hamilton, 1964; Trivers and Hare, 1976). Hence, as with sex allocation, there is a potential kin-selected conflict between queens and workers over male parentage.

Changes in colony kin structure due to either multiple mating by queens (polyandry) or polygyny can alter the expected pattern of conflict because they alter relative relatedness values. For example, under monogyny with an effective queen mating frequency greater than two, a focal reproductive worker remains more closely related to its male eggs ($r = 0.5$) than to those of the queen ($r = 0.25$), but the average worker becomes more closely related to queen-produced males ($r = 0.25$) than to the average worker-produced male ($0.125 < r < 0.25$). Under these conditions, reproductive workers are still predicted to lay male eggs (self-restraint is not favoured), but other workers are predicted to stop these eggs being reared (worker policing is favoured), for example by eating the eggs (Starr, 1984; Woyciechowski and Lomnicki, 1987; Ratnieks, 1988).

Tests of these predictions rely on determining both the exact kin structure of colonies and populations and on measuring male parentage accurately. Since both tasks can be achieved with the necessary accuracy using microsatellite genetic markers (Queller et al., 1993a), there has recently been a great increase in the number of studies analysing this issue. To test specifically for worker policing, there is also a need either for measurements of male parentage at both the egg stage and the adult stage, or for detailed behavioural observations or experiments to determine the fate of worker-produced male eggs.

2.2.1 Interspecific tests

In ants, the prediction that species with monogynous colonies and singly-mated queens should be characterized by high levels of adult male production by workers even in colonies with a queen does not appear to be fulfilled; many ants matching or approximating this kin structure have non-reproductive workers, or workers that reproduce only after the queen has died (e.g. Villesen and Boomsma, 2003; reviewed by: Bourke, 1988b; Choe, 1988; Bourke and Franks, 1995). A possible exception, in which high levels of worker reproduction in colonies with a queen might occur, is found in the slave-making ants (Heinze, 1996; Heinze et al., 1997; Foitzik and Herbers, 2001). Increased reproduction by workers in slave-makers has been predicted on the assumptions that these workers are unable (or, it now appears, only partly able) to bias sex allocation in their favour (Bourke, 1988a) and/or that female slave-maker larvae are more able to determine whether they develop as queens or workers (Nonacs and Tobin, 1992).

In bees, broad comparisons of honeybees (*Apis* spp.), on the one hand, which are characterized by monogyny, high effective queen mating frequencies and low levels of adult male production by workers (e.g. Visscher, 1989; Barron *et al.*, 2001), and stingless bees (Meliponinae) and bumblebees (Bombini), on the other hand, which are characterized by monogyny, low effective mating frequencies and high levels of adult male production by workers (e.g. Peters *et al.*, 1999; Palmer *et al.*, 2002; Tóth *et al.*, 2002a,b; Brown *et al.*, 2003; Paxton *et al.*, 2003), support the predictions of kin selection theory. However, there is considerable unexplained variation in the degree of worker reproduction across and within stingless bee and bumblebee species (e.g. Drumond *et al.*, 2000; Paxton *et al.*, 2001; Brown *et al.*, 2003; Tóth *et al.*, 2003). In addition, worker policing of worker-laid male eggs in honeybees, for which strong evidence exists (Ratnieks and Visscher, 1989; Barron *et al.*, 2001) and which accounts for the lack of adult males derived from workers, has also been found to occur in the Cape honeybee, *A. mellifera capensis*. This lacks a kin structure predicting such policing because workers reproduce by thelytoky (parthenogenetic production of female eggs) (Pirk *et al.*, 2003). Likewise, worker policing by aggression against reproductive workers occurs in the thelytokous ponerine ant, *Platythyrea punctata* (Hartmann *et al.*, 2003). Worker policing conceivably occurs in these cases because it reduces the productivity costs of successful worker reproduction (Ratnieks, 1988; Ratnieks and Reeve, 1992; Hartmann *et al.*, 2003; Pirk *et al.*, 2003). This raises the question of whether, in other *Apis* species, worker policing is driven by relatedness benefits as had previously been assumed, economic benefits, or both.

In polistine wasps, adult male production by workers occurs in queenless colonies but is variable in colonies with a queen, with the estimated fraction of worker-produced adult males ranging from around 0% in *Polistes bellicosus, P. dorsalis* and *P. gallicus* (Arévalo *et al.*, 1998; Strassmann *et al.*, 2003) to 39% in *P. chinensis* (Tsuchida *et al.*, 2003), even though in all these cases the colony kin structure predicted workers should be reproductive. In swarm-founding, polygynous wasps, adult male production by workers is rare; a finding in line with expectations from kin selection theory (Hastings *et al.*, 1998; Henshaw *et al.*, 2002). In the monogynous vespine wasps, patterns of male parentage broadly support kin selection predictions. Across ten species, Foster and Ratnieks (2001b) found an association between high effective queen mating frequencies and the absence of adult male production by workers on the one hand (in three species), and low effective mating frequencies and above-zero levels of adult male production by workers on the other hand (in six species). In the remaining species (the hornet *Vespa crabro*), a low effective mating frequency was coupled with absence of successful worker reproduction, which was later shown to be due to worker policing (Foster *et al.*, 2002). This again suggests that worker policing may occur for its purely economic benefits (Foster *et al.*, 2002). Extensions of kin selection theory predict that, because queen–worker conflict over male parentage is strongest in monogynous colonies with a singly-mated queen, kin-selected worker matricide (killing of the mother queen followed by male production) should be likeliest in these conditions (Trivers and Hare, 1976; Ratnieks, 1988; Bourke, 1994). Foster and

Ratnieks (2001b) found support for this prediction in vespine wasps by demonstrating a significant negative relationship between the frequency of queenless colonies and the effective mating frequency. In the polistine wasp *Polistes gallicus*, Strassmann *et al.* (2003) suggested that the unusually high frequency of queenless nests (74%) also stemmed from worker matricide, since, in these colonies, workers were shown to produce most of the males.

2.2.2 Intraspecific tests

Male parentage among adult males has been genetically investigated in several species with facultative variation in kin structure (e.g. ants: Sundström *et al.*, 1996; Evans, 1998; Herbers and Mouser, 1998; Walin *et al.*, 1998; Foitzik and Heinze, 2000; bees: Paxton *et al.*, 2001; wasps: Arévalo *et al.*, 1998; Hastings *et al.*, 1998; Goodisman *et al.*, 2002; Henshaw *et al.*, 2002). In general, worker production of adult males was found to be absent or rare irrespective of colony kin structure, even when kin selection theory predicted workers to be reproductive. Male parentage has been investigated in detail (genetically or behaviourally) among both male eggs and adult males in two species with facultative variation in kin structure, the wasp *Dolichovespula saxonica* (Foster and Ratnieks, 2000) and the ant *Leptothorax acervorum* (Hammond *et al.*, 2003). In *D. saxonica*, workers under a singly-mated queen laid 70% of male eggs and 70% of adult males reared by the colony were worker-derived. By contrast, workers under a multiply-mated queen laid 25–90% of male eggs and only 7% of adult males were worker-derived. This suggested that, as predicted, workers under a multiply-mated queen facultatively policed worker-laid eggs (Foster and Ratnieks, 2000). However, in *L. acervorum*, the fraction of worker-produced males was low (2–5%) and did not differ either between eggs and adults or between monogynous colonies and polygynous colonies (Hammond *et al.*, 2003). This was contrary to the predictions of kin selection theory, which predicted high frequencies of worker-laid eggs in both colony types and high frequencies of adult worker-derived males in monogynous colonies.

One possible reason for the general lack of success of kin selection theory in intraspecific tests of the male parentage predictions is that, in principle, interactions with sex-ratio splitting may alter the benefits and costs of worker reproduction and policing (Walin *et al.*, 1998; Foster and Ratnieks, 2001a; Hammond *et al.*, 2003). Such effects need considering because species whose facultative variation in colony kin structure makes them likely to be chosen for within-population studies of male parentage are also likely to exhibit split sex ratios. Although these effects were consistent with the absence or rarity of worker reproduction in some cases (Sundström *et al.*, 1996; Walin *et al.*, 1998), in *L. acervorum* patterns of male parentage were still not consistent with predictions of kin selection theory modified to account for the occurrence of sex-ratio splitting in the study population (Hammond *et al.*, 2003). Another possible reason for the lack of match between kin selection predictions and intraspecific measures of male parentage is that there are unmeasured costs of worker reproduction that select for self-restraint among workers in all types of colony (e.g. Arévalo *et al.*, 1998; Foster *et al.*, 2002; Pirk *et al.*, 2003).

Although plausible, this explanation is difficult to test because costs of events that occur infrequently (here, worker egg-laying and the rearing of worker-produced males to adulthood) are hard to measure.

2.3 Caste fate

In the social Hymenoptera, as we have seen, the worker caste is entirely female and, like the queen caste, arises from fertilized, diploid eggs. With a few possible exceptions (Julian *et al.*, 2002; Volny and Gordon, 2002; Helms Cahan and Keller, 2003), the queen-worker dichotomy is non-genetic (Wilson, 1971). Caste determination refers to the process by which a totipotent, female individual (that is, one capable of becoming a member of either caste) develops into an adult queen or worker. In species whose queens and workers differ morphologically, the caste fate of an individual female is of paramount importance in determining her options for realizing evolutionary fitness. Queens are specialized for egg-laying. Workers are specialized for helping and either lack ovaries or, if they have them, usually lack a sperm receptacle and are therefore unable to mate and produce diploid offspring (Wilson, 1971; Bourke, 1988b).

Given its importance for the final fitness of females, it is not surprising that caste fate is also the focus of potential kin-selected conflict. This idea has been raised and discussed by a number of authors over the past 15 years or so (Ratnieks, 1989; Strassmann, 1989; Nonacs and Tobin, 1992; Ratnieks and Reeve, 1992; Keller and Reeve, 1994; Bourke and Franks, 1995; reviewed by Bourke and Ratnieks, 1999). A context in which potential caste conflict occurs that has proved fruitful for testing the theory was described by Bourke and Ratnieks (1999). This involved the case where workers and queens are reared by colonies simultaneously. In this case, there is no temporal factor (i.e. risk of emerging at an unsuitable time in the colony cycle) preventing female larvae from emergence as queens contrary to the interests of other colony members. In this situation, consider the case in which the workers favour rearing one adult queen and one adult worker from *every* pair of diploid larvae. Each individual larva would favour her own development as the queen, because each individual larva is more closely related to her own would-be offspring than to the would-be offspring of the other female larva. However, adult workers should be indifferent as to which of the two larvae became the queen, either because they are equally related to them (e.g. under monogyny with a singly-mated queen) or because workers are assumed to be unable to discriminate between larvae on the basis of their relatedness (e.g. Keller, 1997). Hence a potential kin-selected conflict exists, with each larva favouring her own development as a queen in opposition to the interests of the other larva and, if both succeeded in becoming queens, of the workers.

Bourke and Ratnieks (1999) defined a general condition for the expression of such conflict. This is that female larvae should have some degree of *self-determination*, namely the power to influence their own caste fate. This condition is necessary because, if self-determination is absent, a female larva's caste fate will be determined by the interests of the adult queens or workers and, hence, even if potential conflict exists between her and these parties over her caste fate, it will not be expressed. In general, workers might be expected to control caste fate through their control of the rearing and nutrition of the brood (Bourke and Ratnieks, 1999). Factors that might facilitate self-determination would then include: (i) a low degree of queen–worker size dimorphism (because then just a little extra feeding by a female larva could push her across the size threshold above which she can become a successful queen); and (ii) larvae having some practical control over their own nutrition.

In the stingless bee genus *Melipona*, conditions for actual caste conflict are met. Queens and sexuals are reared simultaneously, queen–worker size dimorphism is absent (i.e. adult queens and workers are the same size and develop in cells of uniform size), and female larvae have practical power over their own nutrition because they are provisioned by workers with food and the cell in which they develop is then sealed. The theory of kin-selected caste conflict then predicts that more queens should emerge than are favoured by the workers, due to female larvae selfishly developing as queens; it likewise predicts worker actions to correct the excess of queens (Bourke and Ratnieks, 1999).

These predictions are fulfilled by *Melipona*. Queen production in this genus is characterized by the emergence of a vast excess of queens (up to 25% of diploid larvae emerge as queens), which the workers then subject to a large-scale cull (Bourke and Ratnieks, 1999; Wenseleers *et al.*, 2003). A factor that probably aggravates this phenomenon is that stingless bees reproduce by colony fission. The colony, which in most species is monogynous, divides into two, with each new colony being headed by a young queen produced in the mother colony. Therefore, only a handful of new queens are required for colony reproduction per fission event, making the unnecessary excess of queens that emerges likely to represent a significant cost to the colony. Trigonine stingless bees (e.g. *Trigona*) and honeybees (*Apis*) share with *Melipona* the simultaneous rearing of queens and workers and reproduction by colony fission. However, in these species there is a relatively large degree of queen–worker size dimorphism, associated with the fact that queens develop in cells that are larger than those of workers (Bourke and Ratnieks, 1999). Female larvae in these species therefore lack self-determination, since a female larva's fate is determined by the type of cell she is reared in. Consistent with this, in these species there is no large-scale overproduction of queens and hence no cull of excess queens (Bourke and Ratnieks, 1999; Wenseleers *et al.*, 2003). Therefore, in stingless and honeybees, the theory of caste conflict successfully predicts actual conflict over caste fate between individual females and workers when self-determination is present, and the lack of actual conflict when conditions are the same except that self-determination is absent.

The theory of caste conflict has been formalized and expanded by Ratnieks (2001) and Wenseleers *et al.* (2003). The formal theory incorporates the key point that the fitness payoff from development as queens by selfish female larvae is frequency-dependent. This is because, although each female larva favours her own development as a queen, if all were to succeed in becoming queens the colony as a whole would suffer a large productivity cost through lack of workers. Ratnieks (2001) and Wenseleers *et al.* (2003) derived the ESS frequency of new queens as a function of the relatedness structure of the colony. For example, under conditions of monogyny, single mating of queens and no worker male-production, the predicted frequency of queens is 20% (Ratnieks, 2001; Wenseleers *et al.*, 2003). This is calculated as the ratio, $1 - r_F$: $r_F + r_M$, where r_F = relatedness to females and r_M = relatedness to males; hence, in the present case, $r_F = 0.75$ (for sisters) and $r_M = 0.25$ (for brothers), so the ESS ratio is 0.25:1, which, expressed as a proportion, is 20%. An intuitive explanation for this result is that each female larva weighs up the net gain to herself of becoming a queen (1 unit of fitness, because relatedness to self is 1, minus the cost she inflicts on a sister by denying it queenhood, 0.75), as against the net benefit of becoming a worker and rearing sisters (0.75) and brothers (0.25) (Wenseleers *et al.*, 2003). The mean value from five *Melipona* species of the frequency of females becoming queens was 21–22% (Ratnieks, 2001). Furthermore, as predicted by the model, this frequency decreases across *Melipona* species as the fraction of worker-produced males (and hence r_M) rises (Wenseleers and Ratnieks, 2004). The predictions of the ESS caste conflict models of Ratnieks (2001) and Wenseleers *et al.* (2003) represent unusual examples of quantitative predictions in kin selection theory outside the context of sex allocation, and moreover ones that are confirmed. Strassmann *et al.* (2002) proposed that a conflict analogous to caste conflict in species with morphological castes occurs in the swarm-founding epiponine wasp *Parachartergus colobopterus*. In this species, an adult female can opt to be a queen by mating and activating her ovaries. Aggression by workers towards recently emerged potential queens suggested that potential queens were in excess and that workers therefore acted to reduce their frequency (Strassmann *et al.*, 2002; Platt *et al.*, 2004).

3. Future Developments

The preceding section demonstrates strong but not universal evidence for an effect of relatedness on social evolution and social behaviour in insects as predicted by kin selection theory. The theory is strongly supported by significantly female-biased population sex investment ratios in monogynous ants, split sex ratios occurring as a function of variation in workers' relatedness asymmetry, and interspecific patterns of variation in male parentage in bees and vespine wasps. The theory is not supported by interspecific patterns of variation in male parentage in ants and non-vespine wasps or, overall, by intraspecific comparisons in male parentage. Other parts of the current evidence base are supportive but not yet conclusive (e.g. as regards worker reproduction in slave-making ants, worker matricide, and caste conflict). This is not because of inconsistent findings but because of the scarcity of relevant studies. Still other parts of the evidence base are currently equivocal (e.g. as regards sex ratios in slave-making ants and bumblebees). These topics all deserve further investigation in order to determine where the balance of the evidence lies.

It seems likely that many anomalies and inconsistencies in the data may in future be resolved by a fuller understanding of: (i) which party has power over relevant reproductive decisions; and (ii) the economic costs and benefits of selfish manipulations. For example, as we have seen, the issue of how power is monopolized or shared could underpin the fact that population sex investment ratios in monogynous ants fall below the values predicted from workers' relatedness asymmetries (e.g. Passera *et al.*, 2001; Reuter and Keller, 2001), the wide variation in population sex investment ratios among slave-making ants (Herbers and Stuart, 1998), and split sex ratio patterns inconsistent with worker control of sex allocation (Helms, 1999; Passera *et al.*, 2001). Likewise, in some cases, costs of worker reproduction and policing may determine, in parallel with or even instead of relatedness differences, the distribution of male parentage and the occurrence of worker policing (e.g. Arévalo *et al.*, 1998; Foster *et al.*, 2002; Hammond *et al.*, 2003; Pirk *et al.*, 2003; Villesen and Boomsma, 2003). Researchers have long recognized the importance of these factors (reviewed by Ratnieks, 1998; Sundström and Boomsma, 2001; Beekman and Ratnieks, 2003; Beekman *et al.*, 2003; Mehdiabadi *et al.*, 2003), but it clearly remains a key task for the future to find appropriate means of analysing and quantifying them. As well as these issues, there are a number of other areas that seem promising for future research in the field, as discussed next.

3.1 Genetic underpinnings

The advent of hypervariable microsatellite markers has made possible extremely detailed investigations of kin structure and parentage in social insects, and hence more rigorous tests of kin selection theory (Queller *et al.*, 1993a; Hughes, 1998; Bourke, 2001b). Microsatellites have also brought other technical benefits, such as a relatively rapid method for sexing eggs and so measuring the primary sex ratio (eggs appearing homozygous at many polymorphic loci are haploid, male eggs) (e.g. Ratnieks and Keller, 1998; Passera *et al.*, 2001; Hammond *et al.*, 2002), although other molecular-genetic techniques for this have also been developed (De Menten *et al.*, 2003). Rapid advances in molecular genetics and genomics promise to bring additional benefits to the study of social evolution in insects. A pioneering study of the fire ant *Solenopsis invicta* by Krieger and Ross (2002) established for the first time in social insects the exact genetic basis of a complex social behaviour. *S. invicta* is a notorious introduced pest in the USA, where it occurs in a monogynous and a polygynous form. Previous work established that queen number variation (gyny) is controlled by allelic variation at a locus, *Gp-9*, with two alleles, *B* and *b*. Specifically, the *b* allele, a recessive lethal, acts as a 'green beard' gene. A 'green beard' gene is one whose bearers treat co-bearers favourably and non-bearers unfavourably on the basis of an external label (here, almost certainly a chemical cue associated with the presence of the *b* allele) (Dawkins, 1976). Heterozygote (*Bb*) workers, but not *BB* workers, tolerate multiple queens, but only if these queens bear the *b* allele; *Bb* workers detect and kill any *BB* queens. The result is that the presence or absence of the *b* allele determines, via effects on workers' social behaviour, the queen number of the colony (Ross, 1992, 1997; Keller and Ross, 1993, 1998, 1999; Ross *et al.*, 1996; Ross and Keller, 1998, 2002). Polymorphism in *Gp-9* was first detected by protein electrophoresis. By electrophoretically separating the *Gp-9* protein and designing degenerate primers based on its amino acid sequence, Krieger and Ross (2002) isolated and sequenced the *Gp-9* gene and, by searching against sequence databases, showed its closest match to be genes encoding pheromone-binding proteins in moths. Because these proteins are used in the chemical recognition of conspecifics, this was consistent with the presumed effect of *Gp-9* in allowing 'green beard' recognition of allelic co-bearers. By sequencing the *Gp-9* gene from queens of both social forms, Krieger and Ross also determined the nature of allelic variation in the gene: the two alleles differ by nine nucleotide substitutions, one of which is synonymous. This means that just eight nucleotide differences in the coding sequences of the two alleles at the *Gp-9* locus determine, via effects on workers' social behaviour, whether a colony has one or several queens (Krieger and Ross, 2002).

This research definitively established that differences at a single genetic locus can determine the presence or absence of a multifaceted social phenotype (Bourke, 2002; Keller and Parker, 2002; Krieger and Ross, 2002). In this respect, it rebuts a criticism sometimes levelled at kin selection models of social evolution, namely that social behaviours are unlikely to be under simple genetic control. In addition, the study pioneered the integrated understanding of the evolution of social behaviour at all levels of the biological hierarchy (Bourke, 2002). It brought the field as close as it has come to a complete understanding of a gene for social behaviour, from knowledge of its sequence identity and protein product, through to effects on individual behaviour, through to the summed outcome for the colony's social structure and the gene's fitness under natural selection.

A future challenge will be, by tracing the causal chain from gene to behaviour at the individual and colony level, to achieve a similarly complete understanding of other genes for social behaviour, including kin-selected genes (*Gp-9* is not strictly a kin-selected gene because workers with the *b* allele favour *b*-bearing queens irrespective of kinship). However, inconveniently, it seems likely that the genes most amenable to this type of analysis will be ones that, like *Gp-9*, are not kin-selected (Bourke, 2002). Kin-selected genes may be relatively 'invisible' because polymorphism at such genes is unlikely to be maintained (as in *Gp-9*) by recessive lethality and their expression should in principle often be facultative (so at fixation there is phenotypic variation but no genetic variation). However, achieving an understanding of the genetic basis of any social phenotype in as much depth as that achieved for queen number variation in *S. invicta* will be valuable regardless of the selective background, because it will deepen our insight into the relationship between genes and behaviour in social evolution. Reaching this goal will be aided by rapid developments in genomics, such as the mass screening of genes using DNA microarrays (Evans and Wheeler, 2001; Robinson, 2002), techniques for experimentally interfering with gene expression (Beye *et al.*, 2002) and, starting with the US National Human Genome Research Institute's sequence of the honeybee genome (http://www.genome.gov/), the sequencing of the entire genomes of social insect species.

3.2 Identifying and investigating unrecognized or neglected arenas of kin-selected conflict

The theory of kin-selected caste conflict (Bourke and Ratnieks, 1999; Ratnieks, 2001; Wenseleers et al., 2003) represents a good example of a case where new predictions from kin selection theory have been generated as a result of identifying and analysing a previously unrecognized arena of kin-selected conflict. Research to identify additional forms of potential conflict is needed. One possible focus of conflict is over allocation to worker production and allocation to sexual production (i.e. of females plus males) in social insect colonies (Pamilo, 1991). This conflict overlaps with caste conflict, but is distinct because it is more concerned with the overall life-history trade-off between growth and reproduction. Although some theories concerning conditions for this conflict have been developed (Pamilo, 1991; Bourke and Chan, 1999; Herbers et al., 2001; Reuter and Keller, 2001), empirical tests have been relatively few (Sundström, 1995; Herbers and Stuart, 1998; Bourke and Chan, 1999; Herbers et al., 2001; Walin and Seppä, 2001; Foitzik et al., 2003). This area is therefore worth expanding, especially since the study of life history evolution in general in social insects has been relatively neglected and deserves to be pursued further (Oster and Wilson, 1978; Bourke and Franks, 1995).

An even more speculative form of potential conflict arises if genomic imprinting of genes for social behaviours occurs (Haig, 1992, 2000). Genomic imprinting describes the situation when alleles have different effects according to the sex of the contributing parent (Moore and Haig, 1991). Say, for example, that genomic imprinting occurs at a locus for sex allocation in workers of monogynous social Hymenoptera with single mating. Paternally derived alleles will be present in sisters but not in brothers and so would favour an all-female sex ratio. Maternally derived alleles are equally represented in both sexes and so would favour an unbiased sex ratio. The outcome would be swings in the sex ratio dependent on which alleles 'win' (Haig, 1992). (The Trivers–Hare prediction of a 3:1 sex ratio is the average of the two extreme predictions of 1:0 and 1:1, because it assumes no genomic imprinting.) Other kin-conflict predictions under conditions of genomic imprinting can also be made in social Hymenoptera (Haig, 1992; Queller and Strassmann, 2002). As has occasionally been noted (Bourke and Franks, 1995; Chapuisat and Keller, 1999), most recently and forcefully by Queller and Strassmann (2002), no empirical tests of this exciting theory have yet been carried out in the social Hymenoptera. They would be well worth devising and performing, especially since the predictions of the theory are truly a priori (Queller and Strassmann, 2002).

It will also be fruitful to re-examine potential areas of conflict that had previously received a lot of attention but since then have been relatively neglected due to negative findings. A prime example involves within-colony kin discrimination. Evidence that social insect workers can discriminate between different classes of relative within colonies with complex kin structures (e.g. in colonies with multiply-mated queens, or in polygynous colonies), and then treat closer relatives more favourably (nepotism), had been judged to be patchy and inconsistent (e.g. Bourke and Franks, 1995; Keller, 1997; Keller and Reeve, 1999). However, a recent study suggests nepotism in rearing sisters in polygynous colonies of the ant *Formica fusca* (Hannonen and Sundström, 2003b). This will revive the issue of within-colony kin discrimination and should also provoke the further development of theoretical models specifying the conditions under which such discrimination arises and is stably maintained (e.g. Ratnieks, 1991; Boomsma *et al.*, 2003).

Finally, an expanding area of theory concerns interactions between different types of kin-selected conflict. There is every reason to suppose that several types of conflict occur simultaneously in social insect colonies. However, to ensure their tractability, conflict models have generally dealt with each type in isolation. Increasingly, more sophisticated models have considered how different types of conflict overlap and interact. Examples include interactions between the conflicts over sex ratio and male parentage (Walin *et al.*, 1998; Foster and Ratnieks, 2001a; Hammond *et al.*, 2003), sex ratio and life-history (Herbers *et al.*, 2001; Reuter and Keller, 2001), sex ratio and reproductive skew (Boomsma, 1996; Bourke, 2001a; Nonacs, 2001, 2002), and worker policing and reproductive skew (Reeve and Jeanne, 2003). Such studies have shown that, for instance, expectations based on comparing relatedness coefficients as to whether or not worker policing of other workers' male eggs occurs may have to be substantially modified if conflict over sex ratios is simultaneously being expressed (Walin *et al.*, 1998; Foster and Ratnieks, 2001a; Hammond *et al.*, 2003). A unified theory of social evolution surely remains the magnificent conceptual prize for which the field as a whole is aiming (e.g. Reeve, 2001). Such a theory will not only have to take into account primary factors such as kin structure, economic costs and benefits, the distribution of power, and environmental variables, but also the second-order factors represented by the interactions between different types of conflict.

3.3 Experimental and phylogenetic approaches

Many predictions in kin selection theory can be tested by a correlative approach. Examples include the association between relatedness asymmetry and sex ratio predicted by split sex ratio theory (Boomsma and Grafen, 1990, 1991), or the association between the level of successful worker reproduction and relative worker relatedness to worker- and queen-produced males (Trivers and Hare, 1976; Ratnieks, 1988). Because the correlations predicted by kin selection theory between reproductive behaviour or allocation patterns and kin structure are often non-intuitive, in the sense that there is no other obvious reason to expect them, and in important cases have also been *a priori* (e.g. the split sex ratio predictions of Boomsma and Grafen 1990, 1991), such correlational tests can be powerful. However, as advocated by Chapuisat and Keller (1999), experimental manipulations are evidently necessary to establish the predicted causal relationships behind the correlations. Largely for technical reasons, few such experiments have been performed. A good example comes from the work of Mueller (1991), who tested Boomsma and Grafen's (1990, 1991) split sex ratio theory by experimentally manipulating the relatedness asymmetry of colonies of the bee *Augochlorella striata* (for other examples, see Evans, 1995; Foitzik *et al.*, 2003). Experimental approaches are also needed to dissect the detailed mechanisms by which queens or workers achieve sex ratio control (e.g. Helms *et al.*, 2000; Passera *et al.*, 2001). Measuring the cost of conflict – a factor that has frequently been invoked to explain patterns of reproductive traits inconsistent with simple kin selection predictions (see previous section) – is also likely to be best performed by experimental methods. For example, a recent study found, by removing aggressive, reproductive workers from colonies and comparing colony productivity with that of control colonies, that there is no detectable cost of worker reproduction to the production of new queens in the bumblebee *Bombus terrestris* (Lopez-Vaamonde *et al.*, 2003).

Finally, studies that map social traits on to robust phylogenies of social insect taxa, and then perform phylogenetically corrected comparative tests of predicted associations (Harvey and Pagel, 1991), are also required (Chapuisat and Keller, 1999). Again, largely through lack of suitable phylogenetic information, phylogenetic approaches to social evolution in insects have been relatively few (e.g. Keller and Genoud, 1997; Hunt, 1999; Schmid-Hempel and Crozier, 1999; Foster and Ratnieks, 2001b; Danforth, 2002; Villesen *et al.*, 2002). As increasing numbers of such studies are performed, they should be sensitive to an important insight derived from the far more advanced, comparative study of life histories, mating systems and social systems in birds (Arnold and Owens, 1998, 1999; Bennett and Owens, 2002). This is that patterns and causes of variation in evolutionary traits in birds may differ across levels in the taxonomic hierarchy (e.g. between families and species), because the ecological factors that determined differences between lineages in the past (that we now see as higher-level taxa) need not be the same as those that determine differences among lower-level taxa or within populations at present (Bennett and Owens, 2002). There seems no reason to suppose that other organisms should differ in this respect, and hence researchers studying the social insects will need to be mindful that differing patterns of association among social, genetic and ecological traits at different taxonomic levels may similarly require a hierarchical interpretation.

References

Alexander, R.D. (1974) The evolution of social behavior. *Annual Review of Ecology and Systematics* 5, 325–383.

Alexander, R.D. and Sherman, P.W. (1977) Local mate competition and parental investment in social insects. *Science* 196, 494–500.

Alonso, W.J. and Schuck-Paim, C. (2002) Sex-ratio conflicts, kin selection, and the evolution of altruism. *Proceedings of the National Academy of Sciences USA* 99, 6843–6847.

Arévalo, E., Strassmann, J.E. and Queller, D.C. (1998) Conflicts of interest in social insects: male production in two species of *Polistes*. *Evolution* 52, 797–805.

Arnold, K.E. and Owens, I.P.F. (1998) Cooperative breeding in birds: a comparative test of the life history hypothesis. *Proceedings of the Royal Society of London, Series B* 265, 739–745.

Arnold, K.E. and Owens, I.P.F. (1999) Cooperative breeding in birds: the role of ecology. *Behavioral Ecology* 10, 465–471.

Aron, S., Campan, E., Boomsma, J.J. and Passera, L. (1999) Social structure and split sex ratios in the ant *Pheidole pallidula*. *Ethology, Ecology and Evolution* 11, 209–227.

Banschbach, V.S. and Herbers, J.M. (1996) Complex colony structure in social insects: II. Reproduction, queen–worker conflict, and levels of selection. *Evolution* 50, 298–307.

Barron, A.B., Oldroyd, B.P. and Ratnieks, F.L.W. (2001) Worker reproduction in honeybees (*Apis*) and the anarchic syndrome: a review. *Behavioral Ecology and Sociobiology* 50, 199–208.

Beekman, M. and Ratnieks, F.L.W. (2003) Power over reproduction in social Hymenoptera. *Philosophical Transactions of the Royal Society of London, Series B* 358, 1741–1753.

Beekman, M. and Van Stratum, P. (1998) Bumblebee sex ratios: why do bumblebees produce so many males? *Proceedings of the Royal Society of London, Series B* 265, 1535–1543.

Beekman, M., Komdeur, J. and Ratnieks, F.L.W. (2003) Reproductive conflicts in social animals: who has power? *Trends in Ecology and Evolution* 18, 277–282.

Bennett, P.M. and Owens, I.P.F. (2002) *Evolutionary Ecology of Birds: Life Histories, Mating Systems, and Extinction*. Oxford University Press, Oxford, UK.

Beye, M., Härtel, S., Hagen, A., Hasselmann, M. and Omholt, S.W. (2002) Specific developmental gene silencing in the honey bee using a homeobox motif. *Insect Molecular Biology* 11, 527–532.

Boomsma, J.J. (1989) Sex-investment ratios in ants: has female bias been systematically overestimated? *American Naturalist* 133, 517–532.

Boomsma, J.J. (1991) Adaptive colony sex ratios in primitively eusocial bees. *Trends in Ecology and Evolution* 6, 92–95.

Boomsma, J.J. (1993) Sex ratio variation in polygynous ants. In: Keller, L. (ed.) *Queen Number and Sociality in Insects*. Oxford University Press, Oxford, UK, pp. 86–109.

Boomsma, J.J. (1996) Split sex ratios and queen–male conflict over sperm allocation. *Proceedings of the Royal Society of London, Series B* 263, 697–704.

Boomsma, J.J. and Grafen, A. (1990) Intraspecific variation in ant sex ratios and the Trivers–Hare hypothesis. *Evolution* 44, 1026–1034.

Boomsma, J.J. and Grafen, A. (1991) Colony-level sex ratio selection in the eusocial Hymenoptera. *Journal of Evolutionary Biology* 4, 383–407.

Boomsma, J.J., Nielsen, J., Sundström, L., Oldham, N.J., Tentschert, J., Petersen, H.C. and Morgan, E.D. (2003) Informational constraints on optimal sex allocation in ants. *Proceedings of the National Academy of Sciences USA* 100, 8799–8804.

Bourke, A.F.G. (1988a) Dominance orders, worker reproduction, and queen–worker conflict in the slave-making ant *Harpagoxenus sublaevis*. *Behavioral Ecology and Sociobiology* 23, 323–333.

Bourke, A.F.G. (1988b) Worker reproduction in the higher eusocial Hymenoptera. *Quarterly Review of Biology* 63, 291–311.

Bourke, A.F.G. (1994) Worker matricide in social bees and wasps. *Journal of Theoretical Biology* 167, 283–292.

Bourke, A.F.G. (1997a) Sex ratios in bumble bees. *Philosophical Transactions of the Royal Society of London, Series B* 352, 1921–1933.

Bourke, A.F.G. (1997b) Sociality and kin selection in insects. In: Krebs, J.R. and Davies, N.B. (eds) *Behavioural Ecology: An Evolutionary Approach.* Blackwell, Oxford, UK, pp. 203–227.

Bourke, A.F.G. (2001a) Reproductive skew and split sex ratios in social Hymenoptera. *Evolution* 55, 2131–2136.

Bourke, A.F.G. (2001b) Social insects and selfish genes. *Biologist* 48, 205–208.

Bourke, A.F.G. (2002) Genetics of social behaviour in fire ants. *Trends in Genetics* 18, 221–223.

Bourke, A.F.G. and Chan, G.L. (1999) Queen–worker conflict over sexual production and colony maintenance in perennial social insects. *American Naturalist* 154, 417–426.

Bourke, A.F.G. and Franks, N.R. (1995) *Social Evolution in Ants.* Princeton University Press, Princeton, New Jersey.

Bourke, A.F.G. and Ratnieks, F.L.W. (1999) Kin conflict over caste determination in social Hymenoptera. *Behavioral Ecology and Sociobiology* 46, 287–297.

Bourke, A.F.G. and Ratnieks, F.L.W. (2001) Kin-selected conflict in the bumble-bee *Bombus terrestris* (Hymenoptera: Apidae). *Proceedings of the Royal Society of London, Series B* 268, 347–355.

Brown, M.J.F., Schmid-Hempel, R. and Schmid-Hempel, P. (2003) Queen-controlled sex ratios and worker reproduction in the bumble bee *Bombus hypnorum*, as revealed by microsatellites. *Molecular Ecology* 12, 1599–1605.

Brown, W.D. and Keller, L. (2000) Colony sex ratios vary with queen number but not relatedness asymmetry in the ant *Formica exsecta. Proceedings of the Royal Society of London, Series B* 267, 1751–1757.

Brown, W.D. and Keller, L. (2002) Queen recruitment and split sex ratios in polygynous colonies of the ant *Formica exsecta. Ecology Letters* 5, 102–109.

Brown, W.D., Keller, L. and Sundström, L. (2002) Sex allocation in mound-building ants: the roles of resources and queen replenishment. *Ecology* 83, 1945–1952.

Bulmer, M.G. (1983) The significance of protandry in social Hymenoptera. *American Naturalist* 121, 540–551.

Chan, G.L., Hingle, A. and Bourke, A.F.G. (1999) Sex allocation in a facultatively polygynous ant: between-population and between-colony variation. *Behavioral Ecology* 10, 409–421.

Chapuisat, M. and Keller, L. (1999) Testing kin selection with sex allocation data in eusocial Hymenoptera. *Heredity* 82, 473–478.

Choe, J.C. (1988) Worker reproduction and social evolution in ants (Hymenoptera: Formicidae). In: Trager, J.C. (ed.) *Advances in Myrmecology.* E.J. Brill, Leiden, The Netherlands, pp. 163–187.

Choe, J.C. and Crespi, B.J. (eds) (1997) *The Evolution of Social Behaviour in Insects and Arachnids.* Cambridge University Press, Cambridge, UK.

Cole, B.J. (1986) The social behavior of *Leptothorax allardycei* (Hymenoptera, Formicidae): time budgets and the evolution of worker reproduction. *Behavioral Ecology and Sociobiology* 18, 165–173.

Crespi, B.J. (2001) The evolution of social behavior in microorganisms. *Trends in Ecology and Evolution* 16, 178–183.

Cronin, A.L. and Schwarz, M.P. (1997) Sex ratios, local fitness enhancement and eusociality in the allodapine bee *Exoneura richardsoni. Evolutionary Ecology* 11, 567–577.

Crozier, R.H. and Pamilo, P. (1996) *Evolution of Social Insect Colonies: Sex Allocation and Kin Selection.* Oxford University Press, Oxford, UK.

Danforth, B.N. (2002) Evolution of sociality in a primitively eusocial lineage of bees. *Proceedings of the National Academy of Sciences USA* 99, 286–290.

Dawkins, R. (1976) *The Selfish Gene.* Oxford University Press, Oxford, UK.

De Menten, L., Niculita, H., Gilbert, M., Delneste, D. and Aron, S. (2003) Fluorescence *in situ* hybridization: a new method for determining primary sex ratio in ants. *Molecular Ecology* 12, 1637–1648.

Deslippe, R.J. and Savolainen, R. (1995) Sex investment in a social insect: the proximate role of food. *Ecology* 76, 375–382.

Drumond, P.M., Oldroyd, B.P. and Osborne, K. (2000) Worker reproduction in *Austroplebeia australis* Friese (Hymenoptera, Apidae, Meliponini). *Insectes Sociaux* 47, 333–336.

Duchateau, M.J. and Velthuis, H.H.W. (1988) Development and reproductive strategies in *Bombus terrestris* colonies. *Behaviour* 107, 186–207.

Duchateau, M.J., Velthuis, H.H.W. and Boomsma, J.J. (2004) Sex ratio variation in the bumblebee *Bombus terrestris. Behavioral Ecology* 15, 71–82.

Elmes, G.W. (1987) Temporal variation in colony populations of the ant *Myrmica sulcinodis.* II. Sexual production and sex ratios. *Journal of Animal Ecology* 56, 573–583.

Evans, J.D. (1995) Relatedness threshold for the production of female sexuals in colonies of a polygynous ant, *Myrmica tahoensis,* as revealed by microsatellite DNA analysis. *Proceedings of the National Academy of Sciences USA* 92, 6514–6517.

Evans, J.D. (1998) Parentage and sex allocation in the facultatively polygynous ant *Myrmica tahoensis. Behavioral Ecology and Sociobiology* 44, 35–42.

Evans, J.D. and Wheeler, D.E. (2001) Gene expression and the evolution of insect polyphenisms. *BioEssays* 23, 62–68.

Fernández-Escudero, I., Pamilo, P. and Seppä, P. (2002) Biased sperm use by polyandrous queens of the ant *Proformica longiseta. Behavioral Ecology and Sociobiology* 51, 207–213.

Field, J., Solís, C.R., Queller, D.C. and Strassmann, J.E. (1998) Social and genetic structure of paper wasp cofoundress associations: tests of reproductive skew models. *American Naturalist* 151, 545–563.

Fisher, R.A. (1930) *The Genetical Theory of Natural Selection.* Clarendon, Oxford, UK.

Fjerdingstad, E.J., Gertsch, P.J. and Keller, L. (2002) Why do some social insect queens mate with several males? Testing the sex-ratio manipulation hypothesis in *Lasius niger. Evolution* 56, 553–562.

Foitzik, S. and Heinze, J. (2000) Intraspecific parasitism and split sex ratios in a monogynous and monandrous ant (*Leptothorax nylanderi*). *Behavioral Ecology and Sociobiology* 47, 424–431.

Foitzik, S. and Herbers, J.M. (2001) Colony structure of a slavemaking ant. I. Intracolony relatedness, worker reproduction, and polydomy. *Evolution* 55, 307–315.

Foitzik, S., Haberl, M., Gadau, J. and Heinze, J. (1997) Mating frequency of *Leptothorax nylanderi* ant queens determined by microsatellite analysis. *Insectes Sociaux* 44, 219–227.

Foitzik, S., Strätz, M. and Heinze, J. (2003) Ecology, life history and resource allocation in the ant, *Leptothorax nylanderi. Journal of Evolutionary Biology* 16, 670–680.

Foster, K.R. and Ratnieks, F.L.W. (2000) Facultative worker policing in a wasp. *Nature* 407, 692–693.

Foster, K.R. and Ratnieks, F.L.W. (2001a) The effect of sex-allocation biasing on the evolution of worker policing in Hymenopteran societies. *American Naturalist* 158, 615–623.

Foster, K.R. and Ratnieks, F.L.W. (2001b) Paternity, reproduction and conflict in vespine wasps: a model system for testing kin selection predictions. *Behavioral Ecology and Sociobiology* 50, 1–8.

Foster, K.R., Gulliver, J. and Ratnieks, F.L.W. (2002) Worker policing in the European hornet *Vespa crabro. Insectes Sociaux* 49, 41–44.

Fournier, D., Aron, S. and Milinkovitch, M.C. (2002) Investigation of the population genetic structure and mating system in the ant *Pheidole pallidula. Molecular Ecology* 11, 1805–1814.

Fournier, D., Keller, L., Passera, L. and Aron, S. (2003) Colony sex ratios vary with breeding system but not relatedness asymmetry in the facultatively polygynous ant *Pheidole pallidula. Evolution* 57, 1336–1342.

Frank, S.A. (1987) Variable sex ratio among colonies of ants. *Behavioral Ecology and Sociobiology* 20, 195–201.

Frank, S.A. (1998) *Foundations of Social Evolution.* Princeton University Press, Princeton, New Jersey.

Goodisman, M.A.D., Matthews, R.W. and Crozier, R.H. (2002) Mating and reproduction in the wasp *Vespula germanica. Behavioral Ecology and Sociobiology* 51, 497–502.

Grafen, A. (1985) A geometric view of relatedness. In: Dawkins, R. and Ridley, M. (eds) *Oxford Surveys in Evolutionary Biology*, Vol. 2. Oxford University Press, Oxford, UK, pp. 28–89.

Haig, D. (1992) Intragenomic conflict and the evolution of eusociality. *Journal of Theoretical Biology* 156, 401–403.

Haig, D. (2000) The kinship theory of genomic imprinting. *Annual Review of Ecology and Systematics* 31, 9–32.

Hamilton, W.D. (1963) The evolution of altruistic behavior. *American Naturalist* 97, 354–356.

Hamilton, W.D. (1964) The genetical evolution of social behaviour I, II. *Journal of Theoretical Biology* 7, 1–52.

Hamilton, W.D. (1972) Altruism and related phenomena, mainly in social insects. *Annual Review of Ecology and Systematics* 3, 193–232.

Hammond, R.L., Bruford, M.W. and Bourke, A.F.G. (2002) Ant workers selfishly bias sex ratios by manipulating female development. *Proceedings of the Royal Society of London, Series B* 269, 173–178.

Hammond, R.L., Bruford, M.W. and Bourke, A.F.G. (2003) Male parentage does not vary with colony kin structure in a multiple-queen ant. *Journal of Evolutionary Biology* 16, 446–455.

Hannonen, M. and Sundström, L. (2003a) Reproductive sharing among queens in the ant *Formica fusca. Behavioral Ecology* 14, 870–875.

Hannonen, M. and Sundström, L. (2003b) Worker nepotism among polygynous ants. *Nature* 421, 910.

Hartmann, A., Wantia, J., Torres, J.A. and Heinze, J. (2003) Worker policing without genetic conflicts in a clonal ant. *Proceedings of the National Academy of Sciences USA* 100, 12836–12840.

Harvey, P.H. and Pagel, M.D. (1991) *The Comparative Method in Evolutionary Biology.* Oxford University Press, Oxford, UK.

Hasegawa, E. (1994) Sex allocation in the ant *Colobopsis nipponicus* (Wheeler). I. Population sex ratio. *Evolution* 48, 1121–1129.

Hasegawa, E. and Yamaguchi, T. (1995) Population structure, local mate competition, and sex-allocation pattern in the ant *Messor aciculatus. Evolution* 49, 260–265.

Hastings, M.D., Queller, D.C., Eischen, F. and Strassmann, J.E. (1998) Kin selection, relatedness, and worker control of reproduction in a large-colony epiponine wasp, *Brachygastra mellifica. Behavioral Ecology* 9, 573–581.

Heinze, J. (1996) The reproductive potential of workers in slave-making ants. *Insectes Sociaux* 43, 319–328.

Heinze, J., Puchinger, W. and Hölldobler, B. (1997) Worker reproduction and social hierarchies in *Leptothorax* ants. *Animal Behaviour* 54, 849–864.

Heinze, J., Hartmann, A. and Rüppell, O. (2001) Sex allocation ratios in the facultatively polygynous ant, *Leptothorax acervorum*. *Behavioral Ecology and Sociobiology* 50, 270–274.

Helms, K.R. (1999) Colony sex ratios, conflict between queens and workers, and apparent queen control in the ant *Pheidole desertorum*. *Evolution* 53, 1470–1478.

Helms, K.R., Fewell, J.H. and Rissing, S.W. (2000) Sex ratio determination by queens and workers in the ant *Pheidole desertorum*. *Animal Behaviour* 59, 523–527.

Helms Cahan, S. and Keller, L. (2003) Complex hybrid origin of genetic caste determination in harvester ants. *Nature* 424, 306–309.

Henshaw, M.T., Queller, D.C. and Strassmann, J.E. (2002) Control of male production in the swarm-founding wasp, *Polybioides tabidus*. *Journal of Evolutionary Biology* 15, 262–268.

Herbers, J.M. (1984) Queen–worker conflict and eusocial evolution in a polygynous ant species. *Evolution* 38, 631–643.

Herbers, J.M. (1990) Reproductive investment and allocation ratios for the ant *Leptothorax longispinosus*: sorting out the variation. *American Naturalist* 136, 178–208.

Herbers, J.M. and Mouser, R.L. (1998) Microsatellite DNA markers reveal details of social structure in forest ants. *Molecular Ecology* 7, 299–306.

Herbers, J.M. and Stuart, R.J. (1998) Patterns of reproduction in slave-making ants. *Proceedings of the Royal Society of London, Series B* 265, 875–887.

Herbers, J.M., DeHeer, C.J. and Foitzik, S. (2001) Conflict over sex allocation drives conflict over reproductive allocation in perennial social insect colonies. *American Naturalist* 158, 178–192.

Hughes, C. (1998) Integrating molecular techniques with field methods in studies of social behavior: a revolution results. *Ecology* 79, 383–399.

Hunt, J.H. (1999) Trait mapping and salience in the evolution of eusocial vespid wasps. *Evolution* 53, 225–237.

Johnstone, R.A. (2000) Models of reproductive skew: a review and synthesis. *Ethology* 106, 5–26.

Julian, G.E., Fewell, J.H., Gadau, J., Johnson, R.A. and Larrabee, D. (2002) Genetic determination of the queen caste in an ant hybrid zone. *Proceedings of the National Academy of Sciences USA* 99, 8157–8160.

Keller, L. (1997) Indiscriminate altruism: unduly nice parents and siblings. *Trends in Ecology and Evolution* 12, 99–103.

Keller, L. (ed.) (1999) *Levels of Selection in Evolution*. Princeton University Press, Princeton, New Jersey.

Keller, L. and Chapuisat, M. (1999) Cooperation among selfish individuals in insect societies. *BioScience* 49, 899–909.

Keller, L. and Genoud, M. (1997) Extraordinary lifespans in ants: a test of evolutionary theories of ageing. *Nature* 389, 958–960.

Keller, L. and Parker, J.D. (2002) Behavioral genetics: a gene for supersociality. *Current Biology* 12, 180–181.

Keller, L. and Reeve, H.K. (1994) Partitioning of reproduction in animal societies. *Trends in Ecology and Evolution* 9, 98–102.

Keller, L. and Reeve, H.K. (1999) Dynamics of conflicts within insect societies. In: Keller, L. (ed.) *Levels of Selection in Evolution*. Princeton University Press, Princeton, New Jersey, pp. 153–175.

Keller, L. and Ross, K.G. (1993) Phenotypic basis of reproductive success in a social insect: genetic and social determinants. *Science* 260, 1107–1110.

Keller, L. and Ross, K.G. (1998) Selfish genes: a green beard in the red fire ant. *Nature* 394, 573–575.

Keller, L. and Ross, K.G. (1999) Major gene effects on phenotype and fitness: the relative roles of *Pgm-3* and *Gp-9* in introduced populations of the fire ant *Solenopsis invicta. Journal of Evolutionary Biology* 12, 672–680.

Krieger, M.J.B. and Ross, K.G. (2002) Identification of a major gene regulating complex social behavior. *Science* 295, 328–332.

Lopez-Vaamonde, C., Koning, J.W., Jordan, W.C. and Bourke, A.F.G. (2003) No evidence that reproductive bumblebee workers reduce the production of new queens. *Animal Behaviour* 66, 577–584.

Maynard Smith, J. and Szathmáry, E. (1995) *The Major Transitions in Evolution.* W.H. Freeman, Oxford, UK.

Mehdiabadi, N.J., Reeve, H.K. and Mueller, U.G. (2003) Queens versus workers: sex-ratio conflict in eusocial Hymenoptera. *Trends in Ecology and Evolution* 18, 88–93.

Michod, R.E. and Roze, D. (2001) Cooperation and conflict in the evolution of multicellularity. *Heredity* 86, 1–7.

Moore, T. and Haig, D. (1991) Genomic imprinting in mammalian development: a parental tug-of-war. *Trends in Genetics* 7, 45–49.

Mueller, U.G. (1991) Haplodiploidy and the evolution of facultative sex ratios in a primitively eusocial bee. *Science* 254, 442–444.

Mueller, U.G., Eickwort, G.C. and Aquadro, C.F. (1994) DNA fingerprinting analysis of parent–offspring conflict in a bee. *Proceedings of the National Academy of Sciences USA* 91, 5143–5147.

Müller, C.B., Shykoff, J.A. and Sutcliffe, G.H. (1992) Life history patterns and opportunities for queen–worker conflict in bumblebees (Hymenoptera: Apidae). *Oikos* 65, 242–248.

Nonacs, P. (1986) Ant reproductive strategies and sex allocation theory. *Quarterly Review of Biology* 61, 1–21.

Nonacs, P. (1993) The effects of polygyny and colony life history on optimal sex investment. In: Keller, L. (ed.) *Queen Number and Sociality in Insects.* Oxford University Press, Oxford, UK, pp. 110–131.

Nonacs, P. (2001) A life-history approach to group living and social contracts between individuals. *Annales Zoologici Fennici* 38, 239–254.

Nonacs, P. (2002) Sex ratios and skew models: the special case of evolution of cooperation in polistine wasps. *American Naturalist* 160, 103–118.

Nonacs, P. and Tobin, J.E. (1992) Selfish larvae: development and the evolution of parasitic behavior in the Hymenoptera. *Evolution* 46, 1605–1620.

Oster, G.F. and Wilson, E.O. (1978) *Caste and Ecology in the Social Insects.* Princeton University Press, Princeton, New Jersey.

Packer, L. and Owen, R.E. (1994) Relatedness and sex ratio in a primitively eusocial halictine bee. *Behavioral Ecology and Sociobiology* 34, 1–10.

Palmer, K.A., Oldroyd, B.P., Quezada-Euán, J.J.G., Paxton, R.J. and May-Itza, W.De.J. (2002) Paternity frequency and maternity of males in some stingless bee species. *Molecular Ecology* 11, 2107–2114.

Pamilo, P. (1982) Genetic evolution of sex ratios in eusocial Hymenoptera: allele frequency simulations. *American Naturalist* 119, 638–656.

Pamilo, P. (1990) Sex allocation and queen–worker conflict in polygynous ants. *Behavioral Ecology and Sociobiology* 27, 31–36.

Pamilo, P. (1991) Evolution of colony characteristics in social insects. I. Sex allocation. *American Naturalist* 137, 83–107.

Pamilo, P. and Seppä, P. (1994) Reproductive competition and conflicts in colonies of the ant *Formica sanguinea. Animal Behaviour* 48, 1201–1206.

Passera, L., Aron, S., Vargo, E.L. and Keller, L. (2001) Queen control of sex ratio in fire ants. *Science* 293, 1308–1310.

Paxton, R.J., Thorén, P.A., Estoup, A. and Tengö, J. (2001) Queen–worker conflict over male production and the sex ratio in a facultatively polyandrous bumblebee, *Bombus hypnorum*: the consequences of nest usurpation. *Molecular Ecology* 10, 2489–2498.

Paxton, R.J., Bego, L.R., Shah, M.M. and Mateus, S. (2003) Low mating frequency of queens in the stingless bee *Scaptotrigona postica* and worker maternity of males. *Behavioral Ecology and Sociobiology* 53, 174–181.

Pearson, B., Raybould, A.F. and Clarke, R.T. (1995) Breeding behaviour, relatedness and sex-investment ratios in *Leptothorax tuberum* Fabricius. *Entomologia Experimentalis et Applicata* 75, 165–174.

Pearson, B., Raybould, A.F. and Clarke, R.T. (1997) Temporal changes in the relationship between observed and expected sex-investment frequencies, social structure and intraspecific parasitism in *Leptothorax tuberum* (Formicidae). *Biological Journal of the Linnean Society* 61, 515–536.

Peters, J.M., Queller, D.C., Imperatriz-Fonseca, V.L., Roubik, D.W. and Strassmann, J.E. (1999) Mate number, kin selection and social conflicts in stingless bees and honeybees. *Proceedings of the Royal Society of London, Series B* 266, 379–384.

Pirk, C.W.W., Neumann, P. and Ratnieks, F.L.W. (2003) Cape honeybees, *Apis mellifera capensis*, police worker-laid eggs despite the absence of relatedness benefits. *Behavioral Ecology* 14, 347–352.

Platt, T.G., Queller, D.C. and Strassmann, J.E. (2004) Aggression and worker control of caste fate in a multiple-queen wasp, *Parachartergus colobopterus*. *Animal Behaviour* 67, 1–10.

Queller, D.C. (2000) Relatedness and the fraternal major transitions. *Philosophical Transactions of the Royal Society of London, Series B* 355, 1647–1655.

Queller, D.C. and Strassmann, J.E. (1998) Kin selection and social insects. *BioScience* 48, 165–175.

Queller, D.C. and Strassmann, J.E. (2002) The many selves of social insects. *Science* 296, 311–313.

Queller, D.C., Strassmann, J.E. and Hughes, C.R. (1993a) Microsatellites and kinship. *Trends in Ecology and Evolution* 8, 285–288.

Queller, D.C., Strassmann, J.E., Solís, C.R., Hughes, C.R. and DeLoach, D.M. (1993b) A selfish strategy of social insect workers that promotes social cohesion. *Nature* 365, 639–641.

Ratnieks, F.L.W. (1988) Reproductive harmony via mutual policing by workers in eusocial Hymenoptera. *American Naturalist* 132, 217–236.

Ratnieks, F.L.W. (1989) Conflict and cooperation in insect societies. Ph.D. thesis, Cornell University, Ithaca, New York.

Ratnieks, F.L.W. (1991) The evolution of genetic odor-cue diversity in social Hymenoptera. *American Naturalist* 137, 202–226.

Ratnieks, F.L.W. (1998) Conflict and cooperation in insect societies. In: Schwarz, M.P. and Hogendoorn, K. (eds) *Social Insects at the Turn of the Millennium*. XIII International Congress of IUSSI, Adelaide, Australia, pp. 14–17.

Ratnieks, F.L.W. (2001) Heirs and spares: caste conflict and excess queen production in *Melipona* bees. *Behavioral Ecology and Sociobiology* 50, 467–473.

Ratnieks, F.L.W. and Keller, L. (1998) Queen control of egg fertilization in the honey bee. *Behavioral Ecology and Sociobiology* 44, 57–61.

Ratnieks, F.L.W. and Reeve, H.K. (1992) Conflict in single-queen Hymenopteran societies: the structure of conflict and processes that reduce conflict in advanced eusocial species. *Journal of Theoretical Biology* 158, 33–65.

Ratnieks, F.L.W. and Visscher, P.K. (1989) Worker policing in the honeybee. *Nature* 342, 796–797.

Ratnieks, F.L.W., Monnin, T. and Foster, K.R. (2001) Inclusive fitness theory: novel predictions and tests in eusocial Hymenoptera. *Annales Zoologici Fennici* 38, 201–214.

Reeve, H.K. (2001) In search of unified theories in sociobiology: help from social wasps. In: Dugatkin, L.A. (ed.) *Model Systems in Behavioral Ecology.* Princeton University Press, Princeton, New Jersey, pp. 57–71.

Reeve, H.K. and Jeanne, R.L. (2003) From individual control to majority rule: extending transactional models of reproductive skew in animal societies. *Proceedings of the Royal Society of London, Series B* 270, 1041–1045.

Reeve, H.K. and Keller, L. (2001) Tests of reproductive-skew models in social insects. *Annual Review of Entomology* 46, 347–385.

Reeve, H.K., Starks, P.T., Peters, J.M. and Nonacs, P. (2000) Genetic support for the evolutionary theory of reproductive transactions in social wasps. *Proceedings of the Royal Society of London, Series B* 267, 75–79.

Reuter, M. and Keller, L. (2001) Sex ratio conflict and worker production in eusocial Hymenoptera. *American Naturalist* 158, 166–177.

Robinson, G.E. (2002) Sociogenomics takes flight. *Science* 297, 204–205.

Ross, K.G. (1992) Strong selection on a gene that influences reproductive competition in a social insect. *Nature* 355, 347–349.

Ross, K.G. (1997) Multilocus evolution in fire ants: effects of selection, gene flow and recombination. *Genetics* 145, 961–974.

Ross, K.G. and Keller, L. (1998) Genetic control of social organization in an ant. *Proceedings of the National Academy of Sciences USA* 95, 14232–14237.

Ross, K.G. and Keller, L. (2002) Experimental conversion of colony social organization by manipulation of worker genotype composition in fire ants (*Solenopsis invicta*). *Behavioral Ecology and Sociobiology* 51, 287–295.

Ross, K.G., Vargo, E.L. and Keller, L. (1996) Simple genetic basis for important social traits in the fire ant *Solenopsis invicta. Evolution* 50, 2387–2399.

Rüppell, O., Heinze, J. and Hölldobler, B. (2002) Intracolonial patterns of reproduction in the queen-size dimorphic ant *Leptothorax rugatulus. Behavioral Ecology* 13, 239–247.

Schmid-Hempel, P. and Crozier, R.H. (1999) Polyandry versus polygyny versus parasites. *Philosophical Transactions of the Royal Society, Series B* 354, 507–515.

Seppä, P., Queller, D.C. and Strassmann, J.E. (2002) Reproduction in foundress associations of the social wasp, *Polistes carolina*: conventions, competition, and skew. *Behavioral Ecology* 13, 531–542.

Stark, R.E. (1992) Sex ratio and maternal investment in the multivoltine large carpenter bee *Xylocopa sulcatipes* (Apoidea: Anthophoridae). *Ecological Entomology* 17, 160–166.

Starr, C.K. (1984) Sperm competition, kinship, and sociality in the Aculeate Hymenoptera. In: Smith, R.L. (ed.) *Sperm Competition and the Evolution of Animal Mating Systems.* Academic Press, Orlando, Florida, pp. 427–464.

Strassmann, J.E. (1989) Early termination of brood rearing in the social wasp, *Polistes annularis* (Hymenoptera: Vespidae). *Journal of the Kansas Entomological Society* 62, 353–362.

Strassmann, J.E. and Hughes, C.R. (1986) Latitudinal variation in protandry and protogyny in polistine wasps. *Monitore Zoologico Italiano (NS)* 20, 87–100.

Strassmann, J.E., Sullender, B.W. and Queller, D.C. (2002) Caste totipotency and conflict in a large-colony social insect. *Proceedings of the Royal Society of London, Series B* 269, 263–270.

Strassmann, J.E., Nguyen, J.S., Arévalo, E., Cervo, R., Zacchi, F., Turillazzi, S. and Queller, D.C. (2003) Worker interests and male production in *Polistes gallicus*, a Mediterranean social wasp. *Journal of Evolutionary Biology* 16, 254–259.

Sumner, S., Casiraghi, M., Foster, W. and Field, J. (2002) High reproductive skew in tropical hover wasps. *Proceedings of the Royal Society of London, Series B* 269, 179–186.

Sundström, L. (1994) Sex ratio bias, relatedness asymmetry and queen mating frequency in ants. *Nature* 367, 266–268.

Sundström, L. (1995) Sex allocation and colony maintenance in monogyne and polygyne colonies of *Formica truncorum* (Hymenoptera: Formicidae): the impact of kinship and mating structure. *American Naturalist* 146, 182–201.

Sundström, L. and Boomsma, J.J. (2001) Conflicts and alliances in insect families. *Heredity* 86, 515–521.

Sundström, L. and Ratnieks, F.L.W. (1998) Sex ratio conflicts, mating frequency, and queen fitness in the ant *Formica truncorum*. *Behavioral Ecology* 9, 116–121.

Sundström, L., Chapuisat, M. and Keller, L. (1996) Conditional manipulation of sex ratios by ant workers: a test of kin selection theory. *Science* 274, 993–995.

Tóth, E., Queller, D.C., Imperatriz-Fonseca, V.L. and Strassmann, J.E. (2002a) Genetic and behavioral conflict over male production between workers and queens in the stingless bee *Paratrigona subnuda*. *Behavioral Ecology and Sociobiology* 53, 1–8.

Tóth, E., Strassmann, J.E., Nogueira-Neto, P., Imperatriz-Fonseca, V.L. and Queller, D.C. (2002b) Male production in stingless bees: variable outcomes of queen–worker conflict. *Molecular Ecology* 11, 2661–2667.

Tóth, E., Strassmann, J.E., Imperatriz-Fonseca, V.L. and Queller, D.C. (2003) Queens, not workers, produce the males in the stingless bee *Schwarziana quadripunctata quadripunctata*. *Animal Behaviour* 66, 359–368.

Trivers, R.L. (1974) Parent–offspring conflict. *American Zoologist* 14, 249–264.

Trivers, R.L. and Hare, H. (1976) Haplodiploidy and the evolution of the social insects. *Science* 191, 249–263.

Tsuchida, K., Saigo, T., Nagata, N., Tsujita, S., Takeuchi, K. and Miyano, S. (2003) Queen–worker conflicts over male production and sex allocation in a primitively eusocial wasp. *Evolution* 57, 2365–2373.

Van der Have, T.M., Boomsma, J.J. and Menken, S.B.J. (1988) Sex-investment ratios and relatedness in the monogynous ant *Lasius niger* (L.). *Evolution* 42, 160–172.

Vargo, E.L. (1996) Sex investment ratios in monogyne and polygyne populations of the fire ant *Solenopsis invicta*. *Journal of Evolutionary Biology* 9, 783–802.

Villesen, P. and Boomsma, J.J. (2003) Patterns of male parentage in the fungus-growing ants. *Behavioral Ecology and Sociobiology* 53, 246–253.

Villesen, P., Murakami, T., Schultz, T.R. and Boomsma, J.J. (2002) Identifying the transition between single and multiple mating of queens in fungus-growing ants. *Proceedings of the Royal Society of London, Series B* 269, 1541–1548.

Visscher, P.K. (1989) A quantitative study of worker reproduction in honey bee colonies. *Behavioral Ecology and Sociobiology* 25, 247–254.

Volny, V.P. and Gordon, D.M. (2002) Genetic basis for queen–worker dimorphism in a social insect. *Proceedings of the National Academy of Sciences USA* 99, 6108–6111.

Walin, L. and Seppä, P. (2001) Resource allocation in the red ant *Myrmica ruginodis* – an interplay of genetics and ecology. *Journal of Evolutionary Biology* 14, 694–707.

Walin, L., Sundström, L., Seppä, P. and Rosengren, R. (1998) Worker reproduction in ants – a genetic analysis. *Heredity* 81, 604–612.

Ward, P.S. (1983) Genetic relatedness and colony organization in a species complex of ponerine ants. II. Patterns of sex ratio investment. *Behavioral Ecology and Sociobiology* 12, 301–307.

Wenseleers, T. and Ratnieks, F.L.W. (2004) Tragedy of the commons in *Melipona* bees. *Proceedings of the Royal Society of London B* (Supplement) 271, S310–S312.

Wenseleers, T., Ratnieks, F.L.W. and Billen, J. (2003) Caste fate conflict in swarm-founding social Hymenoptera; an inclusive fitness analysis. *Journal of Evolutionary Biology* 16, 647–658.

West-Eberhard, M.J. (1975) The evolution of social behavior by kin selection. *Quarterly Review of Biology* 50, 1–33.

Wilson, E.O. (1971) *The Insect Societies.* Belknap Press, Cambridge, Massachusetts.

Woyciechowski, M. and Lomnicki, A. (1987) Multiple mating of queens and the sterility of workers among eusocial Hymenoptera. *Journal of Theoretical Biology* 128, 317–327.

Yanega, D. (1988) Social plasticity and early-diapausing females in a primitively social bee. *Proceedings of the National Academy of Sciences, USA* 85, 4374–4377.

2 Do Insect Sexual Ornaments Demonstrate Heightened Condition Dependence?

SAMUEL COTTON AND ANDREW POMIANKOWSKI

The Galton Laboratory, Department of Biology, University College London, London, UK

1. Introduction

The handicap hypothesis of sexual selection predicts that females prefer exaggerated male sexual ornaments because these traits signal male genetic quality. Males in good genetic condition are assumed to signal their quality through greater sexual trait size or more vigorous courtship display. Males in worse condition are unable to do this because they cannot bear the viability costs associated with such extravagance. Condition (or quality) is viewed as a trait closely related to viability, where higher values confer greater fitness.

Models of the handicap hypothesis show that it can be a potent force in the coevolution of female preferences for exaggerated male sexual ornament (Pomiankowski, 1987; Grafen, 1990; Iwasa et al., 1991; Iwasa and Pomiankowski, 1994). Models show how the sexual ornament size and dependence on male quality respond to female preference. They predict that the handicap process leads to the evolution of heightened condition-dependent expression of the sexual traits. The same logic applies when male quality varies due to environmental conditions and females gain directly (e.g. resources or parenting) from their mate choice, as we expect the cost of ornament exaggeration to have the same dependence on environmental quality as it does on genetic quality (Iwasa and Pomiankowski, 1999). So sexual trait condition dependence can evolve to signal both genetic and environmental qualities of males.

Condition-dependent expression is widely assumed to be a common feature of sexual traits, both in insects and other animals. This view is supported by a number of reviews which list many examples of sexual ornaments and courtship performance being positively correlated with measures of condition (Andersson, 1994; Johnstone, 1995). Most of the evidence reported in these reviews is simply correlational, and whilst this is indicative of an underlying causal relationship, correlation is not cause. It is clear that common indices of condition, such as nutritional state, resources, or energy reserves, are often important components of fitness. But there are no grounds for believing that this will always be the case, or that condition will always equate with an easily measured quantity. For instance, Rolff and Joop (2002) found that fresh weight was a relatively poor predictor of other, more invasive, estimates of physiological condition. There are also many authors who have questioned the biological and statistical justification for size-standardized measures of condition (e.g. Jakob *et al.*, 1996; Kotiaho, 1999; Green, 2000; García-Berthou, 2001). This highlights the need for experimental studies to confirm that sexual ornaments have strong condition-dependent expression. In this chapter we assess the current state of experimental evidence in insects by reviewing studies published over the last decade (for a more general review incorporating other taxa see Cotton *et al.*, 2004a). We ask whether experimental results, like the correlational data, support the condition-dependent sexual trait hypothesis. We also consider the criteria needed for carrying out rigorous experimental studies, and show how improvements in understanding will flow from better experimental design.

2. Experimental Studies of Condition Dependence in Insect Sexual Traits

We surveyed the published literature of experiments using controlled variation of environmental or genetic factors to investigate condition-dependent expression of sexual traits in insects. Johnstone (1995) was used to obtain references for literature published prior to 1995, and we searched an electronic database (Web of Science: http://wos.mimas.ac.uk) for relevant recent articles. We also consulted the reference lists of these papers to identify additional studies. Although not exhaustive, the review is large and represents the state of contemporary literature using insects. In the survey, we do not consider non-experimental studies reporting correlations between sexual ornament size and condition indices or other components of fitness, as these relationships are beyond the scope of this chapter (for recent reviews see Møller and Alatalo, 1999; Jennions *et al.*, 2001).

For each study we inferred an ordinal scale of condition using experimental groups and noted the type and number of treatments in addition to the response of traits to stress. We report where comparisons were made between the response of sexual and non-sexual traits to treatment. Non-sexual traits were taken to be suitable controls if they were similar in dimension and kind (e.g. morphological trait, behaviour) to the sexual traits. We also observed whether the influence of body size on trait expression had been removed, and whether the authors had attempted to investigate the genetic basis of condition dependence.

2.1 Overview

Our review of the experimental literature comprised 21 experiments in 20 studies covering 15 species from ten genera (Table 2.1). It is surprising that so

Table 2.1. Experimental studies investigating condition dependence of insect sexual traits.

Species	Sexual trait	Control trait	Sexual trait CD[1]	Control trait CD	Sexual trait CD > Control trait CD	Controlled for body size	Stress (n)	Genetic design	Genetic effect	Reference
Cricket										
Gryllus campestris	Calling song	Song components	✓	X	✓[2]	✓[3]	A (2)	X	–	Holzer et al., 2003
G. campestris	Calling song	Song components	✓	X	✓[2]	✓[4]	A (3)	✓ (SB)	–	Scheuber et al., 2003a
G. campestris	Calling song/harp size	Song components	✓	X	✓[2]	✓[7]	B (2)	✓ (FS)	✓ (G)	Scheuber et al., 2003b
G. lineaticeps	Calling song	X	X	–	✓[2]	X	B (2)	✓ (SB)	–	Wagner and Hoback, 1999
G. lineaticeps	Courtship song	X	X	–	–	X	B (2)	✓ (SB)	–	Wagner and Reiser, 2000
G. texensis	Courtship song	Song components	X	X	X	X	B (2)	X	–	Gray and Eckhardt, 2001
Damselfly										
Mnais costalis	Wing pigmentation	X	✓	–	–	✓[3]	A (2)	X	–	Hooper et al., 1999
Dung beetle										
Onthophagus taurus	Courtship rate	X	✓	–	–	✓[3]	A (2)	✓ (HS)	✓ (G)	Kotiaho et al., 2001
O. taurus	Horn length	X	✓	–	?[5]	✓[6]	A (4)	X	–	Hunt and Simmons, 1997
O. acuminatus	Horn length	X	✓	–	?[5]	✓[6]	A (2)	✓ (PO)	X (G)	Emlen, 1994
O. taurus, O. binodis, O. australis	Courtship rate	X	✓	–	–	✓[4,7]	A (2)	X	–	Kotiaho, 2002
Fruit fly										
Drosophila grimshawi	Courtship display	X	✓	–	–	X	B (2)	X	–	Droney, 1996
Grain beetle										
Tenebrio molitor	Pheromone	X	✓	–	–	X	A (2)	X	–	Rantala et al., 2003
Stalk-eyed fly										
Cyrtodiopsis dalmanni	Eyespan	Female eyespan	✓	X	✓	✓[7]	A,C (5)	X	–	David et al., 1998
C. dalmanni	Eyespan	Wing, female eyespan	✓	X	✓	✓[8]	B (3)	✓ (HS,FS)	✓ (G×E)	David et al., 2000
C. dalmanni	Eyespan	Wing, female eyespan	✓	X	✓	✓[7]	A (5)	X	–	Cotton et al., 2004b
C. dalmanni	Eyespan	*Sphyracephala beccarii* eyespan	✓	X	✓	✓[7]	A (5)	X	–	Cotton et al., 2004c
Diasemopsis aethiopica	Eyespan	Female eyespan	✓	✓	✓	✓[7]	B (2)	X	–	Knell et al., 1999
Water strider										
Gerris incognitus	Genitalia	Morphological traits	✓	✓	X	X	A,B (2)	✓ (FS,BIP)	✓ (G)	Arnqvist and Thornhill, 1998
Waxmoth										
Achroia grisella	Acoustic signal rate	X	?[9]	–	–	X	A,B,C,D,E (6)	✓ (AS)	✓ (G×E)	Jia et al., 2000
	Acoustic peak amplitude	X	?[9]	–	–	X	A,B,C,D (4)	✓ (AS)	✓ (G)	

Notes:
1 – CD = condition dependence.
2 – Some parameters of calling song show condition dependence with specific stresses whilst others do not. This suggests that different components of the sexual trait show heightened condition dependence under some circumstances.
3 – Repeated measures.
4 – No correlation with body size.
5 – Horn length shows sigmoidal allometry unlike other traits which is interpreted as elevated condition dependence, but horns are thought to be weapons rather than sexual ornaments (Kotiaho, 2002), and under frequency-dependent rather than sexual selection (Moczek and Emlen, 2000).
6 – Via allometry (see note 5)
7 – Body size included as a covariate.
8 – Trait size divided by body size.
9 – Evidence unclear (see text).

Stress:
n – number of stresses
A – diet (quantity)
B – diet (quality)
C – density
D – temperature
E – photoperiod

Genetic design:
SB – split brood
HS – half-sib
PO – parent/offspring
FS – full-sib
AS – artificial selection

Genetic effect:
G – genotype effect
G×E – genotype × environment interaction

few experimental studies have been carried out. All studies measured the effect of environmental stress on one or more sexual characters, whereas only six studies also incorporated a genetic component in the experimental design. This disparity arises for two reasons. First, it is relatively easy to accurately manipulate environmental quality under laboratory conditions. In contrast, it is hard to estimate genetic quality *a priori* or to set up distinct categories of genetic quality. Second, experimental assessment of genetic variation usually requires a breeding design and this is experimentally more complex and difficult to control. However, such studies are needed to assess the genetic component which is assumed to underlie condition-dependent signalling.

All but two of the studies report that sexual traits were condition-dependent; that is, the sexual trait showed decreased size (morphology) or decreased vigour (courtship) in response to experimentally increased stress (but see discussion of Jia *et al.*, 2000, below). The remaining two studies, both in crickets, found that courtship song was not condition-dependent (Wagner and Reiser, 2000; Gray and Eckhardt, 2001).

A comparison with a non-sexual trait is needed to assess the important hypothesis made by the handicap hypothesis that sexual traits have evolved heightened condition dependence. Only 11 studies compared the response of the sexual ornament with suitable non-sexual traits. All but two (Arnqvist and Thornhill, 1998; Gray and Eckhardt, 2001) confirmed that the sexual trait was more sensitive to stress than control traits (David *et al.*, 1998, 2000; Knell *et al.*, 1999; Wagner and Hoback, 1999; Holzer *et al.*, 2003; Scheuber *et al.*, 2003a,b; Cotton *et al.*, 2004b,c). These studies were only in crickets and stalk-eyed flies, so the generality of the result within the Insecta is very constrained. In crickets, comparisons were made between responsive and unresponsive song parameters, whereas in stalk-eyed flies they were made using the homologous trait in females, on the assumption that condition dependence in the female trait approximates that of the ancestral, unexaggerated state. In addition, comparisons were also made with male non-sexual traits to make sure that differences in condition dependence were trait-specific (sexual versus non-sexual) not sex-specific, and to males from other species with the same, but unexaggerated, trait. In other studies, non-sexual trait condition dependence was ignored, which severely limits interpretation of their results.

Only around a half (12/21) of the studies made appropriate adjustments for body size. This lack of control is worrying because correlated change of traits with body size could account for a large part of the condition-dependent response observed. Controlling for body size was attempted in a number of ways. Some experiments simply used repeated measures on the same individuals pre- and post-manipulation. This is appropriate when environmental manipulation was carried out on adults, as this tends to have little or no effect on body size. Another common approach was to use relative trait size (individual trait size divided by body size) or trait size as a percentage of body size. Alternatively, authors have expressed trait size as a function of body size and have statistically accounted for body size as a covariate.

Here we describe studies with the most interesting findings on the condition dependence of sexual traits and evaluate their merits and their shortcomings.

2.2 Crickets

Male crickets produce three types of acoustic sexual signal: a long-range calling song to attract females at a distance, a courtship song to persuade attracted females to mate, and an aggressive song used during encounters with neighbouring males (Alexander, 1961). Wagner and Hoback (1999) investigated the condition dependence of male calling song in *Gryllus lineaticeps* by maintaining adults on high- or low-quality food. They measured five song parameters and found that males called more frequently and had higher chirp rates when reared on a high-quality diet compared with their brothers reared on poorer diets. Chirp and pulse duration, and dominant frequency were unaffected by treatment. Similarly, Scheuber *et al.* (2003a) found that the frequency of calling and chirp rate both declined as adult dietary stress increased in the congener, *G. campestris*. Again, other song characteristics (chirp duration, syllable number, chirp intensity and carrier frequency) were unaffected by adult diet.

The biological relevance of these laboratory-based findings about calling song was confirmed in an experimental field study. Holzer *et al.* (2003) increased male condition in wild *G. campestris* by supplementing food in a confined area close to the burrow. Males with an augmented food supply called more frequently than a group of control males. No effect of treatment was found on any other song characteristic. In addition, food-supplemented males attracted more females than did control males, and this was at least partially attributable to their elevated calling rate.

Stress during nymphal development has also been experimentally investigated in *G. campestris* (Scheuber *et al.*, 2003b). Individuals raised on a poor nymphal food source produced calling song with a higher carrier frequency when adults. This was caused by a disproportionate reduction in the stridulatory harp area with respect to body size. These males were less attractive, as females prefer to mate with males that produce low-frequency calls. No other song characteristic responded to nymphal stress. It is interesting to note that carrier frequency was not affected by adult stress, as this is fixed via harp size at the final moult (Scheuber *et al.*, 2003b). The opposite pattern is seen for chirp rate, which showed no response to stress during the nymphal stage but was strongly affected by adult dietary manipulation (Scheuber *et al.*, 2003a).

In contrast to these reports of condition dependence in calling song, there has been a failure to show condition dependence in courtship song. Gray and Eckhardt (2001) reared *G. texensis* nymphs and adults on high- or low-quality diets. No effect of diet was found at either life-history stage for any of the three courtship song characteristics measured (interphase interval, chirp rate and the number of low-frequency chirps). Since courtship song was also unrelated to two estimates of condition (residual mass and fat reserves), Gray and Eckhardt (2001) concluded that courtship song was not condition-dependent. A similar lack of condition dependence was reported for courtship song in *G. lineaticeps*, although only one aspect of courtship song was measured (chirp rate) and individuals were only stressed in the adult phase (Wagner and Reiser, 2000).

Crickets' acoustic signals are complex multicomponent traits. Calling song was condition-dependent, but courtship song was not. Only certain elements of the calling song showed condition dependence (e.g. frequency of calling, chirp rate and carrier frequency), whilst others did not (e.g. chirp duration, syllable number, chirp intensity). In addition, the timing of stressful conditions (nymphal versus adult) caused different song elements to respond. Taken together, these experimental results suggest that some elements of the call show heightened condition dependence. These heightened responses are contingent on the time at which the traits are expressed. Other call characteristics do not appear to be condition-dependent and probably have different signalling functions, although it remains possible that they signal other aspects of condition that were not tested (e.g. parasite resistance). The strength of these experiments lies in the use of multiple elements of the acoustic signal to identify those that are strongly condition-dependent against non-responding 'controls'.

2.3 Dung beetles

Male dung beetles (*Onthophagus* sp.) have two types of sexually selected trait: courtship display and horns. Male courtship display increases the probability of mating (Kotiaho *et al.*, 2001) and this effect is independent of horn morphology or body size in the absence of male conspecifics (Kotiaho, 2002). Manipulation of environmental quality revealed that courtship is condition-dependent, as adult males kept on plentiful dung had higher courtship display rates than males experimentally deprived of dung (Kotiaho *et al.*, 2001; Kotiaho, 2002). Unfortunately, no comparison with non-sexual behaviour was made. Courtship rate is heritable, and Kotiaho *et al.* (2001) used a half-sib design to show that display rate was genetically correlated to offspring residual mass (a measure of condition) in *O. taurus*. The importance of residual mass in *O. taurus* is unknown, but it exhibits high levels of additive genetic variance either when measured as standardized residual mass (Kotiaho *et al.*, 2001) or, more appropriately, as somatic weight using body size as a covariate (Simmons and Kotiaho, 2002), suggesting that residual mass may make a major contribution to fitness (Houle, 1992; Pomiankowski and Møller, 1995).

Male Onthophagines also exhibit dimorphic horn morphology. Males larger than a critical body size develop disproportionately long horns on their heads, whilst smaller males develop rudimentary horns or none at all (Emlen, 1994; Hunt and Simmons, 1997; Moczek and Emlen, 1999). Manipulation of condition by alteration of food quantity showed that horn expression is condition-dependent (Emlen, 1994; Hunt and Simmons, 1997). Beetles reared as larvae on small amounts of dung had small body size and did not usually develop horns, whereas those given large amounts of dung had large body size with long horns. Although there was no explicit comparison of horn expression with that of non-sexual traits, it is probable that horns have heightened condition dependence because of their unusual (sigmoidal) allometry.

Horn length was deemed unlikely to signal genetic benefits, as neither body size or horn length was found to be heritable in laboratory studies; variation in each trait was entirely attributed to larval dung quantity (Emlen, 1994; Moczek and Emlen, 1999). However, recent research by Kotiaho *et al.* (2003) has suggested that the heritability of male offspring morphology is strongly affected by a sire-mediated maternal component. Female *O. taurus* differentially provision their progeny depending on the phenotype of their mate; mothers provide more resources to offspring when mated with large-horned males. Kotiaho *et al.* (2003) speculated that these differential 'maternal effects' were in fact an indirect sire effect mediated by the transfer of fitness-enhancing seminal products to females during mating, as, contrary to life-history expectations, females had an elevated lifespan and increased reproductive investment when mated to large-horned males.

The absence of female preference for horns (Kotiaho, 2002) suggests that horns are interspecific weapons rather than sexual ornaments (Moczek and Emlen, 2000; Kotiaho, 2002). Horned males aggressively defend tunnels containing a breeding female and contests are usually won by the male with the biggest horns (Moczek and Emlen, 2000). In contrast, hornless males are more agile and adopt a sneaking strategy to gain copulations (Emlen, 1997; Moczek and Emlen, 2000). Thus variation in male size and horn morphology results in part from frequency-dependent selection on different male strategies and is probably not an example of a condition-dependent handicap.

2.4 Waxmoths

The lesser waxmoth, *Achroia grisella*, is a symbiont of honeybee colonies. Its larvae feed on honey and other organic matter within the colony. Upon eclosion, adult moths leave the honeybee colony and mate. Males attract females by producing an ultrasonic signal, the attractiveness of which is predicted by three call characteristics: asynchrony interval (AI), signal rate (SR) and peak amplitude (PA) (Jang and Greenfield, 1996). Jia *et al.* (2000) created genetically distinct lines, which varied in sexual trait expression, through bidirectional selection on SR and PA separately. Larvae from each line were then reared under a suite of non-standard environmental conditions (high and low temperature, high density and short photoperiod) and their ultrasonic call characteristics evaluated.

Significant phenotypic variation in SR and PA was found under some environmental conditions (e.g. temperature in SR lines), but the selection lines did not all respond in the same direction. For example, under high density SR decreased in the high SR line but increased in the low SR line, whereas PA decreased in both the high and low SR lines. This work is difficult to interpret from the perspective of condition dependence, as it is unclear whether many of the non-standard environments were more or less stressful than the standard rearing procedure. The variation observed was also not consistent with strong condition dependence in either SR or PA. Rather, Jia *et al.* (2000) imply that the genetic variation underlying sexual trait expression may reflect local adaptation to particular environmental conditions.

2.5 Stalk-eyed flies

Stalk-eyed flies (Diopsidae: Diptera) show elongation of the head capsule into long processes ('eye-stalks') on to which the eyes and antennae are laterally displaced. Both sexes possess some degree of eye-stalk elongation (Wilkinson and Dodson, 1997; Baker *et al.*, 2001), but in many species male eyespan is much greater than that of females. Some species have no sex differences, however, and eyespan monomorphism is believed to be plesiomorphic, with sexual dimorphism having evolved independently many times (Baker and Wilkinson, 2001).

The main stalk-eyed fly model is *Cyrtodiopsis dalmanni*, a highly sexually dimorphic species inhabiting the forests of South-East Asia. Nocturnal mating aggregations form on root hairs overhanging the banks of streams (Burkhardt and de la Motte, 1985; Wilkinson and Dodson, 1997), and males fight for control of these roosting sites (Fig. 2.1). Contests are usually won by the males with the largest eyespan (Burkhardt and de la Motte, 1983, 1987; Wilkinson and Dodson, 1997; Panhuis and Wilkinson, 1999), and females prefer to roost and mate with males possessing the largest absolute and largest relative eyespan (Wilkinson and Reillo, 1994; Hingle *et al.*, 2001). Male eyespan in *C. dalmanni* is therefore subject to strong inter- and intrasexual selection.

David *et al.* (1998) investigated the condition dependence of male eyespan by rearing larvae from two separate populations on one of five food stress levels by varying the amount of food available to a given number of eggs. Increasing larval density caused male eyespan to decline in both populations.

Fig. 2.1. Male stalk-eyed flies, *Cyrtodiopsis dalmanni*, fight (left) in order to monopolize nocturnal roosting sites containing females (right). The winners usually have the largest eyespan and are preferred as mates by females. In line with the handicap theory, male eyespan is highly condition-dependent, much more so that other traits (see text for details). (Photographs by S. Cotton.)

Larval stress had similar effects on female eyespan as well as male and female wing size. Importantly, David *et al.* (1998) demonstrated that the ornament exhibited *heightened* condition dependence, as male eyespan was significantly more sensitive to changes in larval density than the homologous female trait. Such patterns remained when relative trait size (the ratio of eyespan-to-body size) was investigated and when body size was controlled as a covariate in a general linear model (GLM). However, David *et al.* (1998) used wing size as both a non-sexual trait and a body size index. It is therefore uncertain whether condition dependence arose through changes in the relationship of wing size-to-body size or eyespan-to-body size. A similar result was found across two larval food quality stress regimes in another dimorphic diopsid, *Diasemopsis aethiopica* (Knell *et al.*, 1999). By varying food quality (high versus low), Knell *et al.* (1999) showed that males in high condition produced larger eyespans for their body size than males in low condition. In contrast, females in high condition invested more in both eyespan and body size, suggesting that at least some of the change in female eyespan was the result of correlated change in body size; diet had no effect on male body size. Thus male eyespan appears to be more sensitive to condition than female eyespan in *D. aethiopica*.

A further study in *C. dalmanni* showed that there was genetic variation underlying the response of male eyespan to food stress (David *et al.*, 2000). This experiment used a full- and half-sib design, exposing larvae to three food quality types. Some genotypes maintained large male eyespan under all environments, whilst others became progressively smaller as stress increased. This pattern persisted when body size variation was controlled for using relative trait values. However, female eyespan, and wing length in both sexes, showed no or little genetic condition-dependent response once body size had been controlled for. It was also noted that the sexual trait amplified differences between genotypes under stressful environmental conditions, even though the mean trait size declined (David *et al.*, 2000). Again this pattern persisted when using relative trait values. Non-sexual control traits, female eyespan, and wing length in both sexes, did not show any change in genetic variation across environments.

David *et al.* (2000) overcame their earlier (1998) problem of scaling uncertainty by using thorax length as a measure of body size and looking at the responses of both eyespan and wing traits relative to thorax length. Unfortunately, their use of relative trait size (i.e. dividing by thorax length) did not fully remove the covariance with body size, because traits do not scale isometrically with thorax and exhibit non-zero intercepts (the intercept of male eyespan is more negative than that of other traits). Thus the disproportionate change in relative male eyespan may in part have arisen as an artefact of inappropriate scaling (Packard and Boardman, 1999).

Recently, Cotton *et al.* (2004b) employed a similar experimental approach to that of David *et al.* (1998), but analysed the responses of both eyespan and wing traits using general linear models (GLM) with thorax length as a covariate measure of body size. This analysis confirmed that, after removing the covariance with body size, male eyespan in *C. dalmanni* was more sensitive to changes in condition than the female homologue and male wing length. In addition, Cotton *et al.* (2004b) showed that the body size-independent component of male eyespan variance increased with stress, unlike that of other traits, providing phenotypic support for David *et al.*'s (2000) genetic data.

We have recently investigated condition dependence in *Sphyracephala beccarii*, a species that is sexually monomorphic for absolute eyespan (Cotton *et al.*, 2004c). In *S. beccarii* nocturnal aggregations are not formed and, in contrast to *C. dalmanni*, males exhibit post-copulatory passive mate guarding. There is no evidence of female mate choice based on male eyespan in this or related sexually monomorphic species (Wilkinson and Dodson, 1997; Wilkinson *et al.*, 1998). We found that male eyespan did not exhibit *heightened* condition dependence. Importantly we also found that the reduction in male eyespan caused by food stress in *S. beccarii* was much less than that observed in male eyespan in *C. dalmanni* (Cotton *et al.*, 2004c). This pattern persisted after removing the effects of body size, and suggests that heightened condition dependence is associated specifically with the evolution of *exaggerated* eyespan in stalk-eyed flies, a view corroborated by phylogenetic correlations (Wilkinson and Taper, 1999). It is noteworthy that we found that both male and female eyespan in *S. beccarii* to be generically more sensitive to changes in condition that other traits (Cotton *et al.*, 2004c). This result accords well with Fisher's (1915, 1930) original proposal that sexual selection would be initiated if female preference arose for male traits that conferred a natural selection advantage. Our work suggests that eyespan, even in its unexaggerated state, is a more sensitive indicator of condition than other traits, and that this may have acted as a pre-adaptation to its role in sexual signalling in other species.

3. Discussion and Future Directions

In this chapter, we consider whether insect experimental studies support the prediction that sexual traits have heightened condition dependence as expected by the handicap hypothesis. It is surprising that there are so few experimental studies, which cover a very limited range of insect groups (see Table 2.1). Most report that sexual ornaments are condition-dependent. However, this interpretation is not well supported as the experiments often lack appropriate controls. In this discussion, we consider what criteria need to be met to carry out a rigorous experimental study of condition dependence. We end by making a call for more studies that experimentally investigate the genetic basis of condition-dependent expression, as this is an area that needs far more development.

3.1 Comparison of traits

Statements concerning the condition dependence of sexual traits are of little value unless they refer to other traits against which ornaments can be compared. In particular, it is vital to contrast sexual traits with non-sexually selected or other control traits. The handicap hypothesis assumes that sexual traits are subject to high differential costs, unlike other traits, and that the cost of the ornament varies with male quality; lower-quality individuals pay higher viability costs for larger sexual ornaments (Iwasa *et al.*, 1991; Iwasa and Pomiankowski, 1994, 1999). This theory predicts that condition dependence is proportional to the cost differential (Iwasa and Pomiankowski, 1994, 1999). Sexual ornament expression should therefore show *heightened* condition dependence when compared with other traits. This is the key feature that needs to be addressed in tests of condition dependence and is general, applying to any exaggerated sexual trait whether morphological or behavioural.

In experiments using insects, comparisons have been made between the male sexual trait and a number of others: the homologous trait in females, non-sexual traits in males, other sexual traits and even between the same male trait in different species. All these comparisons have their value, which is best illustrated by particular examples. In stalk-eyed flies, exaggerated male eyespan (sexual trait) has been compared with the homologous trait in females on the assumption that condition dependence in the female trait is typical of the unexaggerated state (David *et al.*, 1998, 2000; Cotton *et al.*, 2004b). In addition, comparison has been made with male wing traits that have not been subject to sexual selection and so do not differ between the sexes (David *et al.*, 1998, 2000; Cotton *et al.*, 2004b). This comparison controls for differences between the sexes. More recently, the male sexual trait has been compared in different species of stalk-eyed fly, which differ in the degree of sexual dimorphism (Cotton *et al.*, 2004c). This allows an assessment of the evolutionary association of exaggeration and condition dependence.

This long list of comparisons in one group brings out the multiple ways in which *heightened* condition dependence of sexual traits can be assessed. Not all of these comparisons can be made in other species. For example, male *Gryllus* crickets make loud calls to attract females (Alexander, 1961), but females do not call, so no female homologous trait exists for comparison. In addition, there are no obvious non-sexual behavioural or acoustic traits in males with which calling can be compared. In the absence of such traits, researchers have examined different aspects of the male call, looking at differences between long-range calls and short-range courtship song, and different elements within these call types. Unfortunately, clear conclusions have not been drawn because of the non-systematic experimental approach, but it does appear that some traits have stronger condition-dependent expression than others. These preliminary results need to be related to the prediction that call elements subject to stronger mate preference, and hence stronger exaggeration, should show heightened condition dependence.

3.2 Control of body size

Many, if not all, traits show some degree of allometric scaling with body size. As a result, many, if not all, traits are likely to show some degree of condition-dependent expression if body size covaries with condition. Therefore removing the effects of body size variation is an important step in comparing the responses of sexual and non-sexual traits to changes in condition.

Sexual ornaments may just be exaggerated representations of body size, with body size being the main condition-dependent trait. In such circumstances we expect to see heightened sensitivity of ornaments to changes in body size, for instance by the evolution of elevated allometry (i.e. a higher value of b in the equation $Y = aX^b$, where Y = trait size and X = body size). Under these conditions, an incremental increase in body size leads to a proportionally greater increase in ornament size. Alternatively, ornaments may reflect a wider range of factors that influence condition, and then we expect part or all of their condition dependence to be independent of body size. In this scenario, ornaments are expected to show greater responses to experimental stress than other traits when the covariance with body size is removed. In general, the need to control for body size has been under-appreciated in the condition dependence literature; assessing the importance of body size scaling will increase our knowledge of sexual signalling by revealing whether ornaments act as indices of body size, some other components of condition, or both.

Controlling for body size has been attempted in a number of ways. Some experiments have simply used repeated measures on the same individuals pre- and post-manipulation (e.g. Holzer et al., 2003) or have shown that stress has no effect on male size (e.g. Kotiaho, 2002). Relative trait size (individual trait size divided by body size; see e.g. David et al., 2000) or trait size as a percentage of body size are also commonly used methods. However, these approaches are ineffective if scaling deviates from isometry, as the covariance with body size remains (Packard and Boardman, 1999); in such cases any 'condition dependence' can be purely artefactual. Although there are no examples in our literature review, numerous authors in other fields and other taxa (e.g. Candolin, 2000) have employed residuals derived from the regression of trait size on body size to control for allometric covariance. However, this procedure also has been the subject of recent criticism, as it introduces assumptions that may not be biologically or statistically justified (Kotiaho, 1999; Green 2000; Darlington and Smulders, 2001; García-Berthoul, 2001). This has led a number of authors to recommend including body size as a covariate in general linear models (Packard and Boardman, 1999; Darlington and Smulders, 2001; García-Berthou, 2001). Such covariance analyses are free from the statistical drawbacks of other methods, and are effective in removing correlations with body size.

3.3 Manipulation of condition

All experimental studies used variation in environmental quality to assess condition dependence of sexual ornaments. Most (13/21) employed two stress treatments, often limiting the analysis to a simple comparison of apparently 'stressful' and 'non-stressful' environments, for example food versus no food (e.g. Kotiaho et al., 2001). This typically allows low and high conditions to be assigned unambiguously to the different treatments, although this is not always the case, as discussed above in the study on waxmoths (Jia et al., 2000).

The use of only two groups is problematic. First, the quantity or quality of stress used may be unrelated to that experienced in nature. In most cases, the range of stress experienced under natural conditions is not known with any degree of accuracy. Truly unstressed animals are probably little more than laboratory artefacts, and many stressed groups are often exposed to environments at the extreme, or beyond, those to which they are adapted. This means that the choice of stress treatments is critical. If animals are exposed to extremely benign and extremely harsh treatments, their response to treatment may be condition-dependent, but biologically unrealistic if the treatment range falls outside that experienced in nature. Such results would therefore be out of context with the trait's true signalling function, and could lead one to declare a false positive with respect to the trait's evolved sensitivity. Alternatively, if there is little biologically relevant difference between the two treatments, there may be insufficient power to detect condition dependence, with the risk of declaring a false negative. We note that the two studies that failed to find condition dependence used only two treatments (Wagner and Reiser, 2000; Gray and Eckhardt, 2001). To firmly establish that traits are not condition-dependent requires further investigation of a wider range of stress levels and even different types of stress.

A simple solution to these problems is to examine a wider range of stress categories. Although this has the cost of increasing sample size, it permits more accurate assessment of the condition-dependent expression of sexual traits. The chances of detecting condition dependence are also enhanced, and it brings increased power for determining whether sexual traits exhibit heightened condition dependence. The timing of stress treatments poses similar problems. Stress can be applied continuously or at specific points during development. These different approaches tend to be appropriate for considering different questions. If continuous stresses are a frequent occurrence in nature, then animals are likely to have an adaptive response to them. So continuous stress experiments are likely to reveal the nature of adaptations to this type of stress. In contrast, brief stress shocks (e.g. extreme temperature) are less predictable events, and may prove useful for determining which parts of development are sensitive to environmental stress.

3.4 Investigating the genetic basis of condition dependence

There is a pressing need for studies that investigate the genetic basis of condition-dependent expression. The paucity of genetic experiments greatly limits how the phenotypic data can be interpreted. Whilst the handicap principle can still work in the absence of a genetic advantage (Price et al., 1993; Iwasa and Pomiankowski, 1999), much of the controversy in sexual selection is concerned with whether male ornaments signal inherited viability benefits. Insects offer our best hope for determining the importance of male genetic quality in mate choice decisions, as they can be cultured with relative ease in the laboratory and their short generation times mean that the large numbers of animals required for such experiments can easily be obtained.

If exaggerated sexual ornaments signal male genetic quality, we expect a genetic basis to condition dependence. Specifically, we expect male genotypes of high quality to possess the largest sexual ornaments. This can be tested directly, where there is independent evidence that different genotypes vary in fitness or in major components of fitness. For instance, if mutation load was experimentally manipulated, then one would expect groups with the most mutations to have smaller ornaments than controls reared in the same environment; non-sexual trait expression would be expected to differ much less between treatments. Alternatively, one could search for genetic variation in condition dependence (e.g. David *et al.*, 2000). In each case, the interaction between environmental and genetic variation is likely to be crucial. If ornaments show heightened condition dependence, then they are expected *a priori* to be greatly influenced by the environment, as condition, like other life-history traits, is likely to exhibit a large component of environmental variance (Price and Schluter, 1991; Houle, 1992). The handicap theory predicts that high-quality male genotypes will produce large ornaments across all environments. We predict that stress will elevate the differential cost of signalling, thereby increasing the differences between genotypes in terms of ornament size; such changes are expected to be absent (or much reduced) in equivalent non-sexual traits. The interaction of genotypes with the environment is therefore critical for our understanding of sexual traits, as it determines to what degree ornaments signal heritable benefits. Future investigations should address this issue.

Acknowledgements

S.C. was supported by a NERC Studentship with additional funding from the Department of Biology, UCL. Janne Kotiaho, Jens Rolff and an anonymous reviewer provided helpful comments on an earlier version of this chapter. We particularly thank Kevin Fowler and Dave Rogers for discussing many of the issues raised.

References

Alexander, R.D. (1961) Aggressiveness, territoriality, and sexual behaviour in field crickets (Orthoptera: Gryllidae). *Behaviour* 17, 130–223.

Andersson, M. (1994) *Sexual selection*. Princeton University Press, Princeton, New Jersey.

Arnqvist, G. and Thornhill, R. (1998) Evolution of animal genitalia: patterns of phenotypic and genotypic variation and condition dependence of genital and non-genital morphology in water strider (Heteroptera: Gerridae: Insecta). *Genetical Research* 71, 192–212.

Baker, R.H. and Wilkinson, G.S. (2001) Phylogenetic analysis of sexual dimorphism and eyespan allometry in stalk-eyed flies (Diopsidae). *Evolution* 55, 1373–1385.

Baker, R.H., Wilkinson, G.S. and DeSalle, R. (2001) The phylogenetic utility of different types of molecular data used to infer evolutionary relationships among stalk-eyed flies (Diopsidae). *Systematic Biology* 50, 87–105.

Burkhardt, D. and de la Motte, I. (1983) How stalk-eyed flies eye stalk-eyed flies: observations and measurements of the eyes of *Cyrtodiopsis whitei* (Diopsidae, Diptera). *Journal of Comparative Physiology A* 151, 407–421.

Burkhardt, D. and de la Motte, I. (1985) Selective pressures, variability, and sexual dimorphism in stalk-eyed flies (Diopsidae). *Naturwissenschaften* 72, 204–206.

Burkhardt, D. and de la Motte, I. (1987) Physiological, behavioural and morphometric data elucidate the evolutive significance of stalked eyes in Diopsidae (Diptera). *Entomologia Generalis* 12, 221–233.

Candolin, U. (2000) Increased signalling effort when survival prospects decrease: male–male competition ensures honesty. *Animal Behaviour* 60, 417–422.

Cotton, S., Fowler, K. and Pomiankowski, A. (2004a) Do sexual ornamants demonstrate heightened condition-dependent expression as predicted by the handicap hypothesis? *Proceedings of the Royal Society of London, Series B* 271, 771–783.

Cotton, S., Fowler, K. and Pomiankowski, A. (2004b) Condition dependence of sexual ornament size and variation in the stalk-eyed fly *Cyrtodiopsis dalmanni* (Diptera: Diopsidae). *Evolution* 58, 1038–1046.

Cotton, S., Fowler, K. and Pomiankowski, A. (2004c) Heightened condition dependence is not a general feature of male eyespan in stalk-eyed flies (Diptera: Diopsidae). *Journal of Evolutionary Biology* 17, 1310–1316.

Darlington, R.B. and Smulders, T.V. (2001) Problems with residual analysis. *Animal Behaviour* 62, 599–602.

David, P., Hingle, A., Greig, D., Rutherford, A., Pomiankowski, A. and Fowler, K. (1998) Male sexual ornament size but not asymmetry reflects condition in stalk-eyed flies. *Proceedings of the Royal Society of London, Series B* 265, 2211–2216.

David, P., Bjorksten, T., Fowler, K. and Pomiankowski, A. (2000) Condition-dependent signalling of genetic variation in stalk-eyed flies. *Nature* 406, 186–188.

Droney, D.C. (1996) Environmental influences on male courtship and implications for female choice in a lekking Hawaiian *Drosophila*. *Animal Behaviour* 51, 821–830.

Emlen, D.J. (1994) Environmental control of horn length dimorphism in the beetle *Onthophagus acuminatus* (Coleoptera: Scarabaeidae). *Proceedings of the Royal Society of London, Series B* 256, 131–136.

Emlen, D.J. (1997) Alternative reproductive tactics and male-dimorphism in the horned beetle *Onthophagus acuminatus* (Coleoptera: Scarabaeidae). *Behavioural Ecology and Sociobiology* 41, 335–342.

Fisher, R.A. (1915) The evolution of sexual preference. *Eugenics Reviews* 7, 184–192.

Fisher, R.A. (1930) *The genetical theory of natural selection.* Clarendon Press, Oxford, UK.

García-Berthou, E. (2001) On the misuse of residuals in ecology: testing regression residuals vs. the analysis of covariance. *Journal of Animal Ecology* 70, 708–711.

Grafen, A. (1990) Sexual selection unhandicapped by the Fisher process. *Journal of Theoretical Biology* 144, 473–516.

Gray, D.A. and Eckhardt, G. (2001) Is cricket courtship song condition-dependent? *Animal Behaviour* 62, 871–877.

Green, A.J. (2000) Mass/length residuals: measures of body condition or generators of spurious results? *Ecology* 82, 1473–1483.

Hingle, A., Fowler, K. and Pomiankowski, A. (2001) Size-dependent mate preference in the stalk-eyed fly *Cyrtodiopsis dalmanni*. *Animal Behaviour* 61, 589–595.

Holzer, B., Jacot, A. and Brinkhof, M.W.G. (2003) Condition-dependent signalling affects male sexual attractiveness in field crickets, *Gryllus campestris*. *Behavioural Ecology* 14, 353–359.

Hooper, R.E., Tsubaki, Y. and Siva-Jothy, M.T. (1999) Expression of a costly, plastic secondary sexual trait is correlated with age and condition in a damselfly with two male morphs. *Physiological Entomology* 24, 364–369.

Houle, D. (1992) Comparing evolvability and variability of quantitative traits. *Genetics* 130, 195–204.

Hunt, J. and Simmons, L.W. (1997) Patterns of fluctuating asymmetry in beetle horns: an experimental examination of the honest signalling hypothesis. *Behavioural Ecology and Sociobiology* 41, 109–114.

Iwasa, Y. and Pomiankowski, A. (1994) The evolution of mate preferences for multiple handicaps. *Evolution* 48, 853–867.

Iwasa, Y. and Pomiankowski, A. (1999) Good parent and good genes models of handicap evolution. *Journal of Theoretical Biology* 200, 97–109.

Iwasa, Y., Pomiankowski, A. and Nee, S. (1991) The evolution of costly mate preferences. II. The 'handicap' principle. *Evolution* 45, 1431–1442.

Jakob, E.M., Marshall, S.D. and Utez, G.W. (1996) Estimating fitness: a comparison of body condition indices. *Oikos* 77, 61–67.

Jang, Y. and Greenfield, M.D. (1996) Ultrasonic communication and sexual selection in wax moths: female choice based on energy and asynchrony of male signals. *Animal Behaviour* 51, 1095–1106.

Jennions, M.D., Møller, A.P. and Petrie, M. (2001) Sexually selected traits and adult survival: a meta-analysis. *Quarterly Review of Biology* 76, 3–36.

Jia, F.Y., Greenfield, M.D. and Collins, R.D. (2000) Genetic variance of sexually selected traits in waxmoths: maintenance by genotype $\times$ environment interaction. *Evolution* 54, 953–967.

Johnstone, R.A. (1995) Sexual selection, honest advertisement and the handicap principle: reviewing the evidence. *Biological Reviews* 70, 1–65.

Knell, R.J., Fruhauf, A. and Norris, K.A. (1999) Conditional expression of a sexually selected trait in the stalk-eyed fly *Diasemopsis aethiopica*. *Ecological Entomology* 24, 323–328.

Kotiaho, J.S. (1999) Estimating fitness: a comparison of body condition indices revisited. *Oikos* 87, 399–400.

Kotiaho, J.S. (2002) Sexual selection and condition dependence of courtship display in three species of horned dung beetles. *Behavioural Ecology* 13, 791–799.

Kotiaho, J.S., Simmons, L.W. and Tomkins, J.L. (2001) Towards a resolution of the lek paradox. *Nature* 410, 684–686.

Kotiaho, J.S., Simmons, L.W., Hunt, J. and Tomkins, J.L. (2003) Males influence maternal effects that promote sexual selection: a quantitative genetic experiment with dung beetles *Onthophagus taurus*. *American Naturalist* 161, 852–859.

Moczek, A.P. and Emlen, D.J. (1999) Proximate determination of male horn dimorphism in the beetle *Onthophagus taurus* (Coleoptera: Scarabaeidae). *Journal of Evolutionary Biology* 12, 27–37.

Moczek, A.P. and Emlen, D.J. (2000) Male horn dimorphism in the scarab beetle, *Onthophagus taurus*: do alternative reproductive tactics favour alternative phenotypes? *Animal Behaviour* 59, 459–466.

Møller, A.P. and Alatalo, R.V. (1999) Good-genes effects in sexual selection. *Proceedings of the Royal Society of London, Series B* 266, 85–91.

Packard, G.C. and Boardman, T.J. (1999) The use of percentages and size specific indices to normalise physiological data for variation in body size: wasted time, wasted effort? *Comparative Biochemistry and Physiology A* 122, 37–44.

Panhuis, T.M. and Wilkinson, G.S. (1999) Exaggerated eyespan influences male contest outcome in stalk-eyed flies. *Behavioural Ecology and Sociobiology* 46, 221–227.

Pomiankowski, A. (1987) Sexual selection: the handicap principle does work – sometimes. *Proceedings of the Royal Society of London, Series B* 231, 123–145.

Pomiankowski, A. and Møller, A.P. (1995) A resolution of the lek paradox. *Proceedings of the Royal Society of London, Series B* 260, 21–29.

Price, T. and Schluter, D. (1991) On the low heritability of life history traits. *Evolution* 45, 853–861.

Price, T., Schluter, D. and Heckman, N.E. (1993) Sexual selection when the female directly benefits. *Biological Journal of the Linnean Society* 48,187–211.

Rantala, M.J., Kortet, R., Kotiaho, J.S., Vainikka, A. and Suhonen, J. (2003) Condition dependence of pheromones and immune function in the grain beetle *Tenebrio molitor*. *Functional Ecology* 17, 534–540.

Rolff, J. and Joop, G. (2002) Estimating condition: pitfalls of using weight as a fitness correlate. *Evolutionary Ecology Research* 4, 931–935.

Scheuber, H., Jacot, A. and Brinkhof, M.W.G. (2003a) Condition dependence of a multicomponent sexual signal in the field cricket *Gryllus campestris*. *Animal Behaviour* 65, 721–727.

Scheuber, H., Jacot, A. and Brinkhof, M.W.G. (2003b) The effect of past condition on a multicomponent sexual signal. *Proceedings of the Royal Society of London, Series B* 270, 1779–1784.

Simmons, L.W. and Kotiaho, J.S. (2002) Evolution of ejaculates: patterns of phenotypic and genotypic variation and condition dependence in sperm competition traits. *Evolution* 56, 1622–1631.

Wagner, W.E. and Hoback, W.W. (1999) Nutritional effects on male calling behaviour in the variable field cricket. *Animal Behaviour* 57, 89–95.

Wagner, W.E. and Reiser, M.G. (2000) The importance of courtship song in female mate choice in the variable field cricket. *Animal Behaviour* 59, 1219–1226.

Wilkinson, G.S. and Dodson, G.N. (1997) Function and evolution of antlers and eye stalks in flies. In: Choe, J. and Crespi, B. (eds) *The Evolution of Mating Systems in Insects and Arachnids*. Cambridge University Press, Cambridge, UK, pp. 310–328.

Wilkinson, G.S. and Reillo, P.R. (1994) Female preference response to artificial selection on an exaggerated male trait in a stalk-eyed fly. *Proceedings of the Royal Society of London, Series B* 255, 1–6.

Wilkinson, G.S. and Taper, M. (1999) Evolution of genetic variation for condition-dependent traits in stalk-eyed flies. *Proceedings of the Royal Society of London, Series B* 266, 1685–1690.

Wilkinson, G.S., Kahler, H. and Baker, R.H. (1998) Evolution of female mating preferences in stalk-eyed flies. *Behavioral Ecology* 9, 525–533.

3 Sperm Competition in Butterflies and Moths

NINA WEDELL

School of Biosciences, University of Exeter in Cornwall, Penryn, UK

1. Introduction

Female choice and competition between males does not necessarily end with mating. Male competition continues if the sperm from more than one male compete for the female's eggs, termed sperm competition (Parker, 1970). By definition, sperm competition occurs whenever females mate more than once and there is overlap of ejaculates from several males. Sperm competition as a research discipline is only about three decades old. It was Geoff Parker, after studying the love-life of yellow dung flies (see Parker, 2001), who first conceptualized sperm competition and its implications, not only for insects but for mating system evolution in general, in a series of influential theoretical and experimental papers (see Parker, 1998, for a summary). Sperm competition has since been found to be virtually ubiquitous throughout the animal kingdom (Smith, 1984; Birkhead and Møller, 1998; Simmons, 2001).

Insects have played a pivotal role in the study of sperm competition for a number of reasons. Female multiple mating, which is a prerequisite for sperm competition, is common in insects. Female insects commonly have specialized structures where sperm is stored for extended periods of time and females often store sperm from several males, ensuring that the ejaculates overlap. Fertilization tends to occur much later than copulation, at the time of egg-laying. These features make sperm competition a particularly important aspect of sexual selection among insect taxa (Simmons and Siva-Jothy, 1998; Simmons, 2001).

Before the discovery of DNA techniques such as the use of microsatellites to determine parentage in broods of multiply inseminated females, most researchers utilized visual phenotypic genetic markers, polymorphic allozymes, or the sterile male technique to determine paternity of offspring of females commonly inseminated by two males. In the sterile male technique, males are rendered infertile by exposing them to a sublethal dose of radiation. This affects sperm viability, so that if sperm from an irradiated male fertilizes a female's egg the developing zygote dies. If a female is mated to two males, one of which is irradiated, all the eggs that fail to hatch are assumed to have been fathered by the irradiated male. When describing the fertilization success of males competing with sperm from a male that has previously inseminated the same female, paternity is commonly measured as P_2: 'the proportion of offspring sired by the second of two males to mate with the same female'.

Studies of sperm competition in a variety of insects have revealed considerable variation in P_2 both between and within species, ranging from a P_2 of 0 where the first male sires all the offspring to a P_2 of 1, where there is a complete second-male sperm precedence. Insects commonly display second-male sperm precedence, where the second male fathers the majority of offspring (Parker, 1970; Simmons and Siva-Jothy, 1998; Simmons, 2001). However, the mechanisms whereby this is achieved are little known, and the majority of these studies come from experiments carried out in the laboratory.

There is little information regarding female mating patterns of insects in the wild. This is because it is virtually impossible to follow a female throughout her lifetime and record the number of times she mates. With the advent of molecular techniques it is possible to determine the number of fathers in a brood of wild-caught females in some species (see Simmons, 2001). However, this only provides us with information about the outcome of sperm competition. It will not tell us about the fertilization success of individual males, or how many males a female has mated with. It is in this area, in particular, that butterflies and moths have made an important contribution to our understanding of female mating patterns and sperm competition.

Butterflies and moths have a unique advantage as model systems for studying sperm competition because it is easy to determine female mating frequency in the wild. In most lepidopterans, empty sperm-packets, one of which is transferred per mating, are retained in the female reproductive tract throughout her life, making it possible to determine the number of times she has mated. Counts of spermatophore remnants have revealed considerable variation in female mating frequency between species (Fig. 3.1), and also within species (Fig. 3.2) (e.g. Pliske, 1973; Drummond, 1984; Svärd and Wiklund, 1989; Wiklund and Forsberg, 1991). In some species, females will only mate once in their lifetime. In other species females mate multiply, as many as 15 times in *Danaus gilippus* (Pliske, 1973) and 13 times in the tiger moth *Utetheisa ornatrix* (LaMunyon, 1994). Few species appear to be exclusively monogamous; there are usually some females that mate more than once (e.g. Torres-Vila *et al.*, 2002; Fig. 3.2). On the other hand, even in polyandrous species, some females are found to never mate more than once (e.g. Wedell *et al.*, 2002a; Fig. 3.2). However, the average female mating frequency of a species (or population) determines the level of sperm competition encountered by males, and hence males' confidence of paternity. If females rarely mate more than once, or if they only remate when sperm supplies are exhausted, then males have high confidence of paternity, whereas if they remate whilst having viable sperm in storage, ejaculates overlap and sperm competition occurs. Therefore, the average mating frequency of a species indicates the risk to males of encountering sperm competition.

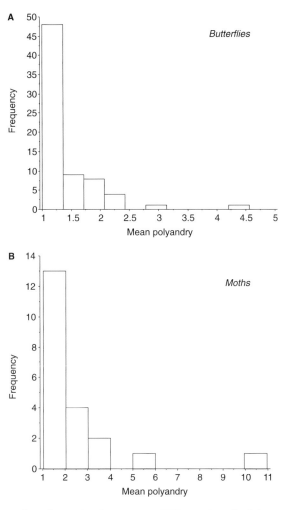

Fig. 3.1. (**A**) Average female mating frequency of 72 species of wild-caught female butterflies as determined by spermatophore counts. (**B**) Average female mating frequency of 23 species of wild-caught female moths as determined by spermatophore counts (data compiled from Drummond, 1984; Svärd and Wiklund, 1989; Wiklund and Forsberg, 1991; Eberhard, 1996).

The reasons why females mate with more than one male (polyandry) are generally not well understood. Females may gain both direct and genetic benefits from mating more than once (Zeh and Zeh, 1996; Vahed, 1998; Arnqvist and Nilsson, 2000; Jennions and Petrie, 2000; Tregenza and Wedell, 2000). In some lepidopterans, males transfer nutrients that females convert into more eggs, or use for somatic maintenance and increased longevity (e.g. Boggs and Gilbert, 1979; Wiklund *et al.*, 1993; Wedell, 1996). In other species, males provide trace elements important to female reproduction, such as sodium (Smedley and Eisner, 1996), or chemicals rendering the eggs and hatchlings unpalatable to predators, increasing their chances of survival (e.g. Dussourd *et al.*, 1988). There is some

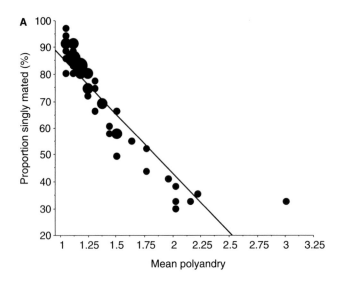

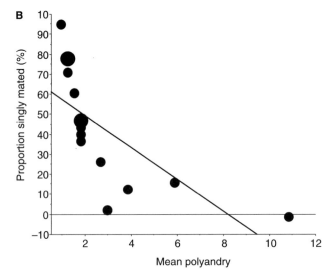

Fig. 3.2. (**A**) The proportion of singly-mated females in relation to average degree of polyandry in 46 species of wild-caught butterflies. (**B**) The proportion of singly-mated females in relation to average degree of polyandry in 17 species of wild-caught moths (data mainly from Drummond, 1984; Eberhard, 1996).

evidence that females mate repeatedly to obtain male-derived nutrients. In Swedish white butterflies (Pieridae), the degree of polyandry is positively correlated with the amount of protein in the spermatophore (Bissoondath and Wiklund, 1995) and with female fecundity (Wiklund *et al.*, 2001). Moreover, male reproductive reserves, measured as abdomen mass relative to total body mass, increase with degree of polyandry (Karlsson, 1995), as does the relative amount of nitrogen in the male's abdomen (Karlsson, 1996), suggesting that males of at least some polyandrous species are adapted to producing large, nutritious spermatophores.

In addition to direct benefits, females may also gain genetic benefits from polyandry. Success in sperm competition may be correlated either with heritable variation in male fitness (Hosken *et al.*, 2003), or sperm competitive success itself may be heritable (Radwan, 1998). Finally, it is possible that variation in female mating frequency is non-adaptive. We now realize that since males and females invest differently in reproduction, there is commonly a conflict between the sexes over female mating rate (Chapman *et al.*, 2003). Males, whose reproductive success tends to increase with the number of matings, are selected to manipulate female reproductive physiology to invest as much as possible in their offspring even if this lowers the female's overall reproductive success (Chapman *et al.*, 1995). In some species, males are able to prevent females from remating or may seduce females to mate at a higher rate than is optimal for the female, since this behaviour favours male reproductive success (see Chapman *et al.*, 2003). There is most probably not one single reason for why female butterflies and moths mate more than once, and the explanations are very likely to differ between species.

The higher the proportion of polyandrous females in the population, the higher the risk of sperm competition for the male. This has resulted in the evolution of various male adaptations to increase fertilization success (Parker, 1998; Simmons and Siva-Jothy, 1998; Simmons, 2001). Male adaptations arising from sperm competition fall into two classes. First, those related to males attempting to avoid having their sperm competing with that of rival males, referred to as *defence* mechanisms. The second set of adaptations are those where males attempt to override the first set of adaptations and outcompete sperm that is already being stored by the female they have mated with, termed *offence* mechanisms (see Simmons and Siva-Jothy, 1998).

2. Adaptations to Reduce Sperm Competition

2.1 Protandry maximizes encounters with virgin females

Males benefit from mating with virgin females for two reasons. First, they do not have to compete with rival males' sperm, ensuring paternity of all the eggs the female lays before she remates. Secondly, fecundity tends to decrease with female age in many insects, and hence virgin females are also the most fecund. The higher returns associated with mating with virgin females exert strong selection on males to become sexually mature before females, termed protandry (Wiklund and Fagerström, 1977). Protandry is also beneficial for females by minimizing the risk of dying unmated, since it reduces the time between emergence and mating (Fagerström and Wiklund, 1982). Female butterflies need to eclose at a time when suitable host plants for egg-laying are available, and males are often selected to eclose before females in order to maximize their mating success. This is most likely to occur in species with female monogamy or where there is a first-male mating advantage (e.g.

Wiklund and Fagerström, 1977; Bulmer, 1983; Iwasa *et al.*, 1983; Wiklund *et al.*, 1992; Wiklund and Kaitala, 1995). Early male emergence can also evolve in response to sperm competition in species where males, by mating with virgin females, fertilize a larger number of eggs than by mating with already mated females (e.g. Wedell, 1992). Protandry may also be the incidental by-product of selection for other life-history traits (e.g. body size), and not be selected *per se* (Wiklund and Solbreck, 1982; Thornhill and Alcock, 1983).

2.2 Mate guarding

One way in which males can reduce the risk of sperm competition is by mate guarding, which is a common adaptation to increase male fertilization success in insects (Alcock, 1994). In butterflies, males guard the female by prolonged mating. In the polyandrous monarch butterfly *Danaus plexippus*, males, unlike in most other butterfly species, can enforce copulations (Pliske, 1975). Copulation in this species only terminates after darkness and males use nightfall as a cue for sperm transfer (Svärd and Wiklund, 1988). In this species males guard the female before sperm transfer, whereas in most insects the guarding phase is initiated after insemination (Alcock, 1994). Guarding benefits the monarch, as the female will use his sperm for fertilization of her eggs on the day after copulation, as there is a last-male sperm precedence in this species. Similar phenomena are suggested to occur in other polyandrous nymphalid butterflies (e.g. Baker, 1972; Bitzer and Shaw, 1979).

2.3 Mating plugs

Another way to reduce the risk of sperm competition is to physically prevent the female from remating. One striking example is found in the production of mating plugs or sphraga in some species, which can be remarkably elaborate. These are formed by the males' accessory glands during copulation, and involves substantial time and energy investment. Male *Heteronympha penelope* butterflies, for example, are only able to produce one or two mating plugs in their lifetime (Orr, 2002). Males tend to produce substantially smaller plugs with number of previous matings; the reduction may be as much as 270% across four consecutive matings in *Atrophaneura alcinous* (Matsumoto and Suzuki, 1992). Similar reductions have been reported in many other species (Orr, 1995, 2002). Elaboration of the sphragis appears to have occurred at the expense of spermatophore size. There is an inverse relationship between spermatophore size and sphragis elaboration (Matsumoto and Suzuki, 1995; Orr, 1995). The function of the plug is most probably to reduce female remating, although females appear capable of remating despite the presence of a plug (Dickinson and Rutowski, 1989; Orr and Rutowski, 1991; Matsumoto and Suzuki, 1992). In one species, males have specialized genitalia capable of removing a rival male's sphragis (Orr, 2002). The presence of sphraga does reduce the probability of female remating. Across species, there is a decline in number of female matings with increasing elaboration or size of the mating plug (Simmons, 2001). It may also function as a visual deterrent to rival males in some species, as the larger the plug the less likely males will be to attempt to mate with the female (Orr and Rutowski, 1991). However, in a study on another butterfly species, mated females with the plug experimentally removed were just as likely to reject courting males as mated females with an intact plug (Dickinson and Rutowksi, 1989), indicating that other factors affect female receptivity in this species.

2.4 Anti-aphrodisiacs

Courting males of the armyworm, *Pseudaletia unipuncta*, release pheromones that deter rival males from approaching receptive females (Hirai *et al.*, 1978). Males also transfer substances that directly reduce the attractiveness of females to rival males. Male *Heliconius erato* butterflies transfer an anti-aphrodisiac pheromone to the female. The female then disseminates the pheromone from special storage organs called 'stink clubs'. The smell of a mated female is highly distasteful to other males (Gilbert, 1976). Interestingly, these odours are race-specific in this species. A similar situation occurs in the green-veined white, *Pieris napi*, where males synthesize and transfer a volatile substance, methyl salicylate (Andersson *et al.*, 2000). This is emitted by mated females and acts as a strong deterrent to courting males. Initially this is beneficial to females as it reduces costly harassment from additional males. However, once the female has laid her eggs this gradually turns into a sexual conflict over remating, as males in this species transfer nutrients to females (Wiklund *et al.*, 1993). Similarly, in the related *P. rapae*, males also synthesize and transfer anti-aphrodisiacs (methyl salicylate and indole) to the female at mating (Andersson *et al.*, 2003). These anti-aphrodisiacs only have a transient effect, as most females will eventually remate.

2.5 Physical presence of a spermatophore

The presence of a spermatophore (a sperm-packet) can reduce female receptivity. During mating the male constructs the spermatophore inside the female's receptacle (bursa copulatrix; see Fig. 3.3). The physical presence of an artificial substance inflating the bursa appears to be sufficient to switch off female receptivity in some species (Sugawara, 1979). The mechanical pressure on stretch receptors in the bursa may be responsible for reduced receptivity, at least initially. There is some support for this. Males of several butterfly species produce larger spermatophores on their first mating, which is associated with longer duration of the female refractory period than those induced by mated males producing smaller spermatophores (Kaitala and Wiklund, 1995; Oberhauser, 1997; Cook and Wedell, 1996; Wedell and Cook, 1999a). In other species, the quality of the spermatophore is more important than its size in influencing female remating behaviour (e.g. Delisle and Hardy, 1997). Females of many lepidopteran species have one or more sets of sclerotized plates (lamina dentata) or spines (signa) in the bursa, which puncture the spermatophore by contraction of the bursal muscles (Rogers and Wells, 1984; Fig. 3.3). There is substantial variation in the number and shape of these 'teeth' between species. There is commonly a conflict between the sexes over female receptivity. This conflict may escalate in an arms race over female mating rate. It is possible that males evolve increasingly tough spermatophores (some higher lepidopteran groups have added chitin in the spermatophore wall; Callahan, 1958), which will remain for longer in the female reproductive tract, resulting in reduced receptivity. Females may counteract this manipulation by investing in increasingly bigger 'teeth', which are required to puncture these tough spermatophores. This idea can be tested by examining spermatophore toughness and the sharpness or size of female signa across species. However, not all species respond to mechanical stimulation of the bursa. In *Manduca sexta*, inflating the bursa artificially has no effect on female behaviour. Spermatophores from castrated males (transferring no testicular products) have a transient effect, but only the presence of a normal spermatophore in the bursa has a lasting effect on reducing female remating behaviour (Sasaki and Riddiford, 1984).

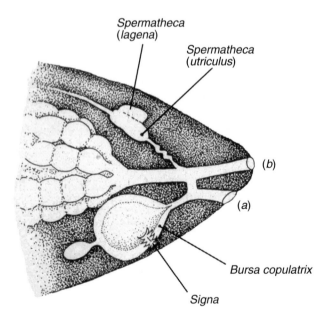

Fig. 3.3. Schematic diagram of the female reproductive tract. Males construct the spermatophore in the bursa copulatrix during copulation through opening (*a*). After mating, the female ruptures the spermatophore (using the signa or lamina dentata in some species) by contracting the bursal muscles, and transports the sperm to the spermatheca where they are stored. The spermatheca consist of two chambers in some species, the utriculus (main chamber) and the lagena. Eggs are fertilized as they travel down the oviduct during oviposition through opening (*b*).

2.6 Seminal factors

Substances transferred in the spermatophore play a role in switching off female receptivity and stimulating oviposition and egg maturation rate (Gillott, 2003). In the moth *Heliothis virescens*, female egg maturation is stimulated by juvenile hormone derived from the males' accessory glands (Park and Ramaswamy, 1998). In addition, it appears that male-derived factors also stimulate the female's own production of juvenile hormone (Park *et al.*, 1998). In *Helicoverpa zea* and *H. armigera* other factors from the male accessory gland stimulate egg maturation and oviposition (Bali *et al.*, 1996; Jin and Gong, 2001). In many moth species, receptive females attract males by releasing pheromones during a characteristic phase of 'calling' behaviour. The regulation of female pheromone production has been intensively studied in moths. A neuropeptide (PBAN) regulates pheromone production (e.g. Raina, 1993; Rafaeli, 2002), mediated by either humoral, hormonal or neural cues. There is large interspecific variation in factors that regulate female pheromone production. In *Helicoverpa zea*, female pheromone production is switched off by a peptide originating from the male accessory glands (Kingan *et al.*, 1995), whilst in the closely related *Heliothis virescens*, a testicular factor is responsible (Ramaswamy *et al.*, 1996). In other moth species, cessation of calling is triggered by a neural signal in the ventral nerve cord (e.g. *Bombyx mori*, Ando *et al.*, 1996; *Lymantria dispar*, Giebultowicz *et al.*, 1991b). In *Helicoverpa zea*, more than one factor controls the switch-off of the pheromone and cessation of calling. Males transferring a spermatophore without accessory gland products do not stop female pheromone production, but do stop the calling behaviour (Kingan *et al.*, 1993). Cessation of calling in female moths appears in general to be triggered by a combination of substances transferred in the ejaculate (i.e. peptides, juvenile hormone) and neural elements (e.g. an intact ventral nerve cord) (Kingan *et al.*, 1995; Marco *et al.*, 1996; Ramaswamy *et al.*, 1996; Delisle *et al.*, 2000). In butterflies, potential seminal factors influencing female receptivity remain largely unexplored.

2.7 Fertile and non-fertile sperm

Sperm present in the female's sperm storage organ (spermatheca, see Fig. 3.3) appear to have a long-term effect on female receptivity. In the butterfly *Pieris rapae*, the presence of sperm in the spermatheca appears to cause neural triggering of female unreceptivity (Obara *et al.*, 1975). Similarly, in several other species of Lepidoptera, the presence of sperm in the spermatheca is required to switch off female receptivity and stimulate oviposition (e.g. Giebultowicz *et al.*, 1991a,b; Karube and Kobayashi, 1999). For example, an inverse relationship between the quantity of sperm in the spermatheca and female remating is reported (e.g. Proshold, 1995; Shantharam *et al.*, 1998). Both the spermatheca and the bursa copulatrix are sensory innervated, suggesting that sperm and seminal fluid may play a role in switching off female receptivity and stimulating oviposition (Sugawara, 1979; Kingan *et al.*, 1995). For example, in *Plodia interpunctella*, females receiving few sperm from a male on his second or third mating are more likely to remate (Cook and Gage, 1995), and in the gypsy moth *Lymantria dispar*, remating is more likely to occur if there are few sperm in the spermatheca (Proshold, 1995).

Male butterflies and moths produce two types of sperm: normal, fertilizing 'eupyrene' sperm, and a large number of non-fertile, anucleate 'apyrene' sperm (Meves, 1902; Friedländer, 1997). Apyrene sperm typically represent 90% or more of the total sperm number (Cook and Wedell, 1996; Solensky, 2003). Fertilizing eupyrene sperm are transferred in the spermatophore to the female in bundles of 256 sperm per bundle (Virkki, 1969; Phillips, 1970; Richard *et al.*, 1975; Witalis and Godula, 1993). Non-fertile sperm represent a significant investment by males, since they are transferred to females in large numbers. Apyrene sperm have different behaviour from eupyrene sperm, being highly active at ejaculation, whilst the eupyrene sperm usually remain in bundles. Apyrene sperm also appear to reach the female's spermatheca before the fertile sperm in both butterflies and moths (Silberglied *et al.*, 1984; Tschudi-Rein and Benz, 1990; Watanabe *et al.*, 2000; Marcotte *et al.*, 2003). Non-fertile sperm seem to be critical to male reproductive success, because males do not decrease investment in apyrene sperm relative to eupyrene sperm when reared on a restricted diet (Gage and Cook, 1994; Cook and Wedell, 1996).

Various hypotheses have been proposed to explain apyrene sperm function (reviewed in Silberglied *et al.*, 1984). Many of these suggest that apyrene sperm have a physiological role, for example in aiding eupyrene sperm transport or activating the eupyrene sperm (e.g. Osanai *et al.*, 1986, 1987). In *Bombyx mori*, they appear to be important for successful fertilization, possibly by aiding transfer of the fertile sperm to the spermatheca and dissociation of the eupyrene sperm bundles (Sahara and Takemura, 2003). Alternatively, they may represent a nutrient source either for the fertile sperm in the spermatheca, or for the female and the developing zygotes. However, Silberglied and co-workers (1984) have argued that these hypotheses do not account for the fact that apyrene sperm reach and persist within the spermatheca and do not appear to be digested. In some species non-fertile sperm may be stored separately in the different parts of the spermatheca (the lagena and utriculus, see Fig. 3.3) (Holt and North, 1970; Miskimen *et al.*, 1983). If apyrene sperm were only involved in eupyrene sperm transport or activation in the spermatophore, it seems unlikely that they would then be stored, and even less likely that they would be sorted and stored in particular areas in the spermatheca. Furthermore, it has been suggested that if apyrene sperm have a physiological role, a given number of non-fertile sperm should be needed for the activation or transport of a single fertile sperm, and therefore the proportion of the two sperm types should be constant within a species (Cook and Gage, 1995). This does not appear to be the case. For example, in at least two species there is a significant increase in the proportion of fertile sperm over the first two matings: in *Plodia interpunctella*, the proportion of eupyrene sperm increases from 7.5% to 10% (Gage and Cook, 1994), and in *Pieris rapae* the increase is from 11% to 15% (Cook and Wedell, 1996).

In their landmark paper, Silberglied *et al.* (1984) suggested that apyrene sperm play a role in sperm competition. They specifically suggested that apyrene sperm may displace or inactivate rival sperm, or, by remaining in the females' spermatheca, they may delay female remating. Both these sperm competition hypotheses predict that apyrene sperm numbers should increase with increased risk of sperm competition. If apyrene sperm displace or inactivate rival males' sperm, they may increase in response to the presence of rival male sperm. In *P. interpunctella*, males provide non-virgin females with more eupyrene, but not apyrene, sperm (Cook and Gage, 1995), whereas in the green-veined white, *Pieris napi*, males provide both a higher number of eupyrene and apyrene sperm to mated females (N. Wedell and P.A. Cook, unpublished data). On the other hand, if apyrene sperm influence female sexual receptivity, we expect their numbers to be related to female remating behaviour. It is, of course, possible that non-fertile sperm may play both these roles.

A recent study on the green-veined white, *P. napi*, suggests that the number of non-fertile sperm in the spermatheca is responsible for reducing female receptivity (Cook and Wedell, 1999). Once-mated females were given the opportunity to remate up to 10 days after their first mating (females rarely remate after this time), dissected upon remating (before the sperm from the second mating reached the spermatheca), and the number of eupyrene and apyrene sperm present in the spermatheca were counted. Females that did not remate within 10 days were also dissected. There was no difference in the number of fertile sperm in the spermatheca of the females that did remate compared with those that did not. However, females that did not remate had significantly more non-fertile sperm in their spermatheca (Fig. 3.4). Moreover, within the females that did remate, there was a positive relationship between the length of time until remating and the number of apyrene sperm in the

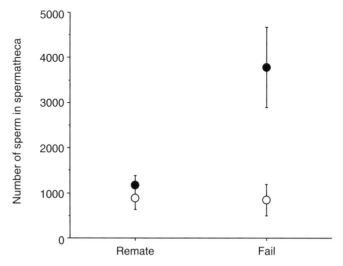

Fig. 3.4. The relationship between number of non-fertile (filled circles) and fertile (open circles) sperm stored by female *Pieris napi* butterflies in relation to probability of remating. Means ± SE (redrawn after Cook and Wedell, 1999).

spermatheca. There was no such relationship with eupyrene sperm number. These results suggest that non-fertile sperm are involved in influencing female sexual receptivity by filling their sperm storage in this species. There is genetic variation between females in their tendency to store non-fertile sperm, which correlates with the duration of their refractory period (Wedell, 2001). Females may have a mechanism to detect the presence of sperm in their spermatheca (e.g. by the presence of mechano-receptors; Lum and Arbogast, 1980) in order to ensure high fertility. Males may have evolved to take advantage of this system: rather than transferring large numbers of nucleate eupyrene sperm, the males transfer large numbers of non-fertile sperm that fill the female's sperm storage organ, thereby delaying female remating and hence reducing sperm competition. It is possible that production of non-nucleated sperm is more efficient than a similar investment in eupyrene sperm to switch off female receptivity, although this is yet to be confirmed.

In *P. napi*, there is a conflict over female remating rate, as males transfer nutrients, increasing female fecundity (Wiklund *et al.*, 1993). Females may have responded to the males' manipulation of their reproductive system by evolving a better detection system to monitor sperm in storage and regain control over their mating rate. This imposes further selection on males to increase the number of non-fertile sperm transferred to females. A coevolutionary arms race between males and females over female receptivity may have resulted in the elaboration of male *P. napi* ejaculates to the point where they now consist of mainly non-fertile sperm, reducing female remating and thus reducing sperm competition for the first male to mate with the female. Similarly, in the armyworm moth, the regaining of female receptivity is associated with a pronounced decline in the number of apyrene, but not eupyrene, sperm, in storage (He *et al.*, 1995), and the presence of motile apyrene sperm in the spermatheca temporarily switches off female receptivity in *Heliothis zea* (Snow *et al.*, 1972).

3. Sperm in Competition

Eventually most females of polyandrous lepidopterans mate again, and sperm compete for fertilization of the female's eggs. The outcome of sperm competition is commonly measured as P_2, the proportion of offspring sired by the second of two males to mate with the same female. The pattern of sperm competition can either be one of sperm mixing, with no obvious mating order effects, or a first- or second-male sperm priority, where one of the two males fertilizes the majority of the female's eggs (Parker, 1970; Simmons, 2001). A common pattern of sperm precedence is for the last male to mate with the female to fertilize most of the eggs (in around 45% of insect species studied so far: Simmons and Siva-Jothy, 1998; Simmons, 2001). Some species of Lepidoptera show complete second-male sperm precedence (e.g. *Trichoplusia ni*, North and Holt, 1968; *Spodoptera litura*, Etman and Hooper, 1979; see Table 3.1). However, there is a more complex pattern of fertilization success that appears to be common in Lepidoptera. Within a species, some females lay eggs fertilized almost exclusively by the first male and some lay eggs fertilized almost exclusively by the second male, whilst very few females lay eggs of mixed paternity. In other words, paternity commonly shows a bimodal distribution (Table 3.1). This is not due to one of the two matings failing to successfully transfer a spermatophore. For most species, identifying the mechanism that underlies the observed pattern of sperm precedence is not straightforward.

Table 3.1. Summary of P_2 studies in butterfly ($n = 8$) and moth ($n = 13$) species.

Species	P_2	Factors influencing P_2	Bimodal distribution	Reference
Papilionidae				
Papilio dardanus	0.67		Yes	Clarke and Sheppard, 1962
Pieridae				
Colias eurytheme	1.00		Yes	Boggs and Watt, 1981
Pieris napi	0.66	Relative male size	Yes	Bissoondath and Wiklund, 1997
Pieris rapae	0.77	Sperm number Male size Female	Yes	Wedell and Cook, 1998
Nymphalidae				
Bicyclus anynana	0.62		Yes	Brakefield *et al.*, 2001
Euphydryas editha	0.72		Yes	Labine, 1966
Limenitis arthemis	0.71		Yes	Platt and Allen, 2001
Danaidae				
Danaus plexippus	0.67	Male size Male mating history	Yes	Oberhauser *et al.* (cited in Simmons, 2001) Solensky, 2003
Arctiidae				
Utetheisa ornatrix	0.52	Relative spermatophore size	Yes	LaMunyon and Eisner, 1993, 1994
Bombicidae				
Bombyx mori	0.34			Omura, 1939
	0.95	Remating interval		Suzuki *et al.*, 1996
Gelechiidae				
Phthorimaea operculella	0.94			Rananavare *et al.*, 1990
Noctuidae				
Heliothis virescens	0.99/ 0.47		Yes	Flint and Kressin, 1968
	0.80			Pair *et al.*, 1977
	0.80	Male age (sperm number) Female size	Yes	LaMunyon, 2000, 2001

continued

Table 3.1. *continued.*

Species	P_2	Factors influencing P_2	Bimodal distribution	Reference
Helicoverpa zea	0.71		Yes	Carpenter, 1992
Pseudaletia separata	0.83			He *et al.*, 1995
P. unipunctata	0.47		Yes	Svärd and McNeil, 1994
Pseudoplusia includens	0.27		Yes	Mason and Pashley, 1991
Spodoptera frugiperda	0.72			Boorman and Parker, 1976
	0.54		Yes	Snow *et al.*, 1970
S. litura	1.00	2nd mating flushes out stored sperm	Yes	Etman and Hooper, 1979
	0.95			Seth *et al.*, 2002a
Trichoplusia ni	0.92			North and Holt, 1968
Pyralidae				
Plodia interpunctella	0.68			Brower, 1978
	0.70	Relative spermatophore size and sperm number	Yes	Cook *et al.*, 1997
Tortricidae				
Choristoneura fumiferana	0.46	Remating interval	Yes	Retnakaran, 1974
Laspeyresia pomonella	0.58			Proverbs and Newton, 1962

3.1 Pattern of sperm storage

During copulation the male constructs the spermatophore within the bursa copulatrix. Sperm transfer usually occurs towards the end of copulation after spermatophore construction (Callahan and Cascio, 1963). Some time after copulation, sperm leave the bursa copulatrix via the ductus seminalis and are transported, probably aided by female reproductive tract movements (Tschudi-Rein and Benz, 1990; LaMunyon and Eisner, 1993), to the spermatheca (Fig. 3.3). This means that there is little scope for males to directly interfere with sperm already stored from previous matings. Females store sperm in the spermatheca until required for fertilization, which occurs at oviposition. Eupyrene sperm leave the spermatheca through the spermathecal duct, and fertilization occurs when the egg passes the spermathecal duct opening as it travels from the oviduct to the ovipositor (Norris, 1932; Callahan and Cascio, 1963). There is therefore scope for ejaculates of more than one male to be stored in the spermatheca and compete for fertilization. In some species, for example noctuids (Callahan and Cascio, 1963; Holt and North, 1970; Miskimen *et al.*, 1983; He *et al.*, 1995), *Manduca sexta* (Sphingidae) (Sasaki and Riddiford, 1984) and *Pieris napi*

(Pieridae) (N. Wedell and P.A. Cook, unpublished observations), the spermatheca has two lobes, the utriculus and lagena (Fig. 3.3). In two noctuids, *Diatraea saccharalis* (Miskimen *et al.*, 1983) and *Trichoplusia ni* (Holt and North, 1970), it has been observed that the two sperm types are stored in the different lobes: both types move into the utriculus and then apyrene sperm separate and move into the lagena. Sperm numbers in the two parts of the spermatheca are reported for *Pseudaletia unipuncta* (Noctuidae) (He *et al.*, 1995). These counts show that, although there is not a complete separation of the two sperm types, there is a higher proportion of non-fertile sperm in the lagena than in the utriculus. However, the pattern of separate apyrene storage does not appear to be the rule (Callahan and Cascio, 1963; LaMunyon, 2000).

It is not known whether sperm from several males mix in the spermatheca or are stored in layers with no mixing, so that the last male's ejaculate is predominantly used. If this is the case we expect a drop in last-male sperm precedence and a shift to increasing use of the first (or previous) male's sperm over time, as the last male's sperm is being used for fertilization. However, the few observations we have to date show that P_2 does not diminish over time (e.g. Cook *et al.*, 1997). Alternatively, sperm storage may be biased so that females store sperm only from certain 'favoured' males. The pattern of sperm storage will give rise to differences in number of sperm stored with consecutive matings. Sperm mixing and layering predict an increase in number of sperm stored with number of matings, whereas preferential sperm storage necessarily does not. For example, in *Spodoptera litura*, Etman and Hooper (1979) report that, prior to the sperm from the second male entering the female's spermatheca, sperm from the first male disappears. In the moth *Heliothis virescens*, despite twice-mated females storing two complements of non-fertile sperm, only one ejaculate's worth of fertile sperm are stored (LaMunyon, 2000). However, this is not the case in all species. In the green-veined white, *Pieris napi*, for example, twice-mated females store about twice as many apyrene and eupyrene sperm as singly mated females, indicating that ejaculates from more than one male are in direct competition (N. Wedell and P.A. Cook, unpublished data).

3.2 Remating interval

The interval between matings by the females can also affect paternity. In the silkworm, the mechanism behind the 'all or nothing' effect is partly understood: if the second mating occurs immediately after the first, the second spermatophore blocks the exit of the bursa and prevents the first male's sperm from leaving; consequently the second male fertilizes all the eggs. If the second mating occurs after 2 hours, then sperm from the first mating already occupies the spermatheca and has the fertilization advantage (Suzuki *et al.*, 1996). A similar explanation has been suggested to operate in the spruce budworm, where there is a shift from first- to last-male sperm precedence when females remate on the same day (Retnakaran, 1974). However, this mechanism is unlikely to explain all incidences of bimodal sperm precedence, since other studies have used standardized remating intervals.

3.3 Sperm number

Theory predicts that an increased number of sperm is advantageous in sperm competition, either because a larger ejaculate volume displaces more rival sperm or due to numerical superiority. In general, sperm competition favouring numerous sperm may have led to the evolution of many tiny sperm and may even be responsible for the evolution of the two sexes (Parker *et al.*, 1972; Bulmer and Parker, 2002). Comparative data in numerous animal groups have corroborated this finding. Males invest more resources in testes and sperm production when they encounter a greater risk of sperm competition (Smith, 1984; Birkhead and Møller, 1998; Simmons, 2001). The same is found in butterflies. In species where females mate more frequently, as determined by spermatophore counts in wild-caught females, males have relatively larger testes (Gage, 1994). As yet, we do not know whether these males also produce more sperm. However, it seems likely, as in other insects, that there is generally a positive relationship between testis size and sperm number (Simmons, 2001). According to theory, we therefore expect that male butterflies and moths should increase the number of sperm when the risk of sperm competition is higher.

In general, we expect that in polyandrous species males should ensure that they are able to mate whenever they encounter a receptive female. In the polyandrous Australian *Hypolimnas bolina* butterfly, females show a facultative adult diapause, possibly due to the unpredictable timing of the tropical season. In contrast, males do not appear to be reproductively dormant, as sperm is present throughout the year. This may be a strategy to cope with the unpredictability of female activity and enable them to mate whenever they encounter receptive females (Pieloor and Seymour, 2001). In highly polyandrous species we also expect males to be able to produce more ejaculates than males of monandrous species, since they, on average, will mate more frequently. This is indeed the case; in butterflies, not only do males of polyandrous species produce bigger spermatophores than males of monandrous species (Svärd and Wiklund, 1989; Bissoondath and Wiklund, 1995), they also have greater spermatophore production capacity. They are capable of producing new spermatophores and remating sooner than males of monogamous species (Svärd and Wiklund, 1989).

3.3.1 Developmental response

One way in which larval experiences may influence adult mating strategy is through the presence of conspecific larvae. The number of conspecifics present may predict the eventual adult density. This in turn is likely to be related to the risk of sperm competition. The armyworm *Pseudaletia separata* has two larval phenotypes depending on rearing densities, a black-coloured gregarious 'crowded' type and a fast growing, pale green 'solitary' type. When armyworm outbreaks occur in the field, the density of adults becomes high and the crowded form is likely to be exposed to a higher risk of sperm competition. Accordingly, crowded type adults produce larger spermatophores (He and Tsubaki, 1992) with more apyrene sperm (He and Miyata, 1997) than those of the solitary type. This increase in apyrene number in the crowded type supports the hypothesis that apyrene sperm play a role in sperm competition, either by actively disrupting or displacing rival sperm or by influencing female sexual receptivity (He and Miyata, 1997). Females receiving spermatophores from males reared under crowded conditions delay remating for longer than females mated to males reared under lower densities (He and Tsubaki, 1995).

In *Plodia interpunctella*, larvae do not have distinct phenotypes depending on rearing density but exhibit a plastic response to variation in rearing density (Gage, 1995). Larval population structure influences adult population structure since, in this species, dispersal appears weak. In turn, adult population structure influences female mating pattern: females are more polyandrous at higher population densities. Males emerging into more dense populations therefore experience a greater risk of sperm competition and mate encounter. Accordingly, males reared at higher larval densities have larger abdomens, relatively bigger testes, and produce ejaculates containing more eupyrene and apyrene sperm. In contrast, males from lower densities have smaller testes and ejaculates but have relatively bigger thoraxes and heads and live longer as adults. These males appear to bias their reproductive strategy towards migration and mate searching. Such plasticity in reproductive strategy is likely to be particularly important for many lepidopteran species because adults are semelparous, short-lived and virtually all resources for reproduction are accrued in the larval stages.

Other species have evolved a fixed strategy to maximize their fertilization success that depends on the risk of sperm competition. In the polyandrous small white butterfly, *Pieris rapae*, the males' second spermatophores are only half the size of their first, but contain almost twice the number of apyrene and eupyrene sperm (Cook and Wedell, 1996). Even males remating on the same day as their first transfer increased the number of sperm on their second mating (Wedell and Cook, 1999a). Why do males have such different strategies on their first and second matings? Cook and Wedell (1996) hypothesized that the pattern of spermatophore allocation can be explained if males are more likely to encounter virgin females when mating for the first time. This seems likely, as there are two to three discrete generations of *P. rapae* per year (Asher *et al.*, 2001), and adults are likely to emerge at similar times. With virgin females, males have complete confidence of paternity, and provide a large nutritious spermatophore to delay female remating and invest in offspring. On males' second matings, encountered females are likely to be non-virgin and males therefore have to compete with sperm from rival males. Males benefit by providing these females with high numbers of sperm, which leads to success in the resulting sperm competition, and reduced nutrient contribution, since they have lower paternity (Wedell and Cook, 1998).

In some species, males transfer an increased number of sperm on their second mating, even when remating on the same day, suggesting that males have a mechanism for reserving sperm for subsequent matings. Male butterflies and moths have a sperm storage organ, the duplex, in which sperm is retained after a mating in some polyandrous species. Male Lepidoptera eclose with a complete complement of sperm (Friedländer, 1997). Sperm release from the testes follows a daily rhythmic pattern and is controlled both by an intrinsic circadian mechanism present in the reproductive tract (Bebas et al., 2001), in combination with photoperiod and temperature (Riemann et al., 1974; Giebultowicz et al., 1989; Proshold, 1997). This results in an increased number of sperm available in the duplex as the number of days until mating increases (LaChance et al., 1977; Proshold, 1991; Giebultowicz and Zdarek, 1996). Males of polyandrous species do not necessarily ejaculate all the sperm they have stored in the duplex in a single mating (Hiroyoshi, 1995; Proshold and Bernon, 1994; Spurgeon et al., 1994; Wedell and Cook, 1999b; Seth et al., 2002b). This may be an adaptation by males to maintain sufficient sperm for additional matings, ensuring that high sperm numbers are available for their second mating, even if it occurs shortly after their first mating. Interestingly, the ability to retain sperm in the duplex varies between species and appears to be associated with variation in degree of polyandry. In the polyandrous P. napi and P. rapae, males reserve sperm in the duplex (N. Wedell and P.A. Cook, unpublished data) and increase the number of sperm provided to females even when remating only 1 hour after their first copulation. In contrast, the monogamous brimstone males ejaculate all available sperm from the duplex (N. Wedell and P.A. Cook, unpublished data), and require several days before being able to produce a new spermatophore, which can result in a female remaining in copula for up to a week if she has the misfortune to encounter a recently mated male (Labitte, 1919).

3.3.2 Ejaculate tailoring

Individual males can strategically ejaculate their limited sperm to maximize their fertilization returns. There is evidence that males assess sperm competition risks at a given mating and respond by increased ejaculate expenditure when risks are elevated. However, spermatogenesis is far from limitless, and males have evolved mechanisms for allocating their finite numbers of sperm optimally to maximize their lifetime reproductive success (Wedell et al., 2002b). Accordingly, males are expected to vary the number of sperm in relation to risk of sperm competition and female fecundity. Males appear to be able to assess mating status and the relative fecundity of females, and to modulate their ejaculate investment (Wedell et al., 2002b). Female reproductive 'quality' may be an important parameter that males have evolved sensitivity to. Female quality can arise from potential fecundity (e.g. body size or age) or female mated status (e.g. risk of sperm competition). When female reproductive quality varies and males are sperm-limited, selection may operate upon males to discriminate between female partners. Indeed, there is some behavioural evidence for male mate choice in butterflies (e.g. in the orange tip butterfly *Anthocaris cardamines*; Wiklund and Forsberg, 1985; and possibly in *Acrea encedon*; Jiggins et al., 2002). However, fewer studies have examined discrimination at the level of the gamete.

FECUNDITY. In *Plodia interpunctella*, female potential fecundity is dictated by adult body size: heavier females have larger ovaries and live longer as adults

(Gage, 1998). Although male body weight does not constrain his first spermatophore size, males are sensitive to female weight and allocate more sperm to heavier females (Gage, 1998). Males may allocate more sperm to heavier females because of greater potential fecundity. They do not ejaculate larger spermatophores into bigger females because they have larger copulatory bursae: bursal size is not related to body size. However, heavier females do have larger spermathecal volumes. In addition, larger females are also more polyandrous than smaller females (perhaps linked to variance in spermathecal volume). Males may therefore also ejaculate more sperm into larger females because of the increased risk of sperm competition associated with such matings. In the gypsy moth, *Lymantria dispar*, males are sensitive to variation in female age and provide older females with less sperm than younger ones (Proshold, 1996). Older females may provide lower fecundity returns and also generate higher risks of sperm competition.

MATING STATUS.   When detectable, female mated status predicts the risk of sperm competition that a male's ejaculate will subsequently face, and hence we expect males to provide more sperm to mated females. In the Indian meal moth, for example, males provide more sperm to non-virgin females (Cook and Gage, 1995). The small white butterfly, *Pieris rapae*, provides a striking suggestion that males indeed tailor the number of sperm provided in relation to risk of sperm competition. In this species, males are not only sensitive to the mating status of the female, and provide more sperm to mated females, but also to the mating status of the female's previous mate. This directly translates to varying sperm competition risk. In this species, mated males provide more sperm than virgin males; hence, when a male mates with a female previously inseminated by a mated male he will have to compete with more sperm than a male mating with a female inseminated with a virgin male. Amazingly, males vary the number of sperm and provide more sperm when the female has previously mated with a mated than a virgin male (Wedell and Cook, 1999a). It is possible that males can assess the mating status, and therefore the number of sperm they will compete with during mating, as remnants of males' spermatophores remain within the female's bursa, where males construct the spermatophore during copulation, and mated males produce smaller spermatophores than do virgin males.

3.4 Male body size

In some species, large male size is advantageous in sperm competition (LaMunyon and Eisner, 1993; Bissoondath and Wiklund, 1997; Wedell and Cook, 1998). This may be explained by larger males producing bigger spermatophores that contain more sperm (Wedell, 1997), or if females favour males with large spermatophores (e.g. LaMunyon and Eisner, 1993, 1994). The effect of male size on sperm precedence has been investigated in seven species. In three, *Pseudaletia unipuncta* (Svärd and McNeil, 1994), *Plodia interpunctella* (Cook *et al.*, 1997) and *Heliothis virescens* (LaMunyon, 2001), there is no effect of male size on male fertilization success. In the monarch, the results are mixed. One study found an effect of male size on fertilization success (Oberhauser *et al.*, cited in Simmons, 2001), whereas another did not (Solensky, 2003). In three others, *Utetheisa ornatrix* (LaMunyon and Eisner, 1993), *Pieris napi* (Bissoondath and Wiklund, 1997) and *P. rapae* (Wedell and Cook, 1998), larger males fertilized more eggs. In these latter species, larger males produced bigger spermatophores; therefore the fertilization advantage may be linked to spermatophore size (LaMunyon and Eisner, 1993; Bissoondath and Wiklund, 1997). This is confirmed in *U. ornatrix*, where males mating for the first time (transferring larger spermatophores) fertilize more eggs when competing with males on their second mating (that transfer smaller spermatophores) (LaMunyon and Eisner, 1994). It has been suggested that in *U. ornatrix*, females bias paternity in favour of males producing large spermatophores (LaMunyon and Eisner, 1993, 1994). However, sperm number (which is predicted to be an important determinant of fertilization success, Parker, 1982, 1990, 1998) could be the cause, if virgin males produce more sperm than mated males. In *P. interpunctella*, males that transfer more sperm (on their first mating) generally have higher fertilization success than males on their second mating (transferring fewer sperm) (Cook *et al.*, 1997). However, in common with most lepidopteran model systems for studying sperm competition, the effects of sperm number and spermatophore size were not separated.

A recent study by Wedell and Cook (1998) was the first to investigate the effects of spermatophore size and sperm number separately. Male *Pieris rapae* transfer a large spermatophore with few sperm on their first mating, and a small spermatophore with many sperm on their second mating (Cook and Wedell, 1996). In their experiment, there were four treatments: females were either mated alternately with mated and virgin males (and *vice versa*); or females were mated with two males of the same mating status (either two virgin or two mated males). Overall, females showed either a first-male or, more commonly, last-male sperm precedence. However, when a mated male competed with a virgin male's sperm, the mated male was more likely to fertilize most of the eggs, even when mating in the less favoured, first-male role. Since mated males produce small spermatophores with many sperm, these findings suggest that sperm number is more important for determining the outcome of sperm competition than spermatophore size. When both males were of the same mating status, mating order seemed to be most important. Similarly, in the moth *Heliothis virescens* a high sperm number results in higher fertilization success. Older males produce more sperm (LaMunyon and Huffman, 2001), and achieve more fertilizations than younger males (LaMunyon, 2001).

3.5 Sperm length

Production of many sperm is not the only way in which males may increase the chances of fertilizing the females' eggs. Another way is investing in bigger sperm. Sperm numbers evidently play a direct role in determining the outcome of sperm competition (Cook *et al.*, 1997; Wedell and Cook, 1998). Less is understood regarding the adaptive significance of sperm size. A popular theory for the evolution of anisogamy is via disruptive selection for either competitive gametes or fecund gametes (Parker, 1982). This ingenious reasoning led to Parker's explanation for the evolution of numerous tiny sperm (Parker, 1982; Bulmer and Parker, 2002). If gamete competition proceeds numerically, then selection would lead to the production of maximal numbers of minimally sized gametes, or sperm. In contrast, selection for gamete fecundity would lead to the evolution of fewer larger gametes, hence ova. This fundamental difference between the sexes carries substantial consequences at all levels of reproduction.

Accordingly, numerical sperm competition explains why there is selection for large numbers of sperm. Males are driven to minimize sperm size to maximize sperm number. On this basis, one might predict across related species which vary in mating pattern, that as sperm competition risk increases then sperm size should either decrease or remain minimal, both of which will enable sperm number to increase as investment in relative testis size increases. Across fish (Stockley *et al.*, 1997) this prediction is satisfied: as sperm competition risk increases, testis size and ejaculate sperm number increase while sperm size decreases. However, across butterflies the opposite is found. Although relative testis size (Gage, 1994) and spermatophore mass (Svärd and Wiklund, 1989) increase with degree of polyandry, there is also a positive increase in relative eupyrene sperm length with level of sperm competition (Gage, 1994). Fertile sperm are longer than would be expected in species which generate higher levels of sperm competition. Longer sperm may swim faster or generate more powerful flagellar forces (Katz and Drobnis, 1990) which might be advantageous if the mechanism of sperm competition demands such active characteristics. It is possible that, while there is fundamental selection on males to produce high numbers of sperm, sperm length is also under selection from sperm competition. In contrast, apyrene sperm lengths show no such associations with level of sperm competition (Gage, 1994). However, if apyrene sperm function as 'cheap fillers' in storage, we predict that their morphometry would be under selection from the dimensions of the female tract.

For moths, a comparative study found a positive association between elongation of the female spermathecal duct and fertile, but not non-fertile, sperm length. However, a positive relationship was found between relative testis volume and female spermathecal volume, suggesting coevolution between male investment in spermatogenesis and female sperm storage capacity (Morrow and Gage, 2000). Since the majority of the ejaculate consists of non-fertile sperm, this may imply that increasing investment in non-fertile sperm number is directly related to female sperm storage capacity.

Within species, sperm length may be under stabilizing selection. In *Plodia interpunctella*, restricting protein in the diet of developing male larvae causes an overall decrease in investment in spermatogenesis. Both testis size and body size were much reduced in dietary-restricted males, as was the number of both sperm types produced (Gage and Cook, 1994). However, there was no evidence for any reduction in investment in individual sperm, which were conserved at the same length as those of males developing on a full-protein diet. This suggests that sperm length may be optimized. Any reduction in investment in sperm length may result in more acute male fitness losses (in, for example, sperm competition or fertilization success) than a reduction towards investment in sperm numbers.

3.6 Female influence

The study of sperm competition has been criticized for taking a male-biased view, allowing little scope for any female influence (Eberhard, 1996). Females were simply seen as dumping grounds for males' sperm and providing an arena on which rival sperm compete. Thankfully, this is no longer our view. It is now apparent that the female exerts strong influence on sperm storage and utilization, which directly affects the outcome of sperm competition.

In the arctiid moth *Utetheisa ornatrix*, relatively larger males sire the majority of offspring in multiply-mated females, which has been interpreted as evidence of female post-copulatory choice (LaMunyon and Eisner, 1993), although the possibility that larger males transfer more sperm cannot be ruled out. Bigger males produce larger spermatophores in this species, and spermatophore size is found to a better predictor of male fertilization success than male size (LaMunyon and Eisner, 1994). Sperm transport appears to be largely dependent on the female's musculature to propel them (Tschudi-Rein and Benz, 1990; LaMunyon and Eisner, 1993), providing scope for females to influence sperm storage. In the moth *Heliothis virescens*, bigger females store more fertile, but not non-fertile, sperm. They also store more fertile sperm from older males and males that provide bigger spermatophores (LaMunyon, 2000). The former may be explained by older males transferring more sperm (LaMunyon and Huffman, 2001), or that females 'favour' high-investing males by storing more of their sperm (Eberhard, 1996). There is no correlation between spermatophore size and sperm number in this species (LaMunyon and Huffman, 2001). In addition, almost half of all females were found to store sperm also in a swollen part of the seminal duct leading from the bursa to the spermatheca. Both sperm types were also found in the lagena (Fig. 3.3).

There may be scope for females to bias sperm use between males. In the noctuid moth *Spodoptera litura*, females may even eject stored sperm when remating (Etman and Hooper, 1979). In *Pieris rapae*, although mated males producing more sperm enjoy higher fertilization success, relative sperm number is not the only factor affecting paternity. At the end of the experiment all females were dissected and the remains of the two used spermatophores were counted (to ensure that each female had achieved her two matings) and weighed. The mass of the remains of the spermatophores was related to paternity: males that had smaller used spermatophores had higher paternity (Wedell and Cook, 1998). This suggests that the spermatophores originating from males with higher paternity may have been used to a greater extent. Since movements of the female tract appear to be involved in draining and transport of sperm to the spermatheca (Tschudi-Rein and Benz, 1990; LaMunyon and Eisner, 1993), it is possible that females influence paternity by draining spermatophores to different extents. Larger spermatophore remains may represent spermatophores from disfavoured males, maybe by affecting the number of sperm reaching the spermatheca.

Females may also affect paternity by varying egg-laying rate. There is some evidence from *Manduca sexta* that a gradual decline in spermatophore mass is linked to egg-laying, since spermatophores within normally mated females show a greater decline in mass than those from females that were either prevented from egg laying or that had mated with castrated males (and therefore received a spermatophore with no sperm) (Sasaki and Riddiford, 1984). Svärd and McNeil (1994) found that in *Pseudaletia unipuncta* (where paternity is also either first or second male), female egg-laying pattern differed depending on whether the first or the second male gained most of the fertilizations. Females were divided into those where the first male fertilized all the eggs and those where the second male fertilized all the eggs, and the number of eggs laid in the 4 days between the first and second matings was compared. When the first male had priority, not only was he more successful in sperm competition, but the females had also laid more eggs prior to remating. Similarly, in the comma butterfly, *Polygonia c-album*, females exercise post-copulatory choice by varying their reproductive investment in relation to spermatophore nutrients received (Wedell, 1996). A similar situation is present in *Pieris napi*, where females increase their immediate egg production after receiving larger male nutrient donations (Wedell and Karlsson, 2003).

4. Conclusions

To date, sperm competition has been examined in 22 species of Lepidoptera. Paternity in the majority of these species shows a bimodal distribution with a strong second-male sperm precedence (Table 3.1). However, the mechanisms whereby this is achieved are still not fully understood, despite nearly 50 years of research! It is clear that high sperm number is advantageous. For the few species where relative sperm numbers are known and have been experimentally manipulated, greater sperm numbers result in increased fertilization success (Cook *et al.*, 1997: Wedell and Cook, 1998; LaMunyon, 2000, 2001). Male size appears at times to affect male fertilization success, but it is not clear how it relates to paternity. In addition, females may influence sperm utilization, most probably through their effect on sperm transport and storage. Clearly, we need to design more ingenious experiments to try and manipulate sperm numbers and circumvent female influences on sperm movements (for instance, by anaesthetizing the female with carbon dioxide; Tschudi-Rein and Benz, 1990; LaMunyon and Eisner, 1993) to gain further insight into the mechanisms determining paternity in the Lepidoptera.

It is also clear that sperm production involves a considerable cost and that sperm is in limited supply. Male Lepidoptera eclose with a full complement of sperm that cannot be replenished. Sperm limitation may have prompted evolution of several male adaptations to reduce the risk of sperm competition in order to conserve sperm (i.e. protandry, mating plugs, anti-aphrodisiacs and pheromones, various substances that switch off female receptivity, and possibly even non-fertile sperm). These adaptations, in combination with mechanisms for partitioning sperm between copulations, allow males to tailor sperm numbers in relation to expected fertilization returns to maximize their lifetime reproductive success.

Sexual conflict over female receptivity is common in Lepidoptera, in particular in species where males provide resources at mating important to female fitness. This may generate a coevolutionary arms race between the sexes, where males try and manipulate female reproductive physiology, and females evolve means to overcome male manipulations and regain control over their receptivity. This conflict may be responsible for rapid divergence in reproductive characters (Chapman et al., 2003). For example, male seminal factors switching off female receptivity in moths show considerable variation between closely related species. More detailed examination of this chemical warfare may reveal sophisticated interactions between the sexes over female receptivity.

Acknowledgements

I would like to thank my colleagues Penny Cook and Christer Wiklund for a most joyous and rewarding collaboration, and The Swedish Natural Science Foundation, NERC and The Royal Society for generous support.

References

Alcock, J. (1994) Post-insemination associations between males and females in insects: the mate-guarding hypothesis. Annual Review of Entomology 39, 1–21.

Andersson, J., Borg-Karlson, A.-K. and Wiklund, C. (2000) Sexual cooperation and conflict in butterflies: a male-transferred anti-aphrodisiac reduces harassment of recently mated females. Proceedings of the Royal Society of London, Series B 267, 1271–1275.

Andersson, J., Borg-Karlson, A.-K. and Wiklund, C. (2003) Antiaphrodisiacs in pierid butterflies: a theme with variations! Journal of Chemical Ecology 29, 1489–1499.

Ando, T., Kasuga, K., Yajima, Y., Kataoka, H. and Suzuki, A. (1996) Termination of sex pheromone production in mated females of the silkworm moth. Archives of Insect Biochemistry and Physiology 31, 207–218.

Arnqvist, G. and Nilsson, T. (2000) The evolution of polyandry: multiple mating and female fitness in insects. Animal Behaviour 60, 145–164.

Asher, J., Warren, M., Fox, R., Harding, P., Jeffcoate, G. and Jeffcoate, S. (2001) The Millennium Atlas of Butterflies in Britain and Ireland. Oxford University Press, Oxford, UK.

Baker, R.R. (1972) Territorial behaviour of the nymphalid butterflies, Aglais urticae and Inachis io. Journal of Animal Ecology 41, 453–469.

Bali, G., Raina, A.K., Kingan, T.G. and Lopez, J.D. (1996) Ovipositional behavior of newly colonized corn earworm (Lepidoptera: Noctuidae) females and evidence for an oviposition stimulating factor of male origin. Annals of the Entomological Society of America 89, 475–480.

Bebas, P., Cymborowski, B. and Giebultowicz, J.M. (2001) Circadian rhythm of sperm release in males of the cotton leafworm, Spodoptera littoralis: in vivo and in vitro studies. Journal of Insect Physiology 47, 859–866.

Birkhead, T.R. and Møller, A.P. (1998) *Sperm Competition and Sexual Selection.* Academic Press, London.

Bissoondath, C.J. and Wiklund, C. (1995) Protein content of spermatophores in relation to monandry/polyandry in butterflies. *Behavioural Ecology and Sociobiology* 37, 365–371.

Bissoondath, C.J. and Wiklund, C. (1997) Effect of male body size on sperm precedence in the polyandrous butterfly *Pieris napi* L. (Lepidoptera: Pieridae). *Behavioral Ecology* 8, 518–523.

Bitzer, R.J. and Shaw, K.C. (1979) Territorial behavior of the red admiral, *Vanessa atalanta. Journal of Research on Lepidoptera* 18, 36–49.

Boggs, C.L. and Gilbert, L.E. (1979) Male contribution to egg production in butterflies: evidence for transfer of nutrients at mating. *Science* 206, 83–84.

Boggs, C.L. and Watt, W.B. (1981) Population structure of pierid butterflies. IV. Genetic and physiological investment in offspring by male *Colias. Oecologia* 50, 320–324.

Boorman, E. and Parker, G.A. (1976) Sperm (ejaculate) competition in *Drosophila melanogaster*, and the reproductive value of females to males in relation to female age and mating status. *Ecological Entomology* 1, 145–155.

Brakefield, P.M., El Filali, E., Van der Laan, R., Breuker, C.J., Saccheri, I.J. and Zwaan, B. (2001) Effective population size, reproductive success and sperm precedence in the butterfly, *Bicyclus anynana*, in captivity. *Journal of Evolutionary Biology* 14, 148–156.

Brower, J.H. (1978) Sperm precedence in the Indian Meal moth, *Plodia interpunctella. Annals of the Entomological Society of America* 68, 78–80.

Bulmer, M.G. (1983) Models for the evolution of protandry in insects. *Theoretical Population Biology* 23, 314–322.

Bulmer, M.G. and Parker, G.A. (2002) The evolution of anisogamy: a game-theoretic approach. *Proceedings of the Royal Society of London, Series B* 269, 2381–2388.

Callahan, P.S. (1958) Serial morphology as a technique for determination of reproductive patterns in the corn earworm, *Heliothis zea* (Boddie). *Annals of the Entomological Society of America* 51, 413–428.

Callahan, P.S. and Cascio, T. (1963) Histology of the reproductive tracts and transmission of sperm in the corn earworm, *Heliothis zea. Annals of the Entomological Society of America* 56, 535–556.

Carpenter, J.E. (1992) Sperm precedence in *Helicoverpa zea* (Lepidoptera: Noctuidae): response to a substerilizing dose of radiation. *Journal of Economic Entomology* 85, 779–782.

Chapman, T., Liddle, L.F., Kalb, J.M., Wolfner, M. and Partridge, L. (1995) Cost of mating in *Drosophila melanogaster* females is mediated by male accessory gland products. *Nature* 373, 241–244.

Chapman, T., Arnqvist, G., Bangham, J. and Rowe, L. (2003) Sexual conflict. *Trends in Ecology and Evolution* 18, 41–47.

Clarke, C.A. and Sheppard, P.M. (1962) Offspring from double matings in swallowtail butterflies. *Entomologist* 95, 199–203.

Cook, P.A. and Gage, M.J.G. (1995) Effects of risks of sperm competition on the numbers of eupyrene and apyrene sperm ejaculated by the moth *Plodia interpunctella* (Lepidoptera: Pyralidae). *Behavioral Ecology and Sociobiology* 36, 261–268.

Cook, P.A. and Wedell, N. (1996) Ejaculate dynamics in butterflies: a strategy for maximizing fertilization success? *Proceedings of the Royal Society of London, Series B* 263, 1047–1051.

Cook, P.A. and Wedell, N. (1999) Non-fertile sperm delay female remating. *Nature* 397, 486.

Cook, P.A., Harvey, I.F. and Parker, G.A. (1997) Predicting variation in sperm precedence. *Philosophical Transactions of the Royal Society of London, Series B* 352, 771–780.

Delisle, J. and Hardy, M. (1997) Male larval nutrition influences the reproductive success of both sexes of the spruce budworm, *Choristoneura fumiferana* (Lepidoptera: Tortricidae). *Functional Ecology* 11, 451–463.

Delisle, J., Picimbon, J.-F. and Simard, J. (2000) Regulation of pheromone inhibition in mated females of *Choristoneura fumiferana* and *C. rosaceana*. *Journal of Insect Physiology* 46, 913–921.

Dickinson, J.L. and Rutowski, R.L. (1989) The function of the mating plug in the chalcedon checkerspot butterfly. *Animal Behaviour* 38, 154–162.

Drummond, B.A., III (1984) Multiple mating and sperm competition in the Lepidoptera. In: Smith, R.L. (ed.) *Sperm Competition and the Evolution of Animal Mating Systems*. Academic Press, London, pp. 291–370.

Dussourd, D.E., Ubik, K., Harvis, C., Resch, J., Meinwald, J. and Eisner, T. (1988) Biparental defensive endowment of eggs with acquired plant alkaloid in the moth *Utetheisa ornatrix*. *Proceedings of the National Academy of Sciences USA* 85, 5992–5996.

Eberhard, W.G. (1996) *Female Control: Sexual Selection by Cryptic Female Choice*. Princeton University Press, Princeton, New Jersey.

Etman, A.A.M. and Hooper, G.H.S. (1979) Sperm precedence of the last mating in *Spodoptera litura*. *Annals of the Entomological Society of America* 72, 119–120.

Fagerström, T. and Wiklund, C. (1982) Why do males emerge before females? Protandry as a mating strategy in male and female butterflies. *Oecologia* 52, 164–166.

Flint, H.M. and Kressin, E.L. (1968) Gamma irradiation of the tobacco budworm: sterilization, competitiveness, and observations on reproductive biology. *Journal of Economic Entomology* 61, 477–483.

Friedländer, M. (1997) Control of the eupyrene–apyrene sperm dimorphism in the Lepidoptera. *Journal of Insect Physiology* 43, 1085–1092.

Gage, M.J.G. (1994) Associations between body size, mating pattern, testes size and sperm lengths across butterflies. *Proceedings of the Royal Society of London, Series B* 258, 247–254.

Gage, M.J.G. (1995) Continuous variation in reproductive strategy as an adaptive response to population density in the moth *Plodia interpunctella*. *Proceedings of the Royal Society of London, Series B* 261, 25–30.

Gage, M.J.G. (1998) Influences of sex, size, and symmetry on ejaculate expenditure in a moth. *Behavioral Ecology* 9, 592–597.

Gage, M.J.G. and Cook, P.A. (1994) Sperm size or numbers? Effects of nutritional stress upon eupyrene and apyrene sperm production strategies in the moth *Plodia interpunctella* (Lepidoptera: Pyralidae). *Functional Ecology* 8, 594–599.

Giebultowicz, J.M. and Zdarek, J. (1996) The rhythms of sperm release from the testis and mating flight are not correlated in *Lymantria* moths. *Journal of Insect Physiology* 42, 167–170.

Giebultowicz, J.M., Riemann, J.G., Raina, A.K. and Ridgway, R.L. (1989) Circadian system controlling release of sperm in the insect testes. *Science* 245, 1098–1100.

Giebultowicz, J.M., Raina, A.K. and Ridgway, R.L. (1991a) Role of the spermatheca in the termination of sex pheromone production in mated gypsy moth females. In: *Proceedings of the Conference on Insect Chemical Ecology*. Academia Prague and SPB Academic Publishers, The Hague, Netherlands, pp. 101–104.

Giebultowicz, J.M., Raina, A.K., Uebel, E.C. and Ridgway, R.L. (1991b) Two-step regulation of sex-pheromone decline in mated gypsy moth females. *Archives of Insect Biochemistry and Physiology* 16, 95–105.

Gilbert, L.E. (1976) Post-mating female odour in *Heliconius* butterflies: a male contributed anti-aphrodisiac? *Science* 193, 419–420.

Gillott, C. (2003) Male accessory gland secretions: modulators of female reproductive physiology and behavior. *Annual Review of Entomology* 48, 163–184.

He, Y. and Miyata, T. (1997) Variations in sperm number in relation to larval crowding and spermatophore size in the armyworm, *Pseudaletia separata*. *Ecological Entomology* 22, 41–46.

He, Y. and Tsubaki, Y. (1992) Variation in spermatophore size in the armyworm, *Pseudaletia separata* (Lepidoptera: Noctuidae) in relation to rearing density. *Applied Entomology and Zoology* 27, 39–45.

He, Y. and Tsubaki, Y. (1995) Effects of spermatophore size on female remating in the armyworm, *Pseudaletia separata*, with reference to larval crowding. *Journal of Ethology* 9, 47–50.

He, Y., Tanaka, T. and Miyata, T. (1995) Eupyrene and apyrene sperm and their numerical fluctuations inside the female reproductive tract of the armyworm, *Pseudaletia separata*. *Journal of Insect Physiology* 41, 689–694.

Hirai, K., Shorey, H.H. and Gaston, L.K. (1978) Competition among courting male moths: male-to-male inhibitory pheromone. *Science* 202, 644–645.

Hiroyoshi, S. (1995) Regulation of sperm quantity transferring to females at mating in the adult male of *Polygonia c-aureum* (Lepidoptera: Nymphalidae). *Applied Entomology and Zoology* 30, 111–119.

Holt, G.G. and North, D.T. (1970) Effects of gamma irradiation on the mechanisms of sperm transfer in *Trichoplusia ni*. *Journal of Insect Physiology* 16, 2211–2222.

Hosken, D.J., Garner, T.W.J., Tregenza, T., Wedell, N. and Ward, P.I. (2003) Superior sperm competitors sire higher-quality young. *Proceedings of the Royal Society of London, Series B* 270, 1933–1938.

Iwasa, I., Odendaal, F.J., Murphy, D.D., Ehrlich, P.R. and Launer, A.E. (1983) Emergence patterns in male butterflies: a hypothesis and a test. *Theoretical Population Biology* 23, 363–379.

Jennions, M.D. and Petrie, M. (2000) Why do females mate multiply? A review of the genetic benefits. *Biological Reviews* 75, 21–64.

Jiggins, F.M., Hurst, G.D.D. and Majerus, M.E.N. (2002) Sex-ratio-distorting *Wolbachia* causes sex-role reversal in its butterfly host. *Proceedings of the Royal Society of London, Series B* 267, 69–73.

Jin, Z.Y. and Gong, H. (2001) Male accessory gland derived factors can stimulate oogenesis and enhance oviposition in *Helicoverpa armigera* (Lepidoptera: Noctuidae). *Archives of Insect Biochemistry and Physiology* 46, 175–185.

Kaitala, A. and Wiklund, C. (1995) Female mate choice and mating costs in the polyandrous butterfly *Pieris napi* (Lepidoptera: Pieridae). *Journal of Insect Behavior* 8, 355–363.

Karlsson, B. (1995) Resource allocation and mating systems in butterflies. *Evolution* 49, 955–961.

Karlsson, B. (1996) Male reproductive investment in relation to mating system in butterflies: a comparative study. *Proceedings of the Royal Society of London, Series B* 263, 187–192.

Karube, F. and Kobayashi, M. (1999) Presence of eupyrene spermatozoa in vestibulum accelerates oviposition in the silkworm moth, *Bombyx mori*. *Journal of Insect Physiology* 45, 947–957.

Katz, D.F. and Drobnis, E.Z. (1990) Analysis and interpretation of the forces generated by spermatozoa. In: Bavister, B.D., Cummins, J. and Roldan, E.R.S. (eds) *Fertilization in Mammals*. Sereno Symposia, Norwell, Massachusetts, pp. 125–137.

Kingan, T.G., Thomas-Laemont, P.A. and Raina, A.K. (1993) Male accessory gland factors elicit change from 'virgin' to 'mated' behaviour in the female corn earworm moth *Helicoverpa zea*. *Journal of Experimental Biology* 183, 61–76.

Kingan, T.G., Bodnar, W.M., Raina, A.K., Shabanowitz, J. and Hunt, D.F. (1995) The loss of female sex pheromone after mating in the corn earworm moth *Helicoverpa zea*: identification of a male pheromonostatic peptide. *Proceedings of the National Academy of Sciences USA* 92, 5082–5086.

Labine, P.A. (1966) The population biology of the butterfly, *Euphydryas editha*. IV. Sperm precedence: a preliminary report. *Evolution* 20, 580–586.

Labitte, M.A. (1919) Observations sur *Rhodocera rhamni*. *Bulletin of the Museum of Natural History Paris* 25, 624–625.

LaChance, L.E., Richard, R.D. and Ruud, R.L. (1977) Movement of eupyrene sperm bundles from the testis and storage in the ductus ejaculatoris duplex of the male pink bollworm: effects of age, strain, irradiation and light. *Annals of the Entomological Society of America* 70, 647–651.

LaMunyon, C.W. (1994) Paternity in naturally-occurring *Utetheisa ornatrix* (Lepidoptera. Arctiidae) as estimated using enzyme polymorphism. *Behavioral Ecology and Sociobiology* 34, 403–408.

LaMunyon, C.W. (2000) Sperm storage by females of the polyandrous noctuid moth, *Heliothis virescens*. *Animal Behaviour* 59, 395–402.

LaMunyon, C.W. (2001) Determinants of sperm precedence in a noctuid moth, *Heliothis virescens*: a role for male age. *Ecological Entomology* 26, 388–394.

LaMunyon, C.W. and Eisner, T. (1993) Postcopulatory sexual selection in an arctiid moth (*Utetheisa ornatrix*). *Proceedings of the National Academy of Sciences USA* 90, 4689–4692.

LaMunyon, C.W. and Eisner, T. (1994) Spermatophore size as determinant of paternity in an arctiid moth (*Utetheisa ornatrix*). *Proceedings of the National Academy of Sciences USA* 91, 7081–7084.

LaMunyon, C.W. and Huffman, T.S. (2001) Determinants of sperm transfer by males of the noctuid moth *Heliothis virescens*. *Journal of Insect Behavior* 14, 187–199.

Lum, P.T.M. and Arbogast, R.T. (1980) Ultrastructure of setae in the spermathecal gland of *Plodia interpunctella* (Hübner) (Lepidoptera: Pyralidae). *International Journal of Insect Morphology and Embryology* 9, 251–253.

Marco, M.-P., Fabriàs, G., Lázaro, G. and Camps, F. (1996) Evidence for both humoral and neural regulation of sex pheromone biosynthesis in *Spodoptera littoralis*. *Archives of Insect Biochemistry and Physiology* 31, 157–167.

Marcotte, M., Delisle, J. and McNeil, J.N. (2003) Pheromonostasis is not directly associated with post-mating sperm dynamics in *Choristoneura fumiferana* and *C. rosaceana* females. *Journal of Insect Physiology* 49, 81–90.

Mason, L.J. and Pashley, D.P. (1991) Sperm competition in the soybean looper (Lepidoptera: Noctuidae). *Annals of the Entomological Society of America* 84, 268–271.

Matsumoto, K. and Suzuki, N. (1992) Effectiveness of the mating plug in *Atrophaneura alcinous* (Lepidoptera: Papilionidae). *Behavioral Ecology and Sociobiology* 30, 157–163.

Matsumoto, K. and Suzuki, N. (1995) The nature of mating plugs and the probability of reinsemination in Japanese Papilionidae. In: Scriber, J.M., Tsubaki, Y. and Lederhauser, R.C. (eds) *Swallowtail Butterflies: Their Ecology and Evolutionary Biology*. Scientific Publishers, Gainesville, Florida, pp. 145–155.

Meves, F. (1902) Ueber oligopyrene und apyrene Spermien und über ihre Entstehung, nach Beobachtungen an *Paludina* und *Pygaera*. *Archiv für Mikroskopische Anatomie* 61, 1–84.

Miskimen, G.W., Rodriguez, N.L. and Nazario, M.L. (1983) Reproductive morphology and sperm transport facilitation and regulation in the female sugarcane borer, *Diatraea saccharalis* (F.) (Lepidoptera: Crambidae). *Annals of the Entomological Society of America* 76, 248–252.

Morrow, E.H. and Gage, M.J.G. (2000) The evolution of sperm length in moths. *Proceedings of the Royal Society of London, Series B* 267, 307–313.

Norris, N.J. (1932) Contributions towards the study of insect fertility. I. The structure and operation of the reproductive organs of the genera *Ephestia* and *Plodia* (Lepidoptera, Phycitidae). *Proceedings of the Zoological Society* 1932, 595–611.

North, D.T. and Holt, G.G. (1968) Genetic and cytogenetic basis of radiation-induced sterility in the adult male cabbage looper *Trichoplusia ni*. In: *Symposium on the Use of Isotopes and Radiation in Entomology (1967)*. IAEA/FAO, Vienna, pp. 391–403.

Obara, Y., Tateda, H. and Kuwabara, M. (1975) Mating behavior of the cabbage white butterfly, *Pieris rapae crucivora* Biosduval. V. Copulatory stimuli inducing changes of female response patterns. *Zoological Magazine* 84, 71–76.

Oberhauser, K.S. (1997) Fecundity, lifespan and egg mass in butterflies: effects of male-derived nutrients and female size. *Functional Ecology* 11, 166–175.

Omura, S. (1939) Selective fertilization in *Bombyx mori*. *Japanese Journal of Genetics* 15, 29–35.

Orr, A.G. (1995) The evolution of sphragis in the Papilionidae and other butterflies. In: Scriber, J.M., Tsubaki, Y. and Lederhauser, R.C. (eds) *Swallowtail Butterflies: Their Ecology and Evolutionary Biology*. Scientific Publishers, Gainesville, Florida, pp. 155–164.

Orr, A.G. (2002) The sphragis of *Heteronympha penelope* Waterhouse (Lepidoptera: Satyridae): its structure, formation and role in sperm guarding. *Journal of Natural History* 36, 185–196.

Orr, A.G. and Rutowski, R.L. (1991) The function of the sphragis in *Cressida cressida* (Fab) (Lepidoptera, Papilionidae): a visual deterrent to copulation attempts. *Journal of Natural History* 25, 703–710.

Osanai, M., Kasuga, H. and Aigaki, T. (1986) Action of apyrene spermatozoa of silkworm on dissociation of eupyrene bundles. *Zoological Sciences* 3, 1031.

Osanai, M., Kasuga, H. and Aigaki, T. (1987) Physiological role of apyrene spermatozoa of *Bombyx mori*. *Experientia* 43, 593–596.

Pair, S.D., Laster, M.L. and Martin, D.F. (1977) Hybrid sterility of the tobacco budworm: effects of alternate sterile and normal matings on fecundity and fertility. *Annals of the Entomological Society of America* 70, 952–954.

Park, Y.I. and Ramaswamy, S.B. (1998) Role of brain, ventral nerve cord, and corpora cardiaca-corpora allata complex in the reproductive behavior of female tobacco budworm (Lepidoptera: Noctuidae). *Annals of the Entomological Society of America* 91, 329–334.

Park, Y.I., Shy, S., Ramaswamy, S.B. and Srinivasan, A. (1998) Mating in *Heliothis virescens*: transfer of juvenile hormone during copulation by male to female and stimulation of biosynthesis of endogenous juvenile hormone. *Archives of Insect Physiology and Biochemistry* 38, 100–107.

Parker, G.A. (1970) Sperm competition and its evolutionary consequences in the insects. *Biological Reviews* 45, 525–567.

Parker, G.A. (1982) Why are there so many tiny sperm? Sperm competition and the maintenance of two sexes. *Journal of Theoretical Biology* 96, 281–294.

Parker, G.A. (1990) Sperm competition games: raffles and roles. *Proceedings of the Royal Society of London, Series B* 242, 120–126.

Parker, G.A. (1998) Sperm competition and the evolution of ejaculates: towards a theory base. In: Birkhead, T.R. and Møller, A.P. (eds) *Sperm Competition and Sexual Selection*. Academic Press, London, pp. 3–54.

Parker, G.A. (2001) Golden flies, sunlit meadows: a tribute to the yellow dungfly. In: Dugatkin, L.A. (ed.) *Model Systems in Behavioral Ecology*. Princeton University Press, Princeton, New Jersey, pp. 3–56.

Parker, G.A., Baker, R.T. and Smith, V.G.F. (1972) The origin and evolution of gamete dimorphism and the male–female phenomenon. *Journal of Theoretical Biology* 36, 181–198.

Phillips, D.M. (1970) Insect sperm: their structure and morphogenesis. *Journal of Cell Biology* 44, 243–277.

Pieloor, M.J. and Seymour, J.E. (2001) Factors affecting adult diapause initiation in the tropical butterfly *Hypolimnas bolina* L. (Lepidoptera: Nymphalidae). *Australian Journal of Entomology* 40, 376–379.

Platt, A.P. and Allen, J.F. (2001) Sperm precedence and competition in doubly-mated *Limenitis arthemis-astyanax* butterflies (Rhopalocera: Nymphalidae). *Annals of the Entomological Society of America* 94, 654–663.

Pliske, T.E. (1973) Factors determining mating frequencies in some New World butterflies. *Annals of the Entomological Society of America* 66, 164–169.

Pliske, T.E. (1975) Courtship behavior of the monarch butterfly, *Danaus plexippus* L. *Annals of the Entomological Society of America* 68, 143–151.

Proshold, F.I. (1991) Number of sperm bundles in duplex of tobacco budworms (Lepidoptera: Noctuidae) as a function of age. *Journal of Economic Entomology* 84, 1485–1491.

Proshold, F.I. (1995) Remating by gypsy moths (Lepidoptera: Lymantriidae) mated with F_1-sterile males as a function of sperm within the spermatheca. *Journal of Economic Entomology* 88, 644–648.

Proshold, F.I. (1996) Reproductive capacity of laboratory reared Gypsy moths (Lepidoptera: Lymantriidae): effect of female age at time of mating. *Journal of Economic Entomology* 89, 337–342.

Proshold, F.I. (1997) Sperm transfer in gypsy moths: effect of constant light or cyclic temperature and constant light or darkness during the pupal stage. *Journal of Entomological Science* 32, 321–331.

Proshold, F.I. and Bernon, G.L. (1994) Multiple mating in laboratory-reared gypsy moths (Lepidoptera: Lymantriidae). *Journal of Economic Entomology* 87, 661–666.

Proverbs, M.D. and Newton, J.R. (1962) Some effects of gamma radiation on the potential of the codling moth, *Carpocapsa pomonella* (L.) (Lepidoptera: Olethreutidae). *Canadian Entomologist* 94, 1162–1170.

Radwan, J. (1998) Heritability of sperm competition success in the bulb mite, *Rhizoglyphus robini*. *Journal of Evolutionary Biology* 11, 321–327.

Rafaeli, A. (2002) Neuroendocrine control of pheromone biosynthesis in moths. *Review of Cytology* 213, 49–91.

Raina, A.K. (1993) Neuroendocrine control of sex pheromone biosynthesis in Lepidoptera. *Annual Review of Entomology* 39, 329–349.

Ramaswamy, S.B., Qiu, Y. and Park, Y.I. (1996) Neuronal control of post-coital pheromone production in the moth *Heliothis virescens*. *Journal of Experimental Zoology* 274, 255–263.

Rananavare, H.D., Harwalkar, M.R. and Rahalkar, G.W. (1990) Studies on the mating behavior of radiosterilized males of potato tuberworm *Phtorimaea operculella* Zeller. *Journal of Nuclear and Agricultural Biology* 19, 47–53.

Retnakaran, A. (1974) The mechanism of sperm precedence in the spruce budworm, *Choristoneura fumiferana* (Lepidoptera: Tortricidae). *Canadian Entomologist* 106, 1189–1194.

Richard, R.D., Lachance, L.E. and Proshold, F.I. (1975) An ultrastructural study of sperm in sterile hybrids from crosses of *Heliothis virescens* and *Heliothis subflexa*. *Annals of the Entomological Society of America* 68, 35–39.

Riemann, J.G., Thorson, B.J. and Ruud, R.L. (1974) Daily cycle of release of sperm from the testes of the Mediterranean flour moths. *Journal of Insect Physiology* 20, 195–207.

Rogers, S.H. and Wells, H. (1984) The structure and function of the bursa copulatrix of the monarch butterfly (*Danaus plexippus*). *Journal of Morphology* 180, 213–221.

Sahara, K. and Takemura, Y. (2003) Application of artificial insemination technique to eupyrene and/or apyrene sperm in *Bombyx mori*. *Journal of Experimental Zoology* 297A, 196–200.

Sasaki, M. and Riddiford, L.M. (1984) Regulation of reproductive behaviour and egg maturation in the tobacco hawk moth, *Manduca sexta*. *Physiological Entomology* 9, 315–327.

Seth, S.K., Kaur, J.J., Rao, D.K. and Reynolds, S.E. (2002a) Sperm transfer during mating, movement of sperm in the female reproductive tract, and sperm precedence in the common cutworm *Spodoptera litura*. *Physiological Entomology* 27, 1–14.

Seth, S.K., Rao, D.K. and Reynolds, S.E. (2002b) Movement of spermatozoa in the reproductive tract of adult male *Spodoptera litura*: daily rhythm of sperm movement and the effect of light regime on male reproduction. *Journal of Insect Physiology* 48, 119–131.

Shantharam, K., Tamhankar, A.J. and Rananavare, H.D. (1998) Impact on calling behavior of *Earias vittella* (Fabricius) females mated to radiation sterilized males. *Journal of Nuclear Agriculture and Biology* 27, 57–60.

Silberglied, R.L., Shepherd, J.G. and Dickinson, J.L. (1984) Eunuchs: the role of apyrene sperm in Lepidoptera? *American Naturalist* 123, 255–265.

Simmons, L.W. (2001) *Sperm Competition and its Evolutionary Consequences in Insects*. Princeton University Press, New Jersey.

Simmons, L.W. and Siva-Jothy, M.T. (1998) Sperm competition in insects: mechanisms and the potential for selection. In: Birkhead, T.R. and Møller, A.P. (eds) *Sperm Competition and Sexual Selection*. Academic Press, London, pp. 341–434.

Smedley, S.R. and Eisner, T. (1996) Sodium: a male moth's gift to its offspring. *Proceedings of the National Academy of Sciences USA* 93, 809–813.

Smith, R.L. (1984) *Sperm Competition and the Evolution of Animal Mating Systems*. Academic Press, London.

Snow, J.W., Young, F.R. and Jones, R.L. (1970) Competitiveness of sperm in female fall armyworms mating with normal and chemosterilized males. *Journal of Economic Entomology* 63, 1799–1802.

Snow, J.W., Young, J.R., Lewis, W.J. and Jones, R.L. (1972) Sterilization of adult fall armyworms by gamma irradiation and its effect on competitiveness. *Journal of Economic Entomology* 65, 1431–1433.

Solensky, M.J. (2003) Reproductive fitness in monarch butterflies (*Danaus plexippus*). PhD thesis, University of Minnesota, St Paul, Minnesota.

Spurgeon, D.W., Raulston, J.R., Lingren, P.D., Shaver, T.N., Proshold, F.I. and Gillespie, J.M. (1994) Temporal aspects of sperm transfer and spermatophore condition in Mexican rice borers (Lepidoptera: Pyralidae). *Journal of Economic Entomology* 87, 371–376.

Stockley, P., Gage, M.J.G., Parker, G.A. and Møller, A.P. (1997) Sperm competition in fishes: the evolution of testis size and ejaculate characteristics. *American Naturalist* 149, 933–954.

Sugawara, T. (1979) Stretch reception in the bursa copulatrix of the butterfly, *Pieris rapae crucivora*, and its role in behaviour. *Journal of Comparative Physiology* 130, 191–199.

Suzuki, N., Okuda, T. and Shinbo, H. (1996) Sperm precedence and sperm movement under different copulation intervals in the silkworm, *Bombyx mori*. *Journal of Insect Physiology* 42, 199–204.

Svärd, L. and McNeil, J.N. (1994) Female benefit, male risk: polyandry in the true armyworm *Pseudaletia unipuncta*. *Behavioral Ecology and Sociobiology* 35, 319–326.

Svärd, L. and Wiklund, C. (1988) Prolonged mating in the monarch butterfly *Danaus plexippus* and nightfall as a cue for sperm transfer. *Oikos* 52, 351–354.

Svärd, L. and Wiklund, C. (1989) Mass and production rate of ejaculates in relation to monandry/polyandry in butterflies. *Behavioral Ecology and Sociobiology* 24, 395–402.

Thornhill, R. and Alcock, J. (1983) *The Evolution of Insect Mating Systems*. Harvard University Press, Cambridge, Massachusetts.

Torres-Vila, L.M., Gragera, J., Rodríguez-Molina, M.C. and Stockel, J. (2002) Heritable variation for female remating in *Lobesia botrana*, a usually monandrous moth. *Animal Behaviour* 64, 899–907.

Tregenza, T. and Wedell, N. (2000) Genetic compatibility, mate choice and patterns of parentage. *Molecular Ecology* 9, 1013–1027.

Tschudi-Rein, K. and Benz, G. (1990) Mechanisms of sperm transfer in female *Pieris brassicae* (Lepidoptera: Pieridae). *Annals of the Entomological Society of America* 83, 1158–1164.

Vahed, K. (1998) The function of nuptial feeding in insects: a review of empirical studies. *Biological Reviews* 73, 43–78.

Virkki, N. (1969) Sperm bundles and phylogenesis. *Zeitschrift für Zellforschung* 101, 13–27.

Watanabe, M., Bon'no, M. and Hacisuka, A. (2000) Eupyrene sperm migrated to the spermatheca after the apyrene sperm in the swallowtail butterfly *Papilio xuthus* L. (Lepidoptera: Papilionidae). *Journal of Ethology* 18, 91–99.

Wedell, N. (1992) Protandry and mate assessment in the wartbiter *Decticus verrucivorus* (Orthoptera: Tettigoniidae). *Behavioral Ecology and Sociobiology* 31, 301–308.

Wedell, N. (1996) Mate quality affects reproductive effort in a paternally investing species. *American Naturalist* 148, 1075–1088.

Wedell, N. (1997) Ejaculate size in bushcrickets: the importance of being large. *Journal of Evolutionary Biology* 10, 315–325.

Wedell, N. (2001) Female remating in butterflies: interaction between female genotype and nonfertile sperm. *Journal of Evolutionary Biology* 14, 746–754.

Wedell, N. and Cook, P.A. (1998) Determinants of paternity in a butterfly. *Proceedings of the Royal Society of London, Series B* 265, 625–630.

Wedell, N. and Cook, P.A. (1999a) Butterflies tailor their ejaculate in response to sperm competition risk and intensity. *Proceedings of the Royal Society of London, Series B* 266, 1033–1039.

Wedell, N. and Cook, P.A. (1999b) Strategic sperm allocation in the Small White butterfly *Pieris rapae* (Lepidoptera: Pieridae). *Functional Ecology* 13, 85–93.

Wedell, N. and Karlsson, B. (2003) Paternal investment directly affects female reproductive effort in an insect. *Proceedings of the Royal Society of London, Series B* 270, 2065–2071.

Wedell, N., Wiklund, C. and Cook, P.A. (2002a) Monandry and polyandry as alternative lifestyles in a butterfly. *Behavioral Ecology* 13, 450–455.

Wedell, N., Gage, M.J.G. and Parker, G.A. (2002b) Sperm competition, male prudence and sperm limited females. *Trends in Ecology and Evolution* 17, 313–320.

Wiklund, C. and Fagerström, T. (1977) Why do males emerge before females? A hypothesis to explain the incidence of protandry in butterflies. *Oecologia* 31, 153–158.

Wiklund, C. and Forsberg, J. (1985) Courtship and male discrimination between virgin and mated females in the orange tip butterfly *Anthocharis cardamines*. *Animal Behaviour* 34, 328–332.

Wiklund, C. and Forsberg, J. (1991) Sexual size dimorphism in relation to female polygamy and protandry in butterflies: a comparative study of Swedish Pieridae and Satyridae. *Oikos* 60, 373–381.

Wiklund, C. and Kaitala, A. (1995) Sexual selection for large male size in a polyandrous butterfly: the effect of body size on male versus female reproductive success in *Pieris napi*. *Behavioral Ecology* 6, 6–13.

Wiklund, C. and Solbreck, C. (1982) Adaptive versus incidental explanations for the occurrence of protandry in a butterfly, *Leptidea sinapsis* L. *Evolution* 36, 55–62.

Wiklund, C., Wickman, P.-O. and Nylin, S. (1992) A sex difference in the propensity to enter direct/diapause development: a result of selection for protandry. *Evolution* 46, 519–528.

Wiklund, C., Kaitala, A., Lindfors, V. and Abenius, J. (1993) Polyandry and its effect on female reproduction in the green-veined butterfly (*Pieris napi* L.). *Behavioral Ecology and Sociobiology* 33, 25–33.

Wiklund, C., Karlsson, B. and Leimar, O. (2001) Sexual conflict and cooperation in butterfly reproduction: a comparative study of polyandry and female fitness. *Proceedings of the Royal Society of London, Series B* 268, 1661–1667.

Witalis, J. and Godula, J. (1993) Postembryonal development of the testes in cotton leaf worm, *Spodoptera littoralis* (Boisd.) (Noctuidae, Lepidoptera). *Acta Biologica Hungarica* 44, 281–295.

Zeh, J.A. and Zeh, D.W. (1996) The evolution of polyandry I: intragenomic conflict and genetic incompatibility. *Proceedings of the Royal Society of London, Series B* 263, 1711–1717.

4 Alternative Mating Tactics and Fatal Fighting in Male Fig Wasps

JAMES M. COOK

Department of Biological Sciences, Imperial College London, Ascot, UK

1. Introduction

In most animal species, conspecific males tend to look quite similar and combat with rival males is not a major source of mortality. However, Hamilton (1979) noted that fig wasps commonly show three traits – male aptery, male dimorphism and lethal male combat – that are all relatively rare in animals. Male dimorphism can be so extreme (Fig. 4.1) that conspecific winged and wingless males have sometimes been mistaken for different species or even genera (Hamilton, 1979). Male fighting can also be extreme and, in some species, 50% of the males may die in lethal combat over mating opportunities.

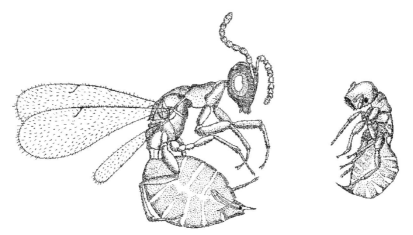

Fig. 4.1. The winged and wingless male morphs of the gall-inducing species *Pseudidarnes minerva* (Sycophaginae). Winged males mate outside the syconium, while wingless males bite into and enter galls containing females in order to mate. (From Cook *et al.*, 1997.) Published with permission.

The dramatic variation in male morphology and behaviour is seen both within and between species and, furthermore, species that do and do not show these strange traits can be found living alongside each other in the same fig fruits (syconia) (Cook *et al.*, 1999). Why are there so many ways to be a male fig wasp? I address this question by reviewing the nature of, and explanations for, male polymorphism and fatal fighting in fig wasps. To begin with, however, I give a brief introduction to male polymorphism and fatal fighting in general.

1.1 Male polymorphism

In many animal species there exist two or more male morphs that differ considerably in behaviour, morphology, or both (Darwin, 1874; Andersson, 1994; Gross, 1996; Shuster and Wade, 2003). Male polymorphism is usually linked to differences in mate acquisition and is common in species where sexual selection is strong, because the mating system generates considerable variance in male reproductive success (Gadgil, 1972; Andersson, 1994; Shuster and Wade, 2003). In theory, male polymorphism can be achieved in different ways. First, different morphs may carry different alleles (Mendelian strategies). Second, they may be genetically similar, but follow different developmental pathways, dependent upon environmental cues (a conditional strategy). Both scenarios have been studied using game theory and phenotypic models (e.g. Maynard Smith, 1982; Parker, 1984; Charnov, 1993), although Shuster and Wade (2003) strongly advocate the use of population genetic and general evolutionary analysis to gain further insights. Since many cases of alternative male tactics involve a dominant and a sneaker/satellite tactic, it is convenient to discuss models in this context. Following Gross (1996) and Shuster and Wade (2003), I will endeavour to use the term 'tactic' for the phenotype and 'strategy' for the underlying genotype. These terms are interchangeable with regard to Mendelian strategies, but a conditional strategy has multiple tactics.

   Game theory has been used to investigate whether there can be an evolutionarily stable strategy (ESS) that permits the coexistence of alternative strategies in a population (Maynard Smith and Price, 1973; Maynard Smith, 1982; Parker, 1984; Charnov, 1993). In some cases, there is an ESS frequency f of type A (e.g. dominant) and $(1 - f)$ of type B (e.g. sneaker) and coexistence is favoured $(0 < f < 1)$, because each phenotype is under frequency-dependent selection, such that it has high fitness below and low fitness above its ESS frequency (Fig. 4.2a). This can be achieved if a fraction f of individuals play A, while $(1 - f)$ adopt B. An important prediction is that the average fitness of A and B is equal at the ESS frequency (ESSf*). Earlier deliberations on alternative mating tactics (e.g. Darwin, 1874; Gadgil, 1972) assumed that these would have equal fitnesses, and indeed a genetically determined alter-native will only invade a population if it has a fitness advantage (relative to the resident tactic) when rare, and will only be maintained if it has equal fitness once it has spread (Darwin, 1874; Maynard Smith, 1982; Parker, 1984; Shuster and Wade, 2003). There is good evidence for Mendelian strategies with equal fitness in a few well-studied cases, such as the lizard *Uta stansburiana* (Sinervo

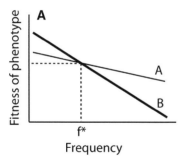 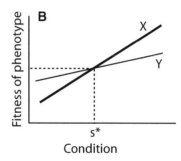

Fig. 4.2. Two ways in which natural selection can maintain alternative tactics in a population: (**A**) Alternative genetic strategies (A and B) are predicted to occur at an ESS frequency (f^*), where each has the same fitness; (**B**) a conditional genetic strategy has two tactics (X and Y) with different mean fitness across all individuals, but the same fitness at the ESS switchpoint, where the optimal tactic to adopt changes from X to Y.

and Lively, 1996) and a marine isopod *Paracerceis sculpta* (Shuster and Wade, 1991, 2003). In theory, the ESS can be achieved in another way (mixed ESS), by each individual adopting tactic A or B stochastically, but with probabilities f and $(1 - f)$ respectively. This is an intriguing possibility for purely behavioural dimorphisms, but no convincing examples are yet known (Gross, 1996).

The majority of empirical studies have reported unequal fitness estimates for different male tactics (Austad, 1984; Gross, 1996; Shuster and Wade, 2003). Such cases could represent conditional strategies, where the tactic employed is condition (e.g. size) and/or status (social context) dependent, such that individuals adopt the tactic most appropriate to maximizing their fitness, given their condition. Theory predicts a single genetic strategy, which allows for condition-dependent choice between tactics, e.g. fighting (X) and sneaking (Y), which are influenced differentially by body size (condition). The average fitness of the two tactics can be different, but each individual is using the tactic that maximizes fitness, given its condition. Models predict an ESS switchpoint (ESSs^*) at which the optimal tactic to adopt changes (Fig. 4.2b). At this switch point, the fitnesses of X and Y are equal, but the average fitnesses of all X and Y individuals are not equal.

A key difference between these two models is that the morphs are predicted to have equal fitness in the Mendelian, but not in the conditional, case. There is good evidence that both scenarios exist in nature, but considerable disagreement as to which predominates. Gross and others (Austad, 1984; Gross, 1996; Gross and Repka, 1998) have argued that few examples support Mendelian strategies and that conditional strategies are common. In contrast, Shuster and Wade (2003) argued that Mendelian alternatives should be common, because essentially the only selection pressure required for one to invade a population is variation in male reproductive success. Good estimates of lifetime reproductive success for different morphs are crucial to these debates, but are generally difficult to obtain. Shuster and

Wade (2003) argued that many studies show higher fitness for the dominant strategy (e.g. territory holder) erroneously, because of a general (unintentional) bias towards omitting unsuccessful dominant males when calculating average fitness.

Another major issue is whether morph determination is primarily genetic or environmental, although it is important to realize that these are not exclusive alternatives (e.g. both effects operate in earwigs; Tomkins, 1999). In some species, this issue is amenable to appropriate rearing experiments, but even then it may be difficult to distinguish between alleles acting in the individual male or in his mother (Greeff, 1995; Shuster and Wade, 2003).

Finally, the accurate detection and description of male morphological polymorphism in itself is not a trivial issue. In some cases (see Fig. 4.1), male morphs are so different that they are very obvious without any quantitative studies. However, in many more cases, detailed measurements may be necessary to test for the presence and number of discrete morphs, to describe the morphological patterns of variation, and to assign the morph of individuals correctly (Eberhard and Gutierrez, 1991; Andersson, 1994; Kotiaho and Tomkins, 2001).

1.2 Fatal fighting

Although fatal fights between conspecifics have been reported from a wide variety of animal species, there are relatively few species in which a high proportion of individuals die as a result of combat with conspecifics (Enquist and Leimar, 1990). Game theory has also played a large role in the theoretical exploration of animal conflicts (Maynard Smith and Price, 1973; Maynard Smith, 1982) and a range of different asymmetries between contestants are predicted to be used to settle disputes without recourse to fatal fights. In many contest situations, individuals do assess each other and rare fatal fights can be attributed largely to the failure in assessment, i.e. they occur when two individuals are actually extremely similar with respect to the key variables. Assessment may involve an impressive stereotyped and escalated series of postures, as in the red deer *Cervus elaphus* (Clutton-Brock et al., 1982) and in the cichlid fish *Nannacara anomala* (Jakobsson et al., 1979), such that one individual can withdraw before serious injury occurs.

Where there is regular fatal fighting, it tends to involve situations in which the current dispute has a massive (overwhelming) impact on the expected lifetime reproductive success of the rivals. For example, new queens of some ant species will cooperate to initiate nests, because a single queen has very low success at this activity, but then fight to the death to become the sole reproductive in the colony. The winner has a high probability of great reproductive success, while the loser(s) gets nothing, but would also have virtually zero chance of success if she dispersed and tried to start a new nest for herself. This line of thought has been formalized in models by Enquist and Leimar (1990) to yield a general prediction that fatal fights should occur when the value of future reproductive opportunities is very low relative to the value of the currently contested reproductive opportunity. More specific models have been tailored to the biology of particular species/taxa, including fig wasps (e.g. Murray, 1987; Reinhold, 2003), and these are discussed later.

2. Fig Wasp Biology

Fig wasps are tiny insects that develop only inside the enclosed inflorescences (syconia or, colloquially, figs) of fig (*Ficus*) trees. There are hundreds of fig species, mostly in tropical and subtropical regions and, typically, each hosts a single species of fig-pollinating wasp and several species of fig-parasitic wasps (Compton *et al.*, 1994; Cook and Rasplus, 2003).

2.1 Fig-pollinating wasps

Fig-pollinating wasps comprise the monophyletic family Agaonidae (Hymenoptera; Chalcidoidea) and are involved in an obligate mutualism with the figs. The pollinating wasps are the only pollen vectors for the figs, while the larvae of the wasps develop only in the female flowers of their host fig species. The wasp life cycle is tied intimately to the biology of the host syconia as follows. A small number (often only one or two) of pollen-carrying wasp foundresses burrow into each receptive syconium through a narrow tunnel called the ostiole. Once inside, they lay eggs into some of the female flowers, while also depositing (actively or passively, depending on species; see Kjellberg *et al.*, 2001) pollen on to many of the female flowers. Since each female flower can give rise to a wasp or a seed, but not both, there is conflict between the insect and plant over the level of wasp oviposition (see Herre, 1989; Bronstein, 1992; Herre and West, 1997; Anstett, 2001; Cook and Rasplus, 2003). The wasp larvae then develop within the female flowers until, several weeks later, the males emerge first into the fig cavity. Males then mate, using their telescopic aedaegi (genitalia), with females that are still within their galls. In at least one, and probably many, species, females are generally unable to leave their galls until males bite exit holes for them, which they do only after mating (Zammit and Schwarz, 2000). In most species, females rely upon males to dig exit tunnels out of the syconium, although the ostiole reopens in a few fig species. Mated females, carrying pollen from the now mature male flowers, then disperse to search for receptive syconia to enter. They must achieve this rapidly, since they live for only a day or two as adults (Compton *et al.*, 1994).

2.2 Fig-parasitic wasps

Fig-parasitic wasps are far more diverse than the pollinating species and a single fig species may host up to 30 parasite species (Compton *et al.*, 1994). Taxonomically, they represent several different lineages that have colonized figs at different times (Rasplus *et al.*, 1998; Cook and Rasplus, 2003). The vast majority are in the Chalcidoidea, but they belong to various different families and lineages within this superfamily (Rasplus *et al.*, 1998). These different sets of interlopers show several convergent adaptations to living in figs. Females in different groups have evolved either long ovipositors to penetrate the wall and lay eggs deep within the syconium or, more rarely (but again repeatedly), the habit of entering the syconium at the same time as the pollinators (Rasplus *et al.*, 1998).

The larvae of fig-parasitic wasps feed in several different ways. Many (e.g. probably all members of subfamilies Epichrysomallinae and Sycophaginae) are gall-inducers (e.g *Anidarnes bicolor*; Bronstein, 1999), while others are known to be kleptoparasites of the pollinators (e.g. *Philotrypesis caricae*; Joseph, 1959), or parasitoids of gall-inducing wasps (e.g. *Physothorax* sp.; West *et al.*, 1996). Nevertheless, the larval ecology of most fig-parasitic wasps is understood poorly and requires dedicated study (Cook and Rasplus, 2003). Most species do not enter the syconium, but use their long ovipositors to lay eggs through the fig wall. Consequently, unlike the pollinators, they are not trapped in the first syconium that they visit. This, combined with longer lifespans than the pollinators (Compton *et al.*, 1994), gives them the opportunity to visit and lay eggs in several different syconia. Brief visits to several syconia may also be selected for by ants and spiders, which often predate fig-parasitic wasps in the act of oviposition (Bronstein, 1988; J.M. Cook, personal observation).

Mating behaviour also varies considerably between species. In general, winged males mate outside and wingless males inside syconia (Hamilton, 1979; Cook *et al.*, 1997). Females of most species appear not to be reliant upon males to assist their exit from galls, and, in some species, males may actually enter the female's gall to mate (Cook *et al.*, 1997). In many species, female parasitic wasps rely upon male pollinators to dig holes through which they can leave the syconium. In other species, notably large gall-inducing wasps like epichrysomallines, both males and females may bite their own way through the syconium wall to disperse.

2.3 Why study fig wasps?

The fig/fig wasp system has been used extensively as a model for both coevolutionary studies (e.g. Herre, 1989, 1993; van Noort and Compton, 1996; Kjellberg *et al.*, 2001; Weiblen and Bush, 2002; Cook *et al.*, 2004) and for studies of sex allocation (e.g. Hamilton, 1967; Frank, 1985a; Herre 1985, 1987; Herre *et al.*, 1997; Greeff, 1997, 2002; West and Herre, 1998; Fellowes *et al.*, 1999) and mating systems (e.g. Hamilton, 1979; Murray, 1987, 1989, 1990; Greeff, 1995; Cook *et al.*, 1997; Greeff and Ferguson, 1999; Bean and Cook, 2001, 2005). But why?

The peculiar natural history of fig wasps creates some unique opportunities, but also imposes some inevitable constraints, upon studies of their mating systems. It is crucial that fig wasps develop only in the syconia of their host fig species, and that this is the arena for most of their mate competition (Herre *et al.*, 1997; Cook *et al.*, 1999). Consequently, despite the fact that they are tiny, short-lived insects, fig wasp mating aggregations can be located precisely and studied in the field. In addition, many key data can be sampled from the well-defined biological unit of the syconium. Thus, one can easily count sex ratios, morph ratios and brood sizes, score injury levels, and, with more effort, observe matings and fights. It all happens quickly, once wasps have eclosed, but there is a limited period of time (around 24 h) in which it is possible to engineer contests between individuals, move wasps between different syconia, etc. (e.g. Frank, 1985b; Zammit and Schwarz, 2000). The downside of syconia is that they occur only on mature fig trees, which makes it difficult to set up experiments that rely on manipulations, controlled matings and multigeneration data. As biological models of small insects, fig wasps contrast strongly with *Drosophila melanogaster*, which is easy to study in the laboratory, but notoriously difficult to study in the field. A major advantage of fig wasps is that one can actually measure key aspects of the real (field) selective regime that influences the evolution of mating systems (Herre *et al.*, 1997). This is no small advantage (see Shuster and Wade, 2003).

A quite different reason to study mating systems in fig wasps is due to phylogeny. Many people have been impressed by the fact that several species, with quite different mate competition systems, can be found developing together in the syconia of even a single fig species (Cook *et al.*, 1999). Although certain traits, such as wing dimorphism and fighting, are found in many species, they often occur due to convergent evolution of distantly related species to similar selection pressures. These convergent adaptations make the taxonomist's job difficult, but represent the independent evolutionary events that comparative biologists require in order to test adaptive hypotheses. Thus, comparative tests of adaptive hypotheses (e.g. Cook *et al.*, 1997; West *et al.*, 1997, 2001) can have considerable power in fig wasps.

3. Aptery and Wing Polymorphism in Male Fig Wasps

3.1 Aptery

The adult males of many fig wasp species are apterous (wingless). In fact, all several hundred species of fig-pollinating wasps (family Agaonidae) have wingless males. Amongst the parasites there is more variation, and an individual species may have wingless, winged or wing-dimorphic males. Male-limited aptery has arisen independently in several different lineages of fig-parasitic wasps, but is relatively rare in chalcidoid wasps in general, suggesting that it is an adaptation to life in fig syconia. Hamilton (1979) suggested that aptery was an adaptation to facilitate movement and mate location within syconia. This seems very likely, especially when one has seen the tangled mass of flowers inside a fig, through which a male must make his way, when searching for mates. Since the inside of a fig is dark, eyes are also of limited value and eye reduction is also common in male fig wasps. Another reduction observed commonly involves the length and number of antennal segments. All these modifications appear to facilitate male function within the syconium. Some features may also be gained. For example, males of some fig wasp species (e.g. *Platyneura* spp.) have evolved abdominal gills, which are beneficial because the male wasps emerge from fig flowers before the resorption of fluids from the syconium interior, i.e. they begin searching for mates in an aqueous environment (Compton and McLaren, 1989).

The diversity of form seen between different types of wingless males is also striking. Some are armoured and soldier-like, adapted for fighting, while others may be flattened, to assist movement between flowers, or dwarf, to facilitate entry into galls (Hamilton, 1979; Murray, 1990; Cook *et al.*, 1997). In addition to overall appearance, particular features may be modified, such as the robust fore and hind legs of pollinator males, which are used to cling on to galls, while the male bites into them to gain access to the female inside. Murray (1989) recognized five basic forms in his study of Malaysian fig wasps and these serve well to illustrate the impressive morphological diversity (Fig. 4.3).

3.2 Wing dimorphism

In general, winged males tend to look very similar to winged females of the same species, but the wingless males look completely different, due to the various adaptations discussed in the previous section. Male wing dimorphism is common in fig wasps, but is still a minority trait. Cook *et al.* (1997) collated data on 114 fig wasp species and found that only ten were wing-dimorphic. This is probably fairly representative, as most underlying studies were not

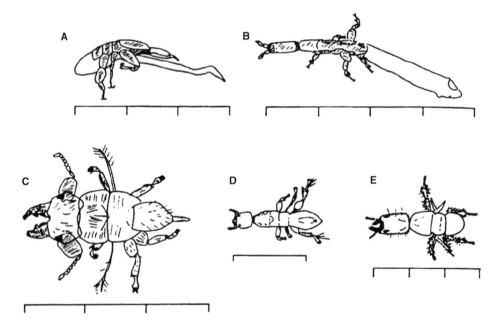

Fig. 4.3. Five types of wingless male recognized by Murray (1990): (**A**) pollinator, (**B**) tubular, (**C**) hog-backed soldier, (**D**) dwarf, (**E**) soldier. Scale bars show 1 mm intervals. From Murray (1990).

targeted towards or against finding male-dimorphic species. Nevertheless, if there are some 750 *Ficus* species, with a rough estimate of ten parasite species per fig, there may be about 850 wing-dimorphic species! It is worth stressing that it is not always a 'fighter-flier' dimorphism. This is true of the species studied by Hamilton (1979), but wing-dimorphic fig wasps do not all have fighting males (Cook *et al.*, 1997). For example, the wingless males of *Pseudidarnes minerva* are better described as precocious dwarf males (Fig. 4.1). They do not fight, but bite into and enter galls containing females. They then mate with the female before she emerges from her gall.

Why do some species have dimorphic males? Hamilton (1979) argued that the type of males found in a species depends on mating opportunities inside and outside the local mating patch (natal syconium). Essentially, wingless males are adapted to find mates within the syconium, while winged males are adapted to find females outside the syconium. Consequently, large syconium population size (SPS) of females selects for (wingless) males adapted to pre-dispersal mating, while small SPS selects for males adapted to post-dispersal mating. This argument leads to the prediction that, across species, wing loss will be correlated with large SPS. Cook *et al.* (1997) used data from 114 species to conduct both phylogeny-free and independent contrast tests of Hamilton's prediction. These tests both show a significant relationship between wing loss and brood size (Fig. 4.4). Across species, the mean SPS is 82 for pollinators, all of which have wingless males. For parasites, mean SPS is 30 for wingless, 7 for dimorphic and 5.5 for winged species. In phylogeny-free tests, winged and dimorphic species are not significantly different. However, in independent contrasts, the sign of the comparison is always in the predicted direction (Cook *et al.*, 1997).

A more quantitative prediction is available for species with dimorphic males. Hamilton proposed a simple model in which all females developing in a patch with wingless males present are mated by those males, while all those developing in patches without wingless males are mated by winged males. Under this scenario, the proportion of winged males should equal the pro-

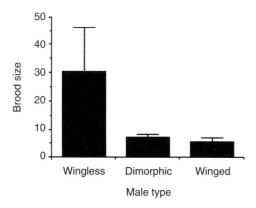

Fig. 4.4. Mean syconium population sizes (± 1 SE) for fig-parasitic wasp species with different types of males. (From Cook *et al.*, 1997.) Published with permission.

portion of females dispersing unmated (termed *j* by Greeff, 1995). However, this model assumed that each female laid only one egg per patch and so removed the possibility of local mate competition (LMC) between wingless males in the same patch. Greeff (1995) developed further models that allowed LMC, which reduces the predicted proportion of wingless males for a given scenario. The two key proportions have been estimated for ten species (Cook *et al.*, 1997) and are compared in Fig. 4.5. There is a clear positive relationship between the two variables and a linear regression explains 71% of the variance. If forced through the origin, as Hamilton's model implies, it still explains 62% of the variation. Looking at individual species values, four differ significantly from the predicted Hamilton value, but only one differs significantly from a value consistent with the Greeff model (Fig. 4.5). This suggests that LMC may have a role in many of these fig-parasitic species, as it does so clearly in the fig-pollinating wasps (Herre *et al.*, 1997). Two other factors might explain, or partly explain, the excess of winged males. First, winged males of some species may mate inside their natal figs before dispersing (Cook *et al.*, 1997; West and Herre, 1998), even if wingless males are present. Second, some females are likely to disperse unmated from patches with wingless males and this may be exacerbated by the occurrence of fatal fights (West *et al.*, 1997). Both factors would increase the proportion of matings available to winged males, leading to them having a higher ESS frequency (Hamilton, 1979; Charnov, 1993).

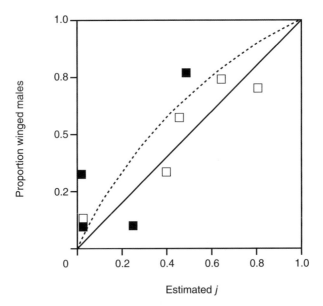

Fig. 4.5. The relationship between proportion of winged males and *j* (proportion of females dispersing unmated) across ten male-dimorphic species. The solid line is the prediction when there is no LMC (Hamilton's 1979 model) and the broken line is when 50% of matings are between siblings (example of Greeff's 1995 model). Shading shows species that do (black) and do not (white) differ significantly from Hamilton's prediction. (From Cook *et al.*, 1997.) Published with permission.

Charnov (1993) provides a lucid general presentation of alternative tactics theory, using a stationary population assumption, and stressing the symmetry (alternative strategies at equilibrium) or asymmetry (condition affecting two tactics differentially) of male beginnings. His approach extends Hamilton's (1979) discrete generations model to continuous generations. One general prediction for the symmetrical case with two tactics (A and B) is that the ESS*f** of A individuals should equal the proportion of eggs (*H*) fertilized by type A males in a given season. This is appealing to empiricists, because *H* may be a measurable parameter. Indeed, the proportion of females dispersing unmated (*j* in Greeff's model, 1995) provides the estimate of *H* in wing-dimorphic fig wasps.

3.3 Polymorphism of apterous males

While male wing dimorphism is more obvious (Fig. 4.1) and more thoroughly studied, there are also cases of polymorphism within species that have only wingless males. The best-studied case involves some African *Otitesella* species, which have two distinctive types, known as the *religiosa* and *digitata* morphs (van Noort and Rasplus, 1997). The *religiosa* morph has typical fighter morphology and mates in the natal syconium, while the *digitata* morph is less armoured and primarily a disperser (Greeff and Ferguson, 1999). In fact, *digitata* males have also been observed to mate within the natal syconium if *religiosa* males are not present (Pienaar and Greeff, 2003a,b), as do the winged males of some wing-dimorphic species if wingless males are absent. This is interesting, but will not influence morph frequency predictions from Hamilton's model, because the females so mated are assumed to be mated by disperser males in any case, i.e. while the mating locality is abnormal, the morph that mates is the expected one. On the other hand, it may also alter the degree of local mate competition, which could have an effect on ESS morph frequencies (Greeff, 1995; Bean and Cook, 2001). Recently, Moore *et al.* (2004) further showed that, in two *Otitesella* species, *religiosa* males below a certain size-related switch point have relatively larger jaws and less sclerotization. Consequently, there is actually a trimorphism in these species. There are certainly many more cases of male polymorphism awaiting detection amongst apterous male fig wasps, and quantitative morphological studies are needed to reveal the patterns of variation, which can then be linked to behaviour (Bean and Cook, 2001; Moore *et al.*, 2004).

3.4 Morph determination

In theory, male morph determination could depend on alternative alleles (Mendelian strategies), or on environmental factors acting on a single conditional strategy. In addition, the genetic and environmental effects can act either in the individual male, or in his mother. The obvious way to investigate these possibilities is through controlled breeding experiments, but these are difficult in fig wasps because of their natural history. Consequently, there have so far been no direct investigations of morph determination in male fig wasps, but a few studies provide indirect insights from survey data on morph ratios and other population variables.

Hamilton (1979) modelled fig wasp male wing dimorphism as Mendelian alternative strategies, with alleles acting in the males, and Cook *et al.* (1997) also supported this as the simplest explanation of the comparative data (see Section 3.2). This interpretation has been accepted widely with fig wasps becoming a textbook example of alternative tactics maintained by frequency-dependent selection (Krebs and Davies, 1993; Shuster and Wade, 2003). However, male dimorphism might result from the expression of maternal alleles. Greeff (1995) developed two theoretical models, in which morph determination depended upon alleles acting in males, or their mothers, and compared the predictions with survey data from six wing-dimorphic species studied by Hamilton (1979). Unfortunately, the predictions are rather similar and did not allow discrimination between the two alternatives. This is an example of a wider problem in that it is often difficult to distinguish between alleles acting in the focal male versus alleles acting in the mother (Shuster and Wade, 2003) and is exacerbated by the fact that the data available are from surveys of natural variation rather than controlled experiments.

In general, the nature of morph determination is expected to depend upon the predictability of environmental variation in mating opportunities and upon what cues are available to developing males or their mothers (Pienaar and Greeff, 2003a,b; Shuster and Wade, 2003). Pienaar and Greeff (2003a,b) have focused on the fit between patterns of morph allocation and environmental variation in mating opportunities to argue that, in the *Otitesella religiosa–digitata* case, male dimorphism is both conditional and under maternal control. They used models incorporating different genetic mechanisms to argue that, in *O. pseudoserrata*, the match between the male morphs produced and the mating opportunities for those males was too good to hold true for an alternative alleles model (Pienaar and Greeff, 2003a). A characteristic of many fig-parasitic wasp species is that the number and sex of individuals per patch (fig) is highly variable. Consequently, if male morph determination depends upon alternative alleles, there should often be a poor match between the male morphs present and their mating opportunities. However, this was not observed. For there to be a good match, morph determination must depend upon reliable cues about mating opportunities. In theory, these might be available to either the ovipositing mother or the developing male offspring.

Pienaar and Greeff (2003b) developed adaptive offspring allocation models for *Otitesella* species that allowed a mother to choose both the sex and morph of each offspring. Field observations suggest that *Otitesella* females typically lay only one or two eggs per fig and so this was assumed in the models (removing LMC effects). Data from two *Otitesella* species agree well with model predictions and show that the proportion of *digitata* (disperser) males decreases as the total number of wasps increases. The effect is significant both at the syconium and at the crop (of syconia on a tree) level, and is emphasized by the fact that over 70% of syconia containing any *digitata* males contain just the one. This is consistent with the idea that the first female to arrive at a syconium lays a *digitata* male, while subsequent females lay *religiosa* males. At the syconium level, females may be able to detect chemical cues of previous oviposition by conspecifics, as occurs in some parasitoid wasps.

Pienaar and Greeff (2003a,b) have argued in favour of conditional morph allocation based upon cues available to ovipositing *Otitesella* females. Another possibility is that developing offspring make use of cues available to themselves, i.e. their developmental fate is influenced by detection of the presence/sex/number of conspecifics in the same fig, or their own body size. For example, the mite *Caloglyphus berlesei* has a fighter/dwarf male dimorphism and morph ratios are influenced by the number of males developing in the colony (Radwan, 1993). Moore *et al.* (2004) used manipulative experiments to test for environmental effects upon wasp size, sex and morph. There were no significant effects of the number of conspecifics in the same patch, or the length of the developmental period. However, wasps developing in the inner layers of flowers (which allow development of larger galls) were larger (see Anstett, 2001, for a similar case with pollinating wasps) and more likely to be female. If male, they were more likely to be *digitata* than *religiosa* males. It was concluded that females may preferentially lay daughters in inner flowers because their fitness is most strongly influenced by body size (Moore *et al.*, 2004).

On balance, it appears likely that a conditional strategy operates in the *Otitesella* species investigated. As long as females can detect whether or not they are the first female to lay in a patch, there is a reliable cue. It seems less likely that developing males use cues about the presence of conspecifics (Pienaar and Greeff, 2003a,b; Moore *et al.*, 2004). They might be able to use a cue from their own body size, but the number of conspecifics present would seem to be a more important determinant of male fitness than any size effect. Given that facultative sex allocation by female fig wasps is well known (e.g. Hamilton, 1967; Frank, 1985a; Herre 1985, 1987; Greeff, 1997, 2002; Herre *et al.*, 1997; West and Herre, 1998; Fellowes *et al.*, 1999), facultative morph allocation might seem unsurprising. However, while haplodiploidy allows fine control of sex allocation via the release of sperm, no equivalent mechanism is known for morph allocation.

3.5 Virginity: a corollary of male aptery

A counterintuitive effect of selection on male mating opportunities is that it may lead to an increase in levels of female unmatedness (virginity) in species with wingless males, because some females develop in patches without males (Godfray, 1988; West *et al.*, 1997). As fig wasps are haplodiploids, these unmated females are then constrained to produce (haploid) sons from unfertilized eggs, but cannot produce (diploid) daughters from fertilized eggs. This is a great disadvantage in fig-pollinating wasps, where most patches contain the offspring of only one, or a few, foundresses, and local mate competition (LMC) generates strong selection for mothers to produce female-biased offspring sex ratios (Godfray, 1988; West *et al.*,

1997). It is probably also a considerable disadvantage in many species of fig-parasitic wasps that experience LMC. Clearly, the risk of a female developing in a patch without any males should increase as SPS decreases and as the sex ratio becomes more female-biased. Empirical data do indeed show a clear effect of SPS on virginity risk (West *et al.*, 1997).

These results are interesting, because there is a 'virginity load' on the population, despite selection on male morphology to match the distribution of mating opportunities. While the virginity risk is very low for most species, it is very high for the few species with wingless males that have the lowest mean SPS (Fig. 4.6). These tend to be in genera (e.g *Sycoscapter*) where male aptery is at, or close to, fixation, and may represent instances of phylogenetic inertia. Thus, there may be selection for male wings, but lack of appropriate genetic variation to allow a response to selection. Should we expect the reappearance of a complex feature like wings anyway? Perhaps it is more likely in fig wasps than in some other cases (e.g. oceanic island insects), because aptery is a sex-limited condition in fig wasps, so it may be only the sex-limitation that requires reversal (Cook *et al.*, 1997; West *et al.*, 1997).

4. Fatal Fighting

Fig wasps include some dramatic and gory examples of lethal male combat, which occurs between the wingless males of some parasitic species. In these species, males engage in fights that can lead to serious injuries (crushing, bruising, limb loss) or even death, often by decapitation (Hamilton, 1979; Murray, 1987, 1989; Bean and Cook, 2001). For example, Bean and Cook

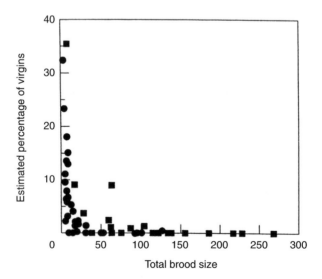

Fig. 4.6. Scatterplots of the estimated percentage of virgin females versus syconium population size. Squares show pollinator and circles non-pollinator species. (From West *et al.*, 1997.)

(2001) found that about 25% of male *Sycoscapter australis* had been fatally injured (most of them decapitated) by the end of the mating period, while Hamilton (1979) referred to 'megadeath', when he estimated that several million deaths may occur in one fruit crop of a large fig tree.

What is the reason for this carnage, and why do fighting and non-fighting species cohabit in the same syconia? It is notable that no pollinator species are known to engage in lethal combat, and Hamilton (1979) argued that general differences in relatedness between pollinators and parasites might provide the key. Pollinating species generally have very low foundress numbers (Herre *et al.*, 1997), such that males are often competing with brothers for mating opportunities. Most parasite species do not enter syconia and so they do not leave behind their cadavers as convenient indices of inbreeding. In addition, they will often oviposit into several different syconia (Herre *et al.*, 1997). Nevertheless, parasite species certainly display a wide variety of population structures, probably including both highly inbred and essentially outbred species. Their sex ratios, which reflect the highly correlated variables of LMC and inbreeding, certainly suggest a broad range of population structures (West and Herre, 1998; Fellowes *et al.*, 1999). In summary, then, most pollinating species are highly inbred, while many non-pollinating species are less inbred (Hamilton, 1979; Herre *et al.*, 1997). This should mean that interactions with brothers are less common in the parasites, so that kin selection (Hamilton, 1964a,b) will not impede the evolution of fatal fighting (Hamilton, 1979).

4.1 Kin selection in viscous populations

One way to conceptualize fatal fights is to consider them as the absence of altruism (e.g. West *et al.*, 2001). Kin selection has been shown to play the major role in the evolution of altruism, but recent theoretical work has shown that it may be sidelined in viscous populations. This occurs because one needs to consider not only the relatedness of interacting individuals, but also the spatial scale of competition (Taylor, 1992; Wilson *et al.*, 1992; Kelly, 1994; Frank, 1998). In essence, when populations are highly viscous, altruism by a male towards his brother can only permit the brother to obtain matings from the same pool of females that is available to himself. Since the male has higher relatedness to himself than to even his brother, it becomes impossible for kin selection to favour altruism in the limiting situation of total viscosity. Given the viscosity of populations of fig wasps with wingless males, and especially pollinating species, theory suggests that relatedness may not be a key factor in explaining variation in the presence of fatal fighting in fig wasps (West *et al.*, 2001).

4.2 The value of present versus future opportunities

A different theoretical approach to the evolution of fatal fighting was developed by Enquist and Leimar (1990), who concluded that fatal fighting should *evolve* only when the expected value of future mating opportunities is small relative to the value of the currently contested opportunity. This simple and general argument fits well with most observations of the occurrence of regular fatal fighting. For example, in some ant species, after the nuptial flight, new queens will cooperate to initiate a nest, but then fight to the death, leaving just one queen. This makes evolutionary sense, because nest initiation has very low success without cooperation, but once the nest is initiated successfully, the decision of who will now be solitary queen has almost total power to set the reproductive success of the individual.

4.3 Contest economics

Another approach to the incidence of fighting in fig wasps was developed by Murray (1987, 1989), who focused on the economics of different mate competition behaviours, building on the biology of the fighting species *Philotrypesis pilosa*. Murray (1987) considered contest competition over mating opportunities and predicted that the intensity of fighting should depend upon the interval between challenges for mating opportunities. At the syconium level, this should mean a positive, but domed, relationship between fighting and the number of males, as well as a positive relationship with the number of females.

4.4 Relatedness revisited

With neither general theory nor existing data supporting a role for relatedness, a new theoretical study by Reinhold (2003) should rekindle interest. Reinhold (2003) constructed models to predict the likelihood of an individual starting a fight in a closed mate competition system like the fig syconium, with and without kin recognition (KR) ability. These models are quite simple and make some interesting and testable predictions. An interesting model assumption is that all males hatch *before* females emerge, so competition is for future matings, rather than for a currently contested female. It turns out that this assumption is not crucial to predictions; however, it is probably correct for many fig wasp species, in which males tend to hatch before females, and where fighting can be intense before female emergence really gets going (Murray, 1987; J.M. Cook, personal observation).

   In Model 1 (no KR), relatedness has no effect on fighting levels, which depend negatively upon the number of males in the fig and positively on the variance in male fighting ability. The first correlation is because killing $1/n$ males improves a male's expected success by the term $[1/(n-1)] - [1/n]$, which gets smaller as n increases. The second correlation stems from the assumption that males can estimate their chance of winning a fight and only start to fight when it increases their expected fitness. Assessment like this may be common in many animals, but has not been investigated in fig wasps. However, it seems certain that there is considerable variation in male fighting ability, given that conspecific males can vary in size by two to as much as seven times in the few species investigated (Bean and Cook, 2001; Moore *et al.*, 2004).

In Model 2 (with KR), males can differentiate related from unrelated competitors and direct aggression primarily at unrelated males, where the threshold value for initiating a fight is lower. There is an increase in predicted fighting levels (relative to Model 1), which are now highest at intermediate levels of relatedness between males (e.g. with two foundresses). Predicted fighting levels still depend in the same way upon the number of males and their variation in fighting ability.

Reinhold (2003) suggested that previous studies may not have found an effect of relatedness on injury levels, because they were testing for a linear, rather than a domed, response. It is also true that all studies to date have used sex ratio as an indirect index of relatedness, instead of estimating relatedness directly from genetic data. Reinhold concluded that a role for kin recognition in fighting can only be resolved by 'detailed intraspecific studies using direct observations in combination with a genetic method to determine relatedness'. This is an open invitation to empiricists.

4.5 Comparative tests of model predictions

Murray (1989) conducted the first detailed comparative analysis of fighting in fig wasps, using data from about 30 Malaysian fig wasp species on injury levels, along with sex ratios, numbers of males and females per syconium, etc. The two factors with significant explanatory power were the number of males, and whether females of a given species entered syconia or oviposited externally. This oviposition behaviour is correlated highly with phylogeny and essentially describes the split between pollinators (all enter) and parasites (most do not). More recently, West et al. (2001) combined Murray's data with a molecular phylogeny to test the contrasting predictions of Hamilton's (1979) and Enquist and Leimar's (1990) models. If, as Hamilton argued, relatedness is important, injury levels should decrease as sex ratios become more female-biased, because biased sex ratios reflect LMC (i.e. interactions between brothers). If, on the other hand, the value of future mating opportunities is important, injury levels should decrease as the number of conspecific females (potential mates) in the patch increases. Using independent contrasts, West et al. (2001) found significant support for the second, but not the first, prediction. Thus, comparative analysis currently supports only Enquist and Leimar's (1990) prediction.

4.6 Constraints on the evolution of fighting morphology

The morphology and behaviour of male fig wasps are adapted to the syconia that they inhabit, as well as to their interactions with conspecifics. This may influence the evolution of fatal fighting in several ways. First, consider pollinator males. These must dig tunnels through the fig wall to enable dispersal and future reproduction of the conspecific females that carry their sperm. In addition, in at least some species, they are required for the females to even get out of the galls in which they have developed (Zammit and Schwarz, 2000). Consequently, there will be selection against any changes that reduce the release of mated females from figs. It is quite likely that mouthparts adapted for fighting are not suitable for release of mates from their galls, or the biting of exit tunnels out of the fig.

Another factor is mating site, which can be divided into four categories: (i) both sexes inside gall, (ii) female in gall, but male in fig cavity, (iii) both sexes in fig cavity, and (iv) both sexes outside fig. Fatal fighting is known only for species in categories ii and iii (Vincent, 1991), where males compete within the fig cavity. Small size and precocious development are characteristics of males that enter galls to mate with females (see Figs 4.1 and 4.3) and may be incompatible with fighting morphology. In addition, the spatial distribution of males and time required to enter seeds and mate may not engender sufficient meetings of males to make fighting an economically viable option in such species (Murray, 1987). Male size may also be constrained in syconia that are tightly packed with flowers at the time when the male wasps emerge and search for mates. There is great variation between fig species in the internal architecture of syconia and a large, soldier morphology may trade-off unfavourably against mobility in cramped conditions. There may be an example of this amongst *Sycoscapter* species on figs of the section *Malvanthera*. Those from figs with loosely packed syconia have fighting males, but those from figs with more tightly packed flowers do not (J.M. Cook, unpublished data). In general, males that are dorsoventrally flattened may be adapted to squeezing between tightly packed flowers.

4.7 Do pollinator males fight?

Fatal fighting is well-documented in the parasites, but are there any pollinator examples? Hamilton (1979) noted that, in fig wasps and other invertebrates (e.g. thrips, beetles), there was an accumulation of evidence that individuals with 'soldier' morphology did, indeed, engage in serious fights. This is useful, because morphological descriptions outnumber detailed observations of contests by a long way. Hamilton (1979) noted that males of one *Alfonsiella* pollinator species showed fighting morphology and suggested that it would actually prove to be a parasite rather than a pollinator. Instead, studies on this and other African species (Greeff *et al.*, 2003) have shown that several pollinators have males with apparently fighting morphology (Fig. 4.7; Greeff *et al.*, 2003). It is also apparent that males of some of these species may disperse and enter nearby syconia, either by biting new holes, or using existing holes (Greeff, 2002; Greeff *et al.*, 2003). There is considerable correlation between fighting morphology and dispersal in these species, and the retractable abdomen in particular is a trait that favours both through increasing walking mobility. Consequently, the morphology observed might reflect dispersal and/or fighting behaviour. Greeff *et al.* (2003) reported the existence of fighting in these species, but without details of the injuries sustained. However, nobody has reported, for pollinators, regular fatal fighting like that seen in some parasites.

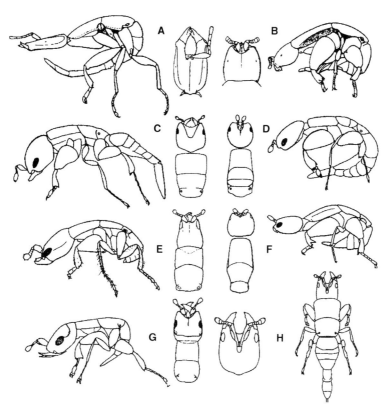

Fig. 4.7. Examples of pollinator males with 'fighting' (F) and 'non-fighting' (NF) morphology: (**A**) *Courtella michaloudi* (F), (**B**) *Courtella armata* (NF), (**C**) *Platyscapa awekei* (F), (**D**) *Platyscapa soraria* (NF), (**E**) *Alfonsiella longiscapa* (F), (**F**) *Elisabethiella stuckenbergi* (NF), (**G**) *Nigeriella excavata* (F), (**H**) *Pegoscapus astomus* (F). Fighting males have falcate mandibles, strong, short thorax and elongate scapes. They can also retract the normally extended gaster. (From Greeff *et al.*, 2003.)

4.8 Operational sex ratios and the intensity of sexual selection

Greeff and Ferguson (1999) studied six wasp species associated with a single fig species (*F. ingens*). The one non-fighting (pollinator) and five fighting (parasite) species showed interesting variations in patterns of emergence. Pollinator females emerged rapidly, over a short period of time, while parasite females emerged slowly over several days. This suggests that, for fig-cavity-mating species, the operational sex ratio (OSR: individuals available to mate) may differ considerably from the syconium sex ratio (SSR: all conspecific males and females present). Importantly, the OSR may generally be highly male-biased, despite a relatively unbiased, or even female-biased, SSR (Greeff and Ferguson, 1999). The presence of several males in the fig cavity, waiting to mate with females who pass through quickly, and at intervals, generates intense sexual selection and may favour fatal fighting as winners can monopolize mating opportunities and make large gains. An essentially similar pattern of emergence (Fig. 4.8), yielding male-biased OSRs in the cavity-mating, fighting species *Sycoscapter australis*, supports the correlation between OSR and fighting (Bean and Cook, 2001). However, there are insufficient data for a full comparison, because of the lack of data on emergence patterns and OSRs in non-fighting, non-pollinating species (Cook *et al.*, 1999). Nevertheless, this approach stresses the likely importance of both the spatial and temporal distribution of receptive females in determining mate competition strategies (Hamilton, 1979; Andersson, 1994; Greeff and Ferguson, 1999; Shuster and Wade, 2003).

4.9 Which individuals fight?

Given that males of a species fight, who fights and when? Is adaptive variation in fighting seen only between species, or is there facultative adjustment within species, as is seen with pollinator sex allocation? This is an important question, because it concerns the precision and flexibility of adaptation by natural selection.

Murray (1987) considered the economics of mate competition in *Philotrypesis pilosa* and developed a model to predict injury levels in this species, dependent upon male encounter rates with rivals and potential mates.

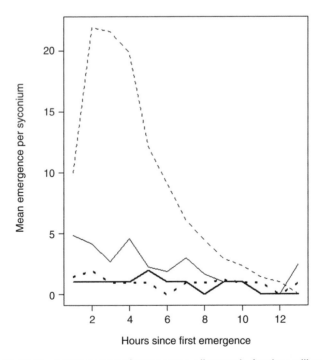

Fig. 4.8. Mean wasp emergence per hour across all syconia for the pollinator *Pleistodontes froggatti* (broken lines) and parasite *Sycoscapter australis* (solid lines) males (thick lines) and females (thin lines). (From Bean and Cook, 2001.)

He found a predicted correlation between fighting intensity and male density, but not other predicted correlations with female density and encounter rates. Murray (1987) argued that the lack of fit between data and models was probably attributable to differences in real and assumed patterns of sperm competition, as well as the ease with which males could switch between searching and guarding behaviours. Lack of data on sperm competition is a limitation for all species studied to date, but the guarding issue may be more species-specific, because males of another fighting species mate females after their emergence from galls and do not guard them subsequently (Bean and Cook, 2001). This emphasizes the importance of knowing the detailed biology of focal species.

The issues of male polymorphism and fatal fighting merge if not all individuals fight and there is morphological variation between individual wingless males. This can be obvious, discrete variation, as in the *religiosa* and *digitata* morphs of *Otitesella* species (van Noort and Rasplus, 1997; Greeff and Ferguson, 1999), or more subtle, even continuous, variation, as in two sycoryctine species (Bean and Cook, 2001, 2005). Size variation is impressive in males of these species – fourfold in *Philotrypesis* sp. and sevenfold in *Sycoscapter australis* (Bean and Cook, 2001, 2005) – but discrete morphs are not immediately obvious. In both species, large males have relatively larger jaws (positive allometry compared with other body parts), and in both cases there is evidence for two subtle, yet discrete, morphs that differ in the allometric relationship between jaw size and head size. In *S. australis,* injury levels are higher in syconia whose wasps have greater mean mandible size, and Bean and Cook (2001) argued that big males were predominantly fighters, while small males sought matings in more concealed positions between the relatively tightly packed flowers of this fig species. There was no evidence that size influenced the possibility of a male being killed, and one possibility is that, while large males are more likely to start fights, they can also sometimes engage and kill small males with relative impunity.

Bean and Cook (2005) collected fighting pairs of males of *Philotrypesis* sp. and compared the size difference of the two individuals fighting with the average size difference of all possible pairs of males in the focal syconium. Fighting pairs differed significantly more in size than randomly chosen males, supporting the impunity idea over the alternatives of failed assessment (predicts size matching) or no role for size (predicts no difference from random pairs). Anecdotal support for impunity killing comes from observations of *Idarnes* spp., in which syconia opened at the end of the mating period typically contain just one live male, which is often the largest (S.A. West, personal communication).

4.10 Fighting: summary

Across species, the intensity of fatal fighting appears to be correlated with future mating opportunities, but not with relatedness of competitors (West *et al.*, 2001). However, other details are also important and fatal fighting appears to occur only in species that mate in the fig cavity, probably because selection upon male morphology to mate elsewhere is not easily compatible with fighting morphology. The spatial and temporal availability of females also seems important and may vary considerably depending on the detailed natural history of species. Pollinators are not known to engage in fatal fights (but see Greeff et al., 2003), and are probably subject to some constraints due to selection on other features of male behaviour (release of females from galls and biting an exit tunnel). Within species that do fight, there may be considerable variation in body size and correlated variation in the use of fighting behaviour (Bean and Cook, 2001, 2005; Moore et al., 2004). Positive allometry of weapon (jaw) size may be linked with bigger individuals who are more likely to fight. However, in some cases, big individuals may be able to kill smaller males with relative impunity (Bean and Cook, 2001, 2005).

5. Conclusion

Hamilton (1979) argued that male aptery in fig wasps evolved to facilitate mate competition inside syconia and was associated with large SPS. Both functional aspects of male morphological changes (e.g. Compton and McClaren, 1989; Murray, 1990) and the repeated loss of wings in species with large SPS (Cook et al., 1997) are supported by comparative studies. Better phylogenies would clarify the number and patterns of change, but the broad picture is already clear.

Male wing dimorphism is probably a case of alternative strategies at equilibrium, as suggested by Hamilton (1979). This conclusion is based upon the association between the proportion of winged males and the proportion of unmated females, both within and across species (Greeff, 1995; Cook et al., 1997). It would be valuable to have further lines of evidence (see Gross, 1996; Shuster and Wade, 2003), especially on the genetic basis of male morph determination, but the difficulty of breeding experiments with fig wasps hampers this line of investigation. Nevertheless, other avenues are more open; for example, studies of spatial and temporal variation in morph ratios within species are less difficult. In addition, genetic markers could be developed and used to study patterns of multiple mating and sperm competition to determine how well these match model assumptions.

Polymorphism also occurs in some species with wingless males and many more cases will surely come to light with the increasing use of quantitative description of male morphological variation (e.g. Bean and Cook, 2001, 2005; Moore et al., 2004). The best-studied cases involve the *Otitesella religiosa* and *digitata* morphs, where the balance of evidence favours a form of conditional strategy (Pienaar and Greeff, 2003a,b; Moore et al., 2004). Again, further study is needed, and it would be valuable to have direct investigation of morph determination, as well as studies of sperm competition and mating success. It is worth stressing that we should expect differences between fig wasp species in many aspects of male polymorphism, including morph determination. First, fig wasps are not a monophyletic group and comprise several diverse wasp lineages that have colonized figs at different times (Rasplus et al., 1998; Cook and Rasplus, 2003). This means that they have different phylogenetic histories and may differ in key traits such as ability to use different cues. Second, although all species develop in figs, there is great variation in population sizes and mating systems that will impact upon the distribution of mating opportunities. Consequently, selection will favour different outcomes in different species and, for example, the situation for *Otitesella* species that typically lay one or two eggs per fig may be quite different from other taxa where individual females lay several eggs per fig.

Fatal fighting in fig wasps, as in many other species, appears to be associated with very limited future reproductive prospects (Enquist and Leimar, 1990; West *et al.*, 2001). Contrary to Hamilton's (1979) original argument, but in line with more recent theory (see Frank, 1998), current evidence does not support a link between fighting and the relatedness of interacting males (West *et al.*, 2001). However, a new model by Reinhold (2003) predicts a domed relationship between fighting and relatedness within fighting species and should spur new work using molecular markers to determine relatedness directly. The number of conspecific females in a syconium is clearly a major determinant of fatal fighting in males (West *et al.*, 2001), but probably not the only one. Syconium architecture and mating site may at least have the potential to constrain the evolution of fighting, while the temporal pattern of female emergence may be another important correlate, through its influence on the ability of males to monopolize mating opportunities.

There is scope for further comparative studies, in which all the key variables are measured for a selected set of species in several lineages. There is also much scope to better understand cryptic variation in morphology and behaviour within species that have fatal fighting (Bean and Cook, 2001; Moore *et al.*, 2004). In asking who is fighting and against who, there is an urgent need to apply molecular markers to estimate relatedness. In addition, while many have been amazed by the behaviour and external appearance of male fig wasps, nobody has ventured within to look at sperm competition patterns. However, molecular approaches have revealed interesting patterns in other male dimorphic taxa, such as pseudoscorpions (Zeh and Zeh, 1994) and dung beetles (Tomkins and Simmons, 2000). In species with male dimorphism (or a wide range of continuous variation), individuals that invest less in fighting or guarding are predicted to invest more in sperm production (Parker, 1990) and such issues could be investigated in fig wasps. A fuller understanding of male variation will come from integration of study at several levels, from phylogeny and comparative studies, through detailed documentation of behaviour and morphology within species, to investigation of reproductive organs and genetic determination of patterns of sperm competition.

Acknowledgements

I am grateful to Stuart West and Jamie Moore for comments on earlier versions of this chapter. My research in this area has received support from the NERC, BBSRC and the Royal Society. Jo Martin kindly drew Fig. 4.1.

References

Andersson, M. (1994) *Sexual Selection*. Princeton University Press, Princeton, New Jersey.

Anstett, M.C. (2001) Unbeatable strategy, constraint and coevolution, or how to resolve evolutionary conflicts: the case of the fig/wasp mutualism. *Oikos* 85, 476–484.

Austad, S.N. (1984) A classification of alternative reproductive behaviours and methods for field testing ESS models. *American Zoologist* 24, 309–320.

Bean, D. and Cook, J.M. (2001) Male mating tactics and lethal combat in the nonpollinating fig wasp *Sycoscapter australis*. *Animal Behaviour* 62, 535–542.

Bean, D. and Cook, J.M. (2005) Cryptic male dimorphism in fighting fig wasps. *Animal Behaviour*, submitted.

Bronstein, J.L. (1998) Predators of fig wasps. *Biotropica* 20, 215–219.

Bronstein, J.L. (1992) Seed predators as mutualists: ecology and evolution of the fig/pollinator interaction. In: Bernays, E.A. (ed.) *Insect–Plant Interactions*, Vol. IV. CRC Press, Boca Raton, Florida, pp. 1–47.

Bronstein, J.L. (1999) Natural history of *Anidarnes bicolor* (Hymenoptera : Agaonidae), a galler of the Florida strangling fig (*Ficus aurea*). *Florida Entomologist* 82, 454–461.

Charnov, E.L. (1993) *Life History Invariants: Some Explorations of Symmetry in Evolutionary Biology*. Oxford University Press, New York.

Clutton-Brock, T.H., Guinness, F.E. and Albon, S.D. (1982) *Red Deer: Behaviour and Ecology of Two Sexes*. University of Chicago Press, Chicago.

Compton, S.G. and McLaren, F.A.C. (1989) Respiratory adaptations in some male fig wasps. *Proceedings of the Koninklijke Nederlandse Akademie Van Wetenschappen, Series C, Biological and Medical Sciences* 92, 57–71.

Compton, S.G., Rasplus, J.Y. and Ware, A.B. (1994) African fig wasps parasitoid communities. In: Hawkins, B. and Sheehan, W. (eds) *Parasitoid Community Ecology*. Oxford University Press, Oxford, UK, pp. 323–348.

Cook, J.M. and Rasplus, J.Y. (2003) Mutualists with attitude: coevolving fig wasps and figs. *Trends in Ecology and Evolution* 18, 241–248.

Cook, J.M., Compton, S.G., Herre, E.A. and West, S.A. (1997) Alternative mating tactics and extreme male dimorphism in fig wasps. *Proceedings of the Royal Society of London, Series B, Biological Sciences* 264, 747–754.

Cook, J.M., Bean, D. and Power, S.A. (1999) Fatal fighting in fig wasps: GBH in time and space. *Trends in Ecology and Evolution* 14, 257–259.

Cook, J.M., Bean, D., Power, S.A. and Dixon, D.J. (2004) Evolution of a complex coevolved trait: active pollination in a genus of fig wasps. *Journal of Evolutionary Biology* 17, 238–246.

Darwin, C.R. (1874) The *Descent of Man and Selection in Relation to Sex*, 2nd edn. Rand, McNally and Co., New York.

Eberhard, W.G. and Gutierrez, E.E. (1991) Male dimorphisms in beetles and earwigs and the question of developmental constraints. *Evolution* 45, 18–28.

Enquist, M. and Leimar, O. (1990) The evolution of fatal fighting. *Animal Behaviour* 39, 1–9.

Fellowes, M.D.E., Compton, S.G. and Cook, J.M. (1999) Sex allocation and local mate competition in Old World non-pollinating fig wasps. *Behavioural Ecology and Sociobiology* 46, 95–102.

Frank, S.A. (1985a) Hierarchical selection theory and sex ratios. II. On applying the theory, and a test with fig wasps. *Evolution* 39, 949–964.

Frank, S.A. (1985b) Are mating and mate competition by the fig wasp *Pegoscapus assuetus* (Agaonidae) random within a fig? *Biotropica* 17, 170–172.

Frank, S.A. (1998) *Foundations of Social Evolution.* Princeton University Press, Princeton, New Jersey.

Gadgil, M. (1972) Male dimorphism as a consequence of sexual selection. *American Naturalist* 106, 574–580.

Godfray, H.C. (1988) Virginity in haplodiploid populations: a study on fig wasps. *Ecological Entomology* 13, 283–291.

Greeff, J.M. (1995) Offspring allocation in structured populations with dimorphic males. *Evolutionary Ecology* 9, 550–558.

Greeff, J.M. (1997) Offspring allocation in externally ovipositing fig wasps with varying clutch size and sex ratio. *Behavioral Ecology* 8, 500–505.

Greeff, J.M. (2002) Mating system and sex ratios of a pollinating fig wasp with dispersing males. *Proceedings of the Royal Society of London, Series B, Biological Sciences* 269, 2317–2323.

Greeff, J.M. and Ferguson, J.W.H. (1999) Mating ecology of the nonpollinating fig wasps of *Ficus ingens. Animal Behaviour* 57, 215–222.

Greeff, J.M., van Noort, S., Rasplus, J.Y. and Kjellberg, F. (2003) Dispersal and fighting in male pollinating fig wasps. *Comptes Rendus de l'Academie Des Sciences, Serie Iii, Sciences de la Vie* 326, 121–130.

Gross, M.R. (1996) Alternative reproductive strategies and tactics: diversity within sexes. *Trends in Ecology and Evolution* 11, 92–98.

Gross, M.R. and Repka, J. (1998) Game theory and inheritance of the conditional strategy. In: Dugatkin, L.A. and Reeve, H.K. (eds) *Game Theory and Animal Behaviour.* Oxford University Press, Oxford, UK, pp. 168–187.

Hamilton, W.D. (1964a) The genetical evolution of social behaviour. I. *Journal of Theoretical Biology* 7, 1–16.

Hamilton, W.D. (1964b) The genetical evolution of social behaviour. II. *Journal of Theoretical Biology* 7, 17–52.

Hamilton, W.D. (1967) Extraordinary sex ratios. *Science* 156, 477–488.

Hamilton, W.D. (1979) Wingless and fighting males in fig wasps and other insects. In: Blum, M.S. and Blum, N.A. (eds) *Reproductive, Competition and Sexual Selection in Insects.* Academic Press, New York, pp. 167–220.

Herre, E.A. (1985) Sex ratio adjustment in fig wasps. *Science* 228, 896–898.

Herre, E.A. (1987) Optimality, plasticity and selective regime in fig wasp sex ratios. *Nature* 329, 627–629.

Herre, E.A. (1989) Coevolution of reproductive characteristics in 12 species of New World figs and their pollinator wasps. *Experientia* 45, 637–647.

Herre, E.A. (1993) Population structure and the evolution of virulence in nematode parasites of fig wasps. *Science* 259, 1442–1445.

Herre, E.A. and West, S.A. (1997) Conflict of interest in a mutualism: documenting the elusive fig wasp seed trade-off. *Proceedings of the Royal Society of London, Series B, Biological Sciences* 264, 1501–1507.

Herre, E.A., West, S.A., Cook, J.M., Compton, S.G. and Kjellberg, F. (1997) Fig-associated wasps: pollinators and parasites, sex-ratio adjustment and male poly-morphism, population structure and its consequences. In: Choe, J.C. and Crespi, B.J. (eds) *The Evolution of Mating Systems in Insects and Arthropods.* Cambridge University Press, Cambridge, UK, pp. 226–239.

Jakobsson, S., Radesater, T. and Jarvi, T. (1979) On the fighting behaviour of *Nannacara anomala* (Pisces, Cichlidae) males. *Zeitschrift fur Tierpsychologie* 49, 210–220.

Joseph, K.J. (1959) The biology of *Philotrypesis caricae* L., parasite of *Blastophaga psenes* L. (Chalcidoidea: Hymenoptera Parasitica). In: *Proceedings of the XVth International Congress of Zoologists, London, 1958,* pp. 662–664.

Kelly, J. (1994) The effect of scale dependent processes on kin selection: mating and density regulation. *Theoretical Population Biology* 46, 32–57.

Kjellberg, F., Jousselin, E., Bronstein, J.L., Patel, A., Yokoyama, J. and Rasplus, J.Y. (2001) Pollination mode in fig wasps: the predictive power of correlated traits. *Proceedings of the Royal Society of London, Series B* 268, 1113–1121.

Kotiaho, J.S. and Tomkins, J.L. (2001) The discrimination of alternative male morphologies. *Behavioural Ecology and Sociobiology* 12, 553–557.

Krebs, J.R. and Davies, N.B. (1993) *An Introduction to Behavioural Ecology*, 3rd edn. Blackwell Scientific, Oxford, UK.

Maynard Smith, J. (1982) *Evolution and the Theory of Games*. Cambridge University Press, Cambridge, UK.

Maynard Smith, J. and Price, G.R. (1973) The logic of animal conflicts. *Nature* 246, 15–18.

Moore, J.C., Pienaar, J. and Greeff, J.M. (2004) Male morphological variation and the determinants of body size in two *Otiteselline* fig wasps. *Behavioral Ecology* 15, 735–741.

Murray, M.G. (1987) The closed environment of the fig receptacle and its influence on male conflict in the Old World fig wasp, *Philotrypesis pilosa*. *Animal Behaviour* 35, 438–506.

Murray, M.G. (1989) Environmental constraints on fighting in flightless male fig wasps. *Animal Behaviour* 38, 186–193.

Murray, M.G. (1990) Comparative morphology and mate competition of flightless male fig wasps. *Animal Behaviour* 39, 434–443.

Parker, G.A. (1984) Evolutionarily stable strategies. In: Krebs, J.R. and Davies, N.B. (eds) *Behavioural Ecology: An Evolutionary Approach*, 2nd edn. Blackwell Scientific, Oxford, UK, pp. 30–61.

Parker, G.A. (1990) Sperm competition games: sneaks and extra-pair copulations. *Proceedings of the Royal Society of London, Series B, Biological Sciences* 242, 127–133.

Pienaar, J. and Greeff, J.M. (2003a) Different male morphs of *Otitesella pseudoserrata* fig wasps have equal fitness but are not determined by different alleles. *Ecology Letters* 6, 286–289.

Pienaar, J. and Greeff, J.M. (2003b) Maternal control of offspring sex and male morphology in the *Otitesella* fig wasps. *Journal of Evolutionary Biology* 16, 244–253.

Radwan, J. (1993) The adaptive significance of male polymorphism in the acarid mite *Caloglyphus beriesi*. *Behavioural Ecology and Sociobiology* 33, 201–208.

Rasplus, J.Y., Kerdelhue, C., Le Clainche, I. and Mondor, G. (1998) Molecular phylogeny of fig wasps Agaonidae are not monophyletic. *Comptes Rendus de l'Academie Des Sciences, Serie Ii, Sciences de la Vie* 321, 517–527.

Reinhold, K. (2003) Influence of male relatedness on lethal combat in fig wasps: a theoretical analysis. *Proceedings of the Royal Society of London, Series B, Biological Sciences* 270, 1171–1175.

Shuster, S.M. and Wade, M.J. (1991) Equal mating success among male reproductive strategies in a marine isopod. *Nature* 350, 608–610.

Shuster, S.M. and Wade, M.J. (2003) *Mating Systems and Strategies*. Princeton University Press, Princeton, New Jersey.

Sinervo, B. and Lively, C. (1996) The rock-scissors-paper game and the evolution of alternative mating strategies. *Nature* 380, 240–243.

Taylor, P.D. (1992) Altruism in viscous populations: an inclusive fitness model. *Evolutionary Ecology* 6, 352–356.

Tomkins, J.L. (1999) Environmental and genetic determinants of the male forceps length dimorphism in the European earwig *Forficula auricularia* L. *Behavioral Ecology and Sociobiology* 47, 1–8.

Tomkins, J.L. and Simmons, L. (2000) Sperm competition games played by dimorphic male beetles: fertilisation gains with equal mating access. *Proceedings of the Royal Society of London, Series B, Biological Sciences* 267, 1547–1553.

van Noort, S. and Compton, S.G. (1996) Convergent evolution of agaonine and sycoecine (Agaonidae, Chalcidoidea) head shape in response to the constraints of host fig morphology. *Journal of Biogeography* 23, 415–424.

van Noort, S. and Rasplus, J.Y. (1997) Revision of the otiteselline fig wasps (Hymenoptera: Chalcidoidea: Agaonidae). I. The *Otitesella digitata* species-group of the Afrotropical region, with a key to Afrotropical species of *Otitesella* Westwood. *African Entomology* 5, 125–147.

Vincent, S.L. (1991) Polymorphism and fighting in male fig wasps. PhD thesis, Rhodes University, Grahamstown, South Africa.

Weiblen, G.D. and Bush, G.L. (2002) Speciation in fig pollinators and parasites. *Molecular Ecology* 11, 1573–1578.

West, S.A. and Herre, E.A. (1998) Partial local mate competition and the sex ratio: a study on non-pollinating fig wasps. *Journal of Evolutionary Biology* 11, 531–548.

West, S.A., Herre, E.A., Windsor, D.M. and Green, P.R.S. (1996) The ecology and evolution of the New World non-pollinating fig wasp communities. *Journal of Biogeography* 23, 447–458.

West, S.A., Herre, E.A., Compton, S.G., Godfray, H.C.J. and Cook, J.M. (1997) A comparative study of virginity in fig wasps. *Animal Behaviour* 54, 437–450.

West, S.A., Murray, M.G., Machado, C.A., Griffin, A.S. and Herre, E.A. (2001) Testing Hamilton's rule with competition between relatives. *Nature* 409, 510–513.

Wilson, D.S., Pollock, G.B. and Dugatkin, L.A. (1992) Can altrusim evolve in purely viscous populations? *Evolutionary Ecology* 6, 331–341.

Zammit, J. and Schwarz, M.P. (2000) Intersexual sibling interactions and male benevolence in a fig wasp. *Animal Behaviour* 60, 695–701.

Zeh, J.A. and Zeh, D.W. (1994) Last-male sperm precedence breaks down when females mate with three males. *Proceedings of the Royal Society of London, Series B* 257, 287–292.

5 Seasonal Plasticity, Host Plants, and the Origin of Butterfly Biodiversity

SÖREN NYLIN, KARL GOTTHARD AND GEORG H. NYGREN

Department of Zoology, Stockholm University, Stockholm, Sweden

1. Introduction

Phenotypic plasticity is the term used when different phenotypes can be produced by the same genotype, in different environments. For many years after the Modern Synthesis, research into this area was largely neglected in favour of strictly genetically determined variation, presumably because of the 'Lamarckian' overtones when dealing with environmentally induced traits (Shapiro, 1976; Sultan, 2003). More recently, however, there has been a renewed interest in plasticity (see e.g. Gotthard and Nylin, 1995; Pigliucci, 2001; West-Eberhard, 2003), following an increasing realization that the dichotomy between genetic and plastic variation is false. Instead genotypes have 'norms of reaction', a repertoire of potential phenotypes that are expressed according to environment, and as a result every phenotype is the result of a genotype-by-environment interaction – phenotypic plasticity (Stearns, 1989; Sultan, 2003). Conversely, plasticity is as much a product of evolutionary processes as genetic differentiation, and adaptive plasticity is shaped by natural selection among genotypes with different reaction norms (Gotthard and Nylin, 1995).

The phenomenon of 'counter-gradient variation' is a good illustration of how genes and the environment interact (Conover and Schultz, 1995). Evidence of local adaptation typically consists of phenotypic differences between environments, but, because genotypes are plastic in their phenotypic expression, natural selection can also result in phenotypic similarity. Selection can favour genotypes with reaction norms that oppose maladaptive effects of the environment. For instance, populations from cold and warm environments may have similar developmental rates in the field, but when they are reared in a common environment (a 'common garden' experiment) individuals of the cold-environment population may grow and develop faster, revealing that the populations are in fact locally adapted. Genetics and the environment thus interact in the truest sense of the word, and for this reason we cannot understand the genetics of evolution without understanding plasticity (Nylin and Gotthard, 1998).

Phenotypic plasticity can be of several different kinds, corresponding to the degree of flexibility in the display of phenotypes. Although the boundaries are not sharp, three types can be recognized.

1. 'Developmental' plasticity is when a certain developmental pathway is taken according to the environment experienced. These environments are known as 'environmental cues' in the case of plasticity, which is thought to represent adaptation to variation in the environment. Because of the cascading effects in ontogeny of going down one such pathway and not the other this is often an irreversible process, of particular importance in short-lived organisms such as insects.

2. Vertebrates and other organisms with lives that span more than a year often more conspicuously show reversible plasticity (e.g. birds and mammals that go white in the winter), also known as 'phenotypic flexibility' (Piersma and Lindstrom, 1997).

3. Finally, when the flexibility of the phenotype increases even more, this grades into 'behaviour'. Although behaviour is usually seen as a field of study separate from plasticity, there is no absolute distinction. Moreover, plasticity is often mediated by behaviour. For instance, plastic growth rates may be due to variations in feeding behaviour and plastic clutch size to variation in oviposition behaviour. Not surprisingly, the methods of behavioural ecology, i.e. predictions from optimality followed by experiments or comparative studies, can be fruitfully applied also to plasticity (Nylin, 1994).

For most insects, not the least in temperate areas where the winter season is wholly unfavourable for growth and development, 'seasonal plasticity' is very important. As noted above, in insects this often takes the form of developmental plasticity, where different phenotypes are seen according to the progress of the season and the whole life cycle is regulated by seasonal cues. There are, however, also more flexible elements of seasonal plasticity and there are many where the actual degree of flexibility is uncertain, such as growth rate regulation by external cues, or seasonal plasticity in host-plant choice.

In the following, we primarily use the comma butterfly (Nymphalidae: *Polygonia c-album* L.) and its close relatives to illustrate the ubiquitous presence of seasonal plasticity, and the optimality approach to understanding its role in the evolution of insect life-cycle regulation and host-plant preference. *P. c-album* is strongly polyphagous, with larvae feeding on at least seven different host-plant families, several of which are not closely related (Nylin, 1988; Janz *et al.*, 2001), namely *Urticaceae, Ulmaceae, Cannabidacae, Salicaceae, Grossulariaceae, Betulaceae* and *Corylaceae*. Related species generally feed on a subset of these plant families, resulting in an interesting model group for studying the plastic effects of larval host plants in combination with other environmental factors.

We will touch upon several important features of the insect life cycle where plasticity is important: seasonal polyphenism, life-cycle regulation by photoperiod, fine-tuning of growth and development, life-history traits, larval host plants and female host-plant preference. For the first two subjects we have included some previously unpublished data, whereas the remaining text reviews other findings from the research project, with a focus on results from the comma butterfly and other *Polygonia*. The final section links phenotypic plasticity to ecological speciation theory.

2. Materials and Methods

The term 'polyphenism' is most appropriately used for discrete phenotypes resulting from plasticity, in an analogy with genetic polymorphism (Shapiro, 1976). Seasonal polyphenism, the occurrence of different phenotypic forms in different seasons, is common in insects and an example of plasticity in its most visual form. There is often a correlation with physiological plasticity in terms of developmental pathway taken: diapause or direct development. Butterflies often show striking seasonal polyphenism (Shapiro, 1976; Brakefield and Larsen, 1984; Koch, 1992) but in most cases the adaptive significance is not clear (Gotthard and Nylin, 1995; Kemp and Jones, 2001). One exception is the seasonal variation in the amount of dark pigment in pierid butterflies, which has been convincingly linked to temperature regulation (e.g. Kingsolver, 1995).

For the previously unreported results on seasonal polyphenism and life-cycle regulation in *Polygonia satyrus*, *P. gracilis zephyrus* and *P. faunus*, methods were as follows. Stock originated from females caught in Washington State, USA. As larvae hatched they were randomly divided among different photoperiod treatments (designed to elucidate the role of photoperiod in the adult stage) and raised in environmental cabinets at 20°C. There were four possible treatments (see Tables 5.1 and 5.2):

1. A short larval daylength of 15 h with adults moved to 22 h at eclosure.
2. A long larval daylength of 22 h with adults moved to 15 h at eclosure.
3. A constant long daylength of 22 h.
4. A shift from 15 h to 22 h after 2 weeks (in the fourth larval instar) with adults retained in 22 h.

Adults were frozen 1 week after the eclosure of the last adult and, in the case of females, were later dissected and checked for mature eggs. For *P. g. zephyrus* and *P. faunus* there were only a few hatchlings available, and these were divided among treatments 1 and 4 only.

Wings were removed and taped on to OH transparencies, which were scanned using an HP ScanJet 5100C flatbed scanner. The resulting JPEG images were converted to 16-bit greyscale and the greyness (% white) measured using Corel PhotoPaint software. This was done in two places on the forewing undersides (Fig. 5.1), one of the darkest areas (the most apical section of the central symmetry system; see Nylin *et al.*, 2001) and one of the lightest (in the area between the basal and central symmetry system, immediately apical of the bend in the former).

Fig. 5.1. The two seasonal forms of *Polygonia c-album*, as illustrated by two female individuals. Top: directly developing form; Bottom: hibernating form. The butterflies have been scanned and the images converted to black and white, using the same threshold for both individuals. (D) Dark spot measured, (L) light spot measured (see text).

3. Seasonal Polyphenism

P. c-album is facultatively multivoltine, with two distinct seasonal forms (Fig. 5.1). Summer-form individuals have light-coloured, yellowish, wing undersides and develop directly to sexual maturation, whereas the darker autumn/spring form spends the winter in reproductive diapause before mating in the spring. Individuals of the spring form may give rise to either form in the next generation, or mixed broods of both forms, depending on the larval environment (Nylin *et al.*, 1989; Nylin, 1992). Temperature regulation seems an unlikely explanation for this polyphenism, as the species warm up at low temperatures by using 'dorsal basking' (wings held horizontal, exposing the upper sides to the sun) and there is little variation in the dorsal wing surface. Interestingly, the North American *P. interrogationis* and *P. comma* do display dorsal polyphenism, but in this case it is the summer form that is darker (Scott, 1986), further indicating that wing pigmentation is not a major factor in *Polygonia* temperature regulation. More probably, the dark form of *P. c-album* is adapted for being cryptic during hibernation (Nylin, 1991). Evidently, the species hibernates fully exposed on trunks and roots of trees (Thomas and Lewington, 1991), where the dark colour would serve it well, coupled with its extremely cryptic appearance in many other respects (mottled coloration with relief effects, jagged outline, etc.). The dark pigment melanin may also have structural importance, making wings more resistant to wear and tear.

Karlsson and Wickman (1989) investigated resource allocation patterns in the two forms of *P. c-album*, which differ dramatically in life history. Whereas the summer form has an expected adult lifespan of at most a few weeks, the hibernating form may live as an adult for up to 10 months. In accordance with predictions from optimality, Karlsson and Wickman (1989) found that the summer form allocates more nitrogen to the abdomen (and hence to reproduction) and the hibernating form more to the thorax and wings (and hence to somatic survival). It has therefore been suggested (Nylin, 1991) that the coloration of the summer form may not be adaptive *per se*, but rather in the sense of saving unnecessary costs of melanization (melanin is rich in nitrogen). In accordance with this idea, Wiklund and Tullberg (2004) have found in experiments with bird predators that the dark form is more efficient at being cryptic than the light form when presented against a 'winter' background consisting of a tree trunk, but that the light form is not superior in this respect against a 'summer' background consisting of stinging nettles.

A very similar polyphenism to that in *P. c-album* is found in the two other Palaearctic species of *Polygonia* for which information is available, namely *P. c-aureum* (Masaki *et al.*, 1989; Koch, 1992) and *P. egea* (Dal, 1981). This demonstrates that a plastic phenotype can be retained over several speciation events (Fig. 5.2). Studies on *P. c-aureum* revealed that the polyphenism is controlled by a hormone which was named 'summer-morph-producing hormone' (Masaki *et al.*, 1989), i.e. independently of hormonal control of adult reproductive diapause, despite a strong correlation. Interestingly, however, the polyphenism is lacking, or not very apparent, in the Nearctic species (see Scott, 1986), although neither the three Palaearctic species, nor the Nearctic species, form monophyletic groups (Nylin *et al.*, 2001; Wahlberg and Nylin, 2003). The simplest explanation would seem to be that this type of polyphenism is ancestral to the genus but was lost twice after representatives of the genus colonized the New World (Fig. 5.2). It could be speculated that this happened because colonization occurred via a northern route, so that all populations of the incipient species were univoltine for many generations and the reaction norm for expressing both seasonal phenotypes was lost.

In accordance with the phylogenetic interpretation depicted in Fig. 5.2 (polyphenism ancestral) it can be noted that both *P. gracilis zephyrus* and

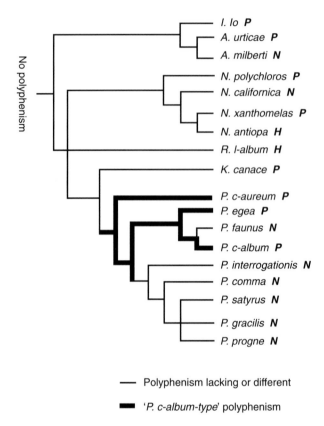

Fig. 5.2. Phylogeny of *Polygonia* and relatives, based on Nylin *et al.* (2001) and Wahlberg and Nylin (2003). Uncertain relationships depicted by unresolved branching patterns. Capital letters after species names refer to biogeographical region: *P* = Palaearctic, *N* = Nearctic, *H* = Holarctic. Lineages with the type of strong ventral polyphenism found in *Polygonia c-album* and two other Palaearctic species shown with thick lines. This trait is reconstructed as being ancestral to *Polygonia* if gains are evolutionarily less likely than losses. Traces of similar polyphenism are found also in *P. satyrus* and *P. gracilis* (see text).

(especially) *P. satyrus* show traces of the Palaearctic form of polyphenism (Table 5.1). Note that the designations of colour forms given in Table 5.1 are rather arbitrary, as the dark and light morphs produced were not as distinct as in, for instance, *P. c-album*. The clearest result was when comparing adults of *P. satyrus* resulting from larvae raised in short or increasing daylengths, respectively, with adults in long daylengths (treatments 1 and 4 in the Materials and Methods section). The first treatment can be expected to produce adults initially destined for hibernation, although the reproductive diapause may subsequently be broken by the long adult daylengths. The second treatment is the only one effective in producing a majority of (light-coloured) directly developing adults in the normally univoltine Swedish population of *P. c-album* (Nylin, 1989, 1992) (see next section).

Table 5.1. Proportion of dark-form adults of *Polygonia* species in different photoperiod treatments. Sequence of photoperiod given for young larvae, older larvae/pupae, adults (see text). Sexes pooled. Number of individuals in parentheses.

Species	Photoperiod treatment			
	15–15–22h	22–22–15h	22–22–22h	15–22–22h
P. satyrus	100% (14)	59% (22)	86% (29)	11% (19)
P. gracilis	100% (6)	–	–	21% (14)
P. faunus	100% (8)	–	–	100% (8)

Most individuals in treatment 1 looked darker than most individuals in treatment 4, and this difference could also be quantified. The measured 'dark' area of treatment-1 individuals was on average darker than the corresponding spot in treatment-4 individuals (males 25.2% white vs. 27.1% white, $n = 8 + 8$, t-test: $P = 0.09$; females 25.0% vs. 29.7%, $n = 7 + 4$, t-test: $P < 0.05$) and the 'light' area was also darker (males 33.5% vs. 37.8%, $n = 8 + 8$, t-test: $P < 0.01$; females 31.0% vs. 38.0%, $n = 7 + 4$, t-test: $P < 0.001$). In most cases, however, the ranges in grey values overlapped slightly. This was even more the case in *P. g. zephyrus* and, because of this and the small sample size, no attempt at quantification was made. In *P. faunus*, the sister species to *P. c-album*, no difference in colour between treatments could be detected ($n = 4$ females and $n = 4$ males per treatment).

For comparison, we measured grey values in four typical adults of *P. c-album* (both forms and sexes, the females are shown in Fig. 5.1), and in these individuals the 'dark' areas differed in amount of white between forms by 5% in the males and 7% in the females, and the light areas by 24% in the males and 28% in the females. In treatments corresponding to those applied here, there were no intermediate individuals (Nylin, 1989, 1992). It is thus questionable whether the weaker polyphenism in the two studied Nearctic species is adaptive, but the fact that it can be detected demonstrates that traces of the ancestral reaction norm are still present.

4. Photoperiodic Control of the Life Cycle

As noted in the Introduction, insects in seasonal areas, especially in the temperate zone, must regulate their life cycles in accordance with the progression of the seasons. Notably, they enter a hormonally controlled diapause in preparation for the winter, well in advance of the onset of unfavourable conditions. This is done using information from the environment, known as seasonal 'cues', of which photoperiod is the most important (Danilevskii, 1965; Tauber *et al.*, 1986). The most common response is that individuals enter diapause (in a species-specific developmental stage) after having experienced short daylengths, signalling the approaching winter, in a preceding stage. The 'critical daylength' is the threshold value below which a majority of the individuals enter diapause.

It is often difficult to determine whether phenotypic plasticity is an adaptation to environmental variation, and not an unavoidable 'direct' effect of the environment, but plasticity in response to token cues such as photoperiod is one of the most convincing cases (Gotthard and Nylin, 1995). This is both because of the existence of physiological 'machinery' specialized for photoperiod detection and response, and because it can be demonstrated that this machinery is locally adapted. Since the amplitude of daylength variation over the year increases with increasing latitude, it can be predicted from optimality that the critical daylength should vary accordingly over the geographical range of a species (Danilevskii, 1965), and this has also repeatedly been found to be the case. In fact, one of the most convincing examples of local adaptation available in any organism is the geographical variation in photoperiodic thresholds in the pitcher-plant mosquito (Bradshaw, 1976). Such examples also provide good illustrations of how plasticity – the response to photoperiod – can be determined by adaptive genetic variation in reaction norms. Variation among families in the propensity for diapause is commonly found, for instance in *P. c-album* (Nylin, 1992), and this is the raw material for natural selection.

The comma butterfly follows the latitudinal pattern, with individuals of the English population developing directly at a larval daylength of 18 h or 20 h, whereas insects of the more northern Swedish population enter hibernation diapause (Nylin, 1989). However, a further complication is that there is also a response to the direction of change in daylength. In the Swedish population, diapause is averted by a change in the larval stage from 16 h to 18 h (Nylin, 1992) or 18 h to 20 h (Nylin, 1989), presumably signalling a date before summer solstice and time for a second generation. English insects enter diapause after a decrease from 20 h to 18 h (Nylin, 1989), i.e. well above the critical constant daylength but indicating a date after summer solstice.

Such true sensitivity to changing daylengths in diapause induction has rarely been demonstrated, and in butterflies it may be associated with the unusual habit of adult hibernation diapause, which results in eggs and larvae early in the summer. Since the same daylengths occur before and after summer solstice, they may provide conflicting information regarding the date in the season if larvae occur early enough to experience both spans of daylength (Nylin, 1989). Misinterpreting a daylength in a late larval stage means several weeks of juvenile development and adult maturation and mating, before the second generation can be initiated, and a risk of failing to complete this generation before winter. This problem can be solved by also using information on the direction of change, or in some cases by postponing the 'decision' on whether to enter diapause or not until the adult stage.

We expected polyphenic species to be relatively insensitive to adult daylengths as a cue for entering diapause or not, since they should have already 'decided' before eclosion, the decision signalled by the irreversible adult morph (Nylin, 1989). This is a potential cost of polyphenism, as it would result in some loss of developmental flexibility. Non-polyphenic species could be predicted to instead make use of the information from adult daylengths. Such a response (where adult daylengths override any cues during the larval stage) is seen in the related non-polyphenic butterfly *Aglais urticae* (Voigt, 1985).

P. gracilis showed a higher proportion of females with eggs after an increase in daylength during the larval stage, as predicted for a polyphenic species. Also as predicted, the non-polyphenic P. faunus did not seem to be sensitive to an increase in daylength during the larval stage (Table 5.2). However, for both of these species, sample sizes were small and missing treatments prevented a real test of the hypotheses outlined above. Concerning P. satyrus, short adult daylengths did not prevent egg maturation. This is as expected for a polyphenic species (Table 5.2), and as found in P. c-album (Nylin, 1989). However, egg maturation was not well correlated with coloration (cf. Tables 5.1 and 5.2), a further demonstration that the observed polyphenism is not necessarily adaptive in this species. Note especially that the constant long-day treatment was the most effective in inducing direct development in females (Table 5.2), but resulted in a high proportion of individuals of the dark morph.

In summary, few clear conclusions could be drawn from this small-scale experiment, but we believe that it may still serve to demonstrate that the research field of comparative studies on insect life-cycle regulation lies open for exploitation.

5. Life-cycle Fine-tuning and Adaptive Growth Rate Plasticity

Once a certain main developmental pathway has been taken, other responses than diapause (or not) and morphological polyphenism may follow. In particular, the optimal development time for a directly developing insect can be expected to be shorter than that for an insect destined to enter diapause, at least in near-critical conditions (e.g. near the critical daylength). This is because the former is aiming to fit a whole extra generation of offspring into the same season, whereas the latter should accordingly have surplus time left to develop up to the stage where it can itself enter diapause. In mixed broods of P. c-album, for instance, development time is shorter for individuals that later eclose as the light form than for the dark form, and this is a result of both shorter larval and pupal developmental times (Nylin, 1992). Such responses may be viewed as a physiological polyphenism, and perhaps also a behavioural polyphenism if shorter larval times are a result of more pronounced larval feeding. However, it is not always clear whether differences in development

Table 5.2. Proportion of females found to contain eggs in different photoperiod treatments. Sequence of photoperiod given for young larvae, older larvae/pupae, adults (see text). Number of individuals in parentheses.

| Species | Photoperiod treatment | | | |
	15–15–22h	22–22–15h	22–22–22h	15–22–22h
P. satyrus	25% (4)	60% (5)	88% (8)	71% (7)
P. gracilis	0% (2)	–	–	88% (8)
P. faunus	50% (4)	–	–	50% (4)

time are purely effects of the developmental pathway taken, or whether they may to some extent be the cause, as poor environments may act as a cue for diapause induction (see Section 9).

In addition to such qualitative differences between pathways, photoperiod and other seasonal cues can also be used to fine-tune growth and development within pathways (Nylin and Gotthard, 1998). In our work, evidence of the adaptive nature of such reaction norms comes primarily from satyrine butterflies. Here, development time is progressively shorter in shorter daylengths when they can be expected to signal progressively later dates in the summer after the summer solstice, in *Pararge aegeria* (Nylin *et al.*, 1989) and *Lasiommata petropolitana* (Nylin *et al.*, 1996a; Gotthard, 1998). In related species hibernating as larvae, such as *L. maera*, the late larval stages instead occur before the summer solstice when daylengths are still on the increase, and here development time is instead shorter in longer daylengths (Nylin *et al.*, 1996a; Gotthard *et al.*, 1999b). Moreover, in the same individuals development time is, as predicted, shorter in shorter daylengths in the early larval stages, before hibernation (Gotthard *et al.*, 1999b). Finally, experiments on *L. maera* show that even the plastic growth rate response to rearing temperature is influenced by photoperiod; individuals experiencing daylengths indicating a higher degree of time-stress increase their growth rates more at higher temperatures compared with individuals under less time-stress (Gotthard *et al.*, 2000a).

In *P. c-album*, studies on the quantitative effects of photoperiod are complicated by the sensitivity to changes in daylength, but there is a clear tendency towards a gradual response, with progressively shorter development times in shorter daylengths (Nylin, 1992).

6. Life-history Patterns and Plasticity

Life-history traits can be defined as those that quantitatively describe the life cycles of organisms, notably juvenile development time, size at sexual maturity, egg weight, clutch size and lifespan (Nylin, 2001).

A common feature in the studies reported in the previous section is that much of the variation in larval development time is achieved by plasticity in larval growth rates, rather than by variation in final pupal or adult weight, the trade-off often assumed in life-history theory (Abrams *et al.*, 1996; Nylin and Gotthard, 1998). Evidently growth rates are typically not maximized in butterflies. This may be because there is often surplus time due to the constraint of species-specific diapause stages, and because high growth rates carry costs such as higher risk of starvation and predation when feeding activity is increased (Gotthard *et al.*, 1994; Gotthard, 2000).

In an experiment designed to investigate the effects of different growth rate levels on the trade-offs between development time and final weight, *P. c-album* was used as a contrast to the presumed growth rate maximizer *Epirrita autumnata* (Tammaru *et al.*, 2004). This geometrid moth feeds in the spring on the foliage of trees, which rapidly declines in quality as food for larvae, and consequently larval growth rates are very high. The Swedish population of *P.*

c-album studied, on the other hand, has plenty of surplus time for its single generation and is known not to maximize its growth rate (Janz *et al.*, 1994). Last-instar larvae of both species were deprived of food for short periods of time. As predicted, pupal weight was little affected by this treatment in the comma butterfly, because larval development time was extended to compensate for the loss of food. In addition, compensatory growth occurred, with faster growth on the day after the starvation treatment. Surprisingly, however, very similar results were obtained from *E. autumnata*, indicating that not even this seasonally constrained species maximizes growth rates. Such universal plasticity in growth rates has profound consequences for life-history theory (Abrams *et al.*, 1996; Nylin and Gotthard, 1998).

One example is the theory on latitudinal and altitudinal patterns of size variation. It has been proposed that insects should often be expected to follow a 'saw-tooth' pattern in development time and (consequently) size (Roff, 1980, 1983). At high latitudes or altitudes the favourable season is short and the optimal development time short. Going south or to lower elevations the optimal time for growth and development becomes longer; hence adult size should increase. Such patterns of increasing size to the south have been documented in crickets (Mousseau and Roff, 1989) and other ectotherms (Mousseau, 1997), including butterflies (Nylin and Svärd, 1990). However, in facultatively multivoltine species an additional generation in the same season can be added when the season becomes long enough to allow it. This effectively cuts the season into two parts and can be predicted to lead to a decrease in size, before it starts increasing again. The best-documented saw-tooth patterns in size are those found in crickets (Mousseau and Roff, 1989), whereas in butterflies they are less apparent but may well be present (Nylin and Svärd, 1990).

Phenotypic plasticity has several roles in creating such geographical patterns, or in preventing them. First, the shift in generation number involves plasticity, since environmental cues are typically used to determine whether an additional generation is possible (see above). However, if the insect is sedentary enough, there may also be a genetic component, a propensity to enter diapause that is locally adapted, i.e. a greater propensity to the north or higher up a mountain than in the long-season areas. In species with strong gene flow any geographical pattern observed must instead be wholly due to plasticity. Second, plasticity determines whether trends in development time will in fact produce trends in size. Roff's theory really predicts variation in development time, and then assumes a direct relationship with adult size. One way that this could happen is by plastically adding extra larval instars when time allows, as suggested by Roff (1983). However, as we have seen, plasticity could also work to prevent the expected size patterns. In *P. c-album*, larvae of the partially bivoltine English population have shorter development times than those of the more northern and univoltine Swedish population – as predicted by saw-tooth theory – but this is achieved by higher growth rates and does not result in decreased adult size (Nylin, 1992).

Another 'classic' life-history trait is lifespan (Stearns, 1992). This trait obviously has a strong plastic component, presumably in most cases representing non-adaptive effects of poor environments. In the speckled wood butterfly *Pararge aegeria*, for instance, levels of drought and food shortage have a strong impact on adult longevity (Gotthard *et al.*, 2000b) (see also next section for an example from *P. c-album* involving host-plant quality). Perhaps more surprisingly, a strong genetic component to variation in ageing and lifespan can also often be demonstrated, a fact that has led to considerable discussion (Stearns, 1992). One approach to explaining such variation is in terms of future reproductive success. From an optimality perspective, it does not pay to invest in a very durable soma if there is little chance of reproduction in late life. In a previous section we saw a result of this equation, with the light form of *P. c-album* investing more in current reproduction and less in a durable thorax and wings than the long-lived dark form; a demonstration that plastic variation in lifespan can sometimes be adaptive.

Extending this reasoning to different populations, with different expectations of future reproduction, we compared *P. aegeria* from Sweden and Madeira. In Sweden there is strong seasonality and populations are synchronized by winter diapause. Females mate only once and there is a strong fitness premium on males being already present when they emerge, but a low premium on long male life. On Madeira there is a relatively non-seasonal climate and female emergence is unpredictable for males over the year and vice versa. As predicted, there is strong protandry in Sweden (i.e. males enter the population earlier), which is lacking in Madeira (Nylin *et al.*, 1993), and virgin butterflies from Madeira are more prone to mate as soon as they meet compared with Swedish ones (Gotthard *et al.*, 1999a). Moreover, male lifespan is shorter in the Swedish population, where there is little chance of a mating once the peak of female emergence has passed (Gotthard *et al.*, 2000b). This reasoning can also be extended to species differences. *P. c-album* is a highly polyandrous species, with females mating 2–3 times on average, whereas females of the peacock butterfly *Inachis io* mate only once. As predicted, males of *P. c-album*, which can expect matings over a longer fraction of the summer than peacock males, live longer relative to females (Wiklund *et al.*, 2003).

7. Responses to Host-plant Quality

For phytophagous insects, the larval host plant constitutes a very important environmental variable, but plastic responses by larvae to host-plant quality are typically not reported as part of the literature on plasticity but rather in the context of insect–plant associations. However, one influential genetic model of the evolution of phenotypic plasticity used examples framed in terms of an insect utilizing one or several host plants (Via and Lande, 1985), which shows the close connection.

Since plants vary in their properties, we can also expect them to vary in quality as host plants. Most often we are concerned with variation among plant species, but sometimes the variation among plant individuals may be greater than among species (Singer and Lee, 2000). What, exactly, is host-plant 'quality'? This is a complicated subject, but it should at least be noted that there are two different aspects of quality that cannot always be separated. One is the 'objective' quality, i.e. the amount of resources contained in the plant, the water content and the amount of material that is hard or impossible to digest, or even toxic. For instance, high nitrogen levels, high water content and low leaf toughness in a plant equals high quality for many insects, unless specific chemistry prevents use of the plant. The other aspect is the 'subjective' quality for the particular insect, taking into account its specific adaptations (or fortunate chance properties) for utilizing the particular resources in the plant, metabolizing toxins and perhaps even sequestering them for its own use. Since one insect's toxin may be another insect's resource, host-plant quality can more unequivocally be defined in terms of a specific insect's 'performance' on that plant.

Performance, in turn, is measured as the fitness of offspring reared on the plant (relative to other potential hosts) summed over the whole life cycle. That is, ideally not only survival, larval development time and final mass (and hence average growth rate) should be measured, but also male and female reproductive success (Nylin *et al.*, 1996b). In the case of the polyphagous comma butterfly, the preferred host plants (the stinging nettle *Urtica dioica* and the hop *Humulus lupulus*) rank highly in terms of most of these fitness correlates. Survival is higher, development time shorter, and growth rate higher than on alternative hosts, which are bushes and trees (Nylin, 1988; Janz *et al.*, 1994; Nylin *et al.*, 1996b). This can probably to some extent be explained in terms of 'objective' quality, as the herb *U. dioica* has higher water and nitrogen contents than the alternative hosts (N. Wedell, N. Janz and S. Nylin, unpublished data; *H. lupulus* not measured). However, a more 'subjective' aspect of quality is also likely to be involved, since larvae also perform well on other (preferred) hosts in the order *Urticales* with lower levels of nitrogen and water, including the tree *Ulmus glabra*. It seems probable that such phylogenetically constrained patterns of responses to host plants have a historical component, and indeed the tribe Nymphalini has a long history of association with this plant order (Janz *et al.*, 2001).

When it comes to adult fitness components, the high nitrogen content of stinging nettles adds another bonus for utilizing this plant as host. Males reared on nettles can produce spermatophores that are richer in nitrogen than males reared on, for instance, sallow *Salix caprea* (Wedell, 1996; N. Wedell, N. Janz and S. Nylin, unpublished data). This nitrogen is transferred to the females, and it has been shown by radioactive labelling that it is incorporated into eggs (Wedell, 1996). Furthermore, females mated to nettle-males can increase their own reproductive effort but still live longer than females mated to sallow-males, and would probably have higher lifetime fecundity if allowed to remate at will (Wedell, 1996). Most aspects of performance that have been measured are poorer on other host plants than on nettle, but larvae reared on sallow pupate at higher weights (after a longer time) and eclose as larger adults, and are also more fecund in the case of females (Janz *et al.*, 1994; Nylin *et al.*, 1996b).

8. Performance and the Evolution of Host-plant Preference

The subject of host-plant preference (in insects where the ovipositing females make the choice of host for their offspring) is traditionally even less tied to plasticity theory than that of larval performance, but since preference and performance are closely connected themes (Thompson, 1988), and 'performance' is just another word for plastic responses to host plants, the preference–plasticity connection is equally strong. To the extent that plant preference is determined by performance on different host plants, it can be said to be determined by reaction norms of offspring, in response to host plants selected by the females. Preference and performance are not always well correlated (Thompson, 1988), but a search for such a correlation is a good starting point in attempting to understand preference. If it is not found, this may be because the most relevant fitness correlates have not been measured, because field characteristics of performance (e.g. predation) are of crucial importance for the preference patterns, because realized fecundity is more important than careful selection of hosts in the study species, or because the insect is in fact not behaving optimally (Nylin et al., 1996b). If a good correlation is found, there is reason to assume that females have been selected to prefer the plants of highest 'quality' in terms of offspring performance. In P. c-album there is a good general agreement between preference and performance (Nylin, 1988; Janz et al., 1994; Nylin et al., 1996b).

Phytophagous insects are often specialized on one or a few host plant species (or genera, or families) (Fig. 5.3; Thompson, 1994), and the reason (or reasons) for this pattern has generated much discussion (Strong, 1988). The traditional explanation is in terms of the 'jack-of-all-trades is a master of none' idea (Pianka, 1983), a general ecological principle which makes intuitive sense. Since plants vary in their properties, optimal performance should not be possible on more than one type of plant simultaneously, and hence the population should specialize to improve performance. But exactly how does this principle apply (if it does)? In the language of plasticity theory there may be a 'negative genetic covariance' across environments (Via and Lande, 1985). That is, a genotype's performance on one plant is not independent of its performance on another plant, but instead they are negatively correlated, perhaps because of physiological trade-offs when larvae are metabolizing plant chemistry. Such negative covariance can have the effect that adaptation for good performance occurs most rapidly to the most frequently used host plant, whereas adaptation to perform well on more rarely used plants is greatly slowed down (Via and Lande, 1985). Ovipositing females, which to a higher degree prefer the already most frequently used host, may then in the meantime be selectively favoured, and hence specialization occurs.

Physiological trade-offs have seldom been demonstrated from genetic evidence; in fact genetic covariance in performance among hosts is more often positive rather than negative, reflecting, for example, generally higher growth rates in some families of offspring (Fox and Caldwell, 1994). This has led interest to shift instead to other aspects of the 'jack-of-all-trades' principle, such as faster and more accurate host-plant choice in specialists (Fox and Lalonde,

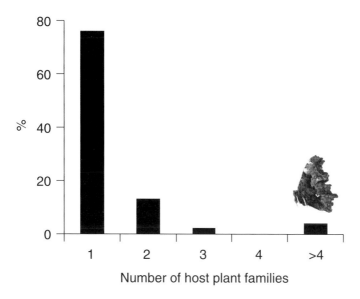

Fig. 5.3. The proportion of Swedish butterflies utilizing one, two, three, four or more than four plant families as larval host plants. *Polygonia c-album* belongs to the last category. Data from sources given in Janz and Nylin (1998).

1993; Bernays and Wcislo, 1994; Janz and Nylin, 1997; Nylin *et al.*, 2000). However, it has been suggested that we can expect to see negative covariance for performance among genotypes only under rather restricted conditions, in populations adapted to use more than one host, close to the genetic equilibrium (Joshi and Thompson, 1995). In other words, negative covariance patterns are evolved traits of generalist populations, as has been experimentally demonstrated using *Drosophila* (Joshi and Thompson, 1997). Consistent with this idea, Ballabeni *et al.* (2003) found positive covariance for performance on two potential hosts in populations of the sedentary beetle *Oreina elongata* that normally use a single host plant, but a lack of covariance among hosts in a two-host population. This population may then be on its way to evolving genotypes that are differentially adapted to showing good performance on one or the other host plant. Clearly, this more restricted view of when to expect to see trade-offs means that the principle can also be less generally applied to the evolution of specialization.

Trade-offs or not, however, the universal existence of performance hierarchies can hardly be denied. Whenever a phytophagous insect has the choice of more than one potential host plant, performance is bound to be better on some of these plants, if all fitness components are summed together. In the case of the comma butterfly, overall performance is good on the hosts in the plant order *Urticales* (especially *Urtica dioica* and *Humulus lupulus*), intermediate on the hosts in the genera *Salix* and *Ribes*, and poor on hosts in the genera *Betula* and *Corylus* (Nylin, 1988; Janz *et al.*, 1994; Nylin *et al.*, 1996b). Other *Polygonia* species, such as *P. gracilis*, instead do better on *Ribes* than on *Urtica* (Janz *et al.*, 2001). It is hard to see why such patterns should exist, unless it is because it *is* difficult to perform equally well on several plant species, especially when they are distantly related and consequently chemically different. Perhaps, then, the most universal reason for specialization is simply that it is best to specialize on the plant (or set of similar plants) where overall performance is best (Nylin and Janz, 1999). One could instead fruitfully ask why there are any generalists at all, and look for explanations in terms of higher realized fecundity or spreading of risks in variable environments.

9. Host Plants, Other Environmental Factors and Life-cycle Regulation

The strong effects of host plants in terms of larval growth rate and development time provide parallels to the sections above on insect life-cycle regulation and fine-tuning by photoperiod. Besides photoperiod, temperature and host plant quality also commonly affect the propensity to enter diapause, and in addition there may be sexual differences. One suggested adaptive interpretation of such patterns is that they arise because the optimal development time under direct development is typically shorter than when the insect is destined for diapause (because in the former case it is aiming to fit an additional generation into the season), and hence optimal growth rates are also typically higher. Low growth rates, and any environmental factors constraining growth rates to be low, may then have been selected to act as cues for diapause development (Wedell *et al.*, 1997; Nylin and Gotthard, 1998).

In several studied butterflies, male and female reaction norms differ in that males are more prone to enter diapause than are females (Wiklund *et al.*, 1992; Nylin *et al.*, 1995; Wedell *et al.*, 1997). This may relate to selection for protandry (Wiklund and Fagerström, 1977). At least in species hibernating as pupae or adults, protandry is achieved by different mechanisms in directly developing generations compared with generations that enter diapause. Protandry in the spring after hibernation diapause can probably be achieved without costs, because males and females have been synchronized by diapause and males simply need to emerge earlier. In order to achieve protandry under direct development, however, males need a shorter development time, which can only be achieved at the cost of either lower pupal weight or higher growth rate. Possibly, males 'choose' diapause at near-critical conditions when females are still developing directly, rather than taking on these costs (Wiklund *et al.*, 1992).

Why does low temperature often increase diapause propensity (Danilevskii, 1965)? Doubtless, temperature often acts as an additional seasonal cue, with low temperatures signalling a late date. Danilevskii (1965) strongly favoured this view, and dismissed the earlier notion that diapause is induced by conditions unfavourable for growth as 'mistaken'. At the time of publication of Danilevskii's seminal book, it was important to establish that diapause is not a consequence of unfavourable conditions, but instead a state that the insect enters in advance of poor conditions in order to survive them. However, there may have been a sense in which the early entomologists were correct. Low temperatures are relatively poor seasonal cues, but they do indicate that high growth rates are costly, at least now and perhaps also in the future. This may tip the balance towards diapause development as the optimal pathway in near-critical conditions (Wedell *et al.*, 1997; Nylin and Gotthard, 1998).

The same type of reasoning holds for host plants: they strongly affect performance and (since the plant is unlikely to change, at least in the butterfly case) is also a good cue for predicting future conditions. Specifically, high growth rates (optimal under direct development) are likely to be costly on poor host plants. Perhaps not surprisingly, effects of host-plant quality on the propensity to enter diapause is often found in the rare cases when this is explicitly investigated, as in *Choristoneura rosaceana* (Lepidoptera: Tortricidae) (Hunter and McNeil, 1997). In the case of the comma butterfly, significant effects of sex, temperature, larval growth rate and host plant, besides photoperiod, were found on the propensity to enter diapause (Wedell *et al.*, 1997). The effect of host plant paralleled the performance hierarchy (highest incidence of direct development on stinging nettle, lowest on birch) and remained after statistically controlling for the effect of larval growth rate itself, suggesting that the plant is truly used as a cue for future conditions (Wedell *et al.*, 1997).

Within developmental pathways, the choice of host plant can act to fine-tune development time. From optimality, it may be predicted that plants permitting fast growth and short development time should be used when the optimal development time is short. Hence, these 'faster' hosts should be used later in the season. In the comma butterfly, the directly developing generation of adults, which always occurs late in the summer and in addition is always associated with a bivoltine life cycle, is more specialized on *U. dioica* than the hibernating generation (Fig. 5.4; Nylin, 1988). It is still unclear whether this really results from an innate higher propensity to prefer *U. dioica* in light-form adults or is an effect of a more drastic seasonal decrease in quality for the alternative hosts, which are trees and bushes rather than herbs, or whether both factors contribute. Experiments to disentangle these factors must be done with the two forms ovipositing simultaneously, when the plants are at the same phenological stage, and this has been difficult to achieve. With the discovery that dark-form adults break reproductive diapause in long days (C. Wiklund, Stockholm, 1999, personal communication), we will be able to establish the respective roles of these factors (G.H. Nygren and S. Nylin, unpublished). In other words, do dark-form females also show an increased preference for nettle later in the season, if manipulated to oviposit, but not to the extent that light-form females do? If so, this adds yet another component to the battery of traits involved in the seasonal polyphenism of *P. c-album*.

Expanding this type of reasoning to variation among populations, we can expect populations under time-stress to be more specialized on 'fast' hosts, even if other hosts are superior in other ways (Nylin, 1988; Scriber and Lederhouse, 1992; Janz *et al.*, 1994). In the comma butterfly, the higher degree of specialization on nettles in the partially bivoltine English population

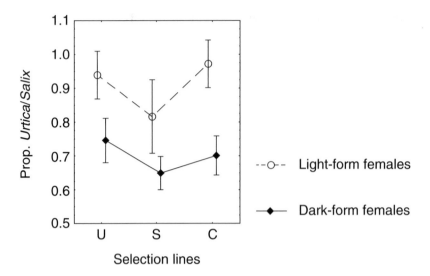

Fig. 5.4. The proportion of eggs laid on *Urtica dioica*, in a choice with *Salix caprea*, by light-form (open symbols) and dark-form (closed symbols) of *Polygonia c-album*. Bars show standard errors from variation among females. Data are from the F$_3$ generation of a selection experiment: U, S = lines selected for *Urtica* and *Salix* preference, respectively. C = unselected control line (Nylin *et al.*, 2005).

than in the univoltine Swedish population has been interpreted in this manner (Nylin, 1988). This body of theory is tightly linked to 'saw-tooth' theory regarding latitudinal variation in development time and size, described above, since preferentially using 'fast' hosts is another way of decreasing development time in offspring. Such preference is expected when the season is very short, as in the extreme north of the distribution, and in areas where there is a shift in generation number, i.e. barely enough time for the additional generation. Achieving a short larval development time in this manner has its advantages. It should typically not carry a cost in either reduced final weight (but here the comma butterfly is an exception if the alternative is to use *Salix caprea*; Janz *et al.*, 1994) or in the form of costs of high growth rates. This is because presumably the 'fast' hosts allow high growth rate because they are plants of high 'quality', i.e. they present a good environment for larval performance.

Other types of costs may be present, however. If the 'fast' hosts are not abundant enough there will be time lost in searching, perhaps even negating the time gained in the next generation. This may explain why females of an extreme short-season population of *P. c-album* from Norway do not specialize on nettles but rather prefer the more abundant sallow (G. Nygren, S. Nylin and R. Krogen, unpublished). In addition, any advantage of a wide host plant range that may originally have promoted its evolution, such as higher realized fecundity or spreading of risks, will be lost by specialization on 'fast' hosts.

10. Speciation and the Origin of Biodiversity

The prevailing view of speciation has been that the most important process is allopatric speciation, often with a high degree of randomness with respect to the ecology of the incipient species. Populations become geographically isolated (for random, climatic or geological reasons), and after some time they have diverged enough, through selection, genetic drift and mutation, so that they are reproductively isolated if they meet again. Recent years have seen an increase in the interest in 'ecological speciation', i.e. processes of speciation where ecological factors play an important role (Schluter, 2000). In the extreme case, ecological specialization of different genotypes may lead directly to a degree of reproductive isolation (e.g. 'host-plant races') and eventually sympatric speciation (Bush and Smith, 1997). However, ecology can play an important role also in allopatric speciation, and in intermediate modes of speciation. If two populations of a species become geographically isolated, gene flow between the populations is interrupted, and they can each adapt to their respective environments. Against a background of the type of results reported above, it is easy to see how, for instance, two insect populations isolated along a north–south gradient could adapt and diverge ecologically, with respect to traits such as life history, life-cycle regulation, seasonal polyphenism and host plant utilization. Co-adapted gene complexes may form, since these traits affect each other in various ways. The populations may not diverge enough to become completely reproductively isolated in case of a secondary contact, but the ecological differences may be enough to promote a process of 'reinforcement'. If hybrids are less fit because they are bad compromises between two ecologies, selection in favour of assortative mating may occur, perhaps simply by divergent choice of habitat.

We have already seen that all of the ecologically relevant traits mentioned in the previous paragraph have strong plastic components; in other words, plasticity cannot be ignored in ecological speciation. It is high time to break with the view of genetics and plasticity as contrasts, where a plastic phenotype is seen as an alternative to genetic differentiation, hence making genetic divergence and speciation less probable. In fact, all genotypes are plastic with respect to most traits, and we must instead concern ourselves with the shape of reaction norms. It has been suggested that, rather than making genetic divergence more difficult, plasticity may provide the necessary raw material of varying phenotypes for natural selection to act upon (West-Eberhard, 1989). After all, imagine two geographically isolated populations of the same species, both fixed for a highly canalized genotype, one that can only produce phenotype A whatever the environment. If the environment changes in the area inhabited by one of the populations, so that phenotype B is now optimal, it must wait for a fortuitous mutation before it can respond. If, instead, both populations have the plastic genotype AB, capable of producing either phenotype, the populations are likely to diverge phenotypically much faster, in response to the environmental change (Fig. 5.5).

Admittedly, it will initially not be a genetic difference, but genotypes within the populations are likely to differ in the exact form of their reaction norms

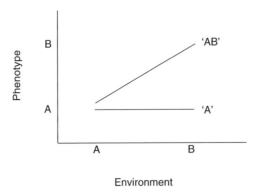

Fig. 5.5. Reaction norms of two genotypes: one ('A') is capable of producing only phenotype A regardless of environment, the other ('AB') is more plastic and produces phenotype A in environment A and phenotype B in environment B. If the plasticity is adaptive, phenotype A leads to higher fitness in environment A than the alternative phenotype, and vice versa.

(genotype-by-environment interactions are ubiquitous). If the environments are stable enough, reaction norms producing phenotype A under prevailing conditions are likely to be favoured by selection in one of the populations, and vice versa for phenotype B in the other population. If this happens by increasing the range of environmental conditions that produces the optimal phenotype, there will be a process of 'genetic assimilation' (Waddington, 1953; Pigliucci and Murren, 2003), in the extreme case leading to canalized phenotypes in the two populations, what we see as a pure 'genetic difference' between them. Even if the potential to produce both phenotypes is retained, 'common garden' experiments can still reveal that the populations now differ genetically in the shape of reaction norms. In other words, upon secondary contact they will display different phenotypes in the same environment, and some form of ecological speciation may then occur.

It is interesting to apply this sort of scenario to insect–host-plant interactions. Phytophagous insects make up a very large proportion of the biodiversity, and this does not seem to be a historical coincidence. Clades of phytophagous insects are consistently more rich in species than the most closely related group of insects feeding on something else (Mitter *et al.*, 1988), suggesting that evolution of the ability to feed on plants somehow opens up a new adaptive zone, promoting diversity. But is it the enormous amount of resources that becomes available by using a lower trophic level that is the key, or is it the diversity of this resource – diversity feeding diversity? This is still unclear, but detailed phylogenetic studies within clades of phytophagous insects may provide clues.

Zooming in on the comma butterfly, one of the examples given by Mitter *et al.* (1988) is the comparison between Lepidoptera and its sister-clade Trichoptera, or perhaps between basal Lepidoptera and the 'higher' forms such as butterflies and large moths (the ancestral diet of Lepidoptera is uncertain). In any case, the phytophagous clade is much more diverse. The phylogenetic relationships among and within 'higher' moth families are still very unclear, but we have some understanding of relationships within the butterflies, an understanding which is rapidly improving through the advent of simple techniques for molecular phylogenetics. Starting with such hypotheses of phylogeny, it is possible to 'map' host-plant utilization on to the phylogenies using the principle of character optimization (Brooks and McLennan, 1991). Simply put, if two closely related species use the same host plant, the simplest explanation is that this host plant was used already by the ancestor of the two species (see Fig. 5.2 for another example of character optimization).

This technique was applied to a butterfly phylogeny, put together from various available sources, by Janz and Nylin (1998). The reconstruction suggested that the ancestral butterfly was a specialist on trees in (or near) *Fabaceae*. Most modern butterflies are also specialists on one (or a few related) plant families (Ehrlich and Raven, 1964), but obviously not all on the same families. This pattern by itself suggests that butterflies sometimes shift plant clades, but then specialize again. There is evidence that shifts to more closely related plants are more likely than shifts to distantly related plants, perhaps because they are chemically more similar and there is some degree of pre-adaptation for good performance (Janz and Nylin, 1998). At the time of shifts, the butterflies must by necessity pass through a phase when they are capable of using at least two clades (the ancestral and the new) as larval food. However, they must respecialize rather quickly, considering the rarity of polyphagous butterfly species.

The comma butterfly belongs to the family Nymphalidae, one of the most diverse butterfly families both in terms of species and host-plant utilization. Relationships within the family are still poorly known, and so we are not yet in a position to reconstruct host-plant utilization at this phylogenetic level. However, hypotheses of phylogeny are available for the tribe Nymphalini (Nylin *et al.*, 2001; Wahlberg and Nylin, 2003). It seems clear from mapping host plants on to these phylogenies (Janz *et al.*, 2001; Wahlberg and Nylin, 2003) that the ancestor of Nymphalini was a specialist on *Urticaceae* (and perhaps other hosts in Urticales), like the basal genera *Hypanartia*, *Antanartia*, *Mynes*, *Symbrenthia* and *Araschnia*. Many of the plant families used in *Polygonia* are shared with the sister-genus *Nymphalis* (Nylin, 1988; Nylin and Janz, 1999). The probable interpretation is that a wider host-plant range evolved in the ancestor of this clade (Janz *et al.*, 2001). In the present context it is interesting to note that the *Polygonia+Nymphalis* clade is more diverse than its sister clade *Aglais+Inachis*, which retains the ancestral specialization on hosts in *Urticales* (Nylin and Janz, 1999).

Possibly, evolution of a wider host-plant range opens up avenues for higher rates of speciation, analogous to evolution of a plastic phenotype. In fact, as pointed out above, an insect that is capable of feeding on more than one plant clade is more plastic compared with a specialist. But how does this ability promote speciation? In some phytophagous insects, one way may be via the formation of host-plant races and sympatric speciation. This could perhaps happen even in some sedentary butterflies (Pratt, 1994), but does not seem likely in strong fliers such as *P. c-album*, where available evidence suggests a very open population structure with high levels of gene flow (Nylin *et al.*, 2005). However, a wide host-plant range may also allow a species to expand its geographical range to areas where the ancestral host is lacking or not very abundant. This sets the scene for speciation, if the species is then over time secondarily broken up into populations with some degree of geographical isolation. Specialization on the host plant which locally provides best performance (also taking into account field characteristics such as relative plant abundance, phenological fit to butterfly life history, predation etc.) may follow. Such emerging differences in host-plant utilization, in turn, have the potential to become an important component in ecological speciation. They may provide some degree of reproductive isolation in time and space, as well as being at the core of co-adapted gene complexes with possibly unfit hybrids between them.

Most of this scenario is at present highly speculative as to how it applies to speciation in *Polygonia* or elsewhere, but there is evidence for one of the necessary components. A dynamic pattern of oscillations in host-plant range is seen in the *Polygonia+Nymphalis* clade, where several species in the clade have respecialized on a subset of the common plant families or, in a few cases, on novel hosts (Janz *et al.*, 2001). Evolutionary dynamics in host-plant range is now emerging as a general property of insect–plant associations (Thompson, 1998; Janz *et al.*, 2001; Nosil, 2002) and it is possible that such oscillations in host-plant range have played a crucial role in the origin of the Earth's biodiversity.

11. Conclusions

Not so long ago the study of plasticity was considered almost a suspect subject in evolutionary biology, but the situation has changed dramatically over the last two decades (West-Eberhard, 2003). Over the coming decades the study of plasticity *per se* will doubtlessly continue and lead to new insights into the evolution of reaction norms and the mechanistic determination of plasticity. Even more importantly, however, it can be hoped that plasticity in traits will increasingly be given the consideration that it deserves as a natural part of any ecological or evolutionary study.

As we hope to have illustrated here, using primarily *Polygonia* butterflies, seasonal plasticity offers a conceptual framework that connects most aspects of insect evolutionary ecology; life-cycle regulation, life-history theory, optimal design, sexual selection, resource use, species interactions and ecological speciation. Some of these are highly theoretical fields, but using the seasonal life cycle of the insect as a starting point helps to ensure that sight is never lost of the organism in its real ecological situation, complete with its constraints and complexities.

Acknowledgements

This study was supported by grants from the Swedish Research Council to S.N. and K.G.

References

Abrams, P.A., Leimar, O., Nylin, S. and Wiklund, C. (1996) The effect of flexible growth rates on optimal sizes and development times in a seasonal environment. *American Naturalist* 147, 381–395.

Ballabeni, P., Gottbard, K., Kayumba, A. and Rahier, M. (2003) Local adaptation and ecological genetics of host-plant specialization in a leaf beetle. *Oikos* 101, 70–78.

Bernays, E.A. and Wcislo, W.T. (1994) Sensory capabilities, information processing, and resource specialization. *Quarterly Review of Biology* 69, 187–204.

Bradshaw, W.E. (1976) Geography of photoperiodic response in a diapausing mosquito. *Nature* 262, 384–385.

Brakefield, P.M. and Larsen, T.B. (1984) The evolutionary significance of dry and wet season forms in some tropical butterflies. *Biological Journal of the Linnean Society* 22, 1–12.

Brooks, D.R. and McLennan, D.H. (1991) *Phylogeny, Ecology, and Behavior: A Research Program in Comparative Biology*. University of Chicago Press, Chicago, Illinois.

Bush, G.L. and Smith, J.J. (1997) The sympatric origin of phytophagous insects. In: Dettner, K., Bauer, G. and Volkl, W. (eds) *Vertical Food Web Interaction*. Springer, Berlin, pp. 3–19.

Conover, D.O. and Schultz, E.T. (1995) Phenotypic similarity and the evolutionary significance of countergradient variation. *Trends in Ecology and Evolution* 10, 248–252.

Dal, B. (1981) *Fjärilar i Naturen 2*. Wahlström and Widstrand, Stockholm.

Danilevskii, A.S. (1965) *Photoperiodism and Seasonal Development of Insects*, English edn. Oliver and Boyd, Edinburgh and London.

Ehrlich, P.R. and Raven, P.H. (1964) Butterflies and plants: a study in coevolution. *Evolution* 18, 586–608.

Fox, C.W. and Caldwell, R.L. (1994) Host-associated fitness trade-offs do not limit the evolution of diet breadth in the small milkweed bug *Lygaeus kalmii* (Hemiptera, Lygaeidae). *Oecologia* 97, 382–389.

Fox, C.W. and Lalonde, R.G. (1993) Host confusion and the evolution of insect diet breadths. *Oikos* 67, 577–581.

Gotthard, K. (1998) Life history plasticity in the satyrine butterfly *Lasiommata petropolitana*: investigating an adaptive reaction norm. *Journal of Evolutionary Biology* 11, 21–39.

Gotthard, K. (2000) Increased risk of predation as a cost of high growth rate: an experimental test in a butterfly. *Journal of Animal Ecology* 69, 896–902.

Gotthard, K. and Nylin, S. (1995) Adaptive plasticity and plasticity as an adaptation: a selective review of plasticity in animal morphology and life history. *Oikos* 74, 3–17.

Gotthard, K., Nylin, S. and Wiklund, C. (1994) Adaptive variation in growth rate: life history costs and consequences in the speckled wood butterfly, *Pararge aegeria*. *Oecologia* 99, 281–289.

Gotthard, K., Nylin, S. and Wiklund, C. (1999a) Mating system evolution in response to search costs in the speckled wood butterfly, *Pararge aegeria*. *Behavioral Ecology and Sociobiology* 45, 424–429.

Gotthard, K., Nylin, S. and Wiklund, C. (1999b) Seasonal plasticity in two satyrine butterflies: state-dependent decision making in relation to daylength. *Oikos* 84, 453–462.

Gotthard, K., Nylin, S. and Wiklund, C. (2000a) Individual state controls temperature dependence in a butterfly (*Lasiommata maera*). *Proceedings of the Royal Society of London, Series B* 267, 589–593.

Gotthard, K., Nylin, S. and Wiklund, C. (2000b) Mating opportunity and the evolution of sex-specific mortality rates in a butterfly. *Oecologia* 122, 36–43.

Hunter, M.D. and McNeil, J.N. (1997) Host-plant quality influences diapause and voltinism in a polyphagous insect herbivore. *Ecology* 78, 977–986.

Janz, N. and Nylin, S. (1997) The role of female search behaviour in determining host plant range in plant feeding insects: a test of the information processing hypothesis. *Proceedings of the Royal Society of London, Series B* 264, 701–707.

Janz, N. and Nylin, S. (1998) Butterflies and plants: a phylogenetic study. *Evolution* 52, 486–502.

Janz, N., Nylin, S. and Wedell, N. (1994) Host plant utilization in the comma butterfly: sources of variation and evolutionary implications. *Oecologia* 99, 132–140.

Janz, N., Nyblom, K. and Nylin, S. (2001) Evolutionary dynamics of host-plant specialization: a case study of the tribe Nymphalini. *Evolution* 55, 783–796.

Joshi, A. and Thompson, J.N. (1995) Trade-offs and the evolution of host specialization. *Evolutionary Ecology* 9, 82–92.

Joshi, A. and Thompson, J.N. (1997) Adaptation and specialization in a two-resource environment in *Drosophila* species. *Evolution* 51, 846–855.

Karlsson, B. and Wickman, P.-O. (1989) The cost of prolonged life: an experiment on a nymphalid butterfly. *Functional Ecology* 3, 399–405.

Kemp, D.J. and Jones, R.E. (2001) Phenotypic plasticity in field populations of the tropical butterfly *Hypolimnas bolina* (L.) (Nymphalidae). *Biological Journal of the Linnean Society* 72, 33–45.

Kingsolver, J.G. (1995) Fitness consequences of seasonal polyphenism in western white butterflies. *Evolution* 49, 942–954.

Koch, P.B. (1992) Seasonal polyphenism in butterflies: a hormonally controlled phenomenon of pattern-formation. *Zoologische Jahrbücher, Abteilung für Allgemeine Zoologie und Physiologie der Tiere* 96, 227–240.

Masaki, T., Endo, K. and Kumagai, K. (1989) Neuroendocrine regulation of the development of seasonal morphs in the Asian comma butterfly, *Polygonia c-aureum* L.: stage-dependent changes in the action of summer-morph producing hormone of the brain-extracts. *Zoological Science* 6, 113–119.

Mitter, C., Farrel, B. and Wiegmann, B. (1988) The phylogenetic study of adaptive zones: has phytophagy promoted insect diversification? *American Naturalist* 132, 107–128.

Mousseau, T.A. (1997) Ectotherms follow the converse to Bergmann's rule. *Evolution* 51, 630–632.

Mousseau, T.A. and Roff, D.A. (1989) Adaptation to seasonality in a cricket: patterns of phenotypic and genotypic variation in body size and diapause expression along a cline in season length. *Evolution* 43, 1483–1496.

Nosil, P. (2002) Transition rates between specialization and generalization in phytophagous insects. *Evolution* 56, 1701–1706.

Nylin, S. (1988) Host plant specialization and seasonality in a polyphagous butterfly, *Polygonia c-album* (Nymphalidae). *Oikos* 53, 381–386.

Nylin, S. (1989) Effects of changing photoperiods in the life cycle regulation of the comma butterfly, *Polygonia c-album* (Nymphalidae). *Ecological Entomology* 14, 209–218.

Nylin, S. (1991) Butterfly life-history adaptations in seasonal environments. PhD thesis, Stockholm University, Sweden.

Nylin, S. (1992) Seasonal plasticity in life history traits: growth and development in *Polygonia c-album* (Lepidoptera: Nymphalidae). *Biological Journal of the Linnean Society* 47, 301–323.

Nylin, S. (1994) Seasonal plasticity and life-cycle adaptations in butterflies. In: Danks, H.V. (ed.) *Insect Life-Cycle Polymorphism*. Kluwer Academic, Dordrecht, The Netherlands, pp. 41–67.

Nylin, S. (2001) Life history perspectives on pest insects: what's the use? *Austral Ecology* 26, 507–517.

Nylin, S. and Gotthard, K. (1998) Plasticity in life history traits. *Annual Review of Entomology* 43, 63–83.

Nylin, S. and Janz, N. (1999) The ecology and evolution of host plant range: butterflies as a model group. In: Drent, R., Brown, V.K. and Olff, H. (eds) *Herbivores, Plants and Predators*. Blackwell, Oxford, UK, pp. 31–54.

Nylin, S. and Svärd, L. (1990) Latitudinal patterns in the size of European butterflies. *Holarctic Ecology* 14, 192–202.

Nylin, S., Wickman, P.-O. and Wiklund, C. (1989) Seasonal plasticity in growth and development of the speckled wood butterfly, *Pararge aegeria* (Satyrinae). *Biological Journal of the Linnean Society* 38, 155–171.

Nylin, S., Wiklund, C., Wickman, P.-O. and Garcia-Barros, E. (1993) Absence of trade-offs between sexual size dimorphism and early male emergence in a butterfly. *Ecology* 74, 1414–1427.

Nylin, S., Wickman, P.-O. and Wiklund, C. (1995) An adaptive explanation for male-biased sex ratios in overwintering monarch butterflies. *Animal Behaviour* 49, 511–514.

Nylin, S., Gotthard, K. and Wiklund, C. (1996a) Reaction norms for age and size at maturity in *Lasiommata* butterflies: predictions and tests. *Evolution* 50, 1351–1358.

Nylin, S., Janz, N. and Wedell, N. (1996b) Oviposition plant preference and offspring performance in the comma butterfly: correlations and conflicts. *Entomologia Experimentalis et Applicata* 80, 141–144.

Nylin, S., Bergström, A. and Janz, N. (2000) Butterfly host plant choice in the face of possible confusion. *Journal of Insect Behavior* 13, 469–482.

Nylin, S., Nyblom, K., Ronquist, F., Janz, N., Belicek, J. and Kallersjo, M. (2001) Phylogeny of *Polygonia, Nymphalis* and related butterflies (Lepidoptera: Nymphalidae): a total-evidence analysis. *Zoological Journal of the Linnean Society* 132, 441–468.

Nylin, S., Nygren, G.H., Windig, J., Janz, N. and Bergström, A. (2005) *Biological Journal of the Linnean Society* 84, in press.

Pianka, E.R. (1983) *Evolutionary Ecology*, 3rd edn. Harper and Row, New York.

Piersma, T. and Lindstrom, A. (1997) Rapid reversible changes in organ size as a component of adaptive behaviour. *Trends in Ecology and Evolution* 12, 134–138.

Pigliucci, M. (2001) *Phenotypic Plasticity: Beyond Nature and Nurture*. Johns Hopkins University Press, Baltimore, Maryland.

Pigliucci, M. and Murren, C.J. (2003) Genetic assimilation and a possible evolutionary paradox: can macroevolution sometimes be so fast as to pass us by? *Evolution* 57, 1455–1464.

Pratt, G.F. (1994) Evolution of *Euphilotes* (Lepidoptera, Lycaenidae) by seasonal and host shifts. *Biological Journal of the Linnean Society* 51, 387–416.

Roff, D.A. (1980) Optimizing development time in a seasonal environment: the 'ups and downs' of clinal variation. *Oecologia* 45, 202–208.

Roff, D.A. (1983) Phenological adaptation in a seasonal environment: a theoretical perspective. In: Brown, V.K. and Hodek, I. (eds) *Diapause and Life Cycle Strategies in Insects*. Dr W. Junk, The Hague, The Netherlands, pp. 253–270.

Schluter, D. (2000) *The Ecology of Adaptive Radiation*. Oxford University Press, Oxford, UK.

Scott, J.A. (1986) *The Butterflies of North America*. Stanford University Press, Stanford, California.

Scriber, J.M. and Lederhouse, R.C. (1992) The thermal environment as a resource dictating geographic patterns of feeding specialization of insect herbivores. In: Hunter, M.R., Ohgushi, T. and Price, P.W. (eds) *Effect of Resource Distribution on Animal–Plant Interactions*. Academic Press, New York, pp. 429–466.

Shapiro, A.M. (1976) Seasonal polyphenism. *Evolutionary Biology* 9, 259–333.

Singer, M.C. and Lee, J.R. (2000) Discrimination within and between host species by a butterfly: implications for design of preference experiments. *Ecology Letters* 3, 101–105.

Stearns, S.C. (1989) The evolutionary significance of phenotypic plasticity. *BioScience* 39, 436–445.

Stearns, S.C. (1992) *The Evolution of Life Histories*. Oxford University Press, Oxford, UK.

Strong, D.R. (1988) Insect host range. *Ecology* 69, 885.

Sultan, S. (2003) The promise of ecological developmental biology. *Journal of Experimental Zoology B (Molecular and Developmental Evolution)* 296B, 1–7.

Tammaru, T., Nylin, S., Ruohomäki, K. and Gotthard, K. (2004) Compensatory responses in lepidopteran larvae: the role of growth rate maximisation. *Oikos* 107, 352–362.

Tauber, M.J., Tauber, C.A. and Masaki, S. (1986) *Seasonal Adaptations of Insects*. Oxford University Press, Oxford, UK.

Thomas, J. and Lewington, R. (1991) *The Butterflies of Britain and Ireland*. Dorling Kindersley, London.

Thompson, J.N. (1988) Evolutionary ecology of the relationship between oviposition preference and performance of offspring in phytophagous insects. *Entomologia Experimentalis et Applicata* 47, 3–14.

Thompson, J.N. (1994) *The Coevolutionary Process*. Chicago University Press, Chicago.

Thompson, J.N. (1998) The evolution of diet breadth: monophagy and polyphagy in swallowtail butterflies. *Journal of Evolutionary Biology* 11, 563–578.

Via, S. and Lande, R. (1985) Genotype–environment interaction and the evolution of phenotypic plasticity. *Evolution* 39, 505–522.

Voigt, W. (1985) Zur Induktion und Termination der reproduktive Diapause des kleinen Fuchses, *Aglais urticae* L. (Lepidoptera, Nymphalidae). *Zoologische Jahrbücher, Abteilung für Systematik, Okologie und Geographie der Tiere* 112, 277–298.

Waddington, C.H. (1953) Genetic assimilation of an acquired character. *Evolution* 7, 118–126.

Wahlberg, N. and Nylin, S. (2003) Morphology versus molecules: resolution of the positions of *Nymphalis, Polygonia* and related genera. *Cladistics* 19, 213–223.

Wedell, N. (1996) Mate quality affects reproductive effort in a paternally investing species. *American Naturalist* 148, 1075–1088.

Wedell, N., Nylin, S. and Janz, N. (1997) Effects of larval host plant and sex on the propensity to enter diapause in the comma butterfly. *Oikos* 78, 569–575.

West-Eberhard, M.J. (1989) Phenotypic plasticity and the origins of diversity. *Annual Review of Ecology and Systematics* 20, 249–278.

West-Eberhard, M.J. (2003) *Developmental Plasticity and Evolution*. Oxford University Press, Oxford, UK.

Wiklund, C. and Fagerström, T. (1977) Why do males emerge before females? A hypothesis to explain the incidence of protandry in butterflies. *Oecologia* 31, 153–158.

Wiklund, C. and Tullberg, B.S. (2004) Seasonal polyphenism and leaf mimicry in the comma butterfly. *Animal Behaviour* 68, 621–627

Wiklund, C., Wickman, P.-O. and Nylin, S. (1992) A sex difference in reaction norms for direct/diapause development: a result of selection for protandry. *Evolution* 46, 519–528.

Wiklund, C., Gotthard, K. and Nylin, S. (2003) Mating system and the evolution of sex-specific mortality rates in two nymphalid butterflies. *Proceedings of the Royal Society of London, Series B* 270, 1823–1828.

6

Life Histories and Parasite Pressure Across the Major Groups of Social Insects

J.J. BOOMSMA,[1] P. SCHMID-HEMPEL[2] AND
W.O.H. HUGHES[1]*

[1]*Institute of Biology, Department of Population Biology, University of Copenhagen, Denmark;* [2]*ETH Zurich, Ecology and Evolution, ETH-Zentrum, Zurich, Switzerland*

*Present address: School of Biological Sciences, A12, University of Sydney, Sydney, Australia.

Thus, whereas ant colonies participate in many symbioses and are sometimes largely dependent on them, honeybee colonies, which are much less permanent in place of abode, have no known symbionts but many parasites.

R. Axelrod and W.D. Hamilton, The evolution of co-operation.
Science 211, 1390–1396 (1981)

1. Introduction

Animal societies are aggregations of cooperating individuals that are isolated from other societies by limitations of dispersal and/or hostile exclusion mechanisms. The individuals within them are more related to the members of their own society than to random individuals in the population at large and quite often this relatedness is high because societies are families or groups of families. For parasites and diseases, however, animal societies are merely patches of suitable hosts to be colonized and exploited and to ultimately produce dispersing propagules to reach other similar patches (Freeland, 1979).

Living in groups or societies has generally been thought to be associated with increased parasitism (Alexander, 1974; Freeland, 1976; Hamilton, 1987; Sherman *et al.*, 1988; Côté and Poulin, 1995; Schmid-Hempel, 1998). However, several recent studies have questioned the generality of this assertion (Watve and Jog, 1997; Lewis, 1998; Naug and Camazine, 2002; Wilson *et al.*, 2003). Others have provided data to show that social behaviour can also be associated with reduced parasite load, due to either behavioural interactions providing an effective defence (Rosengaus *et al.*, 1998; Hughes *et al.*, 2002; Traniello *et al.*, 2002), or density-dependent immune responses (Reeson *et al.*, 1998, Barnes and Siva-Jothy, 2000; Wilson *et al.*, 2003). These discrepancies may result from the exact mode of transmission. In fact, a meta-analysis by Côté and Poulin (1995) has shown that rates of parasitism tend to be positively correlated with group size for parasites that are transmitted by direct contact, whereas these same correlations are negative for parasites that actively find their hosts. Furthermore, while a gregarious habit may increase the risk of intragroup transmission, this may be more than compensated for by a decrease in intergroup transmission (Watve and Jog, 1997), as has been recently demonstrated in a comparative study of Lepidoptera larvae (Wilson *et al.*, 2003).

Ecological epidemiology has developed general theoretical models exploring the conditions under which diseases can co-exist with their hosts. In a series of seminal studies, Anderson and May (1979, 1981, 1982) showed that essentially all diseases need a minimal threshold number of susceptible hosts to maintain themselves. The threshold depends on the rate at which a disease is able to infect new hosts by dispersal, relative to the rate of disappearance of existing infections due to disease-induced mortality and immunity after recovery. When short-range transmission is effective and virulence not too high, diseases may endemically coexist with their hosts in every patch at some expense to the number of hosts and their fitness (Anderson and May, 1979, 1982). Otherwise, extant diseases will have a dynamic metapopulation structure depending on incidental long-range dispersal between patches and epidemic outbreaks hitting different patches at different times (Anderson and May, 1982; Grenfell and Harwood, 1996).

On an evolutionary timescale, parasites often become specialized on a single host species, and coevolve with their hosts in a Red-Queen-like manner (e.g. Jaenike, 1993; Gandon *et al.*, 2002). Other parasites, however, remain generalists exploiting multiple hosts (Johnson *et al.*, 2003a). Host defences have been categorized as avoidance, resistance (before or after recovery) and tolerance, mechanisms that each have their own cost–benefit trade-offs and coevolutionary dynamics with specific pathogens (e.g. Boots and Bowers, 1999; Schmid-Hempel and Ebert, 2003; Weinig *et al.*, 2003). Finally, recent research has increasingly emphasized that virulence of a parasite is a plastic trait, subject to natural selection and sometimes conditionally expressed (Frank, 1996a,b; Herre *et al.*, 1999). Although intermediate virulence often seems to be an evolutionarily stable strategy (ESS), in fact any level of virulence can evolve depending on ecological conditions and evolutionary trade-offs (Ebert and Herre, 1996). This implies that mutualistic and parasitic symbionts are increasingly considered to be two sides of the same coin, with many gradual transitions between them (Frank, 1996a,b; Herre *et al.*, 1999; Bot *et al.*, 2001a).

This chapter concentrates on social insects which, together with our own species, represent extreme examples of group living. We investigate the extent to which non-uniform loads of parasites and diseases can be explained by differences in ecology and life history among the four major groups of social insects: the ants, termites, social bees and social wasps. We realize that comparisons at this large taxonomic scale necessitate crude assumptions and bold generalizations that may not hold for every particular parasite–host interaction. However, we believe that this approach will be helpful to establish an extended general framework for the further investigation of how parasitism affects social species and to clarify what particular elements of coevolution between parasites and social hosts deserve to be emphasized in future studies.

The ants, social bees and social wasps belong to the haplodiploid and holometabolous order of the Hymenoptera and have independently evolved advanced social behaviour (the bees did so multiple times) (Wilson, 1971; Michener, 1974). The termites belong to the hemimetabolous order of the Isoptera, are obligatory social, and share a single common ancestor (Eggleton, 2001). Although the four groups have advanced social behaviour in common, they differ in a series of key traits that will be relevant for the analyses and arguments to be presented in this chapter (Table 6.1). Following Wilson (1971) these can be characterized as follows. Wasps and ants are mostly carnivorous, but many qualify as omnivores as they also collect nectar (wasps) and aphid honeydew (ants). Bees have almost exclusively a vegetarian diet consisting of pollen and nectar, whereas the termites are mostly decomposers of plant-derived biomass. Most bees and wasps have closely related non-social relatives and an annual, semelparous colony cycle, with honeybees, stingless bees and polybiine wasps being important, evolutionarily derived, exceptions to this rule. Ants and termites, on the other hand, are universally eusocial and perennial, with the exception of only very few derived social parasites. Bees and wasps build their nests as an arrangement of brood cells made from freshly collected or manufactured substrates (mud, paper, wax), whereas ants and termites build their nests in

Table 6.1. A comparative overview of the major ecological differences among the four main groups of insects in which eusociality has evolved independently (after Wilson, 1971).

Trait	Bees	Wasps	Ants	Termites
Life cycle	Semelparous/annual[a]	Semelparous/annual[a]	Iteroparous/perennial	Iteroparous/perennial
Diet	Pollen and nectar	Carnivorous/omnivorous	Carnivorous/omnivorous	Decomposer of plant material
Foraging	Flying, foraging areas of colonies typically overlap	Flying, foraging areas of colonies typically overlap	Walking, typically territorial with partly separated foraging grounds	Walking, colonies with strongly separated foraging grounds
Nest building	Constructs of paper or wax, brood arranged in cells	Constructs of paper or wax, brood arranged in cells	Galleries, loose piles of chambers	Galleries, loose piles of chambers
Nest habitat	Closed nests in soil and cavities[b]	Typically open suspended nests[c]	Closed nests, mostly in soil	Closed nests, mostly in wood and soil

[a] Some Apidae and polistine wasps are iteroparous and perennial and some halictine bees are annual but bivoltine.
[b] Some species with open suspended nests (Apidae).
[c] Some Vespinae build nests in underground cavities.

soil or wood by excavating a system of galleries and nest chambers. Finally, there are important differences in worker morphology and foraging. Workers of bees and wasps have wings and a relatively large foraging range, whereas workers of ants and termites are wingless, and normally forage within distinct territories that they may defend against other colonies.

As outlined in Schmid-Hempel (1998), the study of parasites and disease in social insects has been plagued by a general lack of data and a high degree of data skew, with data on honeybees, for example, being much more abundant and accurate than data on other species. Although it has been suspected that social insects do not suffer as much from parasites as their high and aggregated densities would predict, the deficit of data has made it difficult to assess any differences in the prevalence and impact of parasites and diseases across the ants, social bees, social wasps, and termites. Since Schmid-Hempel's (1998) review, theoretical (Naug and Camazine, 2002) and particularly experimental research on social insect diseases has expanded: a number of new parasite–host interactions have been discovered, and experimental studies have clarified detailed mechanisms of defence against diseases (e.g. Baer and Schmid-Hempel, 1999, 2001; Brown et al., 2000; Moret and Schmid-Hempel, 2000, 2001; Lord et al., 2001; Doums et al., 2002; Hughes et al., 2002; Poulsen et al., 2002a, 2003; Traniello et al., 2002). It is timely, therefore, to attempt a specific comparative review of disease pressure across the four major groups of social insects, particularly in connection to the key life-history differences characterizing these groups (Table 6.1), to investigate whether disease pressures are likely to co-vary with these fundamental life-history differences in similar ways as shown for other organisms (Côté and Poulin, 1995; Wilson et al., 2003).

Previous reviews on social insect diseases (Bailey, 1963; Hamilton, 1987; Bailey and Ball, 1991; Macfarlane et al., 1995; Schmid-Hempel, 1995, 1998; Schmid-Hempel and Crozier, 1999) have emphasized the social Hymenoptera, their varying haplodiploidy-induced degrees of relatedness among colony members, and their differences in colony size within each of these groups. This chapter will, without implying any lower importance to relatedness factors, concentrate on differences in ecology between the obligatory perennial ants and termites on one hand and the mostly annual bees and wasps on the other. We offer an updated (compared to Schmid-Hempel, 1998) analysis of comparative data and an explicit life-history framework for comparing insect societies that differ in colony longevity (annual versus perennial), nest building (mud, paper or wax versus galleries in the soil or in wood), and the mode and range of foraging (flying versus walking, and the concomitant range overlap between colonies). Section 2 reviews the various ways in which the different categories of parasites are recruited and transmitted in the four groups of social insects and how their virulence may depend on this. Section 3 evaluates the actual defence mechanisms of individuals and colonies and the ways in which selection is likely to have shaped investment in defences for different social insect life histories. Section 4 presents novel analyses of comparative data to see whether expectations inferred from conceptual considerations in Sections 2 and 3 are supported by evidence.

2. A Comparative Appraisal of the Major Groups of Social Insects as Hosts of Diseases

2.1 The categories of parasites and diseases

Insect societies have a multitude of parasites and diseases. There are microparasites (bacteria, viruses, protozoa, fungi) and macroparasites (mites, nematodes, helminths, insects, many of the latter being true parasitoids). Microparasites are small, have short generation times, and very high rates of reproduction within a host body. Macroparasites are larger, have longer generation times, no or very slow reproduction within a host body, and free-living stages outside the host (Anderson and May, 1979, 1981). Parasites are either endemic or epidemic (mostly restricted to microparasites) and may be specialist or generalist. An extensive review on many aspects of parasitism of insect societies has been provided by Schmid-Hempel (1998). The major characteristics of reproduction, transmission and virulence that can be derived from his review are summarized in Table 6.2. The categorizations of Table 6.2 are obviously generalizations with many exceptions, and only have heuristic value, but they define, together with the host life-history generalizations of Table 6.1, the conceptual framework of this chapter.

In recent years, most work on social insect diseases has been inspired by the idea (Hamilton, 1987; Sherman *et al.*, 1988; reviewed and extended in Schmid-Hempel, 1998) that the dense packing of social insects in nests should have aggravated their problems with diseases over evolutionary time. As 'nature abhors a pure stand' (because it breeds diseases; Hamilton, 1982, 1987), the pressure of rapidly evolving parasites was hypothesized to have been a major driving force behind secondary developments towards genetically less homogeneous societies via multiple queen-mating and multiple queening of colonies (e.g. Crozier and Page, 1985; Keller and Reeve, 1994; Schmid-Hempel, 1994; Boomsma and Ratnieks, 1996). The most straightforward tests of hypotheses of this kind can be done with microparasites. However, on closer inspection, genetic diversity is only one of an entire suite of possible colony-level defence mechanisms against parasites (see e.g. Naug and Camazine, 2002). Before discussing defences though, we need a systematic overview of the assaults that different insect societies may suffer. The factors that determine the frequency and potential severity of challenges by parasites and disease have been listed in Table 6.2. The subsections below evaluate the three processes that are each affected by these factors and which define the impact of parasites for hosts: exposure, intercolony transmission and intracolony transmission.

2.2 Exposure to parasites: nesting ecology, foraging and food

Of the four categories of microparasites, three (bacteria, viruses and protozoa) are usually transmitted orally (*per os*) via the sharing of regurgitated food, ingestion of excrement, etc. (see Table 6.2), whereas the spores of insect

Table 6.2. The categories of social insect diseases and their major characteristics of reproduction, transmission and virulence, the latter assuming that defences are unsuccessful (after Schmid-Hempel, 1998).

	Reproduction within host	Generation time	Transmission stage	Typical infection mode[e]	Transmission route/range	Survival of transmission stage outside host	Virulence for individual[m]	Virulence for colony[p]	Intracolonial epizootic potential
Macroparasites									
Parasitoids	No[a,c]	Very long	Free-living	Active entry	Active dispersed	Moderate, long[i]	High, death inevitable[n]	Low[q]	Absent
Other insects	No[b]	Very long	Free-living	Active entry	Active dispersed	Moderate, long[i]	High, death inevitable[n]	Low[q]	Absent
Mites	No[b,c]	Very long	Free-living	Active entry	Active, local	Short, moderate[k]	Moderate	Moderate	Moderate
Helminths	No[d]	Very long	Free-living, durable eggs	per os	Active dispersed, local[f]	Moderate	Moderate[o]	Low	Absent
Nematodes	Yes	Intermediate	Free-living	per os and other openings	Local	Moderate	Low to high	Moderate	Moderate
Microparasites									
Fungi	Yes	Very short	Spores	Through cuticle	Passive dispersed[g]	Moderate, long[l]	Generally high	High	Very high
Protozoa	Yes	Short	Cells, spores	per os	Local, direct contact[h]	Short	Moderate	Moderate	High
Bacteria	Yes	Very short	Spores	per os	Local, direct contact[h]	Short to long[l]	Low to high	High	Very high
Viruses	Yes	Very short	Spores	per os	Local, direct contact	Very short	Low to high	High	Very high

a Some species with sexual dimorphism and females staying inside host (e.g. Strepsiptera).

b Most species parasitize immatures and grow in brood cells.

c Many mites reproduce on or inside the host (e.g. tracheal mites).

d Social insects are intermediate hosts to helminths (trematodes, cestodes). Reproduction occurs in the final vertebrate host.

e Many exceptions to the typical infection mode are known. Indication given here only as a crude classification.

f Range depends on dispersal by final host.

g Large-scale dispersal by wind, more local dispersal by water or passive transport by other organisms.

h Long-lasting transmission stages also transmitted indirectly through contact with substrates.

i Survival outside hosts depends on life cycle of adult insect, which is typically annual.

k Dispersal stages often do not feed and hence cannot survive for long.

l Some fungi and bacteria are known to have durable stages of extreme longevity.

m Effect on survival or performance of host individual.

n Obligate death sometimes prevented by encapsulation.

o Most helminths do not seriously harm host. Some cause changes that render social insect host more susceptible to predation by final host.

p Effect on survival or reproductive performance of colony.

q In exceptional cases colonies can be devastated (e.g. wax moths).

pathogenic fungi are mostly dispersed passively via wind, rain, etc. and enter hosts via the cuticle or openings such as the trachea (Andreadis, 1987). This implies that fungal spores will be spread out over relatively large areas and will often be associated with the soil, whereas propagules of the other disease categories will be deposited at sites visited by infected individuals. Ant and termite hosts usually start colonies by excavating galleries in the soil or in wood (Table 6.1). Some may erect nest mounds or paper nests when colonies mature, but as a rule the developing brood and foraging workers will remain associated with soil and wood (Brian, 1982). It thus seems obvious to expect that fungal diseases should be common in ants and termites. Most wasps build nests in the open, generally hanging freely from branches, so that they have minimal contact with soil or wood. Bees seem intermediate. Primitively eusocial bees (e.g. halictines) tend to excavate nests in the soil, whereas more advanced taxa (bumblebees, stingless bees, honeybees) mostly use existing cavities under-ground or in hollow trees. This implies that soil contact of nest workers in the latter category of bees is less than that which is normal for ants and termites (Brian, 1982; see also Table 6.1) although probably not quite as low as for wasps. These differences in nesting behaviour among bee taxa have earlier been hypothesized to be associated with different parasite pressure (Michener, 1985), an inference that is supported by data on solitary species (Wcislo, 1996).

Exceptions to these general rules do occur, such as arboreal ants and termites making nests of silk or carton (overviews in Wilson, 1971; Brian, 1982; Hölldobler and Wilson, 1990; Schmid-Hempel, 1998), and allodapine bees nesting in plant stems without making cells (Michener, 1974). However, it seems reasonable to infer that in the course of their social evolution ants and termites have been 'recruiting' their parasites from different habitats than bees and wasps, and that this should be particularly true for parasites and diseases that do not disperse widely. Parasites of ants and termites should thus tend to have close relatives that are soil-borne and relatively intolerant to dehydration, whereas parasites of social wasps and the advanced social bees will tend to be related to organisms that can survive in drier habitats, such as the surfaces of wood and vegetation and the interior of hollow trees. That wet or dry nesting material matters for exposure to microparasites has recently been documented for dampwood and drywood termites (Rosengaus *et al.*, 2003). Similarly, the typical habit of wasps (and some bees) to construct open nests predisposes them to attack by actively searching macroparasites (Keeping and Crewe, 1983; Schmid-Hempel, 1998), whereas the closed nests of most bees, ants, and especially termites, prevent most exposure of this kind. General exposure patterns would thus predict that ants and termites are prone to contracting fungal microparasites and macroparasitic worms (nematodes and helminths), while their underground or otherwise enclosed nesting habits would protect them from assaults by many flying macroparasites. On the other hand, wasps and, to some extent, bees with open nests away from soil may have fewer soil-borne microparasites and relatively more mobile and actively searching macroparasites (Table 6.2). However, these generalizations may not always apply to highly specialized macroparasites, as they may have evolved highly effective host-finding adaptations.

Social insect workers that forage on the wing mostly encounter relatively sterile medium, except when they land on flowers, insect prey or carrion, where disease propagules may be dense, particularly on the latter two. In contrast, social insects whose workers forage on foot will almost never encounter fully sterile habitat, and will thus be likely to have a lower variance in exposure. Also the typical food items collected and ingested differ among the four groups of social insects (Table 6.1) and may incur different risks. Bees almost always forage only on nectar and pollen, which will represent a relatively hygienic food source with very little potential to act as a transmission route for entomopathogens. On the other hand, wasps and, to some extent, ants are largely predatory, and their prey normally consists of other insects. This therefore exposes them frequently to diseases that they will risk contracting via the *per os* route from the animals they eat. Wasps and ants are also predators and scavengers of dead or dying insects infected by disease (Smirnoff, 1959; Tanada and Fuxa, 1987; Baur *et al.*, 1998) and wasp nests and larval faeces commonly contain a very wide range of entomopathogenic microorganisms that they have probably contracted from their food (Morel and Fouillaud, 1992; Rose *et al.*, 1999). Indeed, the life cycles of certain nematode and helminth worms are based upon the infection of their ant or wasp host when it feeds upon another infected arthropod (Kaya, 1987; Molloy *et al.*, 1999), while those of certain protozoan parasites rely upon workers cannibalizing infected pupae and feeding meat from them to larvae (Jouvenaz, 1986; Buschinger and Kleespies, 1999). Finally, a number of wasps and ants are specialized predators on other social insects, which may expose them to diseases that are already adapted to social insect hosts. Overall, therefore, a 'vegetarian' diet (bees and termites) may thus incur a lower risk for general insect diseases than an omnivorous diet (ants and wasps).

2.3 Intercolony interactions, infections and the maintenance of diseases

Although social insects are exposed to diseases via their food, nesting habitat and foraging behaviour as outlined above, a potentially more significant risk of exposure stems from contact with conspecifics, as this transmission route would facilitate the evolution of specialized parasites. A disease can only be maintained if its basic ratio of infection (R_0) remains >1 (Anderson and May, 1979; Grenfell and Harwood, 1996). A detailed treatment of models on transmission between and within social insect colonies is given by Schmid-Hempel (1998), so that here, a verbal summary of the key factors that affect transmission will suffice. For any disease to spread, its growth rate R_0 needs to be >1, i.e. every newly created infection needs to create on average at least one further infection before it disappears with a specific natural death rate (b), a specific disease-induced death rate (α) and a host recovery rate (γ). This requires a minimum threshold number of susceptible hosts N_T, which equals ($\alpha + b + \gamma$) / β (see Anderson and May, 1979, 1982, for the basic theory and Schmid-Hempel, 1998, for more specific elaborations). A key question, therefore, is whether differences in nesting ecology and foraging are likely to affect the intercolony transmission rate (β). Efficient transmission (high β) implies that a disease can maintain itself at relatively low densities of susceptible hosts, whereas inefficient transmission will cause a disease to go extinct, unless the number of susceptible hosts is large. The model sketched out here only applies to microparasites. Macroparasites have free-living stages independent of the host, which complicates their transmission dynamics and implies that the growth rates of populations of macroparasites cannot be simply derived from the rates of gain and loss of infections (Anderson and May, 1979; Schmid-Hempel, 1998). We will restrict our discussion here to horizontal transmission (between individuals within a colony and between individuals from different colonies), as vertical transmission (from a parent colony to a daughter colony via the reproductives) will be dealt with in the next subsection on intracolony transmission.

The intercolony transmission rate is a direct function of the encounter rate of infected and uninfected individuals from different colonies, where encounters are defined as either direct physical contact or indirect contact with an object or patch that contains viable disease propagules. The likelihood of intercolony disease transmission in the four groups of social insects depends to a large degree on how the foraging areas are used. Bees and wasps may cover up to several kilometres on the wing (Spradbery, 1973; Roubik, 1989; Beekman and Ratnieks, 2000; Goulson and Stout, 2001; Dramsted et al., 2003) and typically exploit food sources that are accessible to all colonies in the area. Bees from many different colonies are thus likely to visit the same flowers for nectar and pollen, and wasps from many different colonies may scrape fibres for nest building from the same pieces of dry wood, forage for insect prey in the same patches, or cut pieces of meat from the same carrion. As a consequence, direct or indirect contact between workers from different colonies is frequent, the latter when visits are separated in time but with such short intervals that short-lived transmission propagules survive and can be transmitted between colonies (Durrer and Schmid-Hempel, 1994). The situation in ants and termites is quite different. In both groups, workers forage on foot, cover shorter distances by comparison, and often maintain foraging territories that secure privileged access to food sources (see Brian, 1982; Hölldobler and Wilson, 1990). Termites carry this principle to the extreme. In particular, the 'single site nesters' (Abe, 1987) live 'within their food' and do not forage outside the protected boundaries of their colony in the way that most ants do. As a consequence, contact between neighbouring colonies of termites during the foraging process is extremely limited (Table 6.1). In addition, many bee and wasp species nest in cavities. The availability of such nest sites is limited and the consequent reuse of the same nest sites carries with it the risk of exposure to parasite propagules that may have been left from the previous colony (Ratnieks and Nowakoski, 1989; Roubik, 1989; Greene, 1991; Reeve, 1991; Hansell, 1996). The rates at which colonies are challenged by novel infections may therefore differ considerably across groups of social insects and, once more, the ants and termites seem to be better off than the bees and wasps.

This conclusion is equivalent to one that was reached in a general model showing that clustering of individuals increased within-cluster transmission of diseases, but could also disproportionately decrease between-cluster transmission, making clustering a potentially effective strategy to minimize the overall risk of infection (Watve and Jog, 1997; Wilson *et al.*, 2003). Following Anderson and May (1979, 1981), we would thus expect that the density of susceptible hosts of ants and termites will often be (and has often been over evolutionary time) too low for pathogens to maintain themselves, whereas this constraint would have been much less for pathogens of bees and wasps. In other words, the frequency of extant parasites per average species of ant or termite should be significantly lower than the disease load per average species of bee or wasp. In fact, for ants and termites with large colonies (and thus few colonies per square kilometre of habitat) it seems hard to imagine how selection could maintain any virulence in specialized diseases that would depend on direct or indirect contact with non-nestmates across territory borders. The product βN_T would be low and would still be required to exceed α + b + γ, which would only be possible with very low values of the disease-induced mortality rate α.

A final corollary would be that the few more virulent specialized diseases of ants and termites that do exist would be expected to be epidemic rather than endemic, because the threshold density condition of susceptible hosts is only occasionally met in some host patches. This implies that such diseases will appear in a metapopulation pattern of relatively transient outbreaks (Grenfell and Harwood, 1996; Schmid-Hempel, 1998). Routine searches for infections are unlikely to register such epidemic diseases, in contrast to endemic diseases. Also sexually transmitted diseases will be selected against in social insect hosts, because they require high rates of promiscuous mating (see Hurst *et al.*, Chapter 8, this volume), which does not occur in any social insect (Baer and Boomsma, 2004; Boomsma *et al.*, 2005).

2.4 Intracolony interactions and virulence

The horizontal infectiveness of a diseased colony will depend on the proportion of workers that disperse propagules either while performing their normal foraging activities in spite of being infected, or by dying in places that allow spores to be transmitted. In addition, there is the possibility that a disease is passed on horizontally from workers to the reproductives that they raise, thus resulting in vertical transmission at the colony level (for a model incorporating both, see Schmid-Hempel, 1998). Roughly speaking, there are the following scenarios of intracolony transmission:

- *No transmission*: a diseased individual is recognized as such and 'treated', isolated, or expelled, so that no further nestmates are infected.
- *Only horizontal transmission*: an infection spreads and slows down colony growth and reproductive effort to a degree proportional to the rate of spread; infected workers transmit the disease to other nestmates but not to the reproductives, or infected reproductives are prevented by the disease from founding colonies.

- *Horizontal and vertical transmission*: a disease spreads through the colony and also affects part or all of the colony's reproductives, resulting in direct transmission to the next generation.

Standard epidemiological theory holds that vertical transmission enhances the possibility of a parasite maintaining itself within a fluctuating host population (Anderson and May, 1979, 1981), dynamics that are likely to be typical for many social insect hosts. This is because vertical transmission helps the parasite to survive periods when the minimum threshold density of susceptible hosts is not available. In fact, there is no such threshold for purely vertically transmitted diseases (see Wilson, Chapter 10, this volume). However, vertical transmission may be impossible to maintain in many social insects, when colonies are founded by a single inseminated female (many ants, bees, wasps) or a single mating pair (termites). The demands of raising the first worker brood are likely to be so great that colony foundation is unlikely to succeed when performance is reduced by even mildly negative effects of a disease. Brown *et al.* (2003) have found that the otherwise mild parasite *Crithidia bombi* is apparently vertically transmitted to bumblebee queens and has a severe negative effect on colony founding. In general, therefore, it would seem that vertical transmission can only be maintained when parasites are rather avirulent or in social systems that have multiple queens per colony.

A further general factor that will tend to select against virulence of social insect diseases is the fact that essentially all colonies go through an ergonomic phase of colony growth (a period in which they exclusively produce sterile workers; Oster and Wilson, 1978) before reproducing. For infections that are expressed during this ergonomic phase, there will thus be evolutionary trade-offs between virulence and other fitness components of a disease. For example, an increase in virulence is selected against when the effectiveness of the transmission vehicles is more than proportionally reduced (for a review, see Ebert and Herre, 1996). Selection against virulence should be particularly strong in the perennial ants and termites, where most transmission events will be within the colony. Their ergonomic phase of somatic colony growth normally takes several years and involves numerous subsequent worker cohorts, so that even minor expressions of virulence are likely to terminate the colony in its normal competition with more healthy neighbouring colonies.

Independent of their virulence, social insect diseases gain their short-term fitness by the extent of within-colony transmission that they achieve. Long-term fitness is likewise dependent on within-colony transmission, as the proportion of infected workers ultimately determines the probability that workers or reproductives will carry the parasite to another colony (horizontal transmission) or directly into the next generation (vertical transmission). It is here that social interactions, and in particular liquid food exchange (trophallaxis) between nestmates, are likely to play a key role (Schmid-Hempel, 1998; Naug and Camazine, 2002). Trophallaxis may considerably increase the intracolony transmission rates, thus lowering the threshold density of susceptible hosts needed for the disease to be maintained in the population. Most non-fungal microparasites are transmitted orally and may thus spread through a colony rapidly, because both their arrival in the colony and their subsequent spread via trophallaxis remain unnoticed. High degrees of trophallaxis normally characterize advanced social insects with large colonies. In bees, trophallaxis seems mainly restricted to taxa with progressive provisioning. It is most advanced in honeybees (Wilson, 1971; Seeley, 1985), but also occurs in the allodapine bees (Melna and Schwarz, 1994). Trophallaxis has been shown to occur in stingless bees, but is infrequent and mostly linked to the mass provisioning of brood cells with stored pollen and honey (Sommeijer and De Bruijn, 1994; Hart and Ratnieks, 2002a), in contrast to the honeybee, where it is an almost continuous process. The more primitive bumblebees rarely, if ever, engage in direct food exchange (Michener, 1974) and there is only a single record in carpenter bees (Velthuis and Gerling, 1983). This implies that advanced, long-lived societies could probably only evolve after considerable selection pressure to effectively counter the disease-related negative side-effects of liquid food exchange, i.e. after evolving a series of first-line defences that prevented infections from gaining even a foothold in the colony. Interestingly, the lower termite taxa universally have both oral and anal trophallaxis, partly in connection with the transmission of mutualistic gut flagellates, whereas the higher termites, which have different mutualistic symbionts, no longer have anal trophallaxis (Schmid-Hempel, 1998). It is also interesting to note in this regard, that whereas orally transmitted microsporidian protozoa are the principal parasites of *Solenopsis* fire ants (Jouvenaz, 1983, 1986), which engage in extensive trophallaxis, the non-trophallactic leaf-cutting ant *Acromyrmex octospinosus*, which belongs to the same subfamily (Myrmicinae) but which only rarely engages in trophallaxis, appears to completely lack such parasites (Van Borm *et al.*, 2002).

2.5 Differences in typical disease pressure across the four groups of social insects

In conclusion, we should expect the following pattern in the disease spectra of the four groups of social insects.

● Orally transmitted diseases should be relatively rare, avirulent or transiently epidemic in ants and termites, and more common, up to moderately virulent and often endemic in bees and wasps.
● The soil nesting habit of ants and termites should make them particularly exposed to fungi, nematodes and helminths.
● The occurrence of macroparasites, and especially parasitoids, should be greatest in the wasps, less in the bees, and should be particularly rare in the termites with their cryptic lifestyle.
● Vertical transmission of all but the most avirulent parasites should be absent in ants and termites and rare in bees and wasps, although exceptions may be found in species that always have multiple queen colonies.

Deviations from these overall trends across the four groups of social insects can be expected because wasps and ants have less hygienic food, and thus potentially more parasites than bees and termites, and because perennial societies may have more specialist macroparasites than annual societies. Within each group, comparable species with trophallaxis should have more *per os* transmitted parasites than species without trophallaxis.

3. Defences Against Disease

3.1 The major mechanisms of defence against parasites

A flow diagram of all relevant aspects of the infection process of social insects is shown in Fig. 6.1. The previous section dealt with exposure, infection and transmission, i.e. with the ecological and evolutionary dynamics of disease reproduction, while largely assuming that the individual hosts and their colonies are mostly passive vehicles of disease transmission. This section focuses on the various individual and collective defences that social insects possess and on differences in sophistication and effectiveness of these defences across the major groups of social insects. The part of Fig. 6.1 that addresses defences is marked by the grey frame in the centre of the figure. Essentially there are defences at two different levels: the individual level and the collective level. Each of these in turn also has two components: avoidance (by recognition or expulsion) and the minimization of damage. Successful avoidance at the individual level implies that an exposed individual avoids infection, whereas avoidance at the colony level implies that an individual infection will not spread within the colony. The latter distinction is probably most crucial, because it will determine whether a colony is resistant to a disease, in the sense of not suffering any negative effect on its fitness (i.e. loss of workers, perhaps with the exception of the occasional infected workers that are not admitted back into the colony), or whether a colony can at best be tolerant of a disease (in other words 'accepting' the loss of some significant part of its workforce or brood).

3.2 Individual defences

Individual recognition and avoidance of infection has been documented for pathogenic fungi in ants and termites (Kermarrec *et al.*, 1986; Oi and Pereira, 1993; Rosengaus *et al.*, 1998, 1999a; Jaccoud *et al.*, 1999). Workers are able to recognize spores and are thus often able to prevent them from sticking to their cuticle. As already mentioned, the equivalent recognition of infection sources of *per os* transmitted diseases seems more problematic but has been demonstrated (Drum and Rothenbuhler, 1985). Also in avoidance possibilities, there seems to be a difference between the social insects that forage on the wing and those that forage on foot. Flying workers usually ingest at least part of their food on the spot, bees by drinking nectar (although they collect pollen in external pollen baskets) and wasps by masticating prey or carrion (although they also carry complete prey to the nest). This implies that workers guarding the nest entrance will frequently encounter situations in which it is difficult to

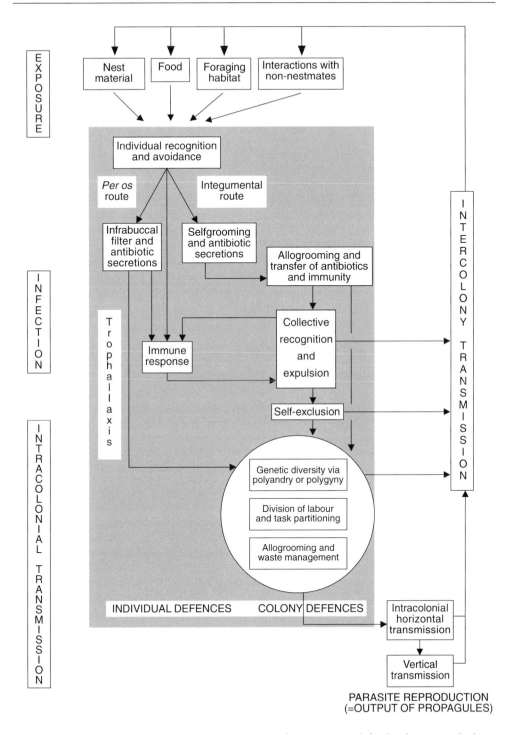

Fig. 6.1. Flow diagram summarizing the dynamics of exposure to, infection by, transmission of, and defence against diseases of individuals and colonies of social insects. See text for details.

detect orally transmitted infections in workers that return to the nest. Recognition of sources of infection thus relies completely on the skills of individual foragers to detect disease propagules before ingesting them. Ants, on the other hand, seem less vulnerable to *per os* transmitted diseases, as they normally carry their prey items wholesale to the nest and only ingest aphid honeydew outside the nest. If recognition at exposure fails, *per os* infections can still be prevented by filtering devices. In ants, this function is carried out by the infrabuccal pocket, which is located at the entrance to the pharynx (Hölldobler and Wilson, 1990). All ingested food is passed through this pocket, with particulate matter being compacted into a pellet for later regurgitation (away from the colony) and thus prevented from entering the digestive system. For example, leaf-cutting ant minor workers can filter out particles at least as small as 10 μm, and probably much smaller (Quinlan and Cherrett, 1978), whereas in fire ants the infrabuccal pocket catches particles as small as 0.88 μm (Glancey *et al.*, 1981). In bees, the proventriculus serves as the filtering mechanism (Seeley, 1985), but appears to be less effective than the mechanism used by ants. Filtering devices seem to be unknown in wasps and termites (for further details, see Schmid-Hempel, 1998).

Closed nests with one or relatively few nest entrances (relative to the size of the worker population) give better possibilities for guard workers to inspect incoming foragers for infections while they are in any case checking their identity as nestmates. In species with closed nests, kin-recognition (e.g. Breed and Bennett, 1987) and disease recognition may thus go hand in hand. On the other hand, open nests have large exposed surfaces that can be accessed from all sides, so that protection from infected kin is necessarily less. This would most strongly affect *per os* infections, as integumental (cuticular) infections with fungal spores will probably be removed by self-grooming behaviour, another individual defence mechanism coming in at this stage and one at which the ants and termites appear particularly adept (Kermarrec *et al.*, 1986; Oi and Pereira, 1993; Rosengaus *et al.*, 1998; Jaccoud *et al.*, 1999; Hughes *et al.*, 2002).

In addition to self-grooming and filtering practices, both *per os* and cuticular infections can be prevented by gland secretions with antibiotic and/or antifungal properties and possibly also by antibiotic cuticular exudates (Boucias and Pendland, 1998). The best-known of these antibiotic defences in social insects are the metapleural glands of ants, which serve as a broad-spectrum defence against unwanted microorganisms (Beattie *et al.*, 1985, 1986; Bot *et al.*, 2002; Poulsen *et al.*, 2002a). These glands are an ancient and unique synapomorphy for the ants and have been retained in almost all genera (Hölldobler and Wilson, 1990) in spite of being metabolically costly (Poulsen *et al.*, 2002a). It thus seems reasonable to infer that disease defence based on the metapleural glands has been of crucial importance for the early evolution and subsequent radiation of the exclusively eusocial ants (Hölldobler and Wilson, 1990). The glands have been secondarily lost on at least two occasions, once in the Oecophyllinae (*Oecophylla*) and once or more in the Camponotini (*Camponotus*, *Dendromyrmex* and *Polyrhachis*) (Johnson *et al.*, 2003b). It is interesting to note that the absence of the glands in these genera appears to be significantly associated with arboreality and the development of nest-weaving, both characters that may reduce exposure to disease (Hölldobler and Engel-Siegel, 1984; Johnson *et al.*, 2003b). Antiseptic glandular compounds, body exudates and faeces have been shown to occur in termites (Rosengaus *et al.*, 1998, 2000), but a convergent equivalent of the metapleural glands is lacking. The salivary glands of wasps and bees have antiseptic properties, which are primarily used to protect the stored honey and nest material (Cane *et al.*, 1983; Gambino, 1993; Schmid-Hempel, 1998). However, these defences seem comparable to defences that ants possess in addition to metapleural gland secretions, for example the ability of *Acromyrmex* leaf-cutting ants to inhibit microbial growth in their infrabuccal pocket by labial gland compounds (Febvay *et al.*, 1984). Overall, it seems therefore that ants have the most sophisticated antibiotic defences, whereas similar defences in the bees and wasps are far more restricted. The termites may well be intermediate.

The final individual defence is the insect immune system (see Rolff and Siva-Jothy, 2003, for a recent review). Studies have clarified functional details of insect immune systems compared with vertebrate ones (reviewed in Schmid-Hempel, 2003; Schmid-Hempel and Ebert, 2003) and experimental work on bumblebees has shown that immune responses are costly and conditionally expressed (König and Schmid-Hempel, 1995; Moret and Schmid-Hempel, 2000, 2001; Lord *et al.*, 2001; Doums *et al.*, 2002). Recent work on termites has revealed individual humoral immune responses to pathogenic fungi (Rosengaus *et al.*, 1999b). Unfortunately, it is at present impossible to evaluate the relative efficiency of individual immune systems across the four groups of social insects, because the studies on bees and termites used different pathogens and comparable data on wasps and ants are lacking. In the absence of such data, one could assume that individual immune defence capacities across the four groups of social insects are similar, and that the major differences are likely to result from variation in collective organization (see below). On the other hand, it cannot be excluded that there may be trade-offs between individual defence and collective defence, and that ants may have reduced humoral defences because they possess metapleural glands. Furthermore, workers in large-sized and thus typically more advanced insect societies tend to be shorter-lived than workers in primitive societies (Schmid-Hempel, 1998), which may imply that investment in individual immune systems has been under selection to be diminished once collective defences improved.

3.3 Colony defences

The various elements of collective (colony-level) defence are summarized at the right-hand side of the defences frame in Fig. 6.1. These second-line defences (after the first-line individual defences) consist of mechanisms such as collective recognition and expulsion of infected individuals, curing of infected individuals by allogrooming or allotransfer of antibiotics before allowing them into the colony or brood chambers, and preventive measures such as the transfer of antibiotic compounds and the induction of faster immune responses in nestmates when re-exposed to the same infection (Rosengaus *et al.*, 1998; Traniello *et al.*, 2002). When successful, alone or in combination, these defence mechanisms effectively prevent a single infected individual from spreading its infection among nestmates and would thus make the colony as a whole resistant to the disease. However, this collective resistance has a price in terms of joint investments in vigilance and allogrooming and in the loss of expelled individuals, expenses that should be added to the individual costs of self-grooming, antibiotics production and the maintenance of immune responses. Finally, infected and expelled individuals may become a source of intercolony transmission if they drift to neighbouring colonies (Fig. 6.1). Defences by collective recognition of sources of infection would probably be most effective if aimed at *per os* infected foragers that return to the colony. However, similar collective screening procedures may also reduce the infection risk from prey items. For example, the wholesale prey items that ants transport to their nest may be processed and screened by a number of other ants before being fed to the larvae, which implies that any other worker would potentially be able to recognize and remove an infected item that the forager and other nestmates might have missed. Social wasps often do not have this option as foragers tend to feed larvae directly (Wilson, 1971).

If individual avoidance, individual defence and collective recognition all fail, a disease brought in by one or several workers will have the possibility of spreading through the colony. When this happens, colonies will suffer more severely and may fail to survive or reproduce because of the infection, unless measures are in place to significantly reduce the impact of the disease. These measures roughly fall into three categories (Fig. 6.1), reducing the rate of spread by:

- Division of labour and/or task partitioning;
- Increasing the genetic diversity of nestmates through polyandry and/or polygyny;
- Hygienic behaviour and waste management.

All three topics have been prominent in disease-related research on social insects during the last decade, and the effects of some of them have recently been modelled (Naug and Camazine, 2002). These studies have significantly improved our understanding of disease tolerance and are briefly summarized below.

Division of labour and task partitioning compartmentalize insect societies in two fundamentally different ways. Division of labour implies that different worker castes specialize on different tasks (Wilson, 1971), whereas task partitioning implies that workers of the same caste split up a complex task into subtasks on which they specialize (Jeanne, 1986; Anderson and Ratnieks, 1999; Ratnieks and Anderson, 1999). Castes can either be permanent, differing in body size (size polymorphism) or genetic inclination to express specific behaviours (genetic polyethism) or temporary, changing with age through the lifetime of a worker (age polyethism). In general, divisions of this kind increase in frequency in the more advanced forms of social organization, i.e. in species with large and long-lived colonies. Although they evolved for reasons of ergonomic efficiency (Oster and Wilson, 1978; Jeanne, 1986), their joint additional effect is a general reduction in interactions between individuals in a society and thus a reduction in intracolony transmission of diseases (Schmid-Hempel and Schmid-Hempel, 1993; Anderson and Ratnieks, 2000; Hart and Ratnieks, 2001; Naug and Camazine, 2002).

The immune resistance of even a fraction of a colony's members may prevent the spread of a disease through the colony, because it makes the local number of susceptible hosts drop below the critical threshold needed for a disease to spread (herd immunity; Anderson and May, 1985). Genetic variation for resistance is well known in many animals and has been demonstrated in several social insect species (Baer and Schmid-Hempel, 2003; Palmer and Oldroyd, 2003; Hughes and Boomsma, 2004). Correlative evidence indicates that genetically more diverse ant colonies have fewer diseases (Schmid-Hempel and Crozier, 1999) and experimental work has shown that genetically diverse colonies of bumblebees (Baer and Schmid-Hempel, 1999, 2001), honeybees (Tarpy, 2002) and leaf-cutting ants (Hughes and Boomsma, 2004) are better able to cope with infections than genetically homogeneous colonies. Recent experimental studies on worker caste allocation and disease resistance in *Acromyrmex* leaf-cutting ants have shown that advantages of genetic diversity may derive from a more flexible allocation to different worker castes (Hughes *et al.*, 2003) or heightened resistance to disease (Hughes and Boomsma, 2004). As the worker castes differ in their exposure to diseases and their effectiveness in defending against them (Hughes *et al.*, 2002; Poulsen *et al.*, 2002b), the two benefits of polyandry will be intertwined. Although patriline-level variation in the chemical mixture of the metapleural gland secretion of *Acromyrmex* workers has not been found (Ortius-Lechner *et al.*, 2003), a better representation of the full spectrum of individual variation in metapleural gland sizes does occur in more genetically diverse colonies (J.J. Boomsma and A.N.M. Bot, unpublished; see also Bot and Boomsma, 1996), suggesting a possible mechanism by which an improvement in disease resistance may occur. The cumulative evidence for social bees and ants (there are no data for social wasps) indicates that genetic diversity for herd immunity is important for disease dynamics, although polyandry may not necessarily have evolved to its present frequencies because of this, with there being a number of other possible benefits that may also apply (Boomsma and Ratnieks, 1996; Crozier and Fjerdingstad, 2001). A direct or indirect causal link between parasite load and relatedness is unlikely if not impossible in termites, because of the almost invariably monogamous colony structure of these social insects (Thorne, 1985).

The most active and flexible forms of behavioural defences can be captured under the category of allogrooming and waste management (Fig. 6.1). Not only do social insects allogroom incoming foragers as discussed above, this behaviour is often routine throughout the colony. However, there are differences in how elaborate and effective this behaviour is across the four groups of social insects. Allogrooming is known to be highly efficient in ants and termites (Kermarrec et al., 1986; Oi and Pereira, 1993; Rosengaus et al., 1998; Hughes et al., 2002), whereas this type of mutual cleaning by nestmates seems to be unknown in the social wasps and bees, except for occasional observations in the honeybee (Naug and Camazine, 2002). A factor connected to this difference may be the extent of hairiness of workers and queens. Bees, in particular, are very hairy so that pollen can be easily collected and trans-ported back to the nest, but this makes grooming for spores of pathogenic fungi difficult. The same applies (but to a lesser extent) to wasps, but the body surfaces of ants and termites are normally smooth enough to make allo-grooming effective. A further factor that makes colony-level behavioural defences more effective in ants and termites is that they have the possibility to abandon sections of the nest (without having to abandon the entire nest) that have an infection that cannot be controlled. This can be done because nests normally consist of a complex network of galleries and brood chambers (Table 6.1). Bees and wasps, however, have single nest units organized as an arrangements of cells in comb-like structures, which makes it much harder if not impossible to abandon sections. Abandoning nest fragments is likely to be costly, but will be far less so than being forced to abandon the entire nest as is otherwise often necessary (Roubik, 1989; Knutson and Murphy, 1990; Williams, 1990; Gadagkar, 1991).

Hygienic behaviours are individually based in social bees and hardly coordinated at the colony level (Trump et al., 1967). The partly genetic determination of this behaviour implies that honeybee colonies may differ considerably in the expression of hygienic behaviour, resulting in different tolerances to disease across colonies (Rothenbuhler, 1964a,b). *Vespula* wasps are surprisingly poor in hygienic behaviour (Greene, 1991; Glare et al., 1996; Harris et al., 2000). In leaf-cutting ants, however, hygienic behaviour and waste management have become a highly integrated colony activity, employing a significant proportion of the worker force (Bot et al., 2001b; Hart and Ratnieks, 2001, 2002b). It has probably been the particular challenges from large amounts of waste due to the fungus agriculture of these ants that have selected for this advanced waste management behaviour that seems to be unmatched in other ants. However, it is also clear that the social characteristics of ants in general (Table 6.1) have predisposed them to evolving elaborate waste management behaviour. They possibly share this predisposition with termites, but not with the social bees and wasps. Again, the honeybee may be an exception, as extreme task specialization of cleaning workers has been observed (Arathi et al., 2000).

3.4 The differences in defence against parasites across the four groups of social insects

Overviewing the above considerations on defences, we conclude that many of them fit and reinforce the conclusions drawn at the end of Section 2. The fundamental differences in our expectations of disease pressure across the four groups of social insects that appeared from analysing differences in exposure and transmission (Section 2) are normally not compensated by opposite differences in individual or collective defence (this section). The annual social bees and wasps face higher risks of introducing infections in their colonies when returning from foraging trips because they are more likely to ingest contaminated food away from the colony and have less effective filtering devices to prevent *per os* infections. Their individual antibiotic defences seem less general and elaborate, and their allogrooming, hygienic behaviour and waste management practices are generally less well developed or less frequent. Although task partitioning probably occurs in all major groups of social insects, physical worker castes that would help in defence against disease are restricted to the perennial ants and termites. In fact, the only factor that is not unambiguously pointing towards a significant advantage in disease defence for the long-lived, perennial societies of ants and termites is intracolonial genetic diversity. Ants are clearly champions in the number of independent transitions to genetically diverse colonies, either because of polygyny or polyandry (Keller and Reeve, 1994; Boomsma and Ratnieks, 1996). They are followed by the social wasps, where polygyny is frequent in the Polistinae (Reeve, 1991) and multiple queen mating occurs in the Vespinae (Foster and Ratnieks, 2001). Next in line are the social bees, where polyandry is almost completely restricted to the honeybees (Palmer and Oldroyd, 2001; Tarpy, 2002) and where polygyny seems less frequent than in the social wasps (Michener, 1974). The termites clearly have the least genetically diverse colonies. The entire group is essentially monogamous, as most documented cases of multiple breeders have been shown to concern offspring reproductives that are on their way to replace parent breeders or to head bud-nests (Thorne, 1985). The only parameter that seems to be correlated with this sequence is the diet: carnivory/omnivory in ants and wasps, followed by a pollen and nectar diet in bees and a decomposer diet in termites (Table 6.1).

We thus expect a number of trends to be apparent in the comparative data. As hosts, we expect ants and termites to be more similar to each other than to the bees and wasps, which should be mutually similar as well. We expect these respective groups of social insect hosts to suffer predominantly from types of parasites and diseases that match their typical nesting and foraging habitats, and we expect these differences to be expressed particularly when grouping parasites and diseases in categories such as the ones proposed in section 2.5.

The next section analyses available comparative data to investigate the extent to which these expectations are supported.

4. An Update and Reappraisal of the Comparative Data

The database of Schmid-Hempel (1998), which was closed in 1996, has been updated with any new host–parasite interactions involving social insect hosts that were not already included in the Schmid-Hempel (1998) database. For the most part, updating was limited to literature that has appeared post-1996. Sampling effort varies dramatically between the social insect groups, with termites and wasps having received considerably less attention than have the ants and bees.

More importantly, the distribution of the effort varies, most notably with a single bee species (*Apis mellifera*) having been studied with very great intensity whereas, for example, a very large number of ant species have been studied, but each relatively little. We used two methods to eliminate the confounding effects of sampling effort. The first method is the same as that applied by Schmid-Hempel (1998) and used the number of studies published on particular host species as an approximation of study effort. Residuals from a regression on these numbers were then analysed. Our second method counted the number of host–parasite interactions for host groups (ants, bees, termites, wasps) as a whole. This means that a large number of parasite species recorded from a single host species counts the same as an identical number of host species recorded for a single parasite species. The method of interaction numbers also allowed the large number of ambiguous records to be included in an estimated, but realistic, fashion. The distribution of such records is not uniform due to the variation in the difficulty of identifying parasites between the different parasite groups. Inclusion of ambiguous records, even in an imprecise manner, helps to reduce the taxonomic bias in the dataset. Although the numbers used for ambiguous records were estimates, the results were reproduced using both the estimated maximum and the absolute minimum number of interactions. In neither case did the patterns observed change significantly.

For the interaction method, data were listed as a column of parasite species and a column of host species, such that each unique parasite–host interaction was represented as an individual row. Where the number of species involved in an interaction was uncertain, due to ambiguous references (for example '*Formica* spp.', '*Camponotus*', 'other ant genera', 'unidentified strepsipteran'), they were estimated by us. Interactions that most probably involved single species, but where the identity of the species was unknown, were counted as being single-species interactions. Interactions that most probably involved multiple species were counted as representing three species (with a minimum of one and an assumed maximum of five), unless the details of the studies allowed a more precise estimate to be made. In as far as it was possible to be certain, only species that had been demonstrated to be parasitic were included and only interactions that were natural, as opposed to those recorded only from laboratory experiments.

Figures 6.2–6.4 show the make-up of the parasites recorded for each of the four social insect groups. In each case, the figures consist of versions produced using each of the two methods described above. The two methods consistently produced extremely similar results, demonstrating that the broad patterns are robust to methodology. The four social insect groups differed significantly in the degree to which they are afflicted by micro- or macroparasites (Fig. 6.2) (Fig. 6.2A: $G_{Het}=392.5$, d.f. $= 3$, $P < 0.001$; Fig. 6.2B: $F_{3, 514} = 76.9$, $P < 0.001$). Wasps are mostly recorded as suffering from macroparasites, while termites suffer predominantly from microparasites. The ants and bees are intermediate. The average ranking of the relative share of microparasites was: 1, termites; 2, ants; 3, bees; 4, wasps, in concordance with the expectations outlined in Sections 2 and 3. Figure 6.3 shows further details of the relative representation of interactions between hosts and ten different categories of parasites and diseases. It shows that wasps are mostly afflicted by parasitoids (particularly

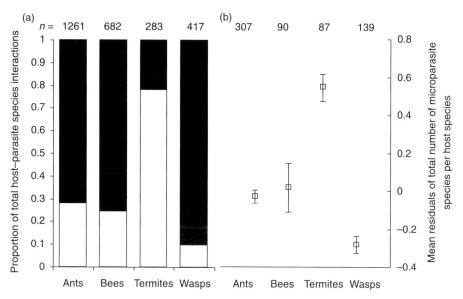

Fig. 6.2. Representation of microparasites (viruses, bacteria, protozoa, fungi) and macroparasites (nematodes, helminths, mites, dipteran and hymenopteran parasitoids and other arthropods) for each social insect group. The graph on the left (a) shows the representation as proportions of the total host–parasite interactions recorded for each social insect group (white: microparasites; black: macroparasites). Interactions were defined as each filled cell in a cross tabulation of parasite species against host species. Sample sizes for each host group are listed above the bars. The graph on the right (b) is based on the method used in Schmid-Hempel (1998) and shows the mean standardized residuals for the arcsine-transformed proportion of parasites (±SE) recorded for individual host species in each social insect group that were microparasites. Sample sizes are given above the means. See text for further details.

Hymenoptera) and other arthropods (which for wasps were mostly Strepsiptera). This is as expected given the highly accessible nests of wasps, which usually hang in free air. Termites are mostly afflicted by fungi and have very few macroparasites. This fits the predictions based on the difficulty for *per os* transmitted microparasites to maintain themselves with termite hosts, the difficulty for mobile macroparasites to find and penetrate their concealed nests, and the hemimetabolous nature of termites, which implies that their immature individuals are not as defenceless as the brood of social Hymenoptera. Bees are afflicted by many mites, while helminths are almost entirely restricted to ants. Finally, Fig. 6.4 tests the expectation that *per os* transmitted diseases should be more common in the bees and wasps, compared with the ants and termites. The results confirm this prediction to a large extent, with the four social insect groups again differing in the make-up of their microparasites (Fig. 6.3A: $G_{\text{Het}}=183.4$, d.f. $= 3$, $P < 0.001$; Fig. 6.3B: $F_{3, 151} = 70.0$, $P < 0.001$). Within the microparasites, bees suffer predominantly from *per os* transmitted parasites, whereas ants and termites suffer mostly from fungi.

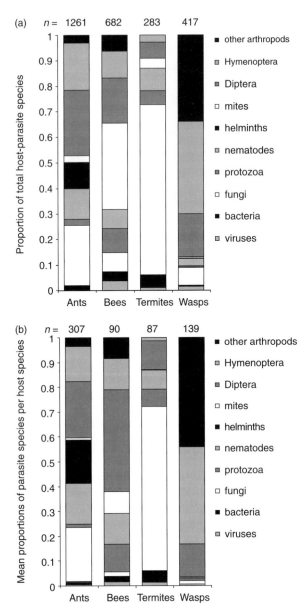

Fig. 6.3. Relative proportions of different parasites among all parasites reported for each social insect group. The top graph (a) shows the proportions of the total host–parasite interactions recorded for each social insect group (parasite groups represented by black, light shading, dark shading and white in a set sequence). Interactions were defined as each filled cell in a cross tabulation of parasite species against host species. Sample sizes for each host group are listed above the bars. The bottom graph (b) is based on the method used in Schmid-Hempel (1998) and shows the mean standardized residuals for the arcsine-transformed proportions of parasites (±SE) recorded for individual host species in each social insect group. Sample sizes are given above the means. See text for further details.

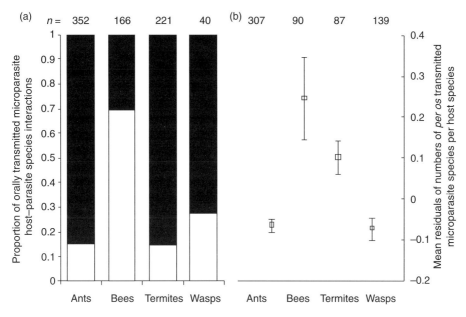

Fig. 6.4. Relative proportions of microparasites that were orally transmitted (viruses, bacteria and protozoa) or fungi. The graph on the left (a) shows the representations of host–microparasite interactions for orally transmitted microparasites (white) and fungal microparasites (black) as proportions of total microparasite interactions recorded for each social insect group. Interactions were defined as each filled cell in a cross tabulation of parasite species against host species. Sample sizes for each host group are listed above the bars. The graph on the right (b) is based on the method used in Schmid-Hempel (1998) and shows the mean standardized residuals for the arcsine-transformed proportion of microparasites (±SE) recorded for individual host species in each social insect group that were orally transmitted. Sample sizes are given above the means. See text for further details.

5. Discussion

5.1 Important issues for future work

In the previous sections we have shown that there are many reasons to expect major differences in disease pressure across the four major groups of social insects. These were expected to be visible as differences in the total number and predominant type of parasites, in the degree of specificity of these parasites, and in their respective virulence. The comparative data largely confirmed the expectations for the overall number and type of parasites, but our large-scale approach does not allow more precise conclusions. Coordinated screening surveys across the major taxa of social insects will be needed to redress the sampling bias of the presently available comparative data and in-depth population-level and experimental studies will have to provide more detailed specific tests. We hope that our analysis will stimulate such work.

Established single-species model systems will remain essential for further experimental studies. However, broader comparative studies will be valuable for uncovering general principles of social evolution and the ways in which parasitism affects social evolution (Schmid-Hempel, 1998; Schmid-Hempel and Crozier, 1999). Further insight into the overall patterns of disease pressure should probably also come from studying paired genera or higher taxonomic units that have much in common but differ in one or two of the key aspects affecting either the infection processes or defences (Fig. 6.1). An obvious case would be to compare disease diversity, prevalence, specificity and virulence in *Apis* honeybees and *Melipona* stingless bees. An advantage of this 'twin' model system would be that virtually all honeybee diseases are known and have been studied (Bailey and Ball, 1991), so that the organization of comparable studies in stingless bees will be straightforward. A difficulty will be to disentangle the effects of genetic diversity (*Apis* always has multiply-mated queens (Palmer and Oldroyd, 2001), whereas *Melipona* queens tend to be singly-mated (Peeters *et al.*, 1999)) and trophallaxis (which occurs at a higher level in *Apis* than in *Melipona*; Sommeijer and De Bruijn, 1994; Hart and Ratnieks, 2002a) on the probability of different diseases to maintain themselves and express virulence. Another promising approach would be to compare the vespid wasp genera *Dolichovespula* and *Vespula* (Foster and Ratnieks, 2001). The former build nests that are freely suspended in the air, whereas the latter use and expand underground cavities (Brian, 1982; Greene, 1991). Multiple queen mating occurs in both genera, obligatory throughout *Vespula* and facultative in *Dolichovespula* (Foster and Ratnieks, 2001), so that this factor can probably be controlled by instrumental insemination (Baer and Schmid-Hempel, 2000). Other good cases would be comparing ants and termites that build nests in trees with sister groups that have remained associated with the soil (see e.g. Johnson *et al.*, 2003b), and comparing ant genera with trophallaxis with sister taxa without.

Studies of virulence and specificity will be particularly rewarding in comparative experimental studies of the type suggested above. Specificity studies are difficult but badly needed, because it is mostly unclear whether what is known as a species of microparasite is indeed a largely panmictic gene pool or in reality a mixture of genetically differentiated lineages (Tibayrenc, 1999). More detailed knowledge about specificity will be crucial for making educated inferences about the extant population size of susceptible hosts and the coevolutionary potential of the interaction between the social host and parasite. The expression of virulence is related to these variables, but will also depend on whether a pathogen is actively or passively dispersed. Selection for or against virulence will generally depend on whether transmission is best served by fit and active foragers who can deposit many propagules at foraging sites, or by dead foragers producing fruiting bodies with passively dispersed (wind, rain, etc.) spores. The microsporidian parasite *Crithidia bombi* infecting bumblebees (Schmid-Hempel, 2001) and the fungus *Metarhizium anisopliae* infecting, for example, leaf-cutting ants (Kermarrec *et al.*, 1986; Jaccoud *et al.*, 1999; Hughes *et al.*, 2002; Poulsen *et al.*, 2002a) illustrate this contrast and underline the fact that any virulence can evolve depending on the ecological conditions that affect transmission and recovery (Anderson and May, 1982;

May and Anderson, 1983; Ebert and Herre, 1996). The trypanosome *Crithidia* is a chronic infection that is iteroparous and relatively mild for both the infected workers (Brown *et al.*, 2000) and established colonies (Shykoff and Schmid-Hempel, 1991), but which severely impedes colony foundation by the queen (Brown *et al.*, 2003). *Metarhizium*, on the other hand, is an 'obligate killer' that reproduces semelparously and has to kill its host in order to do so (Boucias and Pendland, 1998).

A general implication of our findings is that the overall costs of cumulative defences against parasites and diseases are expected to be significantly higher in ants and termites than in bees and wasps. These costs are generally very hard to measure and thus remain largely unknown. However, hints about their magnitude have recently been obtained in *Acromyrmex* leaf-cutting ants. These ants, with their agricultural fungus-rearing societies, have the additional need to control diseases of their fungal crop, so that selection is likely to have promoted consistently high investments in resistance against and tolerance of diseases. The two major components of defence are the metapleural gland secretions, which are generally effective against soil-borne microparasites (Bot *et al.*, 2002; Hughes *et al.*, 2002; Poulsen *et al.*, 2002a), and a cuticular cover of actinomycete bacteria that specifically control a fungal parasite of the mutualistic fungus garden (Currie *et al.*, 1999). Each of these defences has recently been estimated as being equivalent to 10–20% of the basic metabolic rate of *Acromyrmex* workers (Poulsen *et al.*, 2002a, 2003).

5.2 Towards a synthetic life-history theory of disease pressure in social insects

The question as to why ants and termites are obligatory iteroparous, whereas the social bees and social wasps (excepting a few derived lineages that originated and mostly remained in the tropics) have retained the ancestral semelparous life cycle is likely to be directly linked to the type of diseases that affect them and to the prevalence and virulence that these diseases achieve. Iteroparity is generally selected for when adult survival rates are high relative to juvenile survival rates (Stearns, 1977), whereas being parasitized selects for earlier reproduction (Forbes, 1993) and semelparity. Defences favouring colony resistance against, rather than colony tolerance of, parasites may have played a crucial role in this transition that took place early in the social evolution of both ants and termites. In organisms other than social insects, a high life expectancy after the first reproductive effort is generally associated with costly but efficient defences against natural enemies and with elaborate somatic repair mechanisms (Kirkwood, 1981) so that, all else being equal, growth has to be slower. In contrast, life histories characterized by rapid growth and a single early reproduction event are normally characterized by minimal levels of defence and degrees of repair that are only just sufficient to secure survival until the completion of reproduction. Brian (1982) estimated that the typical growth rate of social insect colonies follows the ranking order: wasps > bees > ants > termites, which fits with these expectations.

We therefore submit the hypothesis that, overall, most bee and wasp species have been selected for relatively cheap methods of disease tolerance, whereas the ants and termites have primarily evolved costly mechanisms of colony resistance. The annual bees and wasps may be able to tolerate a fairly high number of diseases of up to moderate virulence, by relying on defences such as individual immune systems, which can be facultatively adjusted according to need, and which are mostly meant to delay the impact of parasites and diseases until the reproductive cycle has been completed. On the other hand, the long-lived fortresses built by ants and termites most probably cannot afford to take such risks once they have grown beyond the colony-founding stage and have thus evolved multiple costly defences to prevent colony infections or to eliminate such infections at an early stage.

The evolution of iterparity in the ants and termites will have been facilitated by the restrictive intercolony transmission dynamics for their parasites (see Section 2). The fact that these taxa made this transition early in their evolution and without reversals in derived clades implies that most of the virulent diseases from which they still suffer are probably general diseases that also use other insect hosts, whereas most of their specialized diseases are likely to be more or less avirulent. Epidemics should thus be rare. Survey data for ants seem to confirm these expectations (Bequaert, 1921; Evans, 1974, 1982, 1989). For example, Evans (1982) argues that diseased army ants and leaf-cutting ants are rarely encountered in the field, suggesting effective defences. The same study reports that the overall number of ants killed by fungal diseases varies little among years, which is compatible with enzootic rather than epizootic prevalences (see also Oi and Pereira, 1993). This implies that the comparative data on a number of diseases may in fact overestimate the disease pressure in ants and termites, as many of the specific pathogens reported may in fact be rather close to being neutral symbionts. Also, vertically transmitted *Wolbachia* symbionts in ants seem to be more neutral in their effects on host reproduction than they often are in non-social insects (Wenseleers *et al.*, 1998; Van Borm *et al.*, 2001, 2003), which might be linked to the long ergonomic phase of colony growth, as argued in Section 3. A recent study documenting that *Wolbachia* infections reduce worker lifespan in ants (Wenseleers *et al.*, 2002) does not necessarily contradict this inference, as the host in question, *Formica truncorum*, starts colonies as a temporary social parasite.

The supposedly moderate effects of diseases in ants are in sharp contrast with the iteroparous honeybees, in which effects of diseases tend to be highly visible, although domestication has undoubtedly aggravated disease problems in this species. The quote at the start of this chapter suggests that Axelrod and Hamilton were aware of this contrast and of a possible explanation along the lines presented here more than 20 years ago. The same quote also suggests that some disease symbionts which were forced into non-virulence by the patterns of clustering and transmission typical for ants or termites, may secondarily have become mutualists.

The arguments and results presented here seem to challenge some established evolutionary concepts about disease pressure as a function of lifespan. To summarize this in a nutshell: Seger and Hamilton (1988) argued that perennials are generally more troubled by parasites than annuals, because they are easier to find (a corollary of the ecological apparency concept; Feeny, 1976) and because they have longer generation times and thus a slower coevolutionary response to parasitic innovations. It is important to realize, however, that this argument implicitly assumes that parasites have unconstrained access to hosts, which is more realistic for plants and other non-social hosts than for insect societies, where social defences are based on active recognition and function as an additional, collective immune system. The coevolutionary part of the argument thus needs a qualifier to be fully transparent. It only says that perennials face more profound challenges because of their longer generation times, but not that they have given up being perennials because of that. The evolution of sophisticated defence systems has, in addition to maintaining sexual reproduction and recombination, allowed perennial organisms to meet these challenges successfully (see Hamilton, 2001, for a review). Social insects, in particular the perennial ones, have also been highly successful in doing exactly that.

Acknowledgements

We thank David Hughes for constructive comments, and the Danish Natural Science Research Council (J.J.B.), Swiss National Science Foundation (P.S.-H.) and Carlsberg Foundation (W.O.H.H.) for providing financial support during the course of this work.

References

Abe, T. (1987) Evolution of life types in termites. In: Kawano, S., Connell, J.J.H. and Hidaka, T. (eds) *Evolution and Coadaptation in Biotic Communities*. University of Tokyo Press, Tokyo, pp. 125–148.

Alexander, R.D. (1974) The evolution of social behavior. *Annual Review of Ecology and Systematics* 5, 325–383.

Anderson, C. and Ratnieks, F.L.W. (1999) Task partitioning in insect societies. I. Effect of colony size on queueing delay and colony ergonomic efficiency. *American Naturalist* 154, 521–535.

Anderson, C. and Ratnieks, F.L.W. (2000) Task partitioning in insect societies: novel situations. *Insectes Sociaux* 47, 198–199.

Anderson, R.M. and May, R.M. (1979) Population biology of infectious diseases. Part I. *Nature* 280, 361–367.

Anderson, R.M. and May, R.M. (1981) The population dynamics of microparasites and their invertebrate hosts. *Philosophical Transactions of the Royal Society* of London, Series B 291, 451–524.

Anderson, R.M. and May, R.M. (1982) Coevolution of hosts and parasites. *Parasitology* 85, 411–426.

Anderson, R.M. and May, R.M. (1985) Vaccination and herd immunity to infectious disease. *Nature* 318, 323–329.

Andreadis, T.G. (1987) Transmission. In: Fuxa, J.R. and Tanada, Y. (eds) *Epizootiology of Insect Diseases*. John Wiley & Sons, New York, pp. 159–176.

Arathi, H.S., Burns, I. and Spivak, M. (2000) Ethology of hygienic behaviour in the honeybee *Apis mellifera* L. (Hymenoptera: Apidae): behavioural repertoire of hygienic bees. *Ethology* 106, 365–379.

Baer, B. and Boomsma, J.J. (2004) Male reproductive investment and queen mating-frequency in fungus-growing ants. *Behavioral Ecology* 15, 426–432.

Baer, B. and Schmid-Hempel, P. (1999) Experimental variation in polyandry affects parasite loads and fitness in a bumblebee. *Nature* 397, 151–154.

Baer, B. and Schmid-Hempel, P. (2000) Applied aspects of the artificial insemination for bumblebees. In: Sommeijer, M.J. and De Ruijter, A. (eds) *Insect Pollination in Greenhouses*. Utrecht University Press, Utrecht, The Netherlands, pp. 31–33.

Baer, B. and Schmid-Hempel, P. (2001) Unexpected consequences of polyandry for parasitism and fitness in the bumblebee, *Bombus terrestris*. *Evolution* 55, 1639–1643.

Baer, B. and Schmid-Hempel, P. (2003) Bumblebee workers from different sire groups vary in susceptibility to parasite infection. *Ecology Letters* 6, 106–110.

Bailey, L. (1963) *Infectious Diseases of the Honey Bee*. Land Books, London.

Bailey, L. and Ball, B.V. (1991) *Honey Bee Pathology*, 2nd edn. Academic Press, London.

Barnes, A.I. and Siva-Jothy, M.T. (2000) Density-dependent prophylaxis in the mealworm beetle *Tenebrio molitor* L. (Coleoptera: Tenebrionidae): cuticular melanisation is an indicator of investment in immunity. *Proceedings of the Royal Society of London, Series B* 267, 177–182.

Baur, M.E., Kaya, H.K. and Strong, D.R. (1998) Foraging ants as scavengers on entomopathogenic nematode-killed insects. *Biological Control* 12, 231–236.

Beattie, A.J., Turnbull, C.L., Hough, T., Jobson, S. and Knox, R.B. (1985) The vulnerability of pollen and fungal spores to ant secretions: evidence and some evolutionary implications. *American Journal of Botany* 72, 606–614.

Beattie, A.J., Turnbull, C.L., Hough, T., Jobson, S. and Knox, R.B. (1986) Antibiotic production: a possible function for the metapleural glands of ants (Hymenoptera: Formicidae). *Annals of the Entomological Society of America* 79, 448–450.

Beekman, M. and Ratnieks, F.L.W. (2000) Long-range foraging by the honey-bee, *Apis mellifera* L. *Functional Ecology* 14, 490–496.

Bequaert, J. (1921) Ants in their diverse relations to the plant world. *Bulletin of the American Museum of Natural History* 45, 333–384.

Boomsma, J.J. and Ratnieks, F.L.W. (1996) Paternity in eusocial Hymenoptera. *Philosophical Transactions of the Royal Society* of *London, Series B* 351, 947–975.

Boomsma, J.J., Baer, B. and Heinze, J. (2005) The evolution of male traits in social insects. *Annual Review of Entomology* 50, 395–420.

Boots, M. and Bowers, R.G. (1999) Three mechanisms of host resistance to microparasites – avoidance, recovery and tolerance – show different evolutionary dynamics. *Journal of Theoretical Biology* 201, 13–23.

Bot, A.N.M. and Boomsma, J.J. (1996) Variable metapleural gland size-allometries in *Acromyrmex* leafcutter ants (Hymenoptera: Formicidae). *Journal of the Kansas Entomological Society* 69 (suppl), 375–383.

Bot, A.N.M., Rehner, S.A. and Boomsma, J.J. (2001a) Partial incompatibility between ants and symbiotic fungi in two sympatric species of *Acromyrmex* leaf-cutting ants. *Evolution* 55, 1980–1991.

Bot, A.N.M., Currie, C.R., Hart, A.G. and Boomsma, J.J. (2001b) Waste management in leaf-cutting ants. *Ethology, Ecology and Evolution* 13, 225–237.

Bot, A.N.M., Ortius-Lechner, D., Finster, K., Maile, R. and Boomsma, J.J. (2002) Variable sensitivity of fungi and bacteria to compounds produced by the metapleural glands of leaf-cutting ants. *Insectes Sociaux* 49, 363–370.

Boucias, D.G. and Pendland, J.C. (1998) *Principles of Insect Pathology.* Kluwer, Norwell, Massachusetts.

Breed, M.D. and Bennett, B. (1987) Kin recognition in highly eusocial insects. In: Fletcher, D.C.J. and Michener, C.D. (eds) *Kin Recognition in Animals.* John Wiley & Sons, Chichester, UK, pp. 243–285.

Brian, M.V. (1982) *Social Insects: Ecology and Behavioural Biology.* Chapman and Hall, London.

Brown, M., Loosli, R. and Schmid-Hempel, P. (2000) Condition-dependent expression of virulence in a trypanosome infecting bumblebees. *Oikos* 91, 421–427.

Brown, M.J.F., Schmid-Hempel, R. and Schmid-Hempel, P. (2003) Strong context-dependent virulence in a host–parasite system: reconciling genetic evidence with theory. *Journal of Animal Ecology* 72, 994–1002.

Buschinger, A. and Kleespies, R.G. (1999) Host range and host specificity of an ant-pathogenic gregarine parasite, *Mattesia geminata* (Neogregariniida: Lipotrophidae). *Entomologia Generalis* 24, 93–104.

Cane, J.H., Gerdin, S. and Wife, G. (1983) Mandibular gland secretions of solitary bees: potential for nest cell disinfection. *Journal of the Kansas Entomological Society* 56, 199–204.

Côté, I.M. and Poulin, R. (1995) Parasitism and group size in social mammals: a meta-analysis. *Behavioral Ecology* 6, 159–165.

Crozier, R.H. and Fjerdingstad, E.J. (2001) Polyandry in social Hymenoptera: disunity in diversity? *Annales Zoologici Fennici* 38, 267–285.

Crozier, R.H. and Page, R.E. (1985) On being the right size: male contributions and multiple mating in social hymenoptera. *Behavioral Ecology and Sociobiology* 18, 105–116.

Currie, C.R., Scott, J.A., Summerbell, R.C. and Malloch, D. (1999) Fungus-growing ants use antibiotic-producing bacteria to control garden parasites. *Nature* 398, 701–704.

Doums, C., Moret, Y., Benelli, E. and Schmid-Hempel, P. (2002) Senescence of immune defense in *Bombus* workers. *Ecological Entomology* 27, 138–144.

Dramsted, W.E., Fry, G.L.A. and Schaffer, M.J. (2003) Bumblebee foraging: is closer really better? *Agriculture, Ecosystems and Environment* 95, 349–357.

Drum, N.H. and Rothenbuhler, W.C. (1985) Differences in non-stinging aggressive responses of worker honeybees to diseased and healthy bees in May and July. *Journal of Apicultural Research* 24, 184–187.

Durrer, S. and Schmid-Hempel, P. (1994) Shared use of flowers leads to horizontal pathogen transmission. *Proceedings of the Royal Society of London, Series B* 258, 299–302.

Ebert, D. and Herre, E.A. (1996) The evolution of parasitic diseases. *Parasitology Today* 12, 96–101.

Eggleton, P. (2001) Termites and trees: a review of recent advance in termite phylogenetics. *Insectes Sociaux* 48, 187–193.

Evans, H.C. (1974) Natural control of arthropods, with special reference to ants (Formicidae), by fungi in the tropical high forest of Ghana. *Journal of Applied Ecology* 11, 37–49.

Evans, H.C. (1982) Entomogenous fungi in tropical forest ecosystems: an appraisal. *Ecological Entomology* 7, 47–60.

Evans, H.C. (1989) Mycopathogens of insects of epigeal and aerial habitats. In: Wilding, N., Collins, N.M., Hammond, P.M. and Webber, J.F. (eds) *Insect–Fungus Interactions: 14th Symposium of the Royal Entomological Society of London in collaboration with the British Mycological Society.* Academic Press, London, pp. 205–238.

Febvay, G., Mallet, F. and Kermarrec, A. (1984) Digestion of chitin by the labial glands of *Acromyrmex octospinosus* (Reich) (Hymenoptera: Formicidae). *Canadian Journal of Zoology* 62, 229–234

Feeny, P. (1976) Plant apparency and chemical defence. *Recent Advances in Phytochemistry* 10, 1–40.

Forbes, M.R.L. (1993) Parasitism and host reproductive effort. *Oikos* 67, 444–450.

Foster, K.R. and Ratnieks, F.L.W. (2001) Paternity, reproduction and conflict in vespine wasps: a model system for testing kin selection predictions. *Behavioral Ecology and Sociobiology* 50, 1–8.

Frank, S.A. (1996a) Host symbiont conflict over the mixing of symbiotic lineages. *Proceedings of the Royal Society of London, Series B* 263, 339–344.

Frank, S.A. (1996b) Models of parasite virulence. *Quarterly Review of Biology* 71, 37–78.

Freeland, W.J. (1976) Pathogens and the evolution of primate sociality. *Biotropica* 8, 12–24.

Freeland, W.J. (1979) Primate social groups as biological islands. *Ecology* 60, 719–728.

Gadagkar, R. (1991) *Belanogaster, Mischocyttarus, Parapolybia*, and independent-founding *Ropalidia*. In: Ross, K.G. and Mathews, R.W. (eds) *The Social Biology of Wasps*. Cornell University Press, Ithaca, New York, pp. 149–187.

Gambino, P. (1993) Antibiotic activity of larval saliva of *Vespula* wasps. *Journal of Invertebrate Pathology* 61, 110.

Gandon, S., Agnew, P. and Michalakis, Y. (2002) Coevolution between parasite virulence and host life-history traits. *American Naturalist* 160, 374–388.

Glancey, B.M., Vander Meer, R.K., Glover, A., Lofgren, C.S. and Vinson, S.B. (1981) Filtration of microparticles from liquids ingested by the red imported fire ant, *Solenopsis invicta* Buren (Hymenoptera: Formicidae*). *Insectes Sociaux* 28, 395–401.

Glare, T.R., Harris, R.J. and Donovan, B.J. (1996) *Aspergillus flavus* as a pathogen of wasps, *Vespula* spp. in New Zealand. *New Zealand Journal of Zoology* 23, 339–344.

Goulson, D. and Stout, J.C. (2001) Homing ability of the bumblebee *Bombus terrestris* (Hymenoptera: Apidae). *Apidologie* 32, 105–111.

Greene, A. (1991) *Dolichovespula* and *Vespula*. In: Ross, K.G. and Mathews, R.W. (eds) *The Social Biology of Wasps*. Cornell University Press, Ithaca, New York, pp. 263–305.

Grenfell, B. and Harwood, J. (1996) (Meta)population dynamics of infectious diseases. *Trends in Ecology and Evolution* 12, 395–399.

Hamilton, W.D. (1982) Pathogens as causes of genetic diversity in their host populations. In: Anderson, R.D. and May, R.M. (eds) *Population Biology of Infectious Diseases*. Springer, Berlin, pp. 269–296.

Hamilton, W.D. (1987) Kinship, recognition, disease and intelligence. In: Ito, Y., Brown, J.L. and Kikkawa, J. (eds) *Animal Societies: Theories and Facts*. Japan Scientific Societies Press, Tokyo, pp. 81–102.

Hamilton, W.D. (2001) *Narrow Roads of Gene Land*. Vol. 2: *Evolution of Sex*. Oxford University Press, Oxford, UK.

Hansell, M.H. (1996) Wasps make nests: nests make conditions. In: Turillazi, S. and West-Eberhard, M.J. (eds) *Natural History and Evolution of Paper-Wasps*. Oxford University Press, Oxford, UK, pp. 272–289.

Harris, R.J., Harcourt, S.J., Glare, T.R., Rose, E.A.F. and Nelson, T.J. (2000) Susceptibility of *Vespula vulgaris* (Hymenoptera: Vespidae) to generalist entomopathogenic fungi and their potential for wasp control. *Journal of Invertebrate Pathology* 75, 251–258.

Hart, A.G. and Ratnieks, F.L.W. (2001) Task partitioning, division of labour and nest compartmentalisation collectively isolate hazardous waste in the leafcutting ant *Atta cephalotes*. *Behavioral Ecology and Sociobiology* 49, 387–392.

Hart, A.G. and Ratnieks, F.L.W. (2002a) Task-partitioned nectar transfer in stingless bees: work organisation in a phylogentic context. *Ecological Entomology* 27, 163–168.

Hart, A.G. and Ratnieks, F.L.W. (2002b) Waste management in the leafcutting ant *Atta colombica*. *Behavioral Ecology* 13, 224–231.

Herre, E.A., Knowlton, N., Mueller, U.G. and Rehner, S.A. (1999) The evolution of mutualisms: exploring the path between conflict and cooperation. *Trends in Ecology and Evolution* 14, 49–53.

Hölldobler, B. and Engel-Siegel, H. (1984) On the metapleural gland of ants. *Psyche* 91, 201–224.

Hölldobler, B. and Wilson, E.O. (1990) *The Ants*. Springer, Berlin.

Hughes, W.O.H. and Boomsma, J.J. (2004) Genetic diversity and disease resistance in leaf-cutting ant societies. *Evolution* 58, 1251–1260.

Hughes, W.O.H., Eilenberg, J. and Boomsma, J.J. (2002) Trade-offs in group living: transmission and disease resistance in leaf-cutting ants. *Proceedings of the Royal Society of London, Series B* 269, 1811–1819.

Hughes, W.O.H., Sumner, S.R., Van Borm, S. and Boomsma, J.J. (2003) Worker caste polymorphism is genetically controlled in a leaf-cutting ant. *Proceedings of the National Academy of Sciences USA* 100, 9394–9397.

Jaccoud, D.B., Hughes, W.O.H. and Jackson, C.W. (1999) The epizootiology of a *Metarhizium* infection in mini-nests of the leaf-cutting ant *Atta sexdens rubropilosa*. *Entomologia Experimentalis et Applicata* 93, 51–61.

Jaenike, J. (1993) Rapid evolution of host specificity in a parasitic nematode. *Evolutionary Ecology* 7, 103–108.

Jeanne, R.L. (1986) The evolution of the organization of work in social insects. *Monitore Zoologico Italiano* (n.s.) 20, 119–133.

Johnson, K.P., Adams, R.J., Page, R.D.M. and Clayton, D.H. (2003a) When do parasites fail to speciate in response to host speciation? *Systematic Biology* 52, 37–47.

Johnson, R.N., Agapow, P.-M. and Crozier, R.H. (2003b) A tree island approach to inferring phylogeny in the ant subfamily Formicinae, with especial references to the evolution of weaving. *Molecular Phylogenetics and Evolution* 29, 317–330.

Jouvenaz, D.P. (1983) Natural enemies of fire ants. *Florida Entomologist* 66, 111–121.

Jouvenaz, D.P. (1986) Diseases of fire ants: problems and opportunities. In: Lofgren, C.S. and Vander Meer, R.K. (eds) *Fire Ants and Leaf-Cutting Ants: Biology and Management*. Westview Press, Boulder, Colorado, pp. 327–338.

Kaya, H.K. (1987) Diseases caused by nematodes. In: Fuxa, J.R. and Tanada, Y. (eds) *Epizootiology of Insect Diseases*. John Wiley & Sons, New York, pp. 453–472.

Keeping, M.G. and Crewe, R.M. (1983) Parasitoids, commensals and colony size in nests of *Belanogaster* (Hymenoptera: Vespidae). *Journal of the Entomological Society of South Africa* 46, 309–323.

Keller, L. and Reeve, H.K. (1994) Genetic variability, queen number, and polyandry in social Hymenoptera. *Evolution* 48, 694–704.

Kermarrec, A., Febvay, G. and Decharme, M. (1986) Protection of leaf-cutting ants from biohazards: is there a future for microbiological control? In: Lofgren, C.S. and Vander Meer, R.K. (eds) *Fire Ants and Leaf-Cutting Ants: Biology and Management*. Westview Press, Boulder, Colorado, pp. 339–356.

Kirkwood, T.B.L. (1981) Repair and its evolution: survival versus reproduction. In: Townsend, A.R.R. and Calow, P. (eds) *Physiological Ecology: An Evolutionary Approach to Resource Use*. Blackwell, Oxford, UK, pp. 165–189.

Knutson, L.V. and Murphy, W.L. (1990) Insects: Diptera (flies). In: Morse, R.A. and Nowogrodzki, R. (eds) *Honey Bee Pests, Predators, and Diseases*. Cornell University Press, Ithaca, New York, pp. 120–134.

König, C. and Schmid-Hempel, P. (1995) Foraging activity and immunocompetence in workers of the bumble bee, *Bombus terrestris* L. *Proceedings of the Royal Society of London, Series B* 260, 225–227.

Lewis, K. (1998) Pathogen resistance as the origin of kin altruism. *Journal of Theoretical Biology* 193, 359–363.

Lord, G.M., Matarese, G., Howard, J.K., Moret, Y. and Schmid-Hempel, P. (2001) The bioenergetics of the immune system. *Science* 292, 855–856.

Macfarlane, R.P., Lipa, J.J. and Liu, H.J. (1995) Bumble bee pathogens and internal enemies. *Bee World* 76, 130–148.

May, R.M. and Anderson, R.M. (1983) Parasite–host coevolution. In: Futuyma, D.J. and Slatkin, M. (eds) *Coevolution*. Sinauer, Sunderland, Massachusetts, pp. 186–206.

Melna, P.A. and Schwarz, M.P. (1994) Behavioral specialization in pre-reproductive colonies of the allodapine bee *Exoneura bicolour* (Hymenoptera, Anthophoridae). *Insectes Sociaux* 41, 1–18.

Michener, C.D. (1974) *The Social Behavior of the Bees*. Harvard University Press, Cambridge, Massachusetts.

Michener, C.D. (1985) From solitary to eusocial: need there be a series of intervening species? In: Hölldobler, B. and Lindauer, M. (eds) *Experimental Behavioural Ecology*. Fisher Verlag, Stuttgart, Germany, pp. 293–305.

Molloy, D.P., Vinikour, W.S. and Anderson, R.V. (1999) New North American records of aquatic insects as paratenic hosts of *Pheromermis* (Nematoda: Mermithidae). *Journal of Invertebrate Pathology* 74, 84–95.

Morel, G. and Fouillaud, M. (1992) Presence of microorganisms and viral inclusion bodies in the nests of the paper wasp *Polistes hebraeus* Fabricius (Hymenoptera, Vespidae). *Journal of Invertebrate Pathology* 60, 210–212.

Moret, Y. and Schmid-Hempel, P. (2000) Survival for immunity: activation of the immune system has a price for bumblebee workers. *Science* 190, 1166–1168.

Moret, Y. and Schmid-Hempel, P. (2001) Immune defence in offspring. *Nature* 414, 506.

Naug, D. and Camazine, S. (2002) The role of colony organization on pathogen transmission in social insects. *Journal of Theoretical Biology* 215, 427–439.

Oi, D.H. and Pereira, R.M. (1993) Ant behavior and microbial pathogens (Hymenoptera: Formicidae). *Florida Entomologist* 76, 63–74.

Ortius-Lechner, D., Maile, R., Morgan, E.D., Petersen, H.C. and Boomsma, J.J. (2003) Lack of patriline-specific differences in chemical composition of the metapleural gland secretion in *Acromyrmex octospinosus*. *Insectes Sociaux* 50, 113–119.

Oster, G.F. and Wilson, E.O. (1978) *Caste and Ecology in the Social Insects*. Princeton University Press, Princeton, New Jersey.

Palmer, K.A. and Oldroyd, B.P. (2001) Very high paternity frequency in *Apis nigrocincta*. *Insectes Sociaux* 48, 327–332.

Palmer, K.A. and Oldroyd, B.P. (2003) Evidence for intracolonial genetic variation in resistance to American foulbrood of honey bees (*Apis mellifera*): further support for the parasite/pathogen hypothesis for the evolution of polyandry. *Naturwissenschaften* 90, 265–268.

Peeters, J.M., Queller, D.C., Imperatriz-Fonseca, V.L., Roubik, D.W. and Strassman, J.E. (1999) Mate number, kin selection and social conflicts in stingless bees and honeybees. *Proceedings of the Royal Society of London, Series B* 266, 379–384.

Poulsen, M., Bot, A.N.M., Nielsen, M.G. and Boomsma, J.J. (2002a) Experimental evidence for the cost and hygienic significance of the antibiotic metapleural gland secretion in leaf-cutting ants. *Behavioral Ecology and Sociobiology* 52, 151–157.

Poulsen, M., Bot, A.N.M., Currie, C.R. and Boomsma, J.J. (2002b) Mutualistic bacteria and a possible trade-off between alternative defence mechanisms in *Acromyrmex* leaf-cutting ants. *Insectes Sociaux* 49, 15–19.

Poulsen, M., Bot, A.N.M., Currie, C.R., Nielsen, M.G. and Boomsma, J.J. (2003) Within colony transmission and the cost of a mutualistic bacterium in the leaf-cutting ants *Acromyrmex octospinosus. Functional Ecology* 17, 260–269.

Quinlan, R.J. and Cherrett, J.M. (1978) Studies on the role of the infrabuccal pocket of the leaf-cutting ant *Acromyrmex octospinosus* (Reich) (Hymenoptera: Formicidae). *Insectes Sociaux* 25, 237–245.

Ratnieks, F.L.W. and Anderson, C. (1999) Task partitioning in insect societies. II. Use of queueing delay information in recruitment. *American Naturalist* 154, 536–548.

Ratnieks, F.L.W. and Nowakoski, J. (1989) Honeybee swarms accept bait hives contaminated with American foulbrood. *Ecological Entomology* 14, 475–478.

Reeson, A.F., Wilson, K., Gunn, A., Hails, R.S. and Goulson, D. (1998) Baculovirus resistance in the noctuid *Spodoptera exempta* is phenotypically plastic and responds to population density. *Proceedings of the Royal Society of London, Series B* 265, 1787–1791.

Reeve, H.K. (1991) *Polistes.* In: Ross, K.G. and Mathews, R.W. (eds) *The Social Biology of Wasps.* Cornell University Press, Ithaca, New York, pp. 99–148.

Rolff, J. and Siva-Jothy, M.T. (2003) Invertebrate ecological immunity. *Science* 301, 472–475.

Rose, E.A.F., Harris, R.J. and Glare, T.R. (1999) Review of pathogens identified from social wasps (Hymenoptera: Vespidae) and their potential as biological control agents. *New Zealand Journal of Zoology* 26, 179–190.

Rosengaus, R.B., Maxmen, A.B., Coates, L.E. and Traniello, J.F.A. (1998) Disease resistance: a benefit of sociality in the dampwood termite *Zootermopsis angusticollis* (Isoptera: Termopsidae). *Behavioral Ecology and Sociobiology* 44, 125–134.

Rosengaus, R.B., Jordan, C., Lefebvre, M.L. and Traniello, J.F.A. (1999a) Pathogen alarm behavior in a termite: a new form of communication in social insects. *Naturwissenschaften* 86, 544–548.

Rosengaus, R.B., Traniello, J.F.A., Chen, T., Brown, J.J. and Karp, R.D. (1999b) Immunity in a social insect. *Naturwissenschaften* 86, 588–591.

Rosengaus, R.B., Lefebvre, M.L. and Traniello, J.F.A (2000) Inhibition of fungal spore germination by *Nasutitermes*: evidence for a possible antiseptic role of soldier defensive secretions. *Journal of Chemical Ecology* 26, 21–39.

Rosengaus, R.B., Moustakas, J.E., Calleri, D.V. and Traniello, J.F.A. (2003) Nesting ecology and cuticular microbial loads in dampwood (*Zootermopsis angusticollis*) and drywood termites (*Incisitermes minor, I. schwarzi, Cryptotermes cavifrons*). *Journal of Insect Science* 3, 31.

Rothenbuhler, W.C. (1964a) Behavior genetics of nest cleaning honeybees. I. Response of four inbred lines to disease-killed brood. *Animal Behaviour* 112, 578–583.

Rothenbuhler, W.C. (1964b) Behavior genetics of nest cleaning honeybees. IV. Responses of F_1 and backcross generations to disease killed brood. *American Zoologist* 4, 111–123.

Roubik, D.W. (1989) *Ecology and Natural History of Tropical Bees.* Cambridge University Press, Cambridge, UK.

Schmid-Hempel, P. (1994) Infection and colony variability in social insects. *Philosophical Transactions of the Royal Society of London, Series B* 346, 313–321.

Schmid-Hempel, P. (1995) Parasites and social insects. *Apidologie* 26, 255–271.

Schmid-Hempel, P. (1998) *Parasites in Social Insects.* Princeton University Press, Princeton, New Jersey.

Schmid-Hempel, P. (2001) On the evolutionary ecology of host–parasite interactions: addressing the questions with bumblebees and their parasites. *Naturwissenschaften* 88, 147–158.

Schmid-Hempel, P. (2003) Variation in immune defence as a question of evolutionary ecology. *Proceedings of the Royal Society of London, Series B* 270, 357–366.

Schmid-Hempel, P. and Crozier, R.H. (1999) Polygyny vs. polyandry vs. parasites. *Philosophical Transactions of the Royal Society of London, Series B* 353, 507–519.

Schmid-Hempel, P. and Ebert, D. (2003) On the evolutionary ecology of specific immune defence. *Trends in Ecology and Evolution* 18, 27–32.

Schmid-Hempel, P. and Schmid-Hempel, R. (1993) Transmission of a pathogen in *Bombus terrestris*, with a note on division of labour in social insects. *Behavioral Ecology and Sociobiology* 33, 319–327.

Seeley, T.D. (1985) *Honeybee Ecology*. Princeton University Press, Princeton, New Jersey.

Seger, J. and Hamilton, W.D. (1988) Parasites and sex. In: Michod, R.E. and Levin, B.R. (eds) *The Evolution of Sex: An Examination of Current Ideas*. Sinauer, Sunderland, Massachusetts, pp. 176–193.

Sherman, P.W., Seeley, T.D. and Reeve, H.K. (1988) Parasites, pathogens, and polyandry in social Hymenoptera. *American Naturalist* 131, 602–610.

Shykoff, J.A. and Schmid-Hempel, P. (1991) Genetic relatedness and eusociality: parasite-mediated selection on the genetic composition of groups. *Behavioral Ecology and Sociobiology* 28, 371–376.

Smirnoff, W.A. (1959) Predators of *Neodiprion swaineri* Midd. (Hymenoptera: Tenthredinidae) larval vectors of virus diseases. *Canadian Entomologist* 91, 246.

Sommeijer, M.J. and De Bruijn, L.L.M. (1994) Intranidal feeding, trophallaxis and sociality in stingless bees. In: Hunt, J. and Nalepa, C. (eds) *Nourishment and Evolution in Insect Societies*. Westview Press, Boulder, Colorado, pp. 391–418.

Spradbery, J.P. (1973) *Wasps. An Account of the Biology and Natural History of Solitary and Social Wasps*. Sidgwick and Jackson, London.

Stearns, S.C. (1977) The evolution of life history traits. *Annual Review of Ecology and Systematics* 8, 145–171.

Tanada, Y. and Fuxa, J.R. (1987) The pathogen population. In: Fuxa, J.R. and Tanada, Y. (eds) *Epizootiology of Insect Diseases*. John Wiley & Sons, New York, pp. 113–158.

Tarpy, D.R. (2002) Genetic diversity within honeybee colonies prevents severe infections and promotes colony growth. *Proceedings of the Royal Society of London, Series B* 270, 99–103.

Thorne, B.L. (1985) Termite polygyny: the ecological dynamics of queen mutualism. In: Hölldobler, B. and Lindauer, M. (eds) *Experimental Behavioural Ecology*. Fisher Verlag, Stuttgart, Germany, pp. 325–341.

Tibayrenc, M. (1999) Toward an integrated genetic epidemiology of parasitic protozoa and other pathogens. *Annual Review of Genetics* 33, 449–477.

Traniello, J.F.A., Rosengaus, R.B. and Savoie, K. (2002) The development of immunity in a social insect: evidence for the group facilitation of disease resistance. *Proceedings of the National Academy of Sciences USA* 99, 6838–6842.

Trump, R.F., Thompson, V.C. and Rothenbuhler, W.C. (1967) Behavior genetics of nest cleaning in honeybees. V. Effect of previous experience and composition of mixed colonies in response to disease-killed brood. *Journal of Apicultural Research* 6, 127–131.

Van Borm, S., Wenseleers, T., Billen, J. and Boomsma, J.J. (2001) *Wolbachia* in leafcutter ants: a widespread symbiont that may induce male killing or incompatible matings. *Journal of Evolutionary Biology* 14, 805–814.

Van Borm, S., Billen, J. and Boomsma, J.J. (2002) The diversity of microorganisms associated with *Acromyrmex* leaf-cutting ants. *BioMed Central Evolutionary Biology* 2, 9.

Van Borm, S., Wenseleers, T., Billen, J. and Boomsma, J.J. (2003) Cloning and sequencing of *wsp* encoding gene fragments reveals a diversity of co-infecting *Wolbachia* strains in *Acromyrmex* leafcutter ants. *Molecular Phylogenetics and Evolution* 26, 102–109.

Velthuis, H.H.W. and Gerling, D. (1983) At the brink of sociality: interactions between adults of the carpenter bee *Xylocopa pubescens* Spinola. *Behavioral Ecology and Sociobiology* 12, 209–214.

Watve, M. and Jog, M.M. (1997) Epidemic diseases and host clustering: an optimum cluster size ensures maximum survival. *Journal of Theoretical Biology* 184, 165–169.

Wcislo, W.T. (1996) Parasitism rates in relation to nest site in bees and wasps (Hymenoptera: Apoidea). *Journal of Insect Behavior* 9, 643–656.

Weinig, C., Stinchcombe, J.R. and Schmitt, J. (2003) Evolutionary genetics of resistance and tolerance to natural herbivory in *Arabidopsis thaliana*. *Evolution* 57, 1270–1280.

Wenseleers, T., Ito, F., Van Borm, S., Huybrechts, R., Volckaert, F. and Billen, J. (1998) Widespread occurrence of the micro-organism *Wolbachia* in ants. *Proceedings of the Royal Society of London, Series B* 265, 1447–1452.

Wenseleers, T., Sundström, L. and Billen, J. (2002) Deleterious *Wolbachia* in the ant *Formica truncorum*. *Proceedings of the Royal Society of London, Series B* 269, 623–629.

Williams, J.L. (1990) Insects: Lepidoptera (moths). In: Morse, R.A. and Nowogrodzki, R. (eds) *Honey Bee Pests, Predators, and Diseases*. Cornell University Press, Ithaca, New York, pp. 96–119.

Wilson, E.O. (1971) *The Insect Societies*. Harvard University Press, Cambridge, Massachusetts.

Wilson, K., Knell, R., Boots, M. and Koch-Osborne, J. (2003) Group living and investment in immune defence: an interspecific analysis. *Journal of Animal Ecology* 72, 133–143.

7

Cascading Effects of Plant Genetic Variation on Herbivore Communities

ROBERT S. FRITZ[1] AND CRIS G. HOCHWENDER[2]

[1]Department of Biology, Vassar College, Poughkeepsie, New York, USA;
[2]Department of Biology, University of Evansville, Evansville, Indiana, USA

1. Introduction

A community consisting of a host plant, the arthropod species that feed on it, and the natural enemies of those herbivores, can be described as a component community (Root, 1973). In addition to plant–herbivore interactions, the structure of a component community could be affected by horizontal interactions among herbivores (e.g. competition and facilitation) and by vertical interactions between herbivores and their natural enemies (e.g. predation, parasitism and mutualism). These communities can be quite large (e.g. Maddox and Root, 1990; Whitham *et al.*, 1994), sometimes including more than 100 species of herbivores and natural enemies, so abundant opportunities exist for both direct and indirect interactions. Even with so many potential interactions, though, a single factor could greatly affect the evolutionary trajectory of a given species, or even the evolution of multiple species, provided that the factor is both ecologically important and temporally consistent. By influencing multiple species in a component community, such a factor might alter the structure of the entire community.

The basis and regulation of community structure in such communities spawned the debate over whether community regulation is most influenced by top-down or bottom-up factors (e.g. Hairston *et al.*, 1960; Hunter and Price, 1992). Hunter and Price (1992) made the case that a bottom-up template is most compelling for terrestrial systems because of the dominance of plants in landscapes and the dependence of many species directly on plants. For many insect herbivores, individual plants act as isolated food resources upon which the insect relies. A plant's nutrient quality, its secondary metabolites and its foliar toughness are just a few factors that may affect the survival and reproductive success of an insect found on that plant. In addition, the conceptual framework of many ecological and evolutionary models (e.g. evolution of preference and performance, plant defence theory, and plant–insect coevolution) also supports a bottom-up perspective; these models either explicitly or implicitly incorporate phenotypic variation among host plants as a part of their design.

While we agree with Hunter and Price's bottom-up perspective, we disagree with their major emphasis on the importance of environmental heterogeneity. Environmental factors, such as nutrient environment, local conditions, or even previous herbivory (Orians and Fritz, 1995; Pilson, 1996; Brown and Gange, 2002, Denno *et al.*, 2002, Moon and Stiling, 2002), can affect insect populations and the structure of herbivores on plants. Nevertheless, genetic variation in plant quality can have dramatic effects on insect herbivores. Plant genetic variation commonly explains differences in insect herbivore preference and performance (Berenbaum and Zangerl, 1992; Karban, 1992; Rausher, 2001). In turn, plant genetic variation can affect the community of herbivores and natural enemies that colonize a plant and the interactions among these species (Fritz and Price, 1988; Boecklen and Spellenberg, 1990; Maddox and Root, 1990; Fritz, 1992; Morrow *et al.*, 1994; Floate and Whitham, 1995; Whitham *et al.*, 1999; Dungey *et al.*, 2000; Hochwender and Fritz, 2004).

In this chapter, we focus our attention on this specific factor of plant quality, plant genetic variation, and we evaluate its impact on the lives of insect herbivores. We hypothesize that the genetic effects of host plants can be expected to 'cascade up' into communities of insects on plants, shaping the structure of communities. The objectives of this chapter, therefore, are threefold:

1. To discuss the evidence of genetic variation in resistance;
2. To examine the effects of genetic variation in structuring communities of herbivores;
3. To consider the influence of genetic variation on the interactions between insects and their natural enemies.

To elaborate upon these points, our discussion centres on studies we have carried out using willow species and their interspecific hybrids.

1.1 Model

The population dynamics and evolution of insects on plants can be influenced by horizontal interactions (competition, facilitation), by vertical interactions from the bottom (environmental and genetic heterogeneity), and by vertical interactions from the top (predation, parasitism and mutualism). As previously mentioned, we think that the 'chutes and ladders' model of Hunter and Price (1992) could be enhanced by shifting focus away from heterogeneity in the host plant caused by environmental variation and toward heterogeneity caused by plant genetic variation. This modified focus would extend the 'chutes and ladders' model to allow for a greater predictive power regarding the evolution of plants, their herbivores, and the natural enemies of those herbivores.

To illustrate this point we present a graphical model of two insect herbivores each with a parasitoid species that utilize a willow host plant (Fig. 7.1). The two outer arrows illustrate the direct effects of the host plant on each herbivore. The symbol h^2_P indicates heritability of genetic variation in resistance of the plant to each herbivore, a well-documented occurrence (Rausher, 2001). The other arrows projecting from the willow plant illustrate the cascading effects of the host plant genetic variation on interactions among the herbivores and natural enemies. The arrow pointing to the interaction between the two herbivores represents the influence of plant genetic variation on the competitive interactions among herbivore species. Two herbivores may compete for plant resources such as oviposition sites or nutrients for larval development. A genotype susceptible to the two species might support greater herbivore densities, leading to greater competition between the herbivores than would be seen for a genotype that has greater resistance to the two herbivore species. In addition, induced resistance and induced susceptibility, which may be genetically variable (Agrawal *et al.*, 2002a,b), to density effects on herbivores may also influence interspecific interactions.

The last two arrows projecting from the willow plant represent the effects of plant genetic variation on interactions between herbivores and their parasitoids. One arrow represents the effect that plants could have on herbivores through their natural enemies; if there exists heritable genetic variation for plant traits that influence the impact of natural enemies, then natural enemies could cause differential mortality on herbivores based on differences in those plant traits. The final arrow represents a more complex indirect effect. If plant genetic variation affects the abundance of one herbivore species, and that herbivore species

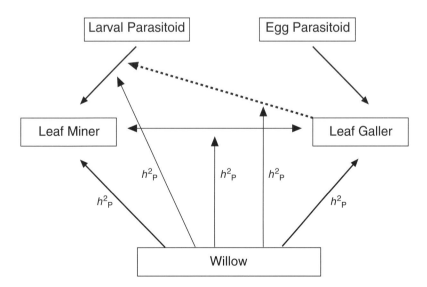

Fig. 7.1. Graphical model illustrating paths in which plant genetic variation can be expressed in communities of insects on plants: h^2_P is the narrow-sense heritability (due to additive genetic variation) attributable to the host plant.

causes greater recruitment of a natural enemy, enemy-mediated competition or apparent competition could occur. In the simplest case, the density of a second herbivore increases searching on the host plant by the parasitoid, and through density-dependent processes parasitism rates increase. Alternatively, an insect that is not a host to the parasitoid could cause the release of plant volatiles that attract parasitoids or mask the presence of the host insect, with opposite outcomes for parasitism rates (Dicke, 1999; Hare, 2002).

In summary, our model suggests that plant genetic variation may influence community structure in three major ways. If plant genetic variation acts as a central framework that structures the component community, plant genetic variation should:

- directly affect the community of herbivores feeding on the plant;
- indirectly affect horizontal interactions among herbivore species;
- indirectly affect vertical interactions between herbivore species and their natural enemies.

Still, to measure the primacy of plant genetic variation over environmental heterogeneity, these patterns should be evaluated for spatial and temporal consistency to determine whether environmental heterogeneity swamps out the effects of plant genetic variation.

2. The Willow Hybrid System

2.1 Willow species

Two willow species, *Salix sericea* Marshall and *S. eriocephala* Michx., and their interspecific hybrids occur naturally at our field site in Otsego County, New York State, USA. *S. sericea* and *S. eriocephala* co-occur in swamps and along streams in central New York, and commonly hybridize throughout their range (Argus, 1986; Mosseler and Papadopol, 1989). These two species have a history of hybridization, with evidence suggesting that introgression has occurred between these species as long ago as the last ice age (Hardig *et al.*, 2000). Although these are two distinct species, genetic exchange regularly occurs between them. At our study site, for example, naturally occurring F_1-like hybrids, F_2 individuals, and backcrosses to both parents occur (Fritz *et al.*, 1996; Hardig *et al.*, 2000). Therefore, arthropod herbivores of this hybrid complex commonly experience a wide range of genotypes at our field site. Moreover, the common nature of hybridization between these two species (Argus, 1986; Mosseler and Papadopol, 1989) suggests that insect herbivores may often be exposed to these hybrid classes in nature.

2.2 Herbivores

A wide array of arthropod herbivores attack these two willow species and their hybrids. These herbivores can be clustered into five different guilds:

1. Five leaf-galling sawfly species: *Eupontania s-gracilis*, *Phyllocolpa eleanorae*, *Phyllocolpa nigrita*, *Phyllocolpa terminalis*, and one we designate as *Phyllocolpa* sp. C (Hymenoptera: Tenthredinidae).

2. Two leaf-mining moths: *Phyllonorycter salicifoliella* and *Phyllocnistis salicifolia* (Lepidoptera: Gracillariidae).

3. Three leaf-folding moths that we call *Caloptilia* sp. T, LF-V, and LF (Lepidoptera: Gracillariidae).

4. Three stem-galling fly species: *Rabdophaga salicisbrassicoides*, *Rabdophaga rigidae*, and one we refer to as *Rabdophaga* sp. G (Diptera: Cecidomyiidae).

5. A leaf-galling mite, *Aculops tetanothrix* (Acarina: Eriophyidae).

Because each of these species creates distinctly different domiciles within which they feed, they can be easily identified and quantified on plants.

2.3 Traits of willow hybrids

Hybrids between *S. sericea* and *S. eriocephala* can usually be distinguished in the field based on morphological traits that are intermediate between those of the parental species (Hardig *et al.*, 2000). The willow species differ in the defensive chemistry of their leaves. *S. sericea* has two phenolic glycosides in its leaves, salicortin and 2'-cinnamoyl salicortin. Their combined concentration is about 15% dry leaf weight. *S. eriocephala* has condensed tannins in its leaves and no phenolic glycosides. The concentration of condensed tannins is about 20% dry weight (C.M. Orians, 2003, personal communication). F_1 hybrids are intermediate in concentrations of condensed tannins and phenolic glycosides, suggesting a primarily additive basis of inheritance of these compounds (Orians and Fritz, 1995). F_2 hybrids have a similar mean concentration of phenolic glycosides and tannins to F_1 hybrids, but the variance among F_2 individuals is greater, as predicted from quantitative genetics theory (Hochwender *et al.*, 2000). Concentrations of tannins in backcrosses to *S. eriocephala* are higher than in F_1s, and likewise, concentrations of phenolic glycosides are higher in backcrosses to *S. sericea*, also as predicted (C.M. Orians, 2003, personal communication). The consequences of hybridization, recombination and back-crossing create substantial genetic variation in traits of hybrid plants which can influence community structure and interaction with natural enemies.

2.4 Effects of hybridization on resistance

Hypothesized effects of hybridization on the resistance of plants compared with parental species assume that resistance is a polygenic trait, influenced by many genes, and that quantitative genetic models best describe the patterns of herbivore and pathogen response to hybrid plants. Some models specify that hybrids are F_1s, but other models do not specify the genetic composition of the hybrids (Fritz, 1999). Providing that tests of resistance are performed in a uniform environment, the outcome of these comparisons can be used to infer the genetic basis of resistance in hybrid plants. Some of the possible patterns are illustrated by herbivores that we have studied in this system. The additive pattern suggests that resistance or host attraction genes act in a dosage-dependent manner (e.g. Aguilar and Boecklen, 1992; Boecklen and Larson, 1994; Eisenbach, 1996; Fritz et al., 1999). This pattern is illustrated in Fig. 7.2A, where the abundance of *Phyllocnistis* sp. on F_1 hybrids was intermediate between densities on the parent species. Dominance of either susceptibility or resistance suggests dominant effects of genes from one parent species. Dominance of susceptibility implies that hybrids have similar attractant or performance traits as the equally susceptible parent (Fig. 7.2B) (Siemens et al., 1994; Mattson et al., 1996; Messina et al., 1996; Hjältén, 1997; Hjältén et al., 2002). Dominance of resistance implies that hybrids have similar repellent or antibiosis traits as the resistant parent (Fig. 7.2C). Hybrid susceptibility occurs when resistance of the hybrid plants is less than that of the parental species (Fig. 7.2D); this response is described as hybrid breakdown (e.g. Boecklen and Spellenberg, 1990; Fritz et al., 1994, 1996; Messina et al., 1996; Orians et al.,

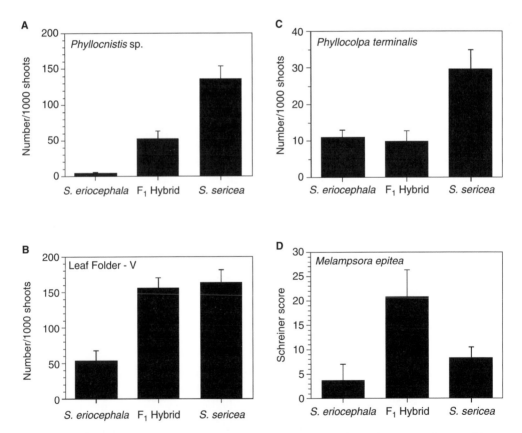

Fig. 7.2. Abundances of three herbivores and infection score of a pathogen on 1-year-old seedling willows of *S. sericea*, *S. eriocephala* and their F_1 hybrids growing in a common garden. **A** *Phyllocnistis* sp., **B** leaf folder – V, **C** *Phyllocolpa terminalis*, **D** *Melampsora epitea*.

1997). In this case, F_1 hybrids were much more susceptible to the rust pathogen, *Melampsora epitea*, than were either parent species. Hybrid resistance occurs when hybrids are more resistant than either parent species; this response is described as heterosis. For this pattern, different resistance genes from each parent may be dominant and act together in their effects on herbivores.

3. Effects of Plant Genetic Variation on Herbivore Abundance

Host plant quality, and hence insect abundance, is commonly affected by plant genotype in natural systems, and a number of studies have demonstrated heritable genetic variation in resistance to herbivores (see reviews by Karban, 1992; Weis and Campbell, 1992; Rausher, 2001). Less resistant plants are often more nutritious or less well defended by toxins or repellents and therefore support larger densities of insect species. For example, gall infestation rates of clones of tall goldenrod (*Solidago altissima*) by the goldenrod ball galler (*Eurosta solidaginis*) are genetically based (Anderson *et al.*, 1989) and are consistent between years in the field (McCrea and Abrahamson, 1987; Abrahamson and Weis, 1997).

Studies of insect herbivores on willows have also demonstrated phenotypic and genetic variation in resistance (Fritz and Price, 1988; Fritz and Nobel, 1989; Roche and Fritz, 1997). Common garden experiments using clones of willows have provided evidence of genetic variation in resistance for many herbivore species, both within populations of willows and for hybrid willow systems (Fritz and Price, 1988, 1990; Fritz, 1995). For half-sib families within a population of *S. sericea*, heritable variation in resistance was detected for two species of herbivores (Roche and Fritz, 1997).

4. Effects of Plant Genetic Variation on Community Structure

Genetic variation can be the underlying cause for differences in community structure of herbivores. The greater the degree of similarity among herbivores in their responses to plant genetic variation, the greater the similarity will be in community structure among plant genotypes (Fritz and Price, 1988; Maddox and Root, 1987, 1990; Fritz, 1992; Weis and Campbell, 1992). If herbivore species respond differently to genetic variation among host plants, then the relative abundances of herbivores will vary greatly among plant genotypes, resulting in distinct communities on each genotype, (Fritz, 1992; Weis and Campbell, 1992). Therefore, genetic correlations of resistance among herbivores will determine the variation in community structure among plants.

Several studies have analysed the correlations among abundances of insect herbivores on plant genotypes in common gardens (Maddox and Root, 1987, 1990; Fritz and Price, 1988; Fritz, 1992; Morrow *et al.*, 1994; Roche and Fritz, 1997). The results of these studies suggest that herbivores respond independently of plant genetic variation (Maddox and Root, 1990; Roche and Fritz, 1997). In addition to suggesting that different genes affect abundances of different herbivore species, these results imply that communities of herbivores on different plant genotypes can vary substantially from each other. In cases where significant genetic correlations are detected, the correlations are typically positive rather than negative (Maddox and Root, 1990; Fritz, 1992; Roche and Fritz, 1997). Positive genetic correlations among herbivore species would favour their co-occurrence on the same plant genotypes, thereby increasing the potential for interactions among these herbivores.

4.1 Genotypic differences in community structure

Mounting evidence from hybrid systems suggests that plant genetic variation can profoundly affect the community structure of insect herbivores (Boecklen and Spellenberg, 1990; Fritz, 1992; Morrow *et al.*, 1994; Floate and Whitham, 1995; Mattson *et al.*, 1996; Whitham *et al.*, 1999; Dungey *et al.*, 2000). In general, these studies suggest that the structure of insect communities reflects genetic differences among plants. For example, the insect community for interspecific F_1 hybrids of *Eucalyptus amygdalina* $\times$ *E. risdonii* is both distinct and unique from the insect community of each parental species (Dungey *et al.*, 2000).

We have recently used the *Salix sericea* $\times$ *S. eriocephala* hybrid complex to examine the importance of plant genetic variation on the structure of an arthropod herbivore community (Hochwender and Fritz, 2004). We examined the arthropod herbivore community using six genetic classes (pure *S. eriocephala*, backcrosses to *S. eriocephala*, F_1 hybrids, F_2 hybrids, backcrosses to *S. sericea*, and pure *S. sericea*) created through controlled crosses. We evaluated the extent to which variation at the level of genetic class affects community structure of arthropod herbivores. Unlike an intraspecific population of plants, these six hybrid genetic classes differed from each other in known ways, so this hybrid system was used to clarify the role of genetic variation on community structure of insect herbivores.

Using a canonical discriminant analysis (CDA) to determine differences in community structure among genetic classes (PROC CANDISC; SAS Institute, 1990), we detected significant variation among genetic classes (Wilk's λ = 0.460, F = 4.7, P < 0.001). Each of the first three linear descriptive functions (LDFs) were significant, with these three LDFs explaining 49%, 29% and 13% of the variance in community structure, respectively (Table 7.1). To further explain the basis for changes in community structure, we evaluated how the herbivore species loaded onto the LDFs (Table 7.1). Based on the loading of each herbivore species for the first LDF, three herbivore species were of greatest importance in explaining the pattern observed: *Phyllocolpa eleanorae*, *Phyllocolpa nigrita* and *Phyllocnistis salicifolia*. For the second LDF, the two most important herbivore species were *Phyllocolpa terminalis* and *Phyllocolpa* sp. C. Again, based on the loadings, *Eupontania s-gracilis*, *Phyllonorycter salicifoliella* and *Rabdophaga salicisbrassicoides* were the three most influential herbivores affecting the third LDF.

We plotted the centroids for each genetic class in three-dimensional space to help interpret the differences in community structure found among genetic

Table 7.1. Results of significance tests of each linear descriptive function from the canonical discriminant analysis (CDA) with the percentage of variation in community structure explained by each variable. Loadings for each linear descriptive function (LDF) on to the ten herbivore species used in the descriptive discriminant canonical analysis. The most influential loadings are in bold (from Hochwender and Fritz (2004) © Springer, Berlin). ***$P < 0.001$, **$P < 0.01$.

Herbivore species	Significance and loadings of the LDFs		
	LDF 1	LDF 2	LDF 3
F-value	4.7***	3.5***	2.3**
% of variance explained	48.6%	28.7%	13.4%
Eupontania gracilis	−0.06	0.01	**0.51**
Phyllocolpa eleanorae	**0.56**	−0.25	−0.09
Phyllocolpa nigrita	**0.38**	−0.02	−0.02
Phyllocolpa terminalis	0.07	**0.54**	−0.04
Phyllocolpa sp. C	−0.10	**0.57**	−0.14
Phyllonorycter salicifoliella	0.23	0.26	**−0.59**
Phyllocnistis salicifolia	**0.49**	0.06	0.41
Caloptilia sp.	−0.21	0.21	−0.17
Leaf folder LF-V	−0.00	0.30	0.33
Rabdophaga salicisbrassicoides	−0.11	0.09	**0.50**

classes (Fig. 7.3). Based on pair-wise comparisons, the location of the centroid for each genetic class was significantly different from the location of the centroids for all other classes, even after sequential Bonferroni corrections were performed ($\alpha = 0.05$). If additive genetic variation was the predominant genetic effect on community structure, the centroids for the hybrid classes should have been located between the centroids of the parental classes. However, the centroids for the four hybrid classes were located well outside of the region between the centroids for the parental species. Furthermore, the distance between the centroids of some hybrid classes was greater than the distance between the two parental classes. For example, the squared distance between the B_E plants and the B_S plants was greater than the squared distance between the two parental species, even though backcross classes are more genetically similar to each other than are parental species. For F_1 and F_2 hybrids, the herbivore communities differed significantly between these two classes even though differences in community structure are due only to recombination following segregation.

To aid in evaluating the underlying basis for differences between the parents, we performed line-cross analysis, a regression technique developed by Cavalli (1952) and Hayman (1958), on each linear descriptive function (i.e. each canonical variable) and its variance. Joint-scaling tests (as described by Mather and Jinks, 1982; Lynch and Walsh, 1998) were used to determine the most appropriate genetic model to fit the data.

Joint-scaling tests showed that a model including additive, dominance and epistatic effects (the ADE model) was required to explain variation in each of the three LDFs (Table 7.2); at least three composite effects were significant for

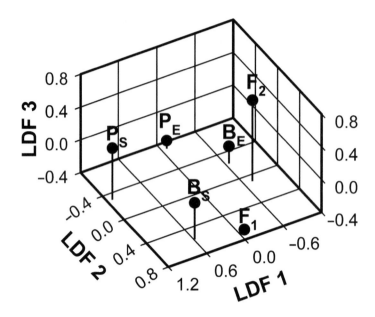

Fig. 7.3. Three-dimensional plot of herbivore community structure on hybrid willows. Each labelled oval represents the centroid of a genetic class. P_E = *Salix eriocephala*, P_s = *Salix sericea*, F_1 = F_1 hybrid, F_2 = F_2 hybrid, B_E = backcross between P_E and F_1, B_S = backcross between P_s and F_1 (from Hochwender and Fritz (2004) © Springer, Berlin).

Table 7.2. Coefficients and significance for mean (μ), additive (α_1), dominance (δ_1), additive $\times$ additive epistasis (α_2), additive $\times$ dominance epistasis ($\alpha*\delta$) and dominance $\times$ dominance epistasis (δ_2) composite effects are shown for the best-fit model for each canonical variable using joint-scaling tests. For each canonical variable, models including additive, dominance and epistasis were sufficient to explain variation among the genetic classes. Blank spaces correspond to parameters dropped from models. *$P < 0.05$, **$P < 0.01$, ***$P < 0.001$ (from Hochwender and Fritz (2004) © Springer, Berlin).

Effect	Model	μ	α_1	δ_1	α_2	$\alpha*\delta$	δ_2
LDF 1	ADE**	− 1.67*	0.69***	2.17	2.22**		− 0.55
LDF 2	ADE	− 0.50***	0.20**	1.24***			1.109*
LDF 3	ADE	2.60**	1.26**	− 5.08**	− 2.64***		2.13*

each LDF. For every LDF, the additive-by-dominance interaction was dropped from the analysis because it did not explain significant variation. The dominance-by-dominance interaction explained significant variation for each of the three LDFs, though, and so was included in all models. For LDF 2, the additive-by-additive interaction was not significant and it was dropped from the final model. For LDF 1, additional, unexplained variation existed, even when all of the parameters were entered into the model. For LDF 2 and LDF 3, the ADE model was sufficient to explain the detected variation.

We found strong support for a plant-based structure to the herbivore community. Similarly, other studies utilizing plant hybrid complexes have found that genetic class can affect the herbivore community (Boecklen and Spellenberg, 1990; Morrow *et al.*, 1994; Whitham *et al.*, 1994; Floate and Whitham, 1995; Mattson *et al.*, 1996; Dungey *et al.*, 2000). For example, Mattson *et al.* (1996) found that communities of arthropods on spruce hybrids differed because of the presence of unique herbivore species that did not colonize either parent species. Whitham *et al.* (1994) and Morrow *et al.* (1994) have shown that species diversity is higher on hybrid plants than on parental species because hybrids support a significant portion of the communities of each parent. Boecklen and Spellenberg (1990), in contrast, found diversity of some species groups on two oak hybrid systems to be lower or intermediate compared to parent species. In all cases, though, the results suggest that underlying genetic differences among plants project upward to the next trophic level, affecting insect herbivore community structure.

4.2 Temporal consistency in community structure

Plant genetic variation can provide a basis for community structure, but temporal fluctuations in insect densities may be large, reducing the predictability of the community structure for insect herbivores (Root and Cappuccino, 1992). Similarly, spatial variation caused by environmental variables such as light, nutrients, salt stress and history of previous herbivory can interact with plant genotype to affect resistance (e.g. Orians and Fritz, 1996; Stiling and Rossi, 1996; Moon and Stiling, 2002; Orians *et al.*, 2003; Sharma *et al.*, 2003).

Consistent patterns of community structure based on plant genetic variation across years would suggest that plant genetic variation is an important organizing force for the structure of herbivore communities. In addition, the influence of plant genetic variation should be detectable in a natural setting, as well as under more controlled environmental conditions, in order for plant genetic variation to be considered an important basis for plant quality. If environmental factors are great enough to mask genetic differences in community structure, plant genetic variation might not be considered as of primary importance. In contrast, if differences in community structure are maintained across years and under natural conditions, plant genetic variation is more likely to have consequential ecological and evolutionary implications.

Insect communities on parental species and intermediate hybrid plants were evaluated for 4 years using field plants. To distinguish genetic variation from environmental variation, we also examined community structure of insect herbivores in a common garden using cuttings from those same individuals. We addressed the following major question: are consistent community differences maintained by genetic differences among plants? Two aspects of this question that we examined were:

- Is consistency in the community structure of insect herbivores maintained on each genetic class across the four years?

• Does environmental variation mask or reduce differences in the community structure of insect herbivores caused by plant genetic variation?

The parental and hybrid plants selected for the experiments were genetically characterized using 20 RAPD (random amplified polymorphic DNA) markers (Hardig et al., 2000). For this study, we used hybrids with scores in the range of 0.4–0.6 to ensure that plants were intermediate hybrids, so these hybrids could be a mixture of F_1 and F_2 hybrids, as well as including more complex pedigrees.

Censuses of herbivore species were conducted on eight S. eriocephala, 15 S. sericea and 12 hybrid field plants during late July and early August 1991, 1992, 1993 and 1994. Galls, leaf mines and leaf folds were counted on 50 shoots per plant for all plants during all years, except for S. sericea in 1991, where 300 shoots were censused. We grouped Phyllocolpa eleanorae with P. nigrita, and we grouped Phyllocolpa terminalis with P. sp. C because we had not recognized them as distinct species for all census periods included in this study. Fortunately, both groups respond similarly to genetic variation in this system (personal observation).

In April 1994, we made cuttings from five S. eriocephala, ten S. sericea, and nine hybrid genotypes from plants that were used in the previous study (191 cuttings with census data that was collected in a comparable way). Plants were randomly placed in four rectangular spatial blocks that were quadrants of a larger rectangle, with equal numbers of clones of each genotype per block. In mid-August 1994 we assessed the densities of most herbivores by counting the total numbers of individuals on plants and total number of shoots per plant to estimate density per shoot. For P. salicifoliella, this censusing procedure was accomplished only for a subset of the plants. For all of the plants, we counted the number of individuals on all leaves on five shoots and divided by the total number of leaves to estimate density per leaf for P. salicifoliella.

The first two linear descriptive functions (LDFs) were significant in explaining variation in community structure for field plants ($P < 0.0001$ in both cases). The three groups of plants censused in the field (S. eriocephala, S. sericea and hybrid plants) clustered into three distinct locations in two-dimensional space in the 4-year study (Fig. 7.4). These locations were significantly different based on pairwise comparisons for the location of their centroids ($P < 0.0001$ in all three cases). Based on the loading of each herbivore species for the first two LDFs, the herbivore species of greatest importance in explaining the pattern observed were: Eupontania s-gracilis, P. eleanorae and P. nigrita, and P. terminalis and Phyllocolpa sp. C. Although genetic differences in community structure were detected between the three groups, the community structure for each genetic class did not change across years (Fig. 7.4); neither census year nor the interaction between year and genetic class had a significant effect for LDF 1 (repeated measures ANOVA: $F_{3, 29} = 0.7$; $P = 0.56$ and $F_{6, 58} = 0.8$; $P = 0.58$, respectively) or for LDF 2 (repeated measures ANOVA: $F_{3, 29} = 0.6$; $P = 0.62$ and $F_{6, 58} = 1.2$; $P = 0.30$, respectively).

To determine whether plants differed in the structure of their community when they occurred naturally versus when placed as cuttings in a common

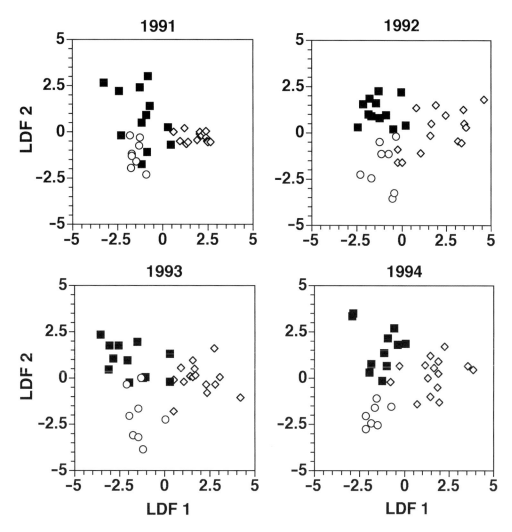

Fig. 7.4. Plots of two linear descriptive functions for parental and hybrid willows for field plants for 4 years. Each symbol represents the position of the insect community for a plant. Genetic classes are represented by different symbols. Circles = *Salix eriocephala*, diamonds = *Salix sericea*, squares = hybrid plants.

environment, we used CDA (PROC CANDISC; SAS Institute, 1990) and MANOVA (PROC GLM; SAS Institute, 1990) to examine whether class-based and treatment-based differences existed in 1994. The MANOVA incorporated both LDFs into a single analysis. For these plants, the first two LDFs were again significant ($P < 0.009$ in both cases), and significant differences were detected in the locations of the centroids for the three genetic classes ($P < 0.002$ in all three cases) (Fig. 7.5). Although the three genetic classes differed in the structure of their insect communities, no significant differences were detected

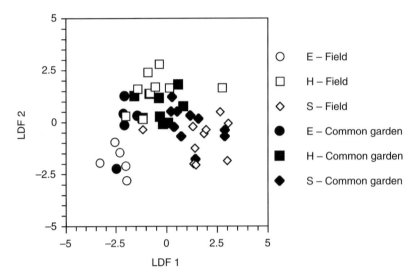

Fig. 7.5. Plot of two linear descriptive functions for parental and hybrid willows from field and common garden plants in 1994. Each symbol represents the centroid of a genetic class. Circles = *Salix eriocephala*, diamonds = *Salix sericea*, squares = hybrid plants. Unfilled symbols represent field plants and filled symbols represent plants from a common garden.

between the field plants and the cuttings grown in a common garden (MANOVA: $F_{2, 41} = 2.2$; $P = 0.13$) (Fig. 7.5).

These results confirm previous findings that the community structure of insect herbivores occurring in this hybrid willow system is shaped by genetic variation among the parental species and hybrid plants. Community structure was strikingly consistent across the 4 years of the field study. The degree to which community patterns are temporally stable for individual plant genotypes has not been well examined in general, but multi-year studies examining the herbivore community of goldenrod have been carried out (Root and Cappuccino, 1992). For *Solidago altissima*, the herbivore community varied greatly over 6 years, with the communities remaining consistent only in that dominant species remained dominant, while rare species remained rare (Root and Cappuccino, 1992). In our analyses of structure of the herbivore community on willow plants, we found that community structure was pre-dictable in the field over a 4-year period. Moreover, the community of insect herbivores demonstrated a similar community structure when clonal individuals were grown in a common garden to remove environmental variation. Based on the results of this experiment, it appears that environmental differences only enhanced the differences due to plant genetic variation. Because neither spatial nor temporal variation negated the effects of plant genetic variation on community structure, these results suggest that plant genetic variation is likely to have influential ecological and evolutionary implications for component communities.

5. Plant Genetic Variation and Interspecific Competition

Clearly, plant genetic variation can create a mosaic of different herbivore communities. This mosaic of communities is likely to create variation in the outcomes of competitive interactions, as well as variation in the interactions of herbivores with their natural enemies. Because the evolution of those interactions may change with hybrid class, and the evolutionary trajectory of interactions may differ among genotypes (Thompson, 1994, 1999). Evidence for plant genetic effects on competition is sparse, but Moran and Whitham (1990) found that competition between a leaf-galling and a root-feeding aphid depended on plant genotype. Competitive effects of the leaf-galling aphid on the root-feeding aphid were severe on susceptible plants. However, plants that were partially resistant to the leaf-galling species provided a refuge for the root-feeding species. Similarly, Fritz (1990) found that the competition coefficients among galling insects on *Salix lasiolepis* varied with plant genotype. The effects ranged from negative effects on densities of three leaf gallers to no significant competitive effects.

6. Plant Genetic Variation and Tritrophic Interactions

Tritrophic level interactions focus on the links between plants, herbivores, and natural enemies of herbivores, and how each trophic level affects the interactions between the other two levels. Natural enemies are thought to affect plant–herbivore interactions so much that Price *et al.* (1980) suggested that they 'must be considered as part of a plant's battery of defences against herbivores'. Natural enemies can provide an indirect benefit to plants by reducing the populations of herbivorous insects that feed on the plants, thus increasing plant fitness. Selection would favour plant genotypes that have traits that enhance the mortality rates by natural enemies for those herbivores that decrease plant fitness. Evidence from crops has demonstrated that genetic differences among cultivars can influence the impact that natural enemies have on herbivores that colonize the plants (see reviews by Hare, 1992, 2002; Bottrell *et al.*, 1998). For natural populations, though, relatively few studies have examined heritable plant genetic variation in enemy impact on herbivores (Boethel and Eikenbary, 1986; Fritz, 1992; Hare, 1992, 2002).

To demonstrate that selection mediated by natural enemies could be responsible for the evolution of a given plant trait, one must first document genetic variation in that trait and determine that a fitness benefit of the trait occurs for plants in the presence of herbivores and their natural enemies (Hare, 2002). Clones of native plants can differ in the effect that natural enemies have on herbivores (Fritz and Nobel, 1990; Fritz and Kaufman, 1993; Quiring and Butterworth, 1994; Fritz, 1995; Fritz *et al.*, 1997), thereby providing some support for genetic variation in the enemy impact. Gómez and Zamora (1994) have shown a substantial fitness benefit to plants due to parasitism of a seed weevil, but have not documented heritable variation in the trait. Recently, Hoballah and Turlings (2001) presented evidence that those maize plants that have herbivore-induced volatiles to attract parasitoids had a fitness benefit when parasitoids reduced feeding and weight gain by *Spodoptera littoralis* larvae.

The effects of plant genotype on tritrophic interactions may also be mediated through other herbivores. In a study of survival and parasitism of *Phyllonorycter salicifoliella*, we found that there was significant variation among cloned willow plants in densities of a non-host herbivore, *Phyllocolpa nigrita*, and that the density of this species was positively correlated with survival and negatively correlated with parasitism of *Phyllonorycter* (Fritz, 1995). Survival of a third species, *Phyllocolpa eleanorae*, was positively associated with densities of *Phyllonorycter* and egg parasitism was negatively correlated with *Phyllonorycter* density (Fritz, 1995) among clones. These data suggest a role of plant genotype in mediating in the interactions of herbivores, their natural enemies, and other non-host herbivores.

We present evidence from pure willow species and from comparisons of pure parents and their interspecific hybrids that plant genetic variation influences the impact of natural enemies on herbivores.

6.1 The system

The following studies focus primarily on *Phyllonorycter salicifoliella* (Chambers) (Lepidoptera: Gracillariidae), an abundant leaf-mining moth on willows at our study site. Eggs are laid on lower leaf surfaces, and mines are formed on the underside of leaves, as the larva passes through sap-feeding and tissue-feeding stages. Causes of larval mortality can often be determined for this species (Pottinger and LeRoux, 1971; Auerbach and Alberts, 1992; Faeth, 1992; Fritz, 1995). One major source of mortality for *Phyllonorycter* is parasitoid attack. *Phyllonorycter* is parasitized by the internal larval parasitoid *Pholetesor salicifolielliae* (Mason) (Hymenoptera: Braconidae) (J. Whitfield, 1990, personal communication). The white silken cocoons of this species, which are strung on silk threads in the mines, identify *Pholetesor*. External larval parasitoids include several genera of Eulophidae (Hymenoptera) (*Sympiesis*, *Pediobius*, *Achrysocharoides*, *Chrysocharis*) (M. Schauff and E. Grissell, 1990, personal communication). Eulophid parasitism was combined, since species could not be reliably distinguished at the time of dissection. We considered host feeding by parasitoids, indicated by a collapsed and empty larval skin adhering to the lower mine surface, to be a separate source of larval mortality (Jervis and Kidd, 1986; Auerbach and Alberts, 1992). This type of mortality on *Phyllonorycter* is usually attributed to eulophids (Van Driesche and Taub, 1983; Barrett and Jorgensen, 1986). Dead larvae usually occurring during the tissue-feeding stage included several unknown causes of death, possibly including nutritional effects of the plant and pathogens (fungi, bacteria, viruses), and this was a further category of mortality. If the mine was opened and there was neither a parasitoid nor a larval body present, and no indication of parasitism or emergence, then the cause of mortality was considered to be unknown. In this last circumstance, it is almost certain that the larva did not survive because it had usually not consumed enough leaf tissue to reach pupation. Densities of *Phyllonorycter* were low so that death due to intraspecific competition seems unlikely.

6.2 Variation in *Phyllonorycter* survival on *S. sericea*

We conducted an experiment on potted *Salix sericea* plants placed in the field in 1990. A factorial crossing design was made between three female (dams) and 13 male (sires) *S. sericea* plants growing at our field site in New York State, USA, in 1989. Seedlings from these crosses were raised in field greenhouses, planted in 3.7 l pots to overwinter, and were transplanted into 7.4 l pots the following spring. In 1990, 1-year-old plants were placed randomly into three spatial blocks (three siblings of each full-sib family per block) in a mown field surrounded by naturally growing willows. In early August, leaves with mines were collected in plastic bags from each plant and returned to the laboratory for dissection.

Heritable (additive) genetic variation in a trait is detected as a significant difference among sire half-sib families. In this study, we tested whether percentage survival and causes of mortality varied significantly among sires, using logistic regression analysis. After testing full models, we eliminated non-significant three-way and two-way interaction terms. We detected significant variation in survival among sires (Table 7.3), which supports the hypothesis of additive genetic variation in *S. sericea* for fitness of this herbivore. Using the logistic regression approach prevented us from estimating the narrow-sense heritability of these traits, but the significance of the statistical test confirms that it is greater than zero. Significant variation in survival and in all causes of mortality also occurred for the sire-by-dam interaction, indicating that dominance genetic variation has a substantial influence on these traits.

Table 7.3. Analysis of survival and sources of mortality (χ^2 values) of *Phyllonorycter salicifoliella* on *Salix sericea* in 1990. Models are presented where non-significant 3-way and 2-way interactions were removed from analyses. Sample size of dissected leaf mines was 10,961.

A

Source	Sire	Dam	S×D	Block	S×B	D×B	S×D×B
d.f.	12	2	20	2	24	4	40
Survival	23.75*	2.82	56.08***	86.79***	43.37**		
Pholetesor parasitism	18.37	4.66	57.76***	26.54***	56.66***		
Eulophid parasitism	6.49	1.11	41.41**	28.51***			
Predation	13.77	1.52	66.77***	4.41	45.52**	2.00	87.03***
Dead larvae	21.42*	9.99**	62.41***	13.42**	37.85*		

B

Block	1	2	3
Survival	60.75***	84.67***	64.56***
Pholetesor parasitism	15.98	22.78*	18.59
Eulophid parasitism	60.77***	10.88	7.41
Predation	23.93*	15.17	24.68**
Dead larvae	27.55**	26.39**	3.52

*$P < 0.05$, **$P < 0.01$, ***$P < 0.005$.

The sire-by-block effects suggest that the environmental conditions that genotypes experienced influenced the heritable genetic variation that existed for survival and for the different causes of mortality. When blocks were analysed separately, we found that survival varied significantly among sires for each block (Table 7.3B). Figure 7.6 shows that survival was much higher in block 1 than in either of the other two blocks (over 30% versus about 10–13%). Another finding from this analysis was that different causes of enemy-related mortality were significant in each block (Table 7.3, Fig. 7.6). In block 1, parasitism by the guild of eulophid parasitoids varied significantly among sires, but caused only 4–17% mortality (Fig. 7.6A). Host-feeding predation and dead larvae also varied significantly among sires in this block (Table 7.3, data not shown). In block 2, parasitism by the braconid, *Pholetesor salicifolielliae*, varied significantly among sires and caused 27–52% mortality (Fig. 7.6B). Dead larvae was also a significant mortality factor in this block. In block 3, host-feeding predation caused 22–39% mortality and was the only mortality factor that varied significantly among sires (Fig. 7.6C). These data suggest that small-scale environmental effects that varied over just a few metres had a substantial influence on the extent of herbivore survival and causes of mortality. These results, however, also demonstrate that heritable plant genetic variation in herbivore fitness and impact of natural enemies is detectable, even though the environment affected natural enemies. Therefore, our results provide evidence that plant traits that recruit natural enemies could evolve.

If plant genotypes vary in specific traits that enhance enemy impact (such as plant volatiles or morphological features), and if the impact of natural enemies on the herbivore improved plant fitness, then these traits could evolve (Hare, 2002). Alternatively, our result could have been due to density-dependent factors. If plants with higher herbivore loads (i.e. susceptible plants) have higher enemy impact because of density-dependent responses of natural enemies, then enhanced enemy impact is unlikely to evolve, since selection should favour resistant plants. Our results do not support a density-dependent pattern of mortality; the pattern was not one where plants with higher densities of *Phyllonorycter* (i.e. lower resistance) had higher enemy impact. Moreover, no significant inverse correlations between density and survival were detected when we examined full-sib means for each block. Furthermore, none of the known causes of enemy mortality (*Pholetesor* parasitism, eulophid parasitism, or host-feeding predation by eulophids) was correlated to density in any block. Thus, our findings lend greater support to the contention that selection may act on plant traits that recruit the third trophic level.

6.3 *Phyllonorycter* on hybrids in common gardens

Only a few studies have examined the effects of interspecific plant hybridization on herbivore susceptibility to natural enemies. If herbivores have decreased mortality by escaping natural enemies when they use hybrid plants, selection could favour herbivore preference for hybrid plants. Herbivores might then

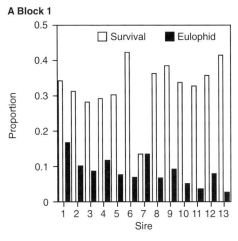

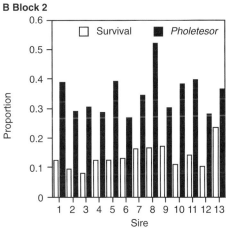

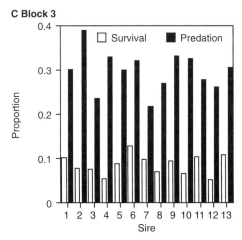

Fig. 7.6. Survival and causes of mortality of *Phyllonorycter salicifoliella* among sire half-sib families in three spatial blocks in 1990. Each factor varied significantly among sires in logistic regression analyses (see Table 7.3).

have a detrimental impact on plant fitness, which could affect the distribution and abundance of hybrids in nature. Some studies have found no effect of hybridization on tritrophic interactions (Moorehead et al., 1993), but other studies have found lower parasitism on hybrids (Preszler and Boecklen, 1994), intermediate parasitism on hybrids (Gange, 1995), or parasitism similar to that of one parental species and not the other (Siemens et al., 1994). Although these studies have been few in number, they suggest that variation in the tritrophic effects of plant hybridization may be as great as the variation in direct effects of hybrid plants on the herbivores (Fritz et al., 1999).

Survival and parasitism differed between hybrid and parental willows in the field in one year; in the following year, predation by host-feeding parasitoids varied significantly, but survival did not vary significantly among taxa (Fritz et al., 1997). However, in a common garden experiment using cloned F_1-type hybrid and parental plants, we found no significant effects of plant taxon on survival or causes of mortality. Still, highly significant differences in survival and causes of mortality occurred among plants within taxa in the common garden. These results support the hypothesis that plant phenotypic and genetic variation influences tritrophic interactions.

We performed a common garden study with an expanded set of genotypes to investigate whether hybrid genetic class, environment, and genotype-by-environment interaction affected survival and causes of mortality of P. salicifoliella. The experimental gardens were laid out in a split-plot design, embedded in three blocks along a local stream. In each block, one garden (size ~150 m^2) was placed less than 5 m from the local stream, representing a wet environment, and another garden was established 30 m from the stream, representing a relatively dry environment. Areas were cleared of willows and other woody vegetation to prevent shading and competition among plants placed in the gardens. Siblings from 5–8 full-sib families of S. sericea, S. eriocephala, and F_1, F_2 and backcrosses (B_E and B_S) to each parent were planted in wet and dry gardens. About 20 individuals of each genetic class were randomly transplanted into each garden (approximately 120 plants per garden). S. eriocephala and hybrids are more common in wet environments, but S. sericea is distributed in both wet and dry environments.

We collected and dissected leaf mines from each plant in the first two spatial blocks to determine miner survival and cause of mortality. We combined data among all plants for each family and genetic class within a garden. We performed logistic regression to determine the significance of genetic class, environment and block. We began with models that included all interactions. If the three-way and any two-way interactions were not significant, they were dropped from subsequent analyses, and data were reanalysed with the reduced models. There were significant interactions, either between genetic class and block or between environment and block, in all cases. Therefore, we performed analyses for each block separately.

Genetic class had a significant effect on survival and on sources of mortality of P. salicifoliella in gardens of block 1, but not in block 2 (Table 7.4). There was also a significant effect of wet versus dry environment and a genetic

Table 7.4. Results of logistic regression analysis of survival and mortality data of *P. salicifoliella* on parental and hybrid genotypes (genetic class = GC) (d.f. = 5) in a common garden experiment in wet and dry environments (E) (d.f. = 1) in two spatial blocks. Effects in the two blocks are analysed separately due to significant interactions of block with genetic class or environment in full models.

Effect	Block	GC	E	G×E	Dry	Wet
Survival	1	13.47*	6.56**	15.66**	19.08**	7.74
	2	8.27	140.54***	2.21		
Eulophid	1	7.63	6.28**	10.61†	11.02†	5.63
	2	8.84	0.00	7.52		
Predation	1	11.77*	2.48	7.53		
	2	8.01	83.81***	0.43		
Dead larvae	1	12.75*	5.49*	1.14		
	2	6.10	45.15***	2.42		

$*P < 0.05$, $**P < 0.01$, $***P < 0.001$, $†P ≈ 0.06$.

class-by-environment interaction for survival in block 1. Analysing the environments separately showed that the significant difference among genetic classes occurred only in the dry environment ($P < 0.01$) (Table 7.4). Parasitism by the guild of eulophid parasitoids also varied with environment and was marginally significant ($P = 0.06$) for the genetic class-by-environment interaction. When analysed separately, the effect of eulophid parasitism was also marginally significant among genetic classes in the dry environment ($P = 0.06$) but not in the wet environment. Leaf miners had the highest survival and the lowest parasitism by eulophid parasitoids on F_2 plants, whereas they had the lowest survival and the highest eulophid parasitism on *S. eriocephala* plants (Fig. 7.7A, C). These data suggest that differential parasitism among genetic classes by eulophids caused variation in survival in the dry environment of block 1.

Wet versus dry environment had a significant effect on survival and eulophid parasitism in block 2 (Table 7.4, Fig. 7.7B, D). Survival was markedly lower in the wet environment. Eulophid parasitism and dead larvae (not shown) were higher in the wet environment. Predation also differed by environment, but was higher in the dry environment (not shown). Density of the leaf miner did not explain variation in survival or causes of mortality.

These data show that genetic variation can have an effect on survival and susceptibility to natural enemies, and suggest that plant genetic variation is an important part of the bottom-up template that influences tritrophic level interactions. These studies also show that environmental variation can be an important factor. The presence of genotype-by-environment interactions in tritrophic level interactions suggests that even if there were fitness differences among plants correlated with enemy impact, directional selection would be unlikely. The ecological effects of both genotype and environment are evident, as is the heritable basis of enemy impact. Still, the evolutionary consequences for the plant remain unclear.

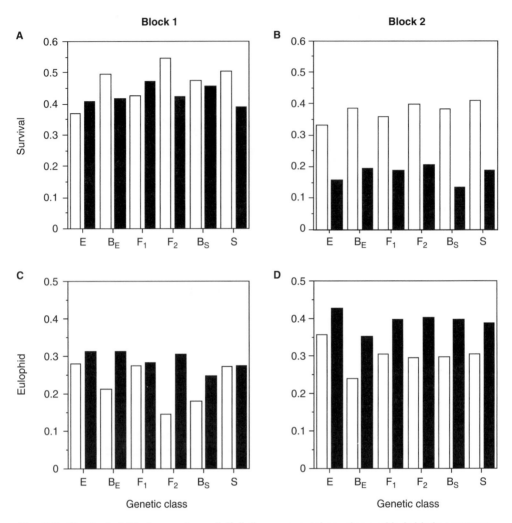

Fig. 7.7. Survival of *Phyllonorycter salicifoliella* on parental species and hybrids in common gardens in two spatial blocks in 2000. Open bars refer to the gardens in the dry environment and filled bars refer to gardens in the wet environment. See Table 7.4 for statistical analysis.

7. Conclusions

Our studies have documented that host-plant genetic variation can have a central role both in structuring the communities of herbivores on plants and in influencing the biotic interactions among herbivores and with natural enemies. While continuing to emphasize the primacy of bottom-up processes in many plant–herbivore systems, our emphasis on plant genetic variation (instead of abiotic variation) places a greater focus on the evolution of biotic interactions in communities. Investing greater attention to genetic variation at all levels when considering plant–animal interactions and community structure will greatly aid in the development of a community genetics paradigm (Antonovics, 1992; Neuhauser *et al.*, 2003; Whitham *et al.*, 2003).

Horizontal and top-down forces can be of primary importance in their own right; evidence demonstrates that predators can greatly influence herbivore communities (Karban, 1986, 1989; Woodman and Price, 1992; Floyd, 1996). For example, differential predation by ants on sawflies can eliminate some gallers from willow plants, dramatically altering community structure on those plants (Woodman and Price, 1992). Interspecific competition can also alter community structure on plants (Fritz *et al.*, 1986; Denno *et al.*, 1995; Hudson and Stiling, 1997). Hudson and Stiling (1997) found that foliar feeding by a chrysomelid beetle on *Baccharis halimifolia* caused a decrease in other common herbivore species. Nevertheless, underlying heterogeneity can be critical to community structure. Although that heterogeneity can stem from abiotic factors, we have documented many examples where plant genetic variation can have a primary role in structuring communities.

Those plant genetic factors may create a complex template within which ecological interactions and evolutionary responses take place when they are expressed under variable environmental conditions. And horizontal and top-down forces may further alter those plant genetic factors. Still, in our system, we found a great degree of constancy in our herbivore communities on plant genetic classes both in time and space. Furthermore, we found that natural enemies responded to plant genetic variation when they attacked herbivores. Thus, we argue that future research should have a greater focus on plant genetics. Those studies that begin and end their investigation with the second and third trophic levels risk overlooking an essential factor when they ignore plant genetic variation. Researchers who view host plants as a monoculture may be creating such simple systems that conclusions regarding how the communities are structured will be over-simplistic at best, and entirely misleading at worst.

Acknowledgements

We are grateful to acknowledge the support of the National Science Foundation grants BSR 89-17752, BSR 96-15038, DEB 99-81406, DEB 01-27369, and the Vassar College Class of '42 Environmental Science Fund. We had assistance from S. Bothwell, E. Boydston, B. Compton, B. Crabb, R.D. Fritz, L. Gedmintas, S. Kaufman, D. Lewkiewicz, S. Manee, N. Murphy, K. Reape, K. Rule, K. Vandenberg and D. Willies. Drs D. Davis, E. Grissell and M. Schauff of the USDA and Dr James Whitfield identified insects. Len and Ellie Sosnowski generously allowed us to conduct research on their land.

References

Abrahamson, W.G. and Weis, A.E. (1997) *Evolutionary Ecology Across Three Trophic Levels: Goldenrods, Gallmakers, and Natural Enemies. Monographs in Population Biology 29*. Princeton University Press, Princeton, New Jersey.

Agrawal, A.A., Conner, J.K., Johnson, M.T. and Wallsgrove, R. (2002a) Ecological genetics of induced plant defense against herbivores: additive genetic variation and costs of phenotypic plasticity. *Evolution* 56, 2206–2213.

Agrawal, A.A., Janssen, A., Bruin, J., Posthumus, M.A. and Sabelis, M.W. (2002b) An ecological cost of plant defense: attractiveness of bitter cucumber plants to natural enemies of herbivores. *Ecology Letters* 5, 377–385.

Aguilar, J.M. and Boecklen, W.J. (1992) Patterns of herbivory in the *Quercus grisea* × *Quercus gambelii* species complex. *Oikos* 64, 498–504.

Anderson, S.S., McCrea, K.D., Abrahamson, W.G. and Hartzel, L.M. (1989) Host genotype choice by the ball gallmaker *Eurosta solidaginis* (Diptera: Tephritidae). *Ecology* 70, 1048–1054.

Antonovics, J. (1992) Toward community genetics. In: Fritz, R.S. and Simms, E.L. (eds) *Plant Resistance to Herbivores and Pathogens: Ecology, Evolution, and Genetics.* University of Chicago Press, Chicago, pp. 240–277.

Argus, G.W. (1986) The genus *Salix* (Salicaceae) in the southeastern United States. *Systematic Botany Monographs* 9, 1–170.

Auerbach, M. and Alberts, J.D. (1992) Occurrence and performance of the aspen blotch miner, *Phyllonorycter salicifoliella*, on three host-tree species. *Oecologia* 89, 1–9.

Barrett, B.A. and Jorgensen, C.D. (1986) Parasitoids of the western tentiform leafminer, *Phyllonorycter elmaella* (Lepidoptera: Gracillariidae), in Utah apple orchards. *Environmental Entomology* 15, 635–641.

Berenbaum, M.R. and Zangerl, A.R. (1992) Genetics of secondary metabolism and herbivore resistance in plants. In: Rosenthal, G.A. and Berenbaum, M.R. (eds) *Herbivores: Their Interactions with Secondary Plant Metabolites*, Vol. II: *Evolutionary and Ecological Processes*, 2nd edn. Academic Press, San Diego, California, pp. 415–438.

Boecklen, W.J. and Larson, K.C. (1994) Gall-forming wasps (Hymenoptera: Cynipidae) in an oak hybrid zone: testing hypotheses about hybrid susceptibility to herbivores. In: Price, P.W., Mattson, W.J. and Baranchikov, Y.N. (eds) *The Ecology and Evolution of Gall-Forming Insects.* North Central Forest Experiment Station, Forest Service USDA, St Paul, Minnesota, pp. 110–120.

Boecklen, W.J. and Spellenberg, R. (1990) Structure of herbivore communities in two oak (*Quercus* spp.) hybrid zones. *Oecologia* 85, 92–100.

Boethel, D.J. and Eikenbary, R.D. (1986) *Interactions of Plant Resistance and Parasitoids and Predators of Insects.* Ellis Horwood, Chichester, UK.

Bottrell, D.G., Barbosa, P. and Gould, F. (1998) Manipulating natural enemies by plant variety selection and modification: a realistic strategy? *Annual Review of Entomology* 43, 347–367.

Brown, V.K. and Gange, A.C. (2002) Tritrophic below- and above-ground interactions in succession. In: Tscharntke, T. and Hawkins, B.A. (eds) *Multitrophic Level Interactions.* Cambridge University Press, Cambridge, UK, pp. 197–222.

Cavalli, L.L. (1952) An analysis of linkage in quantitative inheritance. In: Reeve, E.C.R. and Waddington, C.H. (eds) *Quantitative Inheritance.* Her Majesty's Stationery Office, London, pp. 135–144.

Denno, R.F., McClure, M.S. and Ott, J.R. (1995) Interspecific interactions in phytophagous insects: competition reexamined and resurrected. *Annual Review of Ecology and Systematics* 40, 297–331.

Denno, R.F., Gratton, C., Peterson, M.A., Langellotto, G.A., Finke, D.L. and Huberty, A.F. (2002) Bottom-up forces mediate natural-enemy impact in a phytophagous insect community. *Ecology* 83, 1443–1458.

Dicke, M. (1999) Are herbivore-induced plant volatiles reliable indicators of herbivory identity to foraging carnivorous arthropods? *Entomologia Experimentalis et Applicata* 91, 131–142.

Dungey, H.S., Potts, B.M., Whitham, T.G. and Li, H.F. (2000) Plant genetics affects arthropod community richness and composition: evidence from a synthetic eucalypt hybrid population. *Evolution* 54, 1938–1946.

Eisenbach, J. (1996) Three-trophic-level interactions in cattail hybrid zones. *Oecologia* 105, 258–265.

Faeth, S.E. (1992) Interspecific and intraspecific interactions via plant responses to folivory: an experimental field test. *Ecology* 73, 1802–1813.

Floate, K.D. and Whitham, T.G. (1995) Insects as traits in plant systematics: their use in discriminating between hybrid cottonwoods. *Canadian Journal of Botany* 73, 1–13.

Floyd, T. (1996) Top-down impacts on creosotebush herbivores in a spatially and temporally complex environment. *Ecology* 77, 1544–1555.

Fritz, R.S. (1990) Variation in competition coefficients between insect herbivores on genetically variable host plants. *Ecology* 71, 2008–2011.

Fritz, R.S. (1992) Community structure and species interactions of phytophagous insects on resistant and susceptible host plants. In: Fritz, R.S. and Simms, E.L. (eds) *Plant Resistance to Herbivores and Pathogens: Ecology, Evolution, and Genetics.* University of Chicago Press, Chicago, Illinois, pp. 240–277.

Fritz, R.S. (1995) Direct and indirect of plant genetic variation on enemy impact. *Ecological Entomology* 20, 18–26.

Fritz, R.S. (1999) Resistance of hybrid plants to herbivores: genes, environment, or both? *Ecology* 80, 382–391.

Fritz, R.S. and Kaufman, S.R. (1993) Variable enemy impact at three scales: herbivore species, plant species, and plant genotype. *Oikos* 68, 463–472.

Fritz, R.S. and Nobel, J. (1989) Plant resistance, plant traits, and host plant choice of the leaf-folding sawfly on the arroyo willow. *Ecological Entomology* 14, 393–401.

Fritz, R.S. and Nobel, J. (1990) Host plant variation in mortality of the leaf-folding sawfly on the arroyo willow. *Ecological Entomology* 15, 25–35.

Fritz, R.S. and Price, P.W. (1988) Genetic variation among plants as a cause of community variation in phytophagous insects: willows and sawflies. *Ecology* 69, 845–856.

Fritz, R.S. and Price, P.W. (1990) A field test of interspecific competition on oviposition of gall-forming sawfly species on willow. *Ecology* 71, 99–106.

Fritz, R.S., Sacchi, C.F. and Price, P.W. (1986) Competition versus host plant phenotype in species composition: willow sawflies. *Ecology* 67, 1608–1618.

Fritz, R.S., Nichols-Orians, C.M. and Brunsfeld, S.J. (1994) Interspecific hybridization of plants and resistance to herbivores: hypotheses, genetics, and variable responses in a diverse community. *Oecologia* 97, 106–117.

Fritz, R.S., Roche, B.M., Brunsfeld, S.J. and Orians, C.M. (1996) Interspecific and temporal variation in herbivore responses to hybrid willows. *Oecologia* 108, 121–129.

Fritz, R.S., McDonough, S.E. and Rhoads, A.G. (1997) Effects of plant hybridization on herbivore–parasitoid interactions. *Oecologia* 110, 360–367.

Fritz, R.S., Moulia, C. and Newcombe, G. (1999) Resistance of hybrid plants and animals to herbivores, pathogens, and parasites. *Annual Review of Ecology and Systematics* 30, 565–591.

Gange, A.C. (1995) Aphid performance in an alder (*Alnus*) hybrid zone. *Ecology* 76, 2074–2083.

Gómez, J.M. and Zamora, R. (1994) Top-down effects in a tritrophic system: parasitoids enhance plant fitness. *Ecology* 75, 1023–1030.

Hairston, N.G., Smith, F.E. and Slobodkin, L.B. (1960) Community structure, population control and competition. *American Naturalist* 44, 421–425.

Hardig, T.M., Brunsfeld, S.J., Fritz, R.S., Morgan, M. and Orians, C.M. (2000) Morphological and molecular evidence for hybridization and introgression in a willow (*Salix*) hybrid zone. *Molecular Ecology* 9, 9–24.

Hare, D. (1992) Effects of plant variation on herbivore–natural enemy interactions. In: Fritz, R.S. and Simms, E.L. (eds) *Plant Resistance to Herbivores and Pathogens: Ecology, Evolution, and Genetics.* University of Chicago Press, Chicago, Illinois, pp. 278–298.

Hare, D. (2002) Plant genetic variation in tritrophic interactions. In: Tscharntke, T. and Hawkins, B.A. (eds) *Multitrophic Level Interactions.* Cambridge University Press, Cambridge, UK, pp. 8–43.

Hayman, B.I. (1958) The separation of epistatic from additive and dominance variation in generation means. *Heredity* 12, 371–390.

Hjältén, J. (1997) Willow hybrids and herbivory: a test of hypotheses of phytophage response to hybrid plants using a generalist leaf-feeder *Lochmaea caprea* (Chrysomelidae). *Oecologia* 109, 571–574.

Hjältén, J., Hallgren, P. and Qian, H. (2002) The importance of parent host status for hybrid susceptibility to herbivores: a test with two hybrid lines of willows. *Ecoscience* 9, 339–346.

Hoballah, M.E.F. and Turlings, T.C.J. (2001) Experimental evidence that plants under caterpillar attack may benefit from attracting parasitoids. *Evolutionary Ecology Research* 3, 553–565.

Hochwender, C.G. and Fritz, R.S. (2004) Plant genetic differences influence herbivore community structure: evidence from a hybrid willow system. *Oecologia* 138, 547–557.

Hochwender, C.G., Fritz, R.S. and Orians, C.M. (2000) Using hybrid systems to explore the evolution of tolerance to damage. *Evolutionary Ecology* 14, 509–521.

Hudson, E.E. and Stiling, P. (1997) Exploitative competition strongly affects the herbivorous insect community on *Baccharis halimifolia. Oikos* 79, 521–528.

Hunter, M.D. and Price, P.W. (1992) Playing chutes and ladders: heterogeneity and the relative roles of bottom-up and top-down forces in natural communities. *Ecology* 73, 724–732.

Jervis, M.A. and Kidd, N.A.C. (1986) Host-feeding strategies in hymenopteran parasitoids. *Biological Review* 61, 395–434.

Karban, R. (1986) Interspecific competition between folivorous insects on *Erigeron glaucus. Ecology* 67, 1063–1072.

Karban, R. (1989) Community organization of *Erigeron glaucus* folivores: effects of competition, predation, and host plant. *Ecology* 70, 1028–1039.

Karban, R. (1992) Plant variation: its effects on populations of herbivorous insects. In: Fritz, R.S. and Simms, E.L. (eds) *Plant Resistance to Herbivores and Pathogens: Ecology, Evolution, and Genetics.* University of Chicago Press, Chicago, Illinois, pp. 195–215.

Lynch, M. and Walsh, B. (1998) *Genetics and Analysis of Quantitative Traits.* Sinauer Associates, Sunderland, Massachusetts.

Maddox, G.D. and Root, R.B. (1987) Resistance to sixteen diverse species of herbivorous insects within a population of goldenrod, *Solidago altissima*: genetic variation and heritability. *Oecologia* 72, 8–14.

Maddox, G.D., and Root, R.B. (1990) Structure of the encounter between goldenrod (*Solidago altissima*) and its diverse insect fauna. *Ecology* 71, 2115–2124.

Mather, K. and Jinks, J.L. (1982) *Biometrical Genetics: The Study of Continuous Variation*, 3rd edn. Chapman and Hall, New York.

Mattson, W.J., Haack, R.K. and Birr, B.A. (1996) F_1 hybrid spruces inherit the phytophagous insects of their parents. In: Mattson, W.J., Niemela, P. and Rousi, M. (eds) *Dynamics of Forest Herbivory: Quest for Pattern and Principle. General Technical Report NC-183*. North Central Forest Experiment Station, Forest Service, USDA, St Paul, Minnesota, pp. 142–149.

McCrea, K.D. and Abrahamson, W.G. (1987) Variation in herbivore infestation: historical vs. genetic factors. *Ecology* 68, 822–827.

Messina, F.J., Richards, J.H. and McArthur, E.D. (1996) Variable responses of insects to hybrid versus parental sagebrush in common gardens. *Oecologia* 107, 513–521.

Moon, D.C. and Stiling, P. (2002) The effects of salinity and nutrients on a tritrophic salt-marsh system. *Ecology* 83, 2465–2476.

Moorehead, J.R., Taper, M.L. and Case, T.J. (1993) Utilization of hybrid oak hosts by a monophagous gall wasp: how little host character is sufficient? *Oecologia* 95, 385–392.

Moran, N.A. and Whitham, T.G. (1990) Interspecific competition between root-feeding and leaf-galling aphids mediated by host-plant resistance. *Ecology* 71, 1050–1058.

Morrow, P.A., Whitham, T.G., Potts, B.M., Ladiges, P., Ashton, D.H. and Williams, J.B. (1994) Gall-forming insects concentrate on hybrid phenotypes of *Eucalyptus*. In: Price, P.W., Mattson, W.J. and Baranchikov, Y.N. (eds) *The Ecology and Evolution of Gall-Forming Insects*. North Central Forest Experiment Station, Forest Service, USDA, St Paul, Minnesota, pp. 121–134.

Mosseler, A. and Papadopol, C.S. (1989) Seasonal isolation as a reproductive barrier among sympatric *Salix* species. *Canadian Journal of Botany* 67, 2563–2570.

Neuhauser, C., Andow, D.A., Heipel, G., May, G., Shaw, R. and Wagenius, S. (2003) Community genetics: expanding the synthesis of ecology and genetics. *Ecology* 84, 545–558.

Orians, C.M. and Fritz, R.S. (1995) Secondary chemistry of hybrid and parental willows: phenolic glycosides and condensed tannins in *Salix sericea, S. eriocephala*, and their hybrids. *Journal of Chemical Ecology* 21, 1245–1253.

Orians, C.M. and Fritz, R.S. (1996) Genetic and soil nutrient effects on the abundance of herbivores on willow. *Oecologia* 105, 388–396.

Orians, C.M., Huang, C.H., Wild, A., Dorfman, K.A., Zee, P., Dao, M.T.T. and Fritz, R.S. (1997) Willow hybridization differentially affects preference and performance of herbivorous beetles. *Entomologia Experimentalis et Applicata* 83, 285–294.

Orians, C.M., Lower, S., Fritz, R.S. and Roche, B.M. (2003) The effects of plant genetic variation on secondary chemistry and growth in a shrubby willow, *Salix sericea*: patterns and constraints in the evolution of resistance traits. *Biochemical Systematics and Ecology* 31, 233–247.

Pilson, D. (1996) Two herbivores and constraints on selection for resistance in *Brassica rapa. Evolution* 50, 1492–1500.

Pottinger, R.P. and LeRoux, E.J. (1971) The biology and dynamics of *Lithocolletis blancardella* (Lepidoptera: Gracillariidae) on apple in Quebec. *Memoirs of the Entomological Society of Canada* 77, 1–437.

Preszler, R.W. and Boecklen, W.J. (1994) A three-trophic-level analysis of the effects of plant hybridization on a leaf-mining moth. *Oecologia* 100, 66–73.

Price, P.W., Bouton, C.E., Gross, P., McPheron, B.A., Thompson, J.N. and Weis, A.E. (1980) Interactions among three trophic levels: influence of plants on interactions between insect herbivores and natural enemies. *Annual Review of Ecology and Systematics* 11, 41–65.

Quiring, D.T. and Butterworth, E.W. (1994) Genotype and environment interact to influence acceptability and suitability of white spruce for a specialist herbivore, *Zeiraphera canadensis. Ecological Entomology* 19, 230–238.

Rausher, M.D. (2001) Co-evolution and plant resistance to natural enemies. *Nature* 411, 857–864.

Roche, B.M. and Fritz, R.S. (1997) Genetics of resistance of *Salix sericea* to a diverse community of herbivores. *Evolution* 51, 1490–1498.

Root, R.B. (1973) The organization of a plant–arthropod association in simple and diverse habitats: the fauna of collards, *Brassica oleracea*. *Ecological Monographs* 43, 95–124.

Root, R.B. and Cappuccino, N. (1992) Patterns in population change and the organization of the insect community associated with goldenrod. *Ecological Monographs* 62, 393–420.

SAS Institute (1990) *SAS User's Guide: Statistics, Version 6 Edition.* SAS Institute, Cary, North Carolina.

Sharma, H.C., Venkateswarulu, G. and Sharma, A. (2003) Environmental factors influence the expression of resistance to sorghum midge, *Stenodiplosis sorghicola*. *Euphytica* 130, 365–375.

Siemens, D.H., Ralston, B.E. and Johnson, C.D. (1994) Alternative seed defence mechanisms in a palo verde (Fabaceae) hybrid zone: effects on bruchid beetle abundance. *Ecological Entomology* 19, 381–390.

Stiling, P. and Rossi, A.M. (1996) Complex effects of genotype and environment on insect herbivores and their enemies. *Ecology* 77, 2212–2218.

Thompson, J.N. (1994) *The Coevolutionary Process.* University of Chicago Press, Chicago, Illinois.

Thompson, J.N. (1999) The evolution of species interactions. *Science* 284, 2116–2118.

Van Driesche, R.G. and Taub, G. (1983) Impact of parasitoids on *Phyllonorycter* leafminers infesting apple in Massachusetts, USA. *Protection Ecology* 5, 303–317.

Weis, A.E. and Campbell, D.R. (1992) Plant genotype: a variable factor in insect–plant interactions. In: Hunter, M.D., Ohgushi, T. and Price, P.W. (eds) *Effects of Resource Distribution on Animal–Plant Interactions.* Academic Press, San Diego, California, pp. 75–111.

Whitham, T.G., Morrow, P.A. and Potts, B.M. (1994) Plant hybrid zones as centers of biodiversity: the herbivore community of two endemic Tasmanian eucalypts. *Oecologia* 97, 481–490.

Whitham, T.G., Martinsen, G.D., Floate, K.D., Dungey, H.S., Potts, B.M. and Keim, P. (1999) Plant hybrid zones affect biodiversity: tools for a genetic-based understanding of community structure. *Ecology* 80, 416–428.

Whitham, T.G., Young, W.P., Martinsen, G.D., Gehring, C.A., Schweitzer, J.A., Shuster, S.M., Wimp, G.M., Fischer, D.G., Bailey, J.K., Lindroth, R.L., Woolbright, S. and Kuske, C.R. (2003) Community genetics: a consequence of the extended phenotype. *Ecology* 84, 559–573.

Woodman, R.L. and Price, P.W. (1992) Differential larval predation by ants can influence willow sawfly community structure. *Ecology* 73, 1028–1037.

8

The Role of Parasites of Insect Reproduction in the Diversification of Insect Reproductive Processes

GREGORY D.D. HURST,[1] K. MARY WEBBERLEY[1,2] AND ROBERT KNELL[2]

[1]*Department of Biology, University College London, London, UK;* [2]*School of Biological Sciences, Queen Mary, University of London, London, UK*

1. Introduction

The reproductive processes of insects are remarkably diverse. At a molecular level, genes involved with reproduction show some of the fastest evolutionary rates at both the level of both sequence and expression titre (Civetta and Singh, 1995; Ranz *et al.*, 2003), and networks such as sex determination systems vary between taxa (Schütt and Nöthiger, 2000). At the phenotypic level, genital morphology is highly diversified, as often are rates of mating, secondary sexual characteristics, and the cues used in mate finding and mate choice. This diversification of genotype and phenotype associated with reproduction make it a potential force in reproductive isolation, and thus in higher level processes such as speciation (Arnqvist *et al.*, 2000). It has also led to an intense effort from evolutionary ecologists to discover the factors underlying the diversification of these systems.

Rapid evolution is generally associated with antagonistic coevolution, or, more simply, conflict between parties with different fitness interests in a common process. One obvious conflict that has greatly contributed to diversification of reproductive systems is, of course, the conflict between male and female interests in reproduction (Chapman *et al.*, 2003). This conflict is not the subject of this chapter, and is indeed covered elsewhere in this volume (see Wedell, Chapter 3, this volume). Rather, this chapter concerns another set of conflicts surrounding reproduction, arising from the presence of parasites that either affect the process of reproduction, or are transmitted during it. These clearly have the potential to select for modifiers of reproductive biology that prevent their action (in the case of those affecting the process of reproduction) and prevent their transmission (in the case of parasites transmitted during the process of reproduction).

In this chapter, we examine the potential influence of these parasites on the design of reproductive biology. We deal with three types of parasite. First, there are inherited microorganisms. These are a range of bacteria, viruses and protists that live inside cells and are transmitted from a female to her progeny during reproduction. As well as being intimately involved with testes and ovary tissues, and thus potentially affecting the formation of these tissues, they also commonly manipulate reproduction towards the production of daughters, and this may have a separate impact on the evolution of reproductive processes. We then briefly discuss the potential for parasites that reduce the fecundity of their host to select for changes in reproductive biology, before moving on to the importance of parasites that are transmitted during copulation. In each case, we review the range of parasite–host interactions that exist and assess the incidence of these interactions over insect species. We then examine the potential evolutionary impact of these interactions, and review empirical studies into their impact.

Overall, we argue that the diversification of many aspects of insect (and more widely arthropod) reproduction may be associated with selective pressures deriving from parasites of reproductive processes. The design of oogenesis and spermatogenesis, the number of these sperm used per ejaculate, the number of eggs per clutch, and the way in which male and female are determined, may all partly be a product of selection deriving from parasites that target reproductive processes. There are more conflicts involved with the evolution of reproductive processes than the simple male–female divide.

2. Inherited Microorganisms

2.1 Range of phenotypes and incidence

Many species of insect bear intracellular 'passenger' microorganisms. Because they live within the cytoplasm of host cells, these are passed from a female host to her progeny within the egg. In contrast, male hosts represent a dead end in terms of vertical transmission, as sperm cannot carry these microorganisms.

These passenger microorganisms can be divided into three categories:

1. Microorganisms that are purely vertically transmitted that contribute to host anabolic function.
2. Microorganisms that are purely vertically transmitted and exhibit parasitic 'reproductive manipulation' phenotypes.
3. Microorganisms that are both vertically transmitted through females and horizontally transmitted. These usually do not manipulate reproduction, but may nevertheless be important in the design of reproductive systems if they interfere with processes of oogenesis/spermatogenesis.

We do not deal with the first group in this chapter, as our focus is on parasitic infections.

2.1.1 Microorganisms that are purely vertically transmitted and exhibit parasitic 'reproductive manipulation' phenotypes

In many cases, selection has favoured strains of the microorganism that alter host reproduction with a variety of parasitic manipulations, alike in evolutionary rationale, all being covered by the umbrella term 'reproductive parasites' (see Table 8.1).

The most straightforward of these to understand are the manipulations of host sex ratio. Being transmitted only through females, strains that convert male hosts to female development are favoured. There are two main phenotypes: induction of parthenogenesis and feminization.

Induction of parthenogenesis is already established as a phenotype of *Wolbachia* in haplodiploid species of insects and other arthropods (Huigens and Stouthamer, 2003). Here, the bacterium produces a doubling of chromosome number of unfertilized eggs through modification of early mitosis, changing the

Table 8.1. Major reproductive manipulations induced by inherited bacteria in arthropods, with potential effects on the evolution of host reproductive biology.

Phenotype	Description	Taxa	Known associated microorganisms	Evolutionary impact on reproductive biology
Feminization	Conversion of males to female development	Isopod and amphipod Crustacea; mites	*Wolbachia*, microsporidia, *Cardinium*	Alter behavioural ecology of reproduction; sex determination system evolution
Parthenogenesis induction	Karyotype doubling makes haploid males develop as females in haplodiploid species	Hymenoptera, Thysanoptera, mites	*Wolbachia*, *Cardinium*	Asexuality
'Early' male-killing	Death of male hosts during embryogenesis	Coleoptera, Hymenoptera, Diptera, Hemiptera, Lepidoptera	*Wolbachia*, *Rickettsia*, Flavobacteria, Spiroplasma, Arsenophonus	Behavioural ecology of reproduction; sex determination system evolution
'Late' male-killing	Death of male hosts during 4th instar	Mosquitoes, tortricid moths	microsporidia	
Cytoplasmic incompatibility	Death of zygotes formed from union of sperm from infected fathers and eggs that are uninfected or infected with different strain	Insects, mites, Crustacea	*Wolbachia*, *Cardinium*	Reproductive isolation; development of testes and ovaries to exclude bacteria, polyandry

host from male to female development (Stouthamer and Kazmer, 1994). This phenotype has recently been found to be associated with other bacteria besides *Wolbachia* (Zchori-Fein *et al.*, 2001), and it has also been argued that microorganisms may be responsible for induction of parthenogenesis outside of haplodiploid taxa (Koivisto and Braig, 2003). Persuasive case studies certainly exist in the Collembola (springtails) (Vanderckhove *et al.*, 1999), although Normark (2003) argues that it will be difficult to gain definitive proof of microorganism involvement: without the default of male production in the absence of fertilization, antibiotic treatment is likely to result in sterility in cases of microbe-induced asexuality.

Feminization of hosts has long been known in arthropods other than insects, and is associated with a variety of microorganisms, including *Wolbachia*, *Cardinium*, a member of the Cytophaga–Flexibacter–Bacteroides clade, and a variety of microsporidia (see Bandi *et al.*, 2001, for review; Weeks *et al.*, 2001). Recent work has also suggested the presence of feminizing *Wolbachia* in Lepidoptera (Hiroki *et al.*, 2002).

Aside from these manipulations of primary sex ratio, passenger bacteria in insects most commonly distort the secondary sex ratio, killing male hosts they enter during embryogenesis. Many species of bacteria exhibit this male-killing behaviour, which has evolved on at least seven occasions (Hurst *et al.*, 2003). 'Male-killing' is adaptive for the bacterium where the death of male hosts enhances the reproductive success of sibling females of this host, as then death of the male host, through which the bacterium cannot transmit vertically, enhances the survival of the female sibling. This sibling female bears the same bacterial clone by common descent and can transmit the bacterium on to future generations.

Three main ecological scenarios exist where male host death enhances the survival of sister hosts (Hurst and Majerus, 1993). First, where there is sibling egg consumption. In many coccinellid beetles and lygaeid bugs, the eggs that would have hatched into male larvae are consumed by their sisters. Second, where there are antagonistic interactions between siblings. The intensity of sibling–sibling competition for food or sibling cannibalism is reduced following the death of male hosts, increasing the survival prospects of their sibling female hosts. Finally, male-killing reduces the probability of inbreeding. In host species where inbreeding occurs and is deleterious, females from a male-killed clutch benefit from a reduced rate of copulation with siblings.

These three ecological factors will dictate the incidence of these parasites. They should occur most commonly in species like coccinellid beetles, where there is sibling egg consumption. They should occur, albeit less commonly, in other species where eggs are laid in clutches, by virtue of reduced sibling–sibling competition. They should occur only sporadically in species which lay eggs singly, as here there is just one possible benefit to the bacterium of male-killing: inbreeding avoidance. Current data support these predictions, with a very high incidence (>50% of species infected) in aphidophagous coccinellid beetles (Majerus and Hurst, 1997), moderate incidence in the genus *Acraea* (15% of species infected: Jiggins *et al.*, 2001), with sporadic infection in intensively studied groups such as *Drosophila*. Interestingly, sex determination system does not appear to constrain the presence of male-killers; male-killing bacteria are present in male heterogametic, female heterogametic and haplodiploid species.

In addition to inherited bacteria that kill male hosts during embryogenesis, there are also microsporidial infections that kill male larval hosts (Andreadis, 1985). In this case, the death of males is associated with the dispersal of spores into the environment that are then horizontally transmitted. It has recently been suggested that this 'late male-killing' may be additionally associated with pathogens outside the microsporidia (Hurst *et al.*, 2003).

The final reproductive manipulation conducted by passenger microorganisms is known as cytoplasmic incompatibility, or CI. The most common passenger microorganisms in insects are the *Wolbachia* strains that induce CI. In its simplest form, unidirectional CI in diploid species, crosses between infected males and uninfected females fail due to condensation of the paternal chromosomes during early mitotic events in the new zygote (Breeuwer and Werren, 1990; Callaini *et al.*, 1996). All other cross combinations function, and CI can be described as the paternal chromatin of sperm being marked by the bacterium in males, such that the embryo will die unless it itself is infected with *Wolbachia*. This can be seen as an adaptation in structured populations, where the bacterium in adult males, which cannot transmit the bacterium, is causing the death of uninfected hosts. CI has received much attention because of its ability to induce reproductive isolation between populations bearing different strains of infection. As this is not a direct effect on the evolution of host reproductive processes, it is not covered further here. Interested readers are referred to Hurst and Schilthuizen (1998) and Bordenstein (2003).

Wolbachia strains that induce CI are found in ovaries and testes and are observed in a wide range of insects and other arthropods. Commonly, *Wolbachia* that induce CI reach very high prevalence, being found in over 95% of males and females (Stouthamer *et al.*, 1999). Whilst it is impossible to state precisely how common CI-inducing *Wolbachia* are amongst species, around 20% of species bear the bacterium, and a first guess would be to estimate that the CI phenotype exists in at least half of these cases. The bacterium *Cardinium* has now also been demonstrated to induce CI (Hunter *et al.*, 2003).

2.1.2 Microorganisms showing both vertical and horizontal transmission

In the absence of a direct benefit to infection in female hosts or manipulation of host reproduction, vertically transmitted parasites can only be maintained in a population if they also show some horizontal transmission. A wide variety of vector-borne arboviruses are transmitted horizontally following feeding on an infected host and vertically via the egg (Tesh, 1981), and others are maintained by infectious transmission through the environment in addition to vertical transmission. Intracellular bacteria like *Rickettsia* in fleas can be maintained in a similar way to vector-borne arboviruses (Azad *et al.*, 1992), and *Rickettsia* in Hemiptera may be maintained by joint vertical transmission/horizontal transmission, with horizontal transmission occurring through plant hosts (Davis *et al.*, 1998). Microsporidial infections also commonly show both vertical and horizontal transmission (e.g. Kellen *et al.*, 1965).

In total, the past literature in insect pathology is littered with observations of microbes infecting the ovaries and testes of insects (Afzelius *et al.*, 1989). Some of the bacteria, and a few of the microsporidia, are maintained purely by vertical transmission. A mix of vertical transmission with infectious transmission maintains others. Altogether, the presence of microorganisms in reproductive tissues is highly pervasive.

2.2 Evolutionary impact of passenger microorganisms on reproductive processes

There are five main areas in which passenger microorganisms have influenced the evolution of reproductive processes.

1. They may have been influential in the design of testes and ovary formation.
2. They may dictate the importance of sexual versus asexual reproduction within the population, and may indeed be critical in the transition to obligate asexuality in a variety of insects.
3. They may alter patterns of sexual selection.
4. They may be important in the evolution of polyandry.
5. They may have been influential in the diversification of sex determination systems.

2.2.1 Passenger microorganisms and the design of spermatogenesis and oogenesis

Inherited microorganisms in female hosts need to be within the egg in order to ensure transmission to the next generation. In the case of *Wolbachia* inducing cytoplasmic incompatibility, they must also be in the testes in males to 'mark' the sperm. If their presence in any way disrupts the smooth development of ovary and testes, then adaptation to accommodate them may occur. It is notable that *Wolbachia* that induce CI may be at very high prevalence in both male and females, and thus be a potent coevolutionary force in this regard. It is also possible that adaptation of host reproductive processes to exclude them may also occur.

Involvement of *Wolbachia* with the differentiation of ovaries has now been demonstrated in two cases. First, in the parasitoid wasp *Asobara tabida*, one particular *Wolbachia* infection was found to be necessary for differentiation of ovaries; antibiotic treatment induced female sterility (Dedeine *et al.*, 2001). Notably, other strains of *Wolbachia* that infect this species are not required, and cannot rescue the sterility phenotype. Second, in *Drosophila melanogaster*, mutations in the gene *Sxl* (*Sex lethal*) that produce female sterility through failure of the ovaries to differentiate were rescued by the presence of *Wolbachia* (Starr and Cline, 2001).

It is not clear why these processes happen. One possibility is the ovaries become designed to accommodate the bacterium, and removal of the bacterium results in failure of the ovaries to form properly. This may be a passive structural accommodation, or a molecular accommodation. One could imagine that on original symbiosis, both *Wolbachia* and the host produced some factor that was important in oogenesis, and selection on the host acted to reduce this degeneracy/superfluity by stopping production itself. In the end, this makes oogenesis non-functional without the bacterium (Dedeine *et al.*, 2003). The alternative is that failure of oogenesis in uninfected embryos is a *Wolbachia* strategy, akin to medea in *Tribolium*. The *Wolbachia* places some factor in eggs that prevents oogenesis of the resulting progeny unless it is infected, thus ensuring its representation in the next generation.

A positive accommodation of *Wolbachia* into testis formation and spermatogenesis is less clear. It should be noted that whilst infection of a female may have low cost, infection of male testes may bear higher costs, leading to selection to exclude the bacteria. Nevertheless, *Wolbachia* can persist at high prevalence in males, infecting the testes, and so selection to accommodate their presence is possible. No direct evidence of this exists to date. However, there are two reports of *Wolbachia* improving sperm production rates/ability in sperm competition (Wade and Chang, 1995; Hariri *et al.*, 1998). Explanation of these observations has rather eluded scientists interested in *Wolbachia* in the past, but parallels with oogenesis suggest selection to accommodate infection (and thus loss of some host function) may be a partial explanation.

Selection to accommodate or control passenger microorganisms and their hosts within gonadal tissue might be manifest in microorganism-associated F_1 hybrid sterility (Thompson, 1987). Co-adaptation between passenger microorganisms and their host would break down following hybridization with a population that did not bear the passenger microbe, leading to failure to develop ovaries. The observations of Lee Ehrman on *Drosophila paulistorum* are interesting in this regard: males that are hybrid between geographic races are frequently sterile, due to the growth of bacteria in the testes, with concomitant breakdown of spermatogenesis (Ehrman *et al.*, 1990).

As well as selection to accommodate passenger microorganisms within oogenesis/spermatogenesis, there may also be selection to prevent their access to the germline. In the case of sex-ratio-distorting microorganisms, selection favours modifiers that prevent the microbe entering the ovary, as this prevents the progeny suffering the parasitic phenotype. In the case of *Wolbachia* inducing cytoplasmic incompatibility, selection may favour modifiers that prevent access of the *Wolbachia* to testes. This is for two reasons. First, it may prevent the cost of incompatibility. More importantly, it may prevent the cost to sperm production of having testes densely packed with *Wolbachia*. Snook *et al.* (2000) observed that *D. simulans* males infected with *Wolbachia* strain *wRi* had a lower rate of spermatogenesis, and fertilized fewer eggs per ejaculate.

Empirical studies support, but do not prove, the idea that spermatogenesis and oogenesis have been designed to exclude symbionts. In terms of the access of symbionts to ovaries, there are case studies where there is evidence of host genetic variation for resistance to the transmission of male-killing bacteria in some, but not all, systems studied. The most notable case is in *Drosophila prosaltans*, where a single recessive nuclear gene was associated with refractoriness to the transmission of the male-killer (Cavalcanti *et al.*, 1957).

Whether this occurs at the point of entry of the bacterium to the reproductive tissues (i.e. an alteration in reproductive biology) or at other points in development is unclear. However, the potential is obvious.

In the case of spermatogenesis, it is notable that *Wolbachia* density in sperm cysts varies between species (Veneti *et al.*, 2003). Comparisons of density in transinfected and natural hosts clearly indicate that differences in density are at least partly associated with host genetic differentiation. One hypothesis for this differentiation is selection on the host to prevent infection of the testes in order to safeguard sperm production. These differences would translate into differentiation of the processes underlying testes formation and function. Close examination of the host genetic differences that produce the variation in infection density between *D. melanogaster* and *D. simulans* testes would seem very timely.

2.2.2 Altering the importance of sex as a mode of reproduction

Microbes that induce asexual reproduction clearly have a fundamental impact on the reproductive biology of their host, making sex less important. In many cases, strains of *Wolbachia* that induce parthenogenesis have become fixed in their host population, converting the host from sexual reproduction to asexual (see Huigens and Stouthamer, 2003, for review). This process can be irreversible, as experiments in which asexual populations have been cured of *Wolbachia* have often resulted in the production of non-functional males, with females that are no longer receptive to sex in any case (e.g. Gottleib and Zchori-Fein, 2001). Sexual function has decayed by mutation and, one would expect, selection against aspects of phenotype that are costly in the absence of sex.

It has recently been conjectured that parthenogenesis-inducing *Wolbachia* in haplodiploid species may also select for modifiers of the female host that promote virginity even in the continued presence of males (Huigens and Stouthamer, 2003). The argument runs that sons are more valuable than daughters in these populations, and sons are most likely to be produced if the female is not fertilized. Thus, the sex ratio bias should select for females that resist mating.

2.2.3 Changing the dynamics of male–female sexual conflict through alteration of the population sex ratio

The direction and intensity of sexual conflict is dictated by the relative levels of parental care exhibited by male and females, and the population sex ratio. Microorganisms that bias the sex ratio of the individual that harbours them, such as feminizers and male-killers, have an effect on the population sex ratio of the host, and thus alter this balance.

The degree to which these microorganisms can alter this balance of course depends on the prevalence of the microorganism. The prevalence of male-killers is sometimes low (1–5% of females infected), more often modest (20–50% of females infected), but sometimes very high (>95% of females infected) (Hurst *et al.*, 2003). Effects on patterns of mate competition and sperm competition are expected with a 50% prevalence (gives rise to a 2 females per male population sex ratio), and gross mating system alteration may occur with higher prevalence levels (80–99%, corresponding to a 4:1 to 100:1 population sex ratio).

The example of *Acraea encedon* is a case in point of a host whose reproductive processes have been fundamentally altered by the presence of male-killing bacteria. In Ugandan populations of this species, the population sex ratio bias (14:1) associated with the high-prevalence male-killer infection has led to sex role reversal (Jiggins *et al.*, 2000). Virgin females of this species congregate on hill tops, waiting for males with which to mate.

It is clear that as well as overt mating system differences, these parasites are likely to produce more covert changes in the reproductive biology of the species concerned. In particular, we would expect male ejaculate constitution to change with increasing population sex ratio bias, and corresponding changes in the ability of males to mate multiply. Preliminary evidence for this comes from studies of the nymphalid butterfly *Hypolimnas bolina* (Dyson and Hurst, 2004) (see Fig. 8.1). This species inhabits the island and mainland areas of the Pacific, and the prevalence of a male-killing *Wolbachia* varies between islands. In one island, Independent Samoa, prevalence is extreme (99% of females are infected), and the population sex ratio is 100 females per male. On this island, unmatedness of females is common, indicating that male access to females has become unlimited. The size of spermatophore delivered by males during copulation in Independent Samoa is less than 50% of that found on neighbouring islands with more moderate prevalence. This is consistent with consideration of adaptive male behaviour in the light of male-killer-induced sex ratio changes. It will be informative to ascertain whether the association holds when a wider array of islands with different male-killer prevalence is investigated.

Fig. 8.1. Female *Hypolimnas bolina*. This butterfly is host to a male-killing *Wolbachia* strain. In the population of Independent Samoa, 99% of females are infected, producing a population sex ratio of 100 females per male. This results in high levels of virginity in the field. The consequences of varying male-killer prevalence on host reproductive ecology are currently being investigated.

A second study indicating the capacity of parasite-induced population sex ratio biases to select for change in reproductive biology comes from study of isopod Crustacea. Moreau and Rigaud (2003) measured the ability of males to inseminate females across seven species of woodlice, five of which were naturally infected with a feminizing *Wolbachia*, and two with CI *Wolbachia*. The five feminizer-infected species were homogeneous for mating rate, and this was nearly double that found in the two species that did not bear the feminizer. Whilst not a formal comparative analysis, this result is clearly consistent with the intuitive notion that sex ratio distorters, via the alteration in population sex ratio they produce, select for changes in reproductive tactics.

2.2.4 The evolution of polyandry

There is much debate on the causes of multiple mating (polyandry) in females. Whilst polyandry was historically viewed as something to which males subject females, it is increasingly recognized that females do have control over mating rate. Thus, adaptive explanations of polyandry from the female stance are required.

The concept of genetic incompatibility is one such explanation. Certain sperm–egg combinations result in weak or inviable progeny. In mating multiply, females may either allow selection of compatible sperm or reduce the variance in the rate at with which incompatible matings occur. The benefit of reducing variance in the rate of incompatibility depends on the existence of compensation for the death of inviable progeny, i.e. death of one progeny is associated with increased survival of the remaining individuals. If we imagine that death of one progeny lowers the competitive burden on the remaining individuals, the death of progeny in the polyandrous female from incompatibility is compensated for in increased survival of the remaining progeny. This compensation cannot occur if she mates once, with all siblings dying or none. Thus, genetic incompatibility can select for polyandry without selection of compatible sperm.

Zeh and Zeh (1996, 1997) have proposed that *Wolbachia* inducing CI may be a force producing incompatibility that may favour the evolution of polyandry. Male-killing bacteria could also produce a similar effect, in the case where there are nuclear genes preventing the action of the male-killer. It is currently unclear the extent to which passenger microorganisms have selected for the observed levels of polyandry in nature. Whilst it is clear that they provide some selective pressure for increased polyandry, the quantity of this selective pressure is unclear.

2.2.5 Driving the evolution of sex-determining mechanisms

Sex-determination systems, the developmental processes that underlie the differentiation of male and female, are remarkably variable. This variability is paradoxical when it is considered that the core problem, making male and female, remains constant between taxa. All sexual insects make male and female, but they do so with a variety of systems controlling sex determination. *Sxl*, the major switch gene involved with sex determination in drosophilids, is not involved with sex determination in other flies (Schütt and Nöthiger, 2000). *msl-3*, a gene involved with dosage compensation in *Drosophila*, is present in other fly species, but it is not involved with dosage compensation (Ruiz *et al.*, 2000). At a grosser level, we can observe male heterogametic, female heterogametic and haplodiploid systems of sex determination in insects. Some groups, such as the bark beetles, appear to show transitions on a very regular basis.

This paradox has been subject of much debate, and passenger micro-organisms have been proposed as one factor driving the changes. Hamilton (1993) speculates that haplodiploidy may arise as a consequence of conflict between bacteria that feminize or kill males by attacking or eliminating the Y chromosome. Hamilton conjectures that this selects for the movement of the male determining factor from the Y chromosome to a previous autosome as a mechanism of preventing the parasite from cueing in on sex, with the old Y chromosome no longer determining sex. As the symbiont cues in on the new chromosome, so the sex-determining locus moves, until males are completely haploid, and females diploid, and sex is determined by chromosome dosage (which the bacterium cannot detect before fertilization, at least). This theory has been extended to suggest male-killing bacteria as a driving force behind the evolution of this sexual system (Normark, 2004).

More widely, Werren and Beukeboom (1998) and Hurst and Werren (2001) suggest that changes in the structure of sex determination may be a result of reproductive conflicts. They point to the alteration in sex-determination system in the pill woodlouse, *Armadillidium vulgare*, which have been driven by the feminizing *Wolbachia* bacterium. Feminization first results in the fixation of the Z chromosome in this female heterogametic species, followed by selection for repression of the transmission of the feminizing factor, such that sex is now determined in some populations by presence/absence of *Wolbachia* combined with nuclear genes affecting *Wolbachia* transmission (Rigaud, 1997).

The applicability of this study to insects requires urgent resolution. One thing is certain. Male-killing bacteria and feminizers can produce very strong selection on the host to avoid the action and transmission of male-killing bacteria, and modifiers in the host sex-determination system that prevented recognition by the parasite would spread. What is uncertain is the feasibility of mutation to these modifiers: can a modified system still produce a functional male? Our feeling is that a coevolutionary process, in which mild changes in host sex-determining molecules is followed by 'catch-up' by the parasite, can result in shifts in the protein sequence or titre of sex-determining genes, and these may produce compensatory selection for changes in the structure of sex determination. More studies of the genetic basis of resistance in natural populations and model organisms are required.

2.2.6 Other changes that may be associated with inherited parasites

It has also been speculated that the presence of male-killing bacteria may select on host life history. Theory suggests selection on clutch size (Hurst and McVean, 1998). Male-killing bacteria, it is argued, alter the intensity of sibling–sibling competition, and thus may select for increases in clutch size. It is uncertain how far this theory translates into changes in the field. Some authors have discounted it on discovering male-killers in host species with relatively small clutch sizes (Majerus and Majerus, 2000). However, the required evidence to evaluate the hypothesis has not been obtained, and we await formal comparative analysis between clutch size in populations/species with male-killers and those without.

Beyond clutch size, extreme population sex ratio bias associated with sex-ratio-distorting microorganisms may select for alteration in female life history. If sperm limits egg production, selection on female life history is expected to favour reduction in development time and adult body size in favour of speed of development.

3. Parasite-induced Reduction in Host Fecundity

3.1 Range and incidence

Reduction in the fertility/fecundity of male and female hosts is a relatively common phenotype associated with parasite infection (Hurd, 2001). Reduction in fecundity associated with parasitism may be either an adaptive phenotype from the point of view of the parasite, or of the host, or it may simply be that reproduction is affected by hazard.

These types of rationale are very hard to disentangle. For instance, sexually transmitted pathogens and parasites are a group known particularly to affect reproductive processes, and this is thought to be adaptive (Lockhart *et al.*, 1996). Sexually transmitted pathogens 'require' the host to copulate to transmit infection. In diverting resources from production of viable eggs, rather than maintenance of the soma, the parasite reduces interference with processes that the host requires to stay alive. Whilst a non-reproducing host may still copulate and transmit infection, any resources removed from the soma may reduce longevity, and thus the period over which transmission occurs. Indeed, it has been argued that induction of sterility in female mammals may directly increase copulation rate, and therefore parasite transmission, by preventing pregnancy. However, it is notable that sexually transmitted infections are often (though not exclusively) diseases of reproductive systems. It is thus logical that symptoms of the disease should be most commonly expressed in these tissues as a non-adaptive by-product (blockage of host genitals by sexually transmitted nematode infections certainly appears to be a non-adaptive effect of infection). Thus, by-product and adaptation are hard to disentangle.

In some cases of parasitism-induced fecundity reduction, it is possible to rule out a by-product explanation. For instance, in the case of rat tapeworm (*Hymnolepis diminuta*) infecting the mealworm beetle (*Tenebrio mollitor*), infection reduces male and female host fecundity but increases host longevity (Hurd *et al.*, 2001). Here, relatively small parasite biomass levels have pronounced effects on fecundity, associated with secretion of a 10–50 kDa factor by the cestode that inhibits vitellogenin synthesis (Webb and Hurd, 1999). The observation that this effect is not associated with biomass makes it unlikely to be a by-product. It is still unclear whether the effect is adaptive for host or parasite (or both).

There are, however, situations where the case for parasite-adaptive manipulation is compelling. These centre on cases of parasitic castration, where reproductive function is entirely destroyed. Here, host adaptation is a very unlikely explanation for the data. However, a by-product explanation does need to be ruled out, or an adaptive benefit clearly shown.

An example where a by-product explanation is unlikely is the case of female *Aphidius ervi* that parasitize *Acyrthosiphon pisum*. Female wasps secrete peptides in the venom accompanying the egg that induce cellular alteration, and ultimately degeneration, of the apical germaria of the ovarioles, resulting in cessation of oogenesis (Digilio *et al.*, 2000). Given that these peptides alone are sufficient for castration, it is very unlikely to be a by-product of possession of an energetically costly parasite.

An example of a clear adaptive benefit is Vance's (1996) study of mayflies infected with mermethid worms. Infection causes feminization of male hosts such that they appear externally as intersexes. Most pertinently, they behave as unparasitized females in terms of entering into upstream dispersal and oviposition behaviour. Given that the parasite completes its development in the river bed, this conversion of host to oviposition behaviour (which will return the parasite to the river) rather than the male behaviour of swarming (which takes place at river banks, and may not result in return of the parasite to the water) is adaptive.

3.2 Evolutionary impact

The evolutionary impact of parasite manipulation of fecundity is simply unknown at the moment. In terms of potential effect, it is notable that some of the parasites can reach very high prevalence (sexually transmitted infections are notable in this regard). Combined with a strong phenotype (sterility), selection pressure for modifiers that prevent/ameliorate effects could be extensive.

Whilst there are no empirical studies, selection may act in two ways: first, to alter the signalling processes or physiological processes involved with oogenesis/spermatogenesis to prevent the parasite from effecting its manipulation; second, to reproduce rapidly following infection when infection is followed by inevitable progressive loss of fecundity. There are no empirical tests of either of these ideas in insects, although there is some evidence for the latter in cases of sexually transmitted infection (see next section).

4. Sexually Transmitted Infections

4.1 Range and incidence

Parasites and pathogens that are transmitted during host copulation are a well-known scourge of humans, but their impact on insect populations has only recently been recognized (a full list of known associations can be found in Knell and Webberley, 2004). We can categorize parasites that are transmitted during host copulation into three kinds:

- *Sexual transmission is both necessary and sufficient for maintenance of the pathogen or the parasite.* These are the true sexually transmitted infections. Many of the haematophagous podapolipid mites of insects come into this category, as do several ectoparasitic nematode and fungal infections of insects. The interaction between the mite *Coccipolipus hippodamiae* and its host can be regarded as a canonical sexually transmitted infection, or STI (Fig. 8.2). Simulation studies have shown that sexual transmission is necessary and sufficient to explain the observed dynamics of the parasite, and experiments have found that transmission outside of host copulation in nature is rare (K.M. Webberley and G.D.D. Hurst, personal observation; Webberley and Hurst, 2002). To date, all known 'canonical' STIs in insects are ectoparastic mites, fungi and nematodes.
- *Sexual transmission of the parasite is necessary for parasite maintenance, but not in itself sufficient for it.* Some of the arboviruses, such as the La Crosse virus which infects *Aedes triseriatus*, have been suggested as being in this category, with vertical transmission being required for overwinter

Fig. 8.2. A two-spot ladybird (*Adalia bipunctata*) infected with two sexually transmitted infections. The yellow hyphae on the dorsal surface of the top elytron are *Hesperomyces virescens*, and the elytron has melanized in response to infection. The underside of the bottom elytron bears *Coccipolipus hippodamiae*, an ectoparasitic mite. The infection is visible through the elytron, as is typical in late-stage infections. A cluster of adult mites with the eggs they have produced is visible just anterior to the spot on the elytron; motile larval mites are just visible near the posterior edge of the elytron. Neither of these parasites can reproduce away from the host, and the vast majority of transmission occurs during copulation. This species is also host to three different male-killing bacteria, making it a veritable hotspot for reproductive parasites.

survival of the infection (many of the host species overwinter as eggs), and venereal transmission from male to female through accessory fluid (Thompson and Beaty, 1977, 1978) (*per os* transmission is also known, but is thought to be insufficient for maintenance). Another example in this category is the ectoparasitic mite *Kennethiella trisetosa*, which parasitizes the eumenid wasp *Ancistrocerus antilope*. Mites pass from male host to female during host copulation, as well as vertically, mites dismounting from a female during oviposition and entering the nest into which she has laid an egg (Cowan, 1984).

- *Sexual transmission of the parasite/pathogen occurs, but sexual transmission is not necessary for their maintenance*: transmission outside of copulation is sufficient for this in itself. This type of parasite is well known in birds and mammals (and the parasite may in fact be an insect, such as a louse), but is probably less common in insects. The podapolipid mites of gregarious species such as locusts may be an example (Volkonsky, 1946), as will be the majority of the vector-borne arboviruses, where transmission of the pathogen between insects through mammalian hosts is the principal transmission mode, but low levels of sexual transmission are observed to occur. Some cases of microsporidian infections also fall into this category, showing some transmission during host copulation, but where this route is likely to be a minority of new infections (e.g. Armstrong, 1977).

It is also notable that sexual transmission may become relatively important for many classical 'ordinary' parasites whose host species decrease in density, and for classical parasites that shift to new hosts that are at low density. In hosts that live at low density, copulation may become the most frequent contact event between host individuals, and thus become a major transmission route for a 'classical' contact-transmitted parasite, without any evolutionary change in the parasite. Indeed, reduction in host density may alter an STI from being one in which sexual transmission occurs but is not necessary for maintenance to one in which it is necessary. Low host density may also be important in the evolution of canonical STDs, as selection may then favour more efficient transmission during copulation, even if this reduces non-sexual transmission. In the end, sexual transmission may become both necessary and sufficient for maintenance of the parasite (Thrall *et al.*, 1998).

Canonical sexually transmitted infections are more restricted in their incidence amongst insects than other categories. One restriction is straight-forward to understand: to be maintained purely by sexual transmission, the host must be sufficiently promiscuous to allow spread, and sufficiently long-lived in the adult form to allow the infection to develop from initial infection to becoming infectious. Species that have relatively short adult lives and species that are not very promiscuous clearly cannot maintain an infection that is purely sexually transmitted. The second restriction is that the host species must show reproductive continuity between generations. If the host species exists in cohorts that do not interbreed, sexual transmission can never be sufficient for maintenance, and must be combined with either vertical transmission, or some other form of horizontal transmission to span the gap between generations.

Cohort structures are a feature of many temperate insect species, but also can occur in the tropics associated with 'generation cycles'. Geographical variation in the incidence and prevalence of STIs in ladybirds is likely to be associated with variation in host phenology (Welch *et al.*, 2001).

4.2 Evolutionary impact

The selective importance of sexually transmitted parasites on the evolution of host reproductive biology depends on the prevalence of infection, the frequency of transmission during copulation, and the severity of the disease symptoms. For some parasites, these factors are high, and one can therefore conclude that selection to avoid infection will be strong. In studies of coccinellid beetles, prevalence levels in excess of 90% of adults being infected are commonly recorded (K.M. Webberley and G.D.D. Hurst, personal observation), the parasite is transmitted in over 90% of copulations between infected and uninfected hosts (Webberley *et al.*, 2004), infection causes sterility of female hosts within 2–3 weeks of infection (Hurst *et al.*, 1995; Webberley *et al.*, 2004), and decreases male overwinter survival (Webberley and Hurst, 2002). Other canonical STIs may reach high prevalence, and be transmitted with high efficiency, but produce little selection pressure by virtue of having little impact in terms of disease severity. Laboulbeniales fungal infections of insects are in this category: these are common canonical STIs that are believed to have little impact on their host (Weir and Beakes, 1995). We can therefore conclude that selection to avoid or ameliorate sexually transmitted infections is sometimes strong in insects, and sometimes not.

The most obvious potential evolutionary impacts of sexually transmitted infections on the evolution of reproductive biology are in the areas of mate choice, promiscuity, and sexual conflict.

4.2.1 Mate choice

The involvement of parasites in the evolution of mate choice has been debated now over several years. In the first place, Hamilton and Zuk (1982) suggested that the coevolutionary arms race between parasites and their hosts would maintain genetic variation for fitness in host populations, and that female choice was occurring for indicators of parasite resistance. This idea was later expanded into the general notion of immunocompetence, where females picked males that displayed traits indicative of the ability to ward off infection. It was also noted that host copulation is a time when many parasites are transmitted, and it was suggested that females were choosing males that were parasite-free to obtain the direct benefit of reduced parasite exposure, rather than the indirect benefit of good genes (Borgia and Collis, 1989).

Studies on the role of STIs in the evolution of insect mate choice have produced surprisingly negative results, with no evidence to date for a preference for STI-free partners. In one case, the sexually transmitted mite *Unionicola ypsilophora* that passes from male to female chironomid midges, infected males were in fact overrepresented in mating pairs in the field (McLachlan, 1999). More commonly, no effect of parasitism on mating success is observed. In the *C. hippodamiae–A. bipunctata* system, neither infected females nor infected males were underrepresented in mating pairs collected from the field, and there was no evidence of any reduction in the ability to gain matings when infected males were presented to females in the laboratory (Webberley *et al.*, 2002). This observation contrasts with the background of a female host that does show rejection behaviour when unwilling to mate (e.g. because she has been starved), and a parasite–host interaction where infection is very common and very severe. The system is one in which the conditions are seemingly perfect for the evolution of mate choice, but no choice is observed. In a similar system, Abbot and Dill (2001) found no evidence of preference for uninfected partners in *Labidomera clivicollis* beetles, the host of the sexually transmitted mite *Chrysomelobia labidomera*.

Whilst the lack of discrimination by males can be understood easily in terms of a low relative cost of infection compared to the benefits of multiple mating, the lack of female discrimination between infected and uninfected male partners is surprising, as here there is a large cost to infection compared with a relatively small potential benefit to promiscuity.

Knell (1999) has suggested that the idea that discrimination evolves easily comes from a host-biased perspective. He argues that lack of discrimination is expected because of strong selection on sexually transmitted parasites to be cryptic. He noted that if a parasite is sexually transmitted, it pays to conceal presence, as this maximizes the rate of new infection opportunities. This may be a partial solution to the problem of lack of mate choice, but some questions remain. First, just as it is wrong to decide that the host is all powerful in a coevolutionary game, it may be equally wrong to ascribe full power to the parasite: it does pay the hosts to discriminate if they can, and they may sometimes come out on top in the arms-race. Second, the effect is strongest for the cases of single 'pure' infections within a host individual: superinfection will occur if prevalence is very high and will reduce within host relatedness of the parasite, which may therefore increase the benefits of virulence, with reduced selection for concealment. Third, there may be indicators of infection risk that the parasite cannot remove. Age, for instance, is an excellent 'uncheatable' indicator of infection risk. The probability of a partner being infectious for an STI increases with age for two reasons: greater exposure to the parasite with every subsequent mating, and greater time for the host to become infectious following initial infection. A partner may not be infectious if they are not old enough for the latent period of the infection to pass, and are less likely to be infected when young due to decreased exposure to the STI. Thus, while the parasite may hide, there are certain 'risk factors' that hosts could detect that are beyond parasite control. There may be a certain beauty in youth that is evolutionarily relevant.

An alternative view is that females sometimes simply do not have a strong power to choose and males have a strong desire to mate, making the evolution of mate discrimination difficult. This could be especially true for ectoparasitic infections of beetles detailed above. The initial phases of interaction between male and female ladybirds (and many other beetles) involve the male approaching the female from behind and attempting to mate. During this phase, the female has very little in the way of 'cues' or 'choice' options. These only exist when the male has started trying to mount the female. It is also notable that transfer of the parasite between partners may occur without a successful copulation taking place. For the case of sexually transmitted ectoparasitic fungi, rather short contact is probably sufficient for transmission. For the case of ectoparasitic mites, contact sometimes may not need to be very long before transmission occurs if infection intensity is high (highly infected beetles can be seen to have larval mites crawling over their exposed surfaces). If this is the case, discrimination may be useless, as it will take place after parasite transmission. Thus, whilst in the field of mate choice generally one may accept that a female can exert control over the fertilization process, it is not necessarily the case that the female can discriminate between infected and uninfected mates in time to stop the mate transmitting a parasite.

We are thus left with a somewhat unsatisfactory resolution of the importance of discrimination against sexually transmitted pathogens and parasites in the evolution of mate choice. We have three empirical studies, none of which provides any evidence for the intuitively tempting proposition that these parasites should drive mate choice evolution. It may be that the proposition is naïve, as suggested by Knell (1999), by virtue of selection on the parasite. Alternatively, the validity of the proposition may depend on the timing of parasite transmission. The case studies we have are biologically very similar, and are of interactions where transmission of the parasite without intromission is possible (i.e. female power is reduced). It would be very informative to know whether the same lack of choice of uninfected partners were also true in systems with more 'female power', where transmission occurred only when intromission occurred. Studies of nematodes and mites than inhabit the genital tract would be timely. It would also be very worthwhile investigating whether selection does produce a choice for younger mates in systems with STIs.

Finally, it should be noted that the role of sexually transmitted parasites in the evolution of mate choice may go beyond selection for simple discrimination between infected and uninfected partners, but influence the benefits of female choice more generally. Graves and Duvall (1995) noted that in systems where there is no discrimination between mates on the basis of infection status *per se*, STIs placed a cost on mate choice based on other characters. Preferred mates would have higher prior mating rates, and thus a higher probability of transmitting an infection. In general, the advantages of choice in systems with an STD will be frequency-dependent, and models suggest that STDs are unlikely to eradicate mate choice (Boots and Knell, 2002; Kokko et al., 2002). This prediction awaits data, and insects would appear to be an excellent group in which to perform comparative studies into this question. Are insect systems where female choice is known those in which there are fewer STIs?

4.2.2 Mating rate

The second sphere in which sexual transmission parasites have been suggested to impact on host behaviour is in the evolution of mating rate. In systems with an STI, it is suggested, selection should favour modifiers that reduce mating rate. However, it is very unlikely to select for monogamy. As for the case of mate choice, the advantage gained by less promiscuous individuals is intrinsically frequency-dependent (Thrall *et al.*, 1997, 2000; Boots and Knell, 2002). If there are benefits to promiscuity to either sex, then we would expect either intermediate levels of promiscuity across the population, or a polymorphism with some individuals remaining promiscuous, and others less so.

Knell and Webberley (2004) note that testing for the effect of STIs on promiscuity levels is very difficult, because STI prevalence and promiscuity are intrinsically linked epidemiologically: very promiscuous species are likely to have higher parasite prevalence. This association between promiscuity and prevalence is observed in the field (Webberley *et al.*, 2004), and is likely to be a strong effect that confounds the effect of selection in any comparative analysis. Knell and Webberley (2004) suggest that a selection experiment approach may be the best method to detect whether or not these parasites can select for reduced promiscuity in their hosts.

In a final twist, it is notable that host promiscuity is, of course, beneficial to the parasite, and thus promiscuity increases could potentially occur due to parasite manipulation of the host. In mammalian systems, parasite-induced sterility will of course increase female mating rate (a female will stop mating when she falls pregnant). In insects, a similar effect is possible, if the parasite can divert resources from fertilization and fecundity to increased promiscuity. Definitive experiments need to be undertaken to examine whether this occurs. Knell and Webberley (2004) note some evidence consistent with such effects, although we are a long way from establishing proof.

4.2.3 Sexual conflict

Sexually transmitted infections may also affect the design of reproductive behaviour by altering the intensity of sexual conflict. It is even possible that males use them as a tool to achieve their 'desiderata' in male–female reproductive conflicts.

In terms of STI effects on the intensity of sexual conflict, it is notable that the severity of STI symptoms is often asymmetrical by host sex. In many cases, the effects of infection are more pronounced in females than in males. In this sense, STIs alter the intensity of conflict over mating decisions. In the case of *A. bipunctata*, for instance, sterility is an inevitable consequence of becoming infected if female (Hurst *et al.*, 1995). Thus, while STI presence may produce selection on females to lower mating rate and choose partners, it does not necessarily act on male mating rate, exacerbating any conflicts that already exist.

Indeed, it is possible that STIs may be used as a 'tool' in the sexual conflict between males and females. One adaptive response of a female to becoming infected with a parasite that will reduce future reproduction opportunities is to increase effort into short-term reproduction following parasitism. In a system with multiple mating and sperm competition, this change in behaviour is adaptive with respect to the male interests, and thus the STI can become part of his 'armoury' to induce female oviposition. Indeed, it could be a more powerful force than male accessory proteins that induce oviposition in the female partner. In the case of STI-induced increases in oviposition rate, it is selection on the female that produces the effect, and the male vicariously benefits. In the case of peptides in male accessory fluid that induce oviposition, the female may evolve resistance to any effect, as these are in fact deleterious to the female.

The likelihood of this effect occurring depends crucially on whether females do respond to parasite infection in this manner. This type of response is well known in other cases of parasitization, and Knell and Webberley (2004) note two case studies where there is evidence for increased oviposition following infection with sexually transmitted parasites. First, female *Spodoptera frugiperda* produce more eggs in the short term following infection with the nematode *Noctuidonema guyanese* during copulation (Simmons and Rogers, 1994). Second, infection of females of the earwig *Labidura riparia* with the fungus *Filariomyces forficulae* is associated with more eggs being laid after mating compared with matings where fungus is not transferred (Strandberg and Tucker, 1974). What remains to be seen is the degree to which this interacts with the sexual conflict that occurs in the system, and whether males do ever 'want' to be infected and to infect their partner.

5. Summary

The reproductive biology of insects is fascinating in its diversity. This diversity lies at many levels, from rapid evolution of individual genes, to evolution of developmental systems underlying sex determination, through to differentiation in phenotype (e.g. secondary sexual ornaments and genital structures) and behaviour (tendency to mate multiply, mate preferences). Sexual conflict is one factor that underlies this diversification. Hand in hand with this, we would argue, are the selection pressures produced by reproductive parasites. Many species of insects are affected by these parasites, particularly the passenger microorganisms, and they may affect the design of reproductive systems at many levels. Further study will rely on examining mechanisms in detail (no doubt with the aid of genomics), but will also require continuing work by evolutionary ecologists to reconcile mechanism with diversification.

Acknowledgements

We acknowledge the helpful comments of an anonymous reviewer. We thank the Biotechnology and Biological Sciences Research Council (Grant 31/S15317), The Natural Environment Research Council (Grant GR3/11818) and the Wellcome Trust (Grant 066273/Z/01/Z) for their support of this work.

References

Abbot, P. and Dill, L.M. (2001) Sexually transmitted parasites and sexual selection in the milkweed leaf beetle, *Labidomera clivicollis*. *Oikos* 92, 91–100.

Afzelius, B.A., Alberti, G., Dallai, R., Godula, J. and Witalinski, W. (1989) Virus and *Rickettsia* infected sperm cells in arthropods. *Journal of Invertebrate Pathology* 53, 365–377.

Andreadis, T.G. (1985) Life cycle and epizootiology and horizontal transmission of *Amblyospora* (Microspora: Amblyosporidae) in a univoltine mosquito. *Journal of Invertebrate Pathology* 46, 31–46.

Armstrong, E. (1977) Transmission of *Nosema kingi* to offspring of *Drosophila willistoni* during copulation. *Zeitschrift fur Parasitenkunde* 53, 311–315.

Arnqvist, G., Edvardsson, M., Friberg, U. and Nilsson, T. (2000) Sexual conflict promotes speciation in insects. *Proceedings of the National Academy of Sciences USA* 97, 10460–10464.

Azad, A.F., Sacci, J.B.J., Nelson, W.M., Dasch, G.A., Schmidtmann, E.T. and Carl, M. (1992) Genetic characterization and transovariol transmission of a typhus-like *Rickettsia* found in cat-fleas. *Proceedings of the National Academy of Sciences USA* 89, 43–46.

Bandi, C., Dunn, A.M., Hurst, G.D.D. and Rigaud, T. (2001) Inherited microorganisms, sex-specific virulence and reproductive parasitism. *Trends in Parasitology* 17, 88–94.

Boots, M. and Knell, R.J. (2002) The evolution of risky behaviour in the presence of a sexually transmitted disease. *Proceedings of the Royal Society of London, Series B* 269, 585–589.

Bordenstein, S. (2003) Symbiosis and the origin of species. In: Miller, T. and Bourtzis, K. (eds) *Insect Symbiosis*. CRC Press, Boca Raton, Florida, pp. 283–304.

Borgia, G. and Collis, K. (1989) Female choice for parasite-free male satin bowerbirds and the evolution of bright male plumage. *Behavioral Ecology and Sociobiology* 25, 445–454.

Breeuwer, J.A.J. and Werren, J.H. (1990) Microorganisms associated with chromosome destruction and reproductive isolation between two insect species. *Nature* 346, 558–560.

Callaini, G., Riparbelli, M.G., Giordano, R. and Dallai, R. (1996) Mitotic defects associated with cytoplasmic incompatibility in *Drosophila simulans*. *Journal of Invertebrate Pathology* 67, 55–64.

Cavalcanti, A.G.L., Falcao, D.N. and Castro, L.E. (1957) 'Sex-ratio' in *Drosophila prosaltans*: a character due to interaction between nuclear genes and cytoplasmic factors. *American Naturalist* 91, 327–329.

Chapman, T., Arnquist, G., Bangham, J. and Rowe, L. (2003) Sexual conflict. *Trends in Ecology and Evolution* 18, 41–47.

Civetta, A. and Singh, R.S. (1995) High divergence of reproductive tract proteins and their association with postzygotic reproductive isolation in *Drosophila melanogaster* and *Drosophila virilis* group species. *Journal of Molecular Evolution* 41, 1085–1095.

Cowan, D.P. (1984) Life history and male dimorphism in the mite *Kennethiella trisetosa* (Acarina: Winterschmidtiidae) and its symbiotic relationship with the wasp *Ancistrocerus antilope* (Hymenoptera: Eumendiae). *Annals of the Entomological Society of America* 77, 725–732.

Davis, M.J., Ying, Z., Brunner, B.R., Pantoja, A. and Ferwerda, F.H. (1998) Rickettsial relative associated with papaya bunchy top disease. *Current Microbiology* 36, 80–84.

Dedeine, F., Vavre, F., Fleury, F., Loppin, B., Hochberg, M.E. and Boulétreau, M. (2001) Removing symbiotic *Wolbachia* bacteria specifically inhibits oogenesis in a parasitic wasp. *Proceedings of the National Academy of Sciences USA* 98, 6247–6252.

Dedeine, F., Bandi, C., Bouletreau, M. and Kramer, L.H. (2003) Insights into *Wolbachia* obligatory symbiosis. In: Miller, T. and Bourtzis, K. (eds) *Insect Symbiosis*. CRC Press, Boca Raton, Florida, pp. 267–282.

Digilio, M.S., Isidoro, N., Tremblay, E. and Pennachio, F. (2000) Host castration by *Aphidus ervi* venom proteins. *Journal of Insect Physiology* 46, 1041–1050.

Dyson, E.M. and Hurst, G.D.D. (2004) Persistence of an extreme sex ratio bias in a natural population. *Proceedings of the National Academy of Sciences USA* 101, 6520–6523.

Ehrman, L., Somerson, N.L. and Kocka, J.P. (1990) Induced hybrid sterility by injection of streptococcal L-forms into *Drosophila paulistorum*: dynamics of infection. *Canadian Journal of Zoology* 68, 1735–1740.

Gottleib, Y. and Zchori-Fein, E. (2001) Irreversible thelytokous parthenogenesis in *Muscidifurax uniraptor*. *Entomologia Experimentalis et Applicata* 100, 271–278.

Graves, B.M. and Duvall, D. (1995) Effects of sexually transmitted diseases on heritable variation in sexually selected systems. *Animal Behaviour* 50, 1129–1131.

Hamilton, W.D. (1993) Inbreeding in Egypt and in this book: a childish perspective. In: Thornhill, N.W. (ed.) *The Natural History of Inbreeding and Outbreeding*. University of Chicago Press, Chicago, Illinois, pp. 429–450.

Hamilton, W.D. and Zuk, M. (1982) Heritable true fitness and bright birds: a role for parasites? *Science* 218, 384–387.

Hariri, A.R., Werren, J.H. and Wilkinson, G.S. (1998) Distribution and reproductive effects of *Wolbachia* in stalk eyed flies (Diptera: Diopsidae). *Heredity* 81, 254–260.

Hiroki, M., Kato, Y., Kamito, T. and Miura, K. (2002) Feminization of genetic males by a symbiotic bacterium in a butterfly, *Eurema hecabe* (Lepidoptera: Pieridae). *Naturwissenschaften* 89, 167–170.

Huigens, T. and Stouthamer, R. (2003) Parthenogenesis associated with *Wolbachia*. In: Miller, T. and Bourtzis, K. (eds) *Insect Symbiosis*. CRC Press, Boca Raton, Florida, pp. 247–266.

Hunter, M.S., Perlman, S.J. and Kelly, S.E. (2003) A bacterial symbiont in the Bacteroidetes induces cytoplasmic incompatibility in the parasitoid wasp *Encarsia pergandiella*. *Proceedings of the Royal Society of London, Series B* 270, 2185–2190.

Hurd, H. (2001) Host fecundity reduction: a damage limitation strategy? *Trends in Parasitology* 17, 363–368.

Hurd, H., Ward, E. and Polwart, A. (2001) A parasite that increases host lifespan. *Proceedings of the Royal Society of London, Series B* 268, 1749–1753.

Hurst, G.D.D. and Majerus, M.E.N. (1993) Why do maternally inherited microorganisms kill males? *Heredity* 71, 81–95.

Hurst, G.D.D. and McVean, G.A.T. (1998) Parasitic male-killing bacteria and the evolution of clutch size. *Ecological Entomology* 23, 350–353.

Hurst, G.D.D. and Schilthuizen, M. (1998) Selfish genetic elements and speciation. *Heredity* 80, 2–8.

Hurst, G.D.D. and Werren, J.H. (2001) The role of selfish genetic elements in eukaryotic evolution. *Nature Reviews Genetics* 2, 597–606.

Hurst, G.D.D., Sharpe, R.G., Broomfield, A.H., Walker, L.E., Majerus, T.M.O., Zakharov, I.A. and Majerus, M.E.N. (1995) Sexually transmitted disease in a promiscuous insect, *Adalia bipunctata*. *Ecological Entomology* 20, 230–236.

Hurst, G.D.D., Jiggins, F.M. and Majerus, M.E.N. (2003) Inherited microorganisms that selectively kill males. In: Bourtzis, K. and Miller, T.A. (eds) *Insect Symbiosis*. CRC Press, Boca Raton, Florida, pp. 177–197.

Jiggins, F.M., Hurst, G.D.D. and Majerus, M.E.N. (2000) Sex ratio distorting *Wolbachia* causes sex role reversal in its butterfly host. *Proceedings of the Royal Society of London, Series B* 267, 69–73.

Jiggins, F.M., Bentley, J.K., Majerus, M.E.N. and Hurst, G.D.D. (2001) How many species are infected with *Wolbachia*? Cryptic sex ratio distorters revealed by intensive sampling. *Proceedings of the Royal Society of London, Series B* 268, 1123–1126.

Kellen, W.R., Chapman, H.C., Clark, T.B. and Lindegren, J.E. (1965) Host–parasite relationships of some Thelohania from mosquitoes (Nosematidae: Microsporidia). *Journal of Invertebrate Pathology* 7, 161–166.

Knell, R.J. (1999) Sexually transmitted disease and parasite mediated sexual selection. *Evolution* 53, 957–961.

Knell, R.J. and Webberley, K.M. (2004) Sexually transmitted diseases of insects: distribution, evolution, ecology and host behaviour. *Biological Reviews of the Cambridge Philosophical Society* 79, 557–581.

Koivisto, R.K.K. and Braig, H.K. (2003) Microorganisms and parthenogenesis. *Biological Journal of the Linnean Society* 79, 43–58.

Kokko, H., Ranta, E., Ruxton, G. and Lundberg, P. (2002) Sexually transmitted disease and the evolution of mating systems. *Evolution* 56, 1091–1100.

Lockhart, A.B., Thrall, P.H. and Antonovics, J. (1996) Sexually transmitted diseases in animals: ecological and evolutionary implications. *Biological Reviews of the Cambridge Philosophical Society* 71, 415–471.

Majerus, M.E.N. and Hurst, G.D.D. (1997) Ladybirds as a model system for the study of male-killing endosymbionts. *Entomophaga* 42, 13–20.

Majerus, M.E.N. and Majerus, T.M.O. (2000) Female-biased sex ratio due to male-killing in the Japanese ladybird *Coccinula sinensis*. *Ecological Entomology* 25, 234–238.

McLachlan, A. (1999) Parasites promote mating success: the case of a midge and a mite. *Animal Behaviour* 57, 1199–1205.

Moreau, J. and Rigaud, T. (2003) Variable male potential for reproduction: high male mating capacity as an adaptation to prevent parasite-induced excess of females. *Proceedings of the Royal Society of London, Series B* 270, 1535–1540.

Normark, B.B. (2003) The evolution of alternative genetic systems in insects. *Annual Review of Entomology* 48, 397–423.

Normark, B.B. (2004) Haplodiploidy as an outcome of coevolution between male-killing cytoplasmic elements and their hosts. *Evolution* 58, 790–798.

Ranz, J.M., Castillo-Davis, C.I., Meiklejohn, C.D. and Hartl, D.L. (2003) Sex-dependent gene expression and evolution of the *Drosophilai* transcriptome. *Science* 300, 1742–1745.

Rigaud, T. (1997) Inherited microorganisms and sex determination of arthropod hosts. In: O'Neill, S.L., Hoffmann, A.A. and Werren, J.H. (eds) *Influential Passengers: Inherited Microorganisms and Arthropod Reproduction*. Oxford University Press, Oxford, UK, pp. 81–102.

Ruiz, M.F., Esteban, M.R., Donoro, C., Goday, C. and Sánchez, L. (2000) Evolution of dosage compensation in Diptera: the gene *maleless* implements dosage compensation in *Drosophila* (Brachycera suborder) but its homolog in *Sciara* (Nematocera suborder) appears to play no role in dosage compensation. *Genetics* 156, 1853–1865.

Schütt, C. and Nöthiger, R. (2000) Structure, function and evolution of sex determining systems in dipteran insects. *Development* 127, 667–677.

Simmons, A.M. and Rogers, C.E. (1994) Effects of an ectoparasitic nematode, *Noctuidonema guyanese* on adult longevity and egg fertility in *Spodoptera frugiperda* (Lepidoptera: Noctuidae). *Biological Control* 4, 285–289.

Snook, R.R., Cleland, S.Y., Wolfner, M.F. and Karr, T.L. (2000) Offsetting effects of *Wolbachia* infection and heat shock on sperm production in *Drosophila simulans*: analysis of fecundity, fertility and accessory gland proteins. *Genetics* 155, 167–178.

Starr, D.J. and Cline, T.W. (2001) A host–parasite interaction rescues *Drosophila* oogenesis defects. *Nature* 418, 76–79.

Stouthamer, R. and Kazmer, D. (1994) Cytogenetics of microbe-associated parthenogenesis and its consequence for gene flow in *Trichogramma* wasps. *Heredity* 73, 317–327.

Stouthamer, R., Breeuwer, J.A.J. and Hurst, G.D.D. (1999) *Wolbachia pipientis*: microbial manipulator of arthropod reproduction. *Annual Review of Microbiology* 53, 71–102.

Strandberg, J.O. and Tucker, L.C. (1974) *Filariomyces forficulae*: occurrence and effects on the predatory earwig, *Labidura riparia*. *Journal of Invertebrate Pathology* 24, 357–364.

Tesh, R.B. (1981) Vertical transmission of athropod-borne viruses of vertebrates. In: McKelvey, J.J., Eldridge, B.F. and Maramosch, K. (eds) *Vectors of Disease Agents*. Praeger, New York, pp. 122–137.

Thompson, J.N. (1987) Symbiont-induced speciation. *Biological Journal of the Linnean Society* 32, 385–393.

Thompson, W.H. and Beaty, B.J. (1977) Venereal transmission of La Crosse (California encephalitis) arbovirus in *Aedes triseriatus* mosquitoes. *Science* 196, 530–531.

Thompson, W.H. and Beaty, B.J. (1978) Venereal transmission of La Crosse virus from male to female *Aedes triseratus*. *American Journal of Tropical Medicine and Hygiene* 27, 187–196.

Thrall, P.H., Antonovics, J. and Bever, J.D. (1997) Sexual transmission of disease and host mating systems: within-season reproductive success. *American Naturalist* 149, 485–506.

Thrall, P.H., Antonovics, J. and Wilson, W.G. (1998) Allocation to sexual and non-sexual disease transmission. *American Naturalist* 151, 29–45.

Thrall, P.H., Antonovics, J. and Dobson, A.P. (2000) Sexually transmitted disease in polygynous mating systems: prevalence and impact on reproductive success. *Proceedings of the Royal Society of London, Series B* 267, 1555–1563.

Vance, S.A. (1996) Morphological and behavioural sex reversal in mermethid-infected mayflies. *Proceedings of the Royal Society of London, Series B* 263, 907–912.

Vanderckhove, T.T.M., Watteyne, S., Willems, A., Swings, J.G., Mertens, J. and Gillis, M. (1999) Phylogenetic analysis of the 16S rDNA of the cytoplasmic bacterium *Wolbachia* from the novel host *Folsomia candida* (Hexapoda: Collembola), and its implications for Wolbachial taxonomy. *FEMS Microbiological Letters* 180, 279–286.

Veneti, Z., Clark, M.E., Zabalou, S., Karr, T.L., Savakis, B. and Bourtzis, K. (2003) Cytoplasmic incompatibility and sperm cyst infection in different *Drosophila*/host associations. *Genetics* 164, 545–552.

Volkonsky, M. (1946) *Podapolipus diander* n. sp. acarien hétérostymate parasite du criquet migrateur (*Locusta migratoria*). *Archives de l'Institut Pasteur d'Algerie* 18, 321–340.

Wade, M.J. and Chang, N.W. (1995) Increased male fertility in *Tribolium confusum* beetles after infection with the intracellular parasite *Wolbachia*. *Nature* 373, 72–74.

Webb, T.J. and Hurd, H. (1999) Direct manipulation of insect reproduction by agents of parasite origin. *Proceedings of the Royal Society of London, Series B* 266, 1537–1541.

Webberley, K.M. and Hurst, G.D.D. (2002) The effect of aggregative overwintering on an insect sexually transmitted parasite system. *Journal of Parasitology* 88, 707–712.

Webberley, K.M., Hurst, G.D.D., Buszko, J. and Majerus, M.E.N. (2002) Lack of parasite-mediated sexual selection in a ladybird/sexually transmitted disease system. *Animal Behaviour* 63, 131–141.

Webberley, K.M., Hurst, G.D.D., Husband, R.W., Schulenburg, J.H.G., Sloggett, J.J., Isham, V., Buszko, J. and Majerus, M.E.N. (2004) Host reproduction and a sexually transmitted disease: causes and consequences of *Coccipolipus hippodamiae* distribution on coccinellid beetles. *Journal of Animal Ecology* 73, 1–10.

Weeks, A.R., Marec, F. and Breeuwer, J.A.J. (2001) A mite species that consists entirely of haploid females. *Science* 292, 2479–2482.

Weir, A. and Beakes, G. (1995) An introduction to the Laboulbeniales: a fascinating group of entomogenous fungi. *Mycologist* 9, 6–10.

Welch, V.L., Sloggett, J.J., Webberley, K.M. and Hurst, G.D.D. (2001) Short-range clinal variation in the prevalence of a sexually transmitted fungus associated with urbanisation. *Ecological Entomology* 26, 547–550.

Werren, J.H. and Beukeboom, L.W. (1998) Sex determination, sex ratios and genetic conflict. *Annual Review of Ecology and Systematics* 29, 233–261.

Zchori-Fein, E., Gottlieb, Y., Kelly, S.E., Brown, J.K., Wilson, J.M., Karr, T.L. and Hunter, M.S. (2001) A newly discovered bacterium associated with parthenogenesis and a change in host selection behavior in parasitoid wasps. *Proceedings of the National Academy of Sciences USA* 98, 12555–12560.

Zeh, J.A. and Zeh, D.W. (1996) The evolution of polyandry. I. Intragenomic conflict and genetic incompatibility. *Proceedings of the Royal Society of London, Series B* 263, 1711–1717.

Zeh, J.A. and Zeh, D.W. (1997) The evolution of polyandry. II. Post-copulatory defences against genetic incompatibility. *Proceedings of the Royal Society of London, Series B* 264, 69–75.

9 The Evolution of Imperfect Mimicry

Francis Gilbert

School of Biology, Nottingham University, Nottingham, UK

1. Ideas About Mimicry

Apart from some notable exceptions, mimicry and brightly coloured aposematic patterns have been discussed by biologists mainly from three very different points of view, each making unrealistic assumptions about aspects of the other two (Mallet and Joron, 1999). The most obvious in the voluminous literature (see Komarek, 1998) is the insect natural history approach, used by both naturalists and professional biologists, which has simplistic ideas about the ways in which predators behave and of their evolutionary impact on their prey. The second is a modelling approach, that of evolutionary dynamics: this virtually ignores predator behaviour and any details of the interactions between predators and their prey. The final viewpoint is centred on the details of predator behaviour, but this is often simplistic about the evolutionary dynamics, and can make unrealistic assumptions about the psychological processes of learning and forgetting. A gradual synthesis is taking place between these viewpoints, partly in response to the inadequacy of older theory to explain the phenomenon of imperfect mimicry.

In this chapter, I outline the basic ideas of mimicry theory, and show how they fail to account for the commonly imperfectly mimetic patterns of the main taxonomic group in the Holarctic that contains mimics, the hoverflies (Diptera, Syrphidae). I review the relevant information about this group, and assess a variety of new theories of imperfect mimicry, which have been put forward largely to account for the evolution of their colour patterns. I conclude that only one of these recent ideas – Sherratt's (2002) multiple-model theory – accounts for all the facts.

1.1 Basics

Traditionally the main forms of mimicry are Batesian and Müllerian, formulated in 1862 and 1878, respectively, and still thought to provide some of the most easily understandable examples of the way in which natural selection operates (Poulton, 1890; Carpenter and Ford, 1933; Malcolm, 1990; Joron and Mallet, 1998; Mallet and Joron, 1999). In this chapter, the word 'mimic' without any qualifying adjective refers to Batesian mimics only: the word 'model' includes Müllerian mimics and Müllerian mimicry rings.

Batesian mimicry is thought to occur when a rare harmless species evolves to resemble closely an abundant 'unprofitable' model. A Batesian mimic gains protection from predators which cannot tell the difference between model and mimic, and since they tend to encounter models rather than mimics when searching for food, they associate the colour pattern of the model with a nasty experience, and tend to avoid it in future. Since the more closely a mimic resembles its model, the more protection it gets, there is constant selection for mimetic perfection, which results in mimics evolving to be indistinguishable from models to their predators. The basis of the unprofitability is usually assumed to be unpalatability, or more generally, noxiousness; however, there is no reason why other bases might not be important, such as difficulty of capture ('escape' mimicry: van Someren and Jackson, 1959; Hespenheide, 1973).

Müllerian mimicry occurs when several noxious species evolve to resemble each other, and hence they all benefit by a reduction in predation: Nicholson's (1927) analysis of Australian insects is a very clear example of this for an entire fauna. Since predators need only recognize resemblance rather than identity, there is no reason to suppose that protection is proportional to similarity, and Müllerian mimics are therefore not indistinguishable from one another: they are 'imperfect'.

In Batesian mimicry a new mimetic form will be advantageous since it is rare, but at a high frequency it loses mimetic protection and is selected against. This frequency dependence generates 'diversifying' (i.e. disruptive) selection for different morphs, and hence for polymorphism. In contrast, there is no such diversifying selection on models (Nur, 1970). Different morphs of a model will be disadvantageous, because they will not be identified as inedible; this 'purifying' (i.e. stabilizing) selection maintains a single colour pattern in the population of the model (and in a group of Müllerian mimics).

The study of Batesian mimicry has traditionally involved imagining a one-to-one correspondence between model and mimic species (e.g. Howarth et al., 2000), but this may be incorrect as a general rule. Many insect models are bound up in Müllerian complexes, and typically many harmless insects mimic each complex (see Nicholson, 1927). European work has rarely acknowledged this aspect of mimicry, preferring to identify a single model species for each apparent mimic. Because each mimic usually resembles a Müllerian mimicry ring, a whole suite of models, this has resulted in different models being cited in different studies, generating a great deal of confusion. The situation may be different in the tropics, where models are usually much more diverse (M. Edmunds, personal communication), but knowledge about tropical mimetic relationships outside the Lepidoptera is very scarce.

1.2 What maintains imperfect mimicry?

With some notable exceptions (Schmidt, 1960; Duncan and Sheppard, 1965; Holling, 1965; Ford, 1971; Pilecki and O'Donald, 1971; Dill, 1975; Goodale and Sneddon, 1977; Schuler, 1980; Greenwood, 1986; Hetz and Slobodchikoff, 1988; Lindström *et al.*, 1997; Rowe *et al.*, 2004), consideration of the evolution of mimicry has been mostly confined to the evolution of good mimics. However, these views ignore the obvious fact that most mimicry is of rather poor quality (Getty, 1985), and such imperfect mimicry is widespread in many Batesian mimetic systems (e.g. salticid spiders mimicking ants: Edmunds, 1993, 2000). 'Imperfect mimicry' in a Batesian mimic is defined here as being obviously different (to us) from its putative model, usually visually.

The prevailing opinion was (and still is) that since individuals with closer resemblance to the model gain more protection, Batesian mimics should be under constant selection to improve their mimetic resemblance, and hence eventually only good mimics would exist. Some models are very noxious, and others less so. The degree of protection afforded to mimics will depend upon just how noxious a model is, and predators will generalize more widely to poorer and poorer mimics as the noxiousness of the model increases (Duncan and Sheppard, 1965; Goodale and Sneddon, 1977). This effect is measurable in experiments (e.g. Lindström *et al.*, 1997). However, there would still be selection, however slight, for improvement of mimetic resemblance and thus there would still be constant selection for perfect mimicry.

Very few, and only recent, authors have considered what kind of resistance there might be to directional selection for improving mimicry (apart from the usual 'genetic constraints' possibility). If we view the current colour patterns of models and their mimics as existing in some kind of equilibrial state, then for poor mimicry to be stable, such opposing or balancing forces must exist (Grewcock, 1992; Sherratt, 2002). What might such forces be? Only the costs of producing mimetic colours (Grewcock, 1992), kin selection opposing natural selection (Johnstone, 2002) and the existence of multiple models (Sherratt, 2002) have been suggested. These ideas are described in detail in Section 6.9.

Of course, it is possible that the colour patterns are not at equilibrium, but are constantly evolving. Models do suffer extra mortality when a mimic is present, because by chance some predators encounter mimics rather than models, learn that the pattern is edible, and consequently attack models in error. This generates 'advergence' (Brower and Brower, 1972), i.e. where the colour pattern of a Batesian mimic evolves towards that of its model, but the model evolves away: this is the 'coevolutionary chase' process of Turner (1987). However, selection is always stronger on the mimic, which always catches the model in this evolutionary process, resulting in stability. Simulations of multispecies systems (Franks and Noble, 2004) show that Müllerian mimicry only evolves if there is some initial resemblance among models; however, the additional presence of Batesian mimics strongly promotes the formation of Müllerian mimicry rings by causing models to move in colour space, and hence converge. Despite this, the end result is stasis once again, and an equilibrium state. Thus I do not think there is any empirical or theoretical evidence for a non-equilibrial view of the evolution of mimetic colour patterns.

1.3 Other elements of mimicry theory: relative abundance and predator behaviour

Traditional mimicry theory suggests that mimics should be rarer than models, so that predators tend to meet unprofitable rather than profitable examples, and hence associate the pattern with the unprofitability. The rarity of mimics relative to models has been a constantly repeated feature of the way in which Batesian mimicry has been described. Protection can extend to commoner mimics when their models are also common, and the close correlation between model and mimic abundances was shown in butterflies by Sheppard (1959). It also may be that it is not that mimics are rare, but that models are common, since abundant warningly coloured models will gain more mimics than rare ones (Turner, 1984). More sophisticated models and thinking have altered this viewpoint considerably: mimics can still be protected even when more abundant than their model (e.g. Brower, 1960), when that model is really nasty, or if the mimic has low nutritional value, or if distributions are clumped. This is also true when profitable alternative prey are present (which encourage the predator to drop both model and mimic from its diet). Holling (1965) reached this conclusion explicitly almost 30 years ago:

> The greater the proportion of distasteful models to mimics the greater is the protection in each of the four cases simulated, although some protection is afforded even if the distasteful models are very rare. … Thus the often expressed belief that the advantages of mimicry collapse when the palatable mimic outnumbers its model is no longer tenable, if it ever was.

Furthermore, what matters is the relative abundances as perceived by predators, and these are very difficult to measure realistically (see below for an example).

The recognition, discrimination and generalization of prey by predators are features hardly considered in the early theoretical treatment of mimicry. Errors (real or apparent) must be made in order for the whole system to work, since mimics must be avoided in case they turn out to be models. Most of the discussions in the literature focus only upon perfect mimics, and therefore assume that predators are unable to distinguish them from the models. We now know more, but still not enough, about the way in which predators generalize from a colour pattern associated with unprofitability to other similar, but not identical, colour patterns (see Rowe et al., 2004). The few experiments done on this actually show clearly that even a vague resemblance can be protective: just a bit of black or red is often good enough (e.g. Schmidt, 1960). Many factors probably affect the extent of generalization, the degree of noxiousness of the model being the most obvious: the more dangerous or toxic the model, the more likely it is that even a partial resemblance will afford protection to a mimic. This interacts with model–mimic relative frequencies, as was shown by a clever experiment using birds feeding on mealworm larvae experimentally modified to form models with two levels of nastiness (Lindström et al., 1997; Mappes and Alatalo, 1997). Mimics survived best when the model was relatively common and highly distasteful. When the model was highly distasteful, the birds clearly did not bother to discriminate, whatever the relative abundances, and poor mimics gained while models suffered slightly reduced protection. However, when the model was only slightly distasteful, birds discriminated between model and mimic when models were common, but did not bother when models were rare: the mimics still gained some protection, but not as much (see also Alcock, 1970). The results of these experiments show that discrimination is often perfectly possible, but when correct decisions are less beneficial (prey less profitable), or mistakes are more likely (relative frequencies) or more costly (noxiousness), birds may prefer not to risk it (see Dill, 1975). It seems logical that this decision may also depend on how hungry the bird is in relation to its perception of the availability of prey (including alternatives to the model and mimic). In a marvellous early work of computer-based modelling, Holling (1965) included all of these features in a simulation predicting that even vague resemblances would be protective; thus, he concluded, mimicry should be a very pervasive feature of natural communities.

The perceptual problem of an experienced predator encountering a mimic is one of signal detection. Psychologists have developed signal detection theory for measuring the way in which decisions are made between a desired objective and undesirable 'noise'. It quantifies the fundamental trade-off between making correct choices sufficiently often while keeping the cost of making mistakes to a minimum (Shettleworth, 1998: 61–69). The appearance of the insect is often the only information the predator has in order to make the discrimination, and the more similar the model and mimic are, the more likely it is that the predator will make a mistake. The probabilities of making a correct decision and making a mistake cannot be varied independently, since mimics are by definition sometimes or always confused with models. All the predator can do is to set a threshold value of prey 'appearance', using whatever clues can be obtained at reasonable cost (usually time). Exceeding this threshold determines whether the predator attacks or not (see below, and Fig. 9.3). A predator with a low threshold attacks more often, is correct more often, but also makes more mistakes. A conservative predator makes fewer mistakes, but also makes fewer correct decisions. Despite its obvious applicability, only Oaten et al. (1975), Getty (1985), Greenwood (1986), Johnstone (2002) and Sherratt (2002) have used this approach to analyse the way in which predators affect model–mimic complexes.

Holling (1965) argued that because the distance of perception is so large in visual vertebrate predators (i.e. those that can learn associations), the potential for regulating even low-density populations is present, and the evolution of mimicry results in an increase in the equilibrial population density (cf. Sherratt, 2002). However, current ecological opinion today considers it very unlikely that populations of insects in general, let alone mimics, are regulated by visually based predators of adults.

The ways in which the evolution of aposematism and mimicry affect other aspects of the life history are only just begining to be explored. The evolution of effective defence is costly, and these costs should be measurable in terms of fitness components. As with Bates' original observations that led to the idea of mimicry, such studies have involved the South American heliconiid butterflies. Marden and Chai (1991) and Srygley (1999) found a real dichotomy between palatable, non-mimetic versus unpalatable or mimetic species: palatable non-mimics maintained higher body temperatures, and had larger flight muscles, allowing them to accelerate faster in flight, and had smaller digestive tracts (in males) and smaller ovaries (in females). Thus there were measurable reproductive costs to the need to evade predators effectively, some of which mimics could avoid paying. However, there are also measurable aerodynamic costs to evolving a mimetic flight pattern (Srygley, 2004).

1.4 Variation in noxiousness/unprofitability at all levels

The clear and simple distinction between Batesian and Müllerian mimicry is currently under scrutiny, using theories of the psychological processes of predators. These 'receiver psychology' models make a variety of assumptions about the processes of learning, forgetting and extinction (a learned erasing of a previously memorized association), and can lead to a great variety of different forms of mimicry (see Speed and Turner, 1999). In particular, where Müllerian mimics differ in the degree of noxiousness, the existence of a more palatable one can result in an increase in attack rate on the less palatable, leading to a parasitic form of Müllerian mimicry called 'quasi-Batesian' (Speed, 1993, 2001). This kind of Müllerian mimicry would allow the evolution of polymorphism, which we know occurs in heliconiine butterflies (although there are other explanations: see Mallet and Gilbert, 1995).

In this context it is interesting that the classic case of Batesian mimicry, between Monarch and Queen butterfly models (*Danaus* spp.) and the Viceroy mimic (*Limenitis archippus*) is now known to be much more complex (Brower, 1988; Mallet, 1999). Some, but not all, individuals in monarch populations sequester poisons (cardenolides, cardiac glycosides) from their larval food plants, the milkweeds (*Asclepiadaceae* – whose members vary in their glycoside content), which render the butterfly distasteful and cause vomiting in a number of bird predators. Evidently a great deal of variation in palatability exists within and among populations of monarch butterflies, a veritable palatability spectrum. Whilst initial experiments mainly appeared to show that the viceroy was palatable, we know now that this butterfly can be just as unpalatable as the monarch (Ritland and Brower, 1991).

2. What Phenomena Does Theory Need to Explain?

I have outlined some of the more traditional ideas of mimicry theory, and some of their deficiencies: I shall introduce the two newest theories later on, when I have described what it is they are trying to explain. Thus I turn now to the question of whether traditional theory can account for the data. In order to do this manageably, I concentrate on a single taxonomic group that contains a very large proportion of the Holarctic Batesian mimics: the hoverflies (Diptera, Syrphidae). If theory cannot explain the colour patterns of these insects, then it is clearly deficient. I consider hoverfly colour patterns, ecology and behaviour in detail to show that classical mimicry theory copes poorly in explaining them, as do most of the more recent additions. In my opinion, only one very recent theory (Sherratt, 2002: described below) based on predator signal detection in a world of multiple models has the potential to explain all the characteristics of hoverfly mimicry.

3. Mimicry Complexes in the Holarctic

There are few general overviews of mimetic relationships in particular faunas, except for Australia (Nicholson, 1927) and the UK (Brown, 1951). Most models have conspicuous aposematic patterns, often involving sharp contrasts of two or more different colours. In Britain, the following aposematic patterns occur amongst models: yellow and black (wasps, hornets, many bumblebees), red and black (many beetles, some bugs, moths and a few bumblebees), red (beetles), black (beetles) and metallic shining colours (beetles). There are some non-aposematic mimics, for example of honeybees, and ants (e.g. spiders). Not all possible models are made use of by Batesian mimics: for example, the red-and-black burnet moths and other similar distasteful insects have not been copied. Two sets of models and their Batesian mimics make up significant proportions of the fauna: hymenopteran models with dipteran mimics, and unpalatable coleopteran models with palatable coleopteran mimics. Numerically the Hymenoptera form the most important group of models, and the Diptera the largest group of mimics. Diptera are exclusively mimics of Hymenoptera, and most of the mimics belong to one family, the hoverflies (Syrphidae). This is one of the largest and most diverse of all dipteran families, with a worldwide distribution and more than 5600 species described (see Rotheray, 1993; Rotheray and Gilbert, 1999). The literature contains information about the mimetic status of some 279 species of hoverfly (F. Gilbert, unpublished), an astonishingly high proportion relative to other insect groups: clearly mimicry is a dominant theme of the evolution of this group of flies. Especially in Europe, most of the models for hoverfly mimics appear to be social insects: the bumblebees, honeybees and social wasps.

4. The Models of Hoverfly Mimics

4.1 Bumblebees

Bumblebees are well known both taxonomically (Williams, 1998) and biologically (Prŷs-Jones and Corbet, 1991; Goulson, 2003), occurring largely in boreal or high-altitude habitats. Most of the 239 recognized species are Holarctic, but some extend down the Andes into South America. While morphologically very uniform, some species of bumblebee can be extraordinarily variable in the colour patterns of their body hairs, with several different morphs and broad geographic trends within a single species (see Williams, 1991; von Hagen and Aichhorn, 2003); different judgements about the significance of this variation have resulted in more than 2800 names (Williams, 1998). However, the colour patterns are not 'all possible combinations' (as Drees, 1997, thought), but unrelated species have converged in the colour patterns of their morphs into just a few Müllerian mimicry rings (Nicholson, 1927; Vane-Wright, 1978; Plowright and Owen, 1980; Williams, 1991). Thus virtually all eastern Nearctic bumblebees have broad bands of yellow and black, often with a black spot in the middle of the yellow thorax (see Table 9.1: group G); western Nearctic species are predominantly black with narrow yellow bands (group B); in between in the Rocky Mountains there are two mimicry groups, yellow-and-black banded (group D), and yellow-and-black banded with an extensively red abdomen (group H) (Gabritschevsky, 1924, 1926). In Europe they are more diverse, forming four main Müllerian mimicry rings (groups A, E, F and J: see Prŷs-Jones and Corbet, 1991). These groupings are inevitably not always clear-cut. For example, some individual workers of *Bombus pascuorum* have dense tawny hairs all over the thorax and abdomen, whereas in others the abdominal hairs are thin and do not obscure the dark cuticle beneath, creating the appearance of a dark abdomen. The former pattern would be classified into group J of Table 9.1, whereas the latter would fall either into group G (which are virtually all Nearctic bumblebees) or even into the honeybee group (M), where there are many species with the pattern of a hairy thorax and bare dark abdomen.

Why is there more than one Müllerian ring? Why have all models not converged into one massive Müllerian complex? Holling (1965) suggested the evidence implied a limit to the number of species within a ring, but could not think of a mechanism other than a vague recourse to other general density-dependent features of their biology. It is possible that there are subtle differences in habitat segregation in heliconiine mimicry rings (Mallet and Gilbert, 1995), but this is unlikely in bumblebees. Mallet (1999) suggested that while it might theoretically be because of quasi-Batesian processes caused by different degrees of noxiousness, the evidence is very weak (really just the fact that some species are polymorphic, with each morph belonging to a different mimicry ring, and some emerge later than others). This deserves detailed study in bumblebees, where a huge amount of information is available, especially in Europe. A fascinating recent simulation (Franks and Noble, 2004) suggests that it is the presence of Batesian mimics and the 'coevolutionary chase' that causes models to converge into a smaller number (typically three) of Müllerian mimicry rings than would otherwise form. However, too many Batesian mimetic forms break up Müllerian mimicry rings or prevent them from evolving.

How noxious are bumblebees to their predators? With the exception of the specialist red-backed shrike (*Lanius collurio*), adults of the 19 species of birds in Mostler's (1935) amazingly comprehensive trials only ate 2% of the 646 bumblebees that were presented to them, rejecting all the rest without even attempting to attack, whatever the species involved (mainly *Bombus terrestris*, *B. lapidarius*, *B. hortorum* and *B. ruderarius*). The question arises as to what may be the source of the aversion, and most authors (e.g. Stiles, 1979; Plowright and Owen, 1980) assume that it is their sting. However, the evidence is only convincing in one case: naïve toads ate *Bombus pennsylvanicus* that had had their stings removed, but having attempted to eat an intact bumblebee, from then on strongly avoided them (Brower et al., 1960), a pattern repeated with honeybees (Brower and Brower, 1962, 1965). The evidence that birds are

Table 9.1. Holarctic mimicry rings from assessments in the literature (full data of all the hoverfly species and their morphs are provided in F. Gilbert (unpublished)). Some species have both good and poor morphs, and hence only summary numbers of each are given.

Ring	Colour pattern	Species	Number of hoverfly mimics		Polymorphic species	Species with sex-linked morphs
			Morphs			
			Good	Poor		
Bumblebees						
A	All black with a variable-sized red or orange tip to the abdomen		12	2		
B	Black with a yellow collar and thin yellow posterior band		2	0		
C	Black with a yellow collar and white or whitish tail		1	1		
D	Broad yellow-and-black bands with a black tail		7	0		
E	Broad yellow-and-black bands with a white tail		5	2		
F	Broad yellow-and-black bands with a red/orange/yellow/brown tail		18	3		
G	Yellow/brown anterior (± a central black spot on the thorax), dark brown-black posterior		5	2		
H	Yellow thorax with a black central band, and yellow/orange abdomen		13	0		
J	All tawny		14	3		
K	Broad white-and-black bands with a red/orange/yellow tail		3	1		
L	All black		2	0		
	Total	62	82	14	25	14
Honeybee						
M	Light-haired thorax, dark abdomen with ± orange anterior and thin whitish bands, **total**	28	16	12	0	6?
Wasps						
I	'Eumenids': black with 1–3 narrow, widely spaced bright yellow bands on the abdomen		22	2		
II	Mainly polistines: alternating, more or less equal-sized bright yellow and black bands		6	26		
III	Small *Vespula*: black with paired yellow spots or lunules on each segment		1	15		
IV	Large *Vespula*: mostly bright yellow with narrow black triangular bands ± spots		19	12		
V	Hornets: large, mostly dull yellow with reddish brown to blackish markings		4	0		
VI	*Dolichovespula*: large, mostly dark with white bands at the abdomen tip		2	0		
VII	Larger solitary wasps: strongly 'waisted', brown/orange with poorly defined yellow markings		0	0		
VIII	Small to medium, black with a red/orange abdomen and black tip		5	2		
IX	Small to medium, narrow abdomen with small whitish/yellow/orange/red side spots/bands		9	32		
	Total	157	68	89	0	4?

also deterred by the sting is weak and unconvincing. Mostler (1935) recorded no stings suffered by experienced adult birds, and of 70 prolonged contacts between bumblebees and young naïve birds trying to eat them, there were only three stings. He found that bumblebee tissues were highly palatable, never evoking any of the unpalatability reactions that were so typical of contact with wasp and honeybee abdominal tissues. He attributed the noxiousness of bumblebees to the difficulty of handling: a hand-reared young female whitethroat (*Sylvia communis*), feeding independently for the first time, took 18 min to kill, dismember and eat a bumblebee, after which it was completely exhausted. The equivalent handling time for houseflies, mealworms or beetles was a fraction of a minute, even for young birds. Mostler suggested three reasons why bumblebees were aversive; these were, in order of importance: (i) the tough chitin and hairy bodies of these insects made it necessary for birds to have to expend great efforts in subduing, dismembering and swallowing them; (ii) bumblebee tissues were not easily digestible; and (iii) the sting. Likewise Evans and Waldbauer (1982) thought that the sting of *Bombus pennsylvanicus americanorum* was not the main protection against birds. Only two of their birds were stung; the others avoided eating bumblebees only after having eaten the 'middle segments of the abdomen', presumably with the venom sac. In this case unpalatability may be due to distasteful venom.

Different bumblebee species are not equally noxious, partly because they vary a lot in aggressiveness, although nothing systematic seems to have been studied about this: the subgenus *Fervidobombus* is supposed to be particularly aggressive (Kearns and Thomson, 2001: 70). Such differences should have an impact on the effectiveness of Batesian mimicry, and the occurrence of quasi-Batesian processes. Rupp (1989) noted differences in attack readiness among the German and Swiss bumblebees with which he worked – the workers of *B. terrestris* and *B. lucorum* reacted with particularly fierce attacks to irritation of the nest, whilst he could dig out the nests of *B. pascuorum* and *B. wurfleini mastrucatus* without any special protection being required. He found an unexpectedly low proportion of the black–red *bombylans* morph of the mimetic syrphid *Volucella bombylans* in Switzerland, and attributed this to the fact that the relevant model in the lowlands (*B. lapidarius*) was replaced by a far less aggressive visual counterpart (*B. wurfleini mastrucatus*) in the mountains, which therefore provided less protection. Alford (1975) too mentions the different levels of aggressiveness amongst species of bumblebee: most species in the UK are benign and easy to handle, and their nests are simple to collect, but *B. terrestris* and *B. muscorum* are noticeably much more aggressive and difficult to deal with.

4.2 Honeybees

Honeybees are one of the best known of all insects (Seeley, 1985). *Apis mellifera* probably originally had an African distribution, together with all but northernmost Europe and western Asia. Whilst all individuals have an obviously tawny-haired thorax, most workers either have a dark-brown/black abdomen or carry transverse orange bands on the abdominal tergites. The extent of the banding is mainly a racial difference, but is also sensitive to temperature, and hence varies seasonally. A wide variety of solitary bees and some other Hymenoptera also belong to this mimicry ring.

Mostler (1935) conducted about 480 feeding trials on honeybees with his insectivorous birds. Spotted flycatchers (*Muscicapa striata*) were perfectly willing to take honeybees as prey. There was individual variation in two species: pied flycatchers (*Ficedula hypoleuca*) (where one individual regularly fed on honeybees), and redstarts (*Phoenicurus phoenicurus*) (where females, but not males, were willing to feed on honeybees). In the other birds, the insects were scarcely even looked at in most trials. When tasted or eaten, honeybees induced the same unpalatability reactions as for wasps (see below), but these were more limited in degree, and appeared less frequently. Using mealworms smeared with abdominal tissues, Mostler showed that unpalatability was the main cause of the rejection response. As in the case of wasps, Liepelt (1963) demonstrated that the bad taste of the abdomen derived from the venom. The removal of the entire sting apparatus, including the venom sac, rendered honeybees completely palatable, and all were eaten.

Do honeybees sting predators more or less readily than wasps do? It has been reported that 25–100% of toads were stung during feeding attempts (Cott, 1940; Brower and Brower, 1965), but birds appear to be rarely, if ever, stung (Mostler, 1935; Liepelt, 1963), and probably the sting is not a significant deterrent (Liepelt, 1963). Unpalatability arising from the taste of the venom must be the main cause of avoidance by birds. It is probable that the beak of an insectivorous bird is a useful defence against stings, since it is hard and non-living, but the fleshy mouths of toads offer no protection. In some birds the feeding method exposes them to multiple stings, and they must be immune (e.g. swifts: M.F. Johannsmeier, personal communication).

There are substantial differences in noxiousness among honeybee races (Seeley, 1985: 139–149). A European beekeeper is astonished at the defensive ferocity of African bees, receiving 6–10 times more stings for the same hive manipulation. This is probably a consequence of the much greater level and longer history of nest predation in Africa, especially by humans. We might predict from mimicry theory that these highly noxious bees would therefore be used more frequently as models for palatable insects than other races: although hardly studied at all, honeybee mimicry is said to be a dominant theme of the Diptera of southern Africa (B. Stuckenberg, personal communication), much more so than in Europe.

Given that its members are noxious, and birds learn to avoid them, why is the honeybee complex not aposematically coloured (Holloway, 1976)? Perhaps their high abundance and gregarious foraging render bright colours unnecessary in promoting memorability (J. Mallet, personal communication).

4.3 'Wasps'

The 'wasps' as models for hoverflies contain three main groups: social wasps (Vespidae) and two groups of solitary wasps, the potter (Eumenidae) and digger

(Sphecidae) wasps. It is probable that other related taxa can also serve as models, including spider wasps (Pompilidae) and sawflies (Tenthredinidae), although little is known of their noxiousness. Many different species have been identified as possible models for mimetic syrphids. Whether birds really differentiate among these species is the critical point. Workers in North America have grouped models into a few Müllerian complexes (e.g. Evans and Eberhard, 1970: 245; Waldbauer, 1988), but this sort of classification of wasp colour patterns has hardly ever been done by European workers, although the existence of Müllerian pattern groups has been mentioned occasionally (e.g. Heal, 1979).

Females of all species have a sting, but the sting of the social vespids is often thought to be much more painful. Like the other model groups, the basis of the noxiousness of these models has generally been assumed to be their sting (e.g. Edmunds, 1974: 62, 82). It comes as rather a surprise, therefore, to read the work of those (Mostler, 1935; Steiniger, 1937a,b; Liepelt, 1963) who actually offered wasps experimentally to birds, and who discovered that, as with honeybees, the stings are only rarely used against birds, and that the taste of the venom sac is responsible for their noxiousness. Adults of these birds attacked fewer than 10% of the wasps, eating fewer than 3% of them. Hand-reared naïve young birds ate a somewhat higher proportion overall, but this average masked an initial willingness to attack, followed by rapid learned avoidance. There was no sign of any innate avoidance of black-and-yellow colour patterns in these studies. Mostler (1935) only recorded four birds being stung by wasps in the 1082 presentations in his extensive experiments, and Liepelt (1963) states definitively that 'no wasp stings occurred' during his 99 wasp presentations, although during another series of experiments, two redstarts were stung. Although not fatal, the redstarts spent 6–7 h recovering from their experience, which in nature might easily have been critical to their chances of surviving the night (Birkhead, 1974). From these studies it seems clear that, although having a dramatic effect when used, birds only rarely get stung by wasps, and therefore the sting cannot be the primary source of wasp noxiousness. This seems to be a classic case of risk versus hazard: the risk is low but the hazard great (C.J. Barnard, personal communication). Mostler considered the unpalatability of the abdomen to be the major source of noxiousness for wasps, and the sting being only secondary: subsequently Liepelt (1963) found that venom-free abdominal tissue evoked none of the typical unpalatability reactions. It is the terrible taste that the venom imparts to the abdomen that is the main deterrent for birds.

4.4 Comparison among models

Based upon the data available at present, I conclude therefore that all three main Holarctic groups of aposematic insects form Müllerian mimicry rings, one of them (bumblebees) consisting of subsets of Müllerian rings based on particular colour patterns, and involving some polymorphic species whose morphs are members of different rings. Any theory that predicts that such polymorphic mimicry rings should be rare cannot therefore be correct. For most bird predators, wasps are the most noxious models, and bumblebees are the least noxious: bumblebees seem to be classified as unprofitable food by small insectivorous birds, whereas honeybees and especially wasps are categorized as noxious food.

5. The Hoverfly Mimics

5.1 Resemblance

Hoverfly colour patterns have often been labelled as mimetic, but only some species resemble their models closely, whereas others resemble their supposed models only vaguely, so are at best rather poor mimics (see, for example, the assessments in Howarth *et al.*, 2000). There is a clear distinction in the literature between bumblebee mimics, which are usually accepted as such without question, and honeybee and wasp mimics, where a large proportion are generalized or imperfect to the human eye. Furthermore, there have been many conflicts among writers about the supposed models of particular species. For example, *Criorhina asilica* was labelled as a bumblebee mimic by Verrall (1901), but as a perfect or almost perfect honeybee mimic by most authors (e.g. Dlusskii, 1984; Röder, 1990), although Drees (1997) called it 'cryptic'.

Table 9.1 counts all the Holarctic hoverfly species that have been named as mimics in the literature, organized by mimicry ring (F. Gilbert, unpublished). The striking thing is their sheer number, 256 species from a total of 2334 Holarctic species (11%). The world totals cited above (279 from 5600) demonstrate that outside Europe the available information is very fragmentary and unsystematic, hence these numbers are almost certainly an underestimate. European insects have been studied much more intensively, and in Europe there are 138 mimics out of a total of about 630 species (22%).

Virtually all the model identifications made by the authors concerned were purely on the basis of visual similarity according to our own human perception, with no experimental or any other kind of evidence. Of course, in natural circumstances predators are required to deal with potential prey in a wide variety of circumstances, including as fast-moving evasive insects, and some potential prey represent a significant threat to well being. Identifications based upon our own perceptions may not correspond to the perceptual confusions between models and mimics generated by the eyes of predators, and this might distort our view of biological reality. One element that has been highlighted is the UV-component of colour patterns (Cuthill and Bennett, 1993; Church *et al.*, 2004), invisible to mammalian predators, but possibly conspicuous to UV-sensitive birds or insects. *A priori* an unsuspected and different UV-component to the colour pattern is unlikely in Diptera, since their black colours are indole-based eumelanins: melanins strongly absorb in the UV, and therefore syrphids are unlikely to have UV patterns superimposed on any black part of their body. Photographs of social wasps and some of their hoverfly mimics in both visible and UV light have no UV patterns evident in either, nor in non-mimetic *Sarcophaga* flies (L. Gentle, personal communication). Similarly, Nickol (1994) took photographs of the hoverflies *Volucella inanis*, *V. zonaria* and their models (social wasps and hornets), and also found both models and mimics to appear entirely black under UV light. Thus the ability of birds but not humans to see UV light does not seem to be a serious problem in assessing the model–mimic relationships of hoverflies. In principle, if they did exist, these kinds of distortions are simple to remove, providing that we have realistic predator-based assessments of the degree of model–mimic confusion (see Green et al., 1999); in practice, such assessments are difficult to obtain. The fact that we are able to classify some mimics as 'extremely accurate' probably implies that our perception is rather similar to that of at least some other predators.

Amongst the hoverfly mimics of bumblebees, most model identifications are reasonably obvious, and the lists of bumblebee models generated from the various suggestions by different authors are generally very similar in their colour patterns. Usually there is little ambiguity, since the quality of the mimicry is very high. The distribution of mimics among the various Müllerian complexes is very different between the Nearctic and the Palaearctic. No form seems to mimic the black bumblebees with thin yellow bands (complex B) in the western Nearctic, and this complex is absent from the Palaearctic. A large proportion of Palaearctic mimics are either black with red tails like B. lapidarius (complex A), or all tawny-coloured like B. pascuorum (complex I), or yellow-banded with a white tail like B. lucorum (complex E); all of these complexes and their mimics are largely absent from the Nearctic. In contrast, the Nearctic complex of syrphid mimics with a pattern of a yellow anterior and a black posterior, like B. impatiens (complex G), is absent from the Palaearctic, although the inconspicuous white tail of Criorhina berberina berberina and possibly one morph of Cheilosia illustrata are rather similar (and which therefore lack a closely corresponding model pattern in the Palaearctic). The white-patterned bumblebees of the Caucasus are paralleled by the white-patterned mimetic Diptera there. These distributional correspondences themselves constitute powerful corroborative evidence for the reality of mimetic relationships.

Only four Palaearctic hoverfly species have a quality of bumblebee mimicry that can be regarded as poor or unclear. In the Nearctic, very little work on models and their mimics has been done, except for the series of papers by Waldbauer and colleagues (see Waldbauer, 1988). In all their work, Waldbauer et al. decided to consider all bumblebees as members of a single Müllerian complex, and hence clearly regarded the differences among their colour patterns as irrelevant. Mimetic flies were labelled merely as generalized bumblebee mimics, without noting any closer resemblance to particular species. The authors were then able to assume that Mallota bautias was a general mimic of bumblebees, although in fact it resembles a particular group of eastern Nearctic bumblebee species rather closely. The context dependency of mimicry is highlighted, however, by the fact that M. bautias was for decades regarded as conspecific with the Palaearctic M. cimbiciformis, so closely do they resemble one another morphologically. However, M. cimbiciformis is uniformly interpreted as 'a particularly fine mimic of the honeybee' (Stubbs and Falk, 1983), and to my knowledge has never been identified as a mimic of any Palaearctic bumblebee. In the Nearctic, where honeybees were only introduced in the 19th century, the identical colour pattern can operate as a bumblebee mimic: there are even some good experimental data showing that this bumblebee mimicry is effective in protecting the fly from predation (Evans and Waldbauer, 1982).

The mimicry of honeybees by some hoverflies (mostly *Eristalis* species, commonly called droneflies) has been commented upon for a very long time (Osten-Sacken, 1894), and even experts can be fooled. Benton (1903) exhibited a photograph published in an apicultural journal of 'Bees working on Chrysanthemums' which were in fact *Eristalis tenax*. He also recounted his role in 'the famous Utter trial' (whatever that was!), where the prosecution could not distinguish between honeybees and droneflies, and therefore were unable to prove positively that bees were the cause of some alleged damage. Even experienced beekeepers were unable to make the same discrimination. However, other entomologists have been less impressed with the match between the honeybee model and *Eristalis* species, and Mostler (1935) attributed their lower protection in his experiments to their lesser resemblance to the model. Nicholson (1927) agreed that *E. tenax* was 'somewhat like the common hive-bee', but insisted that it was 'one of the least convincing cases of mimicry I know'. I suspect that most entomologists would agree with Mostler (1935). There is a range of different mimics that correspond to the colour variants of honeybees: for example, *Eristalis tenax* and female *E. arbustorum* are like the darker varieties, and male *E. arbustorum* resemble the lighter varieties. There are also a number of bee-like *Eristalis* species in North America, but honeybees are not native to the New World and we have little idea about which of the native bee fauna might have led to the evolution of mimetic colour patterns among Nearctic *Eristalis* species.

It is the apparently wasp-mimetic syrphids that cause the greatest difficulties in assessing the extent of mimicry among the Syrphidae. They are freely quoted as examples of mimicry, but are often unsatisfactory under critical consideration. The resemblance is often not particularly close, and the quality of mimicry varies from good to bad: many authors have made this point (e.g. Brown, 1951; Dlusskii, 1984; Waldbauer, 1988; Dittrich *et al.*, 1993). Such problems led Waldbauer to define wasp mimicry to include only specialists that also mimic the long antennae and folded wings of vespoid wasps. However, this still does not mean that there is a one-to-one correspondence between species of models and these mimics, since often there is no particular exact replica of the mimic among available models. Some of the morphological adaptations for mimicry in this group are truly remarkable. For example, there are at least ten independently evolved solutions to the problem of mimicking wasp antennae (Waldbauer, 1970), three of which involve using the front legs. Species of the genera *Spilomyia* and *Temnostoma* and *Volucella bombylans* have only the normal short cyclorrhaphan type of antennae, but instead the anterior half of the forelegs is darkened, and the flies hold them up and wave them about in front of the head to create an amazingly good illusion of wasp-like antennae. Interestingly, not one bumblebee mimic has evolved elongated antennae, and only *V. bombylans* uses behaviour to mimic having them (although only females do this, in their final stealthy approach to the bumblebee nest in which they oviposit: Fincher, 1951; Rupp, 1989). This difference must tell us something about the salience of such features to predators; presumably long antennae are important features of identifying wasps, but the coat of hairs dominates when identifying bumblebees. Wasps do indeed wave their antennae about conspicuously, and bumblebees do not.

One characteristic of aposematic models and their mimics is that they often have harder, more durable bodies than other insects, toughened to withstand attack by predators so that the predators taste them but the prey still survive (Rettenmeyer, 1970: 58). The abdomen of many syrphine species is 'emarginate', i.e. each tergite is compressed just before the lateral margin, creating a narrow ridge or beading along the edge; this feature may have arisen in order to toughen the abdomen, since it occurs only in mimics. Specialized mimics have gone much further, and have the entire abdomen arched and convex, or cylindrical; the cuticle is punctate and hence greatly strengthened; and the joints between the overlapping tergites are very strong. If possession of an abdomen of this type is taken to define which of the black-and-yellow syrphids are truly wasp mimics, then rather few temperate species pass the test – in the northern hemisphere, only those of the genera *Ceriana*, *Chrysotoxum*, *Sphecomyia*, *Spilomya* and *Temnostoma*. There are, however, many genera with this type of abdomen in the tropics, perhaps an indication of a longer period of evolution among models and mimics there compared with the Holarctic.

The possibility that large Müllerian complexes of many wasp species together constitute a single model for Batesian mimics has hardly been addressed by anyone since Nicholson's (1927) largely uncited paper, except by Waldbauer and his colleagues in the eastern USA; this is an especially surprising omission among Palaearctic workers. Only Nickol (1994) has really identified this property clearly in his discussion of mimicry in *Volucella zonaria*, although it was also implicit in Dlusskii's (1984) important paper. Despite this omission, many such complexes appear to exist amongst black-and-yellow noxious insects and their mimics. The Müllerian complexes themselves are much less homogeneous than those of bumblebees, and overlap so that the boundaries are less distinct; presumably this is a consequence of their noxiousness.

5.2 Overall features of resemblance

Two patterns are very striking across these major model–mimic groups. The first is the great difference in the incidence of polymorphism among the groups (Table 9.1). There are 71 species of bumblebee mimics, at least 25 of which are polymorphic (35%), with each morph mimicking a different bumblebee species. This number of polymorphic species may in fact be even higher, since several species are only represented by a few specimens, or are poorly known taxonomically (e.g. some of the magnificent *Criorhina* species from western North America): there may still be multiple instances of 'species' which in fact represent different colour morphs of a single species. Among the better-known European species, more than 50% are polymorphic (18 of 35 species). Many of the polymorphisms are different between the sexes, either by each sex having a different (overlapping or non-overlapping) range of morphs, or mimicry being limited to only one sex (there are 17 male-only and 11 female-only morphs recorded); this may indicate a role for apostatic (non-mimetic) mechanisms in these polymorphisms (B. Clarke, personal communication). In stark contrast, of all the other mimetic hoverfly species, there is not a single example of a truly polymorphic species. A few species have a small degree of sexual dimorphism in their patterns, which some authors have then interpreted as mimicking different models, but none is very convincing.

The second striking pattern is the difference in specificity of mimicry (Table 9.1), to which I have already alluded. Virtually all authors who describe bumblebee mimics comment on their amazing similarity to their models. In contrast, many of the supposed wasp mimics are not very similar at all to their models. Honeybee mimics are somewhat intermediate, with some species being described as very good mimics, but others as rather poor.

An obvious objection to the claim of differences in the specificity of mimicry is that human perception is not the same as that of predatory birds, and perhaps they perceive the 'poor' mimics differently. Dittrich et al. (1993) used operant conditioning to test whether a representative bird would make the same sort of mistakes that humans do when presented with models and their hoverfly mimics. They chose pigeons (*Columba livia*) to represent a generalized avian visual system. The birds were trained to discriminate between images of wasps and non-mimetic flies, and then tested to see how they would respond to images of wasp mimics. One group (fly$^+$) were trained to peck at the images of non-mimetic flies for food, wasp images being unrewarded. A second group (wasp$^+$) were trained to peck at wasp images for food, with the non-mimetic flies being unrewarded. Both the fly$^+$ and wasp$^+$ groups of pigeons learned to discriminate between the two sets of images equally quickly, after only two training sessions. During the 20-s projection time of a rewarding stimulus, the pigeons pecked 50–60 times, whereas they hardly pecked at all at non-rewarding images. This suggests that there is no inherent bias of the pigeon visual system against black-and-yellow wasp-like patterns (perhaps not surprisingly, since it is not insectivorous).

The pigeons were then tested using images of hoverflies chosen to represent a range of mimetic quality as perceived by humans. The pigeons were extremely consistent in their responses, ranking the hoverflies in more or less the same order as did the humans, with the fly$^+$ group being more or less the mirror image of the wasp$^+$ group. Thus the main conclusion from this work is that pigeons do appear to see hoverfly mimics in roughly the same way as we do: they seem to rank the images in the same way, and make the same sort of category mistakes. The result is not an artefact of the lack of UV in the photographic images used for the experiment, as claimed by Cuthill and Bennett (1993), since using real specimens produces the same result (Green et al., 1999).

There were two interesting exceptions to this general pattern, which may contain pointers to some aspects of the evolution of mimicry. *Syrphus ribesii* was classified by the pigeons as the most wasp-like of all the images they looked at, and *Episyrphus balteatus* was also classified as extremely wasp-like:

these are not at all like the decisions made by humans, who are generally not impressed by their mimicry. Perhaps these species have exploited some peculiar aspect of the avian visual system or the psychology of learning so that they get classified as very wasp-like, even though to our eyes they are not; or pigeons use rather different features of insects than humans use in the classification process; or there may be some entirely different explanation.

5.3 Abundance

Considering how many people are interested in mimicry, it is amazing how little is known about mimic:model ratios in nature, even for butterflies. There are several hundred studies of syrphid communities, often using Malaise traps that also catch Hymenoptera, but not one presents any estimates of these ratios. Even vague estimates are very rare. Occasionally authors will comment on perceived ratios: for example, Heal (1982) stated that *Eristalis tenax* was more abundant than its honeybee model at many sites in the autumn, but he did not say what the actual ratios were. Heal did measure the relative abundance of light and dark morphs of *E. tenax* in relation to the light and dark forms of honeybees: over all his sites the light morph of the mimic was at a frequency of 44%, and of the model 45%, which is certainly consistent with mimicry maintaining the morph frequencies.

The quantitative data we do have to reconstruct these ratios consist of the extensive Malaise-trap sampling from an urban garden in Leicester (Owen, 1991) and a similar but more restricted dataset from the ancient forest of Bernwood in Oxfordshire (Watts, 1983; Archer, 1988), and a set of observer censuses from the USA (Waldbauer and Sheldon, 1971; Waldbauer *et al.*, 1977; Waldbauer and LaBerge, 1985), the UK (Grewcock, 1992; Howarth and Edmunds, 2000), European Russia (Dlusskii, 1984) and the Massane forest in the French Pyrenees (a fragment of the original wildwood of Europe: Grewcock, 1992). All methods of systematic sampling introduce some sort of bias (Southwood and Henderson, 2000), especially among species, and Malaise traps are poor at catching certain species, rendering the trapping data potentially misleading. In contrast, observer censuses could be considered to be a fairly close approximation of the hunting behaviour of predators (but see below).

Owen's (1991) data on a yearly basis show generally low mimic:model ratios, with mimics usually less or much less common than their models. This is always true for the bumblebee and honeybee mimicry complexes found in the garden, and for all the good wasp mimics; only the poor mimics of social wasps (complexes III and IV of Table 9.1) are much more abundant than their models. In Bernwood Forest the social wasps (43%) and bumblebees (39%) made up most of the hymenopteran catch, with solitary bees (10%), solitary wasps (6%) and honeybees (2%) being much less common. Poor wasp mimics were more than four times commoner than their models, and both good and poor honeybee mimics were also commoner than their models. In contrast, all the bumblebee mimics were rarer than their models, the commonest (complex J of Table 9.1, mainly *Criorhina berberina*) being only 32% as common as its model (*Bombus pratorum*).

From observer censuses of only good mimics in Illinois (USA), Waldbauer and colleagues found mimic:model ratios were all consistently at or below 1.00, meaning that mimics were always about as common, or more usually less common than their models. All the wasp complexes had higher ratios of mimics to models than the bumblebee complex, which had consistently low ratios in all areas. In ancient woodland sites in the UK, Grewcock and Howarth also showed that these ratios were low (<1) for the good bumblebee and wasp mimics, but poor wasp mimics were much more abundant than their models (by a factor of 4–19). In Massane, wasp mimics were more abundant than their models, but only by a factor of 2. Honeybee mimics were also much more abundant than their models in the UK (by a factor of up to 50), less so in Massane (4.5). Thus as in the Malaise-trap data, excess mimic:model ratios are a feature of poor wasp mimics, and also sometimes of honeybee mimics.

The relative abundance of models should also probably be an important feature of mimicry theory, especially if these vary in aversiveness (which could imply quasi-Batesian systems): what is the evidence from field studies? Among insects visiting flowers in dry grasslands of Germany (Kratochwil, 1983), bumblebee mimicry rings were very common (complex A of Table 9.1 – 90, E – 220, F – 188, J – 185), as were honeybee-like forms (222, but none of them *Apis*), but social wasps were rather rare (23). In Howarth's (1998) census walks in UK ancient woodlands the same pattern was evident, with bumblebee rings (complexes of Table 9.1: A – 61, F – 3271, J – 663) and honeybees (1475) being common relative to social (complexes II to IV of Table 9.1 – 880) or solitary wasps (I – 312). Waldbauer's group working in forests in the USA also found that bumblebee models were between 1.5 and 3.2 times as common as all the wasp models combined. Only in Bernwood (Archer, 1988) were social wasps (complexes II to IV of Table 9.1 – 1846) a bit commoner than bumblebees (complexes of Table 9.1: A – 1, E – 1021, F – 142, J – 549) and much more common than honeybees (224, including 74 *Apis*). The pattern seems very clear: the least noxious bumblebee models are also normally substantially more common than the most noxious wasp models.

It is possible that simultaneous model:mimic ratios are not what matters. In much-cited studies of Waldbauer and colleagues (see Waldbauer, 1988), the phenologies of mimics in Illinois showed a gap in the mid- to late-summer period, which the researchers concluded was timed to coincide with the period when young fledglings were learning about what was good and not good to eat. Furthermore, early-emerging mimics did not necessarily coincide with the appearance of their supposed models, whose flight period was much later in the season. Putting these phenological patterns together with the finding that some individual birds can, under some circumstances, remember aposematic colour patterns for long periods of time (months), they concluded that mimics must be protected by these memories, and therefore models and mimics do not have to coincide in space and time in order for mimicry to be effective. There are several problems with these conclusions. The first is that even non-mimetic hoverflies show the same mid-season lull in numbers, which seems to be more connected with the availability of aphids than with the number of fledgling birds. Many host-alternating aphids switch hosts in mid-season, creating a mid-season gap in aphid numbers well known to applied entomologists (e.g. Bombosch, 1963). A second problem is that the predator education process is not simply one of the rates of learning and forgetting, but also that of extinction, the learned alteration of a previously established association (see Shettleworth, 1998). Very little is known about the rates of these three processes in birds in any realistic setting, but Holling's (1965) experiments with shrews and Mostler's (1935) with birds generally showed a much faster process of losing the association than Waldbauer envisaged. Although rather easy to test experimentally, the hypothesis has only been tested observationally by repeated phenological studies in different areas; more critical testing is needed before these ideas can be accepted as established. Most other natural history information indicates that mimicry is only effective when models and mimics can be experienced simultaneously by a predator.

Heal (1995) noted that the main model changes from month to month. In spring, the best model to copy is the honeybee, because workers start foraging in the very first days of spring, and indeed some of the common honeybee mimics are active at that time. The great increase in bumblebee workers from growing colonies occurs in June and July, and many bumblebee mimics are on the wing then. Wasps are most obvious in late summer, when wasp mimicry peaks. Thus there is a regular shape to the phenology of mimetic complexes in the UK. These speculations were broadly supported by the quantitative data of Howarth and Edmunds (2000).

Howarth *et al.* (2004) predicted further that there would be an hour-to-hour dependency between the numbers of models and their Batesian mimics because of behavioural convergence in responses to habitat and weather conditions. Testing for covariation among models and their mimics, over and above the effects of month, site, weather and other general conditions, they found some very striking patterns. There were nine significantly positive relationships out of the 17 model–mimic pairs tested, and furthermore all 17 (bar one) were positive, itself a highly non-random pattern. Only one relationship was significantly negative, between *Helophilus* and their poorly mimicked social wasp models. In four out of seven poor mimics, and five out of ten good mimics, there was a significant positive relationship with numbers of the presumed model: this pattern is not significantly different between good and poor mimics. Interestingly, though, six of the seven common or abundant species showed relationships with their models, whereas this was true for only three of the ten rarer species, a significantly non-random pattern; thus common mimics are more likely to co-vary with their models. This result is very different from Waldbauer's since it involves intimate simultaneous co-occurrences between models and mimics, rather than their phenologies being completely different because of long predator memories. Possibly predators require constant reminding of noxious patterns (Rothschild's, 1984, *aide-mémoire* mimicry), but we need much more study of predator behaviour so that realistic models of their learning processes can be used in mimicry theory.

We do not have to reject the idea that common syrphids have evolved to be mimics merely because they are common. This crucial point was made by Nicholson as long ago as 1927. He pointed out that mimicry evolves when some individuals have a slightly higher probability of surviving than others because their variant of the colour pattern happens to provide some protection from predation via its greater resemblance to a noxious model. As long as that is true, then mimetic colour patterns will increase in the population relative to normal patterns. As alluded to in the introduction, ecologists would be very surprised if syrphid populations were regulated by birds feeding on the adult stages.

It is probably true that if the population is low, the selective advantage to mimicry will be greater, and hence there may be stronger selection to perfect the mimetic pattern. Thus, as Glumac (1962) suggested, low relative abundance may lead to mimicry, rather than the other way round. He pointed out also that many of the morphological and behavioural components of good mimics are also present in non-mimics, but acquire a new significance when selection acts to improve mimetic resemblance. Thus 'wasp-waisted' abdomens occur in *Baccha*, *Sphegina* and *Neoascia* without much or any mimetic coloration, but this morphology is particularly effective when combined with other morphological components of mimicry, such as elongated antennae and a darkened anterior sector of the wing. Glumac regarded only certain rare syrphids as mimetic; indeed, he really thought that mimicry was completely incidental in syrphid ecology. Most of the features normally interpreted as mimetic, he regarded as the result of convergent evolution, and his thinking was really part of the group-selectionist mind-set of biologists before the Darwinian revolution in behaviour and ecology.

5.4 Overall characteristics of hoverfly mimicry

Table 9.2 gives a picture of the range of mimetic quality within one of the major model types, mimics of social wasps. The hoverflies are ordered along a spectrum of mimetic quality, indicated by the arrow, as assessed by both

Table 9.2. Spectrum of mimetic quality among the hoverfly mimics of social wasps. Assessments of variability are from Holloway *et al.* (2002); mimetic quality estimates come partly from Dittrich *et al.* (1993), Howarth *et al.* (2000); all other data are from F. Gilbert (unpublished).

	Model	Hoverfly mimics			Non-mimics
Palatability	Noxious				Palatable
Mimetic accuracy	n/a	Accurate	Less accurate	Poor quality	n/a
Pattern variability	Variable	Variable	Quite variable	Less variable	Variable
Abundance	Common	Rare	Common	Abundant	Common
Structure	Armoured	Often armoured	Rarely armoured	Not armoured	Not armoured
Behavioural mimicry	n/a	Absent	Present	Present	n/a

Gradient of mimetic quality

human and pigeon eyes (see Dittrich *et al.*, 1993). Other axes of variation correlated with this spectrum are pattern variability, relative abundance and the occurrence of behavioural mimicry. Like the models, good mimics have very variable colour patterns, consistent with a relaxation of selection pressure caused by their accurate mimetic appearance: poor-quality mimics have much less variable patterns, but non-mimics are just as variable as the models (Holloway *et al.*, 2002). Visually poor mimics are much commoner than their models, and may compensate for these factors by evolving behavioural mimicry such that they tend to co-occur much more frequently with the models at particular times and places where predators encounter them (Howarth *et al.*, 2004). In the model, the tergites of the abdomen are joined together extremely strongly with overlapping sclerites, making it very difficult for a predator to grasp hold and dismember it. This kind of structure is often also present in the rare, highly accurate mimics, but is absent in common poor mimics.

Across the major types of model (Table 9.3a) there are also some very strong patterns. I have ordered the groups along a gradient of increasing noxiousness (justified above). Unlike the other two groups, the bumblebees contain a diverse set of Müllerian complexes, and the overwhelming characteristics of its hoverfly mimics are: a matching diversity, including a very high degree of polymorphism in both models and mimics; relative rarity; and highly accurate visual mimicry. The contrast with the social wasp group is very clear. Their hoverfly mimics are not polymorphic, are often visually very poor matches to the model, and are often many times more abundant than their models.

Thus polymorphism in certain Müllerian models (bumblebees) is common, and polymorphism among their apparently Batesian mimics is also common. Where the model has racial colour variation (honeybees), its mimics match this. Finally, in the wasp complex where there is no polymorphism within any of the model species (even though variation among the colour patterns of its constituent species certainly exists), no polymorphism is seen in the mimics.

6. Can Mimicry Theory Account for the Spectrum of Hoverfly Mimicry?

The widespread occurrence of poor-quality mimicry has given rise to a variety of different hypotheses to try to account for the evolution of these 'mimetic' patterns, many of them based to some extent upon the consideration of syrphid colours. Many are not really part of standard mimicry theory, but are additions or alternatives to some of its postulates. I have listed the main elements of these ideas in Table 9.3b, with an indication of the way in which each idea accounts for the spectrum of mimicry identified in Table 9.3a. These ideas are not necessarily mutually exclusive, since in the main they involve different types of trade-offs, costs and benefits, and several are almost certainly operating at the same time.

Table 9.3. (a) Overall characteristics of hymenopteran-model:hoverfly-mimic complexes; and (b) their possible causes. The thick arrow indicates the gradient of model noxiousness (see text for justification), and the possible explanations each creates a spectrum correlated with this gradient.

(a) **Models**	**Bumblebees**	**Honeybees**	**Wasps**
Size	Large	Medium-sized	Medium-sized
Coloration	Aposematic	Non-aposematic	Aposematic
Number of Müllerian rings	Usually 1–4 in any one region	One	One large diffuse ring
Polymorphism	Present	Racial variation	Absent
Nature of defence	Heavily armoured	Armoured	Armoured
	Palatable	Unpalatable	Highly unpalatable
	Rarely sting	Rarely sting	Sometimes sting
Hoverfly mimics			
Size	Large	Medium-sized	Small to medium-sized
Physical defence	Often armoured	Rarely armoured	Rarely armoured
Abundance	Rare	Often common	Often abundant
Polymorphism	Polymorphic; often sex-linked	Variation matching model	Non-polymorphic
Visual similarity	Usually visually accurate	Reasonable quality	Occasionally good, usually poor

(b) **Possible explanations**	**Bumblebees**	**Honeybees**	**Wasps**
1 Occurrence of mimicry	Mimics		Non-mimics
2 Nature of predator perception	Same as human		Different from human
	Birds		Invertebrates
3 Nature of mimicry	Batesian		Müllerian
4 Nature of model	Model only mildly noxious		Terribly noxious
5 Type of mimicry	Mainly visual		Mainly behavioural
6 Effect on predator	Deceived		Reminded/confused
7 Speed of evolution	Fast, complete		Slow, incomplete
8 Disturbance by man	Forest habitats, undisturbed		Habitats disturbed by man
9 Resistance to evolving perfection:			
(a) cost of mimetic pattern	High cost, high benefit		Low cost, low benefit
	Hair colour easily modified	Slight changes to tergite colours possible	Major changes to tergite colours not possible
(b) attack threshold + kin selection	Attack threshold low, no kin-structured distribution		Attack threshold high, kin-structured distribution
(c) number of models	Single model	Different races of model	Multiple models

6.1 The occurrence of mimicry: 'poor mimics' are not mimics at all

An obvious tack is to deny that mimicry can possibly be of poor quality, and hence to claim that poor 'mimics' simply are not mimics at all. Before the advent of more rigorous quantitative testing of mimicry theory by the Browers and others, there were several who claimed that because some mimics were eaten by some predators sometimes, this meant that mimicry did not 'work', and hence was invalid as a theory (Heikertinger, 1918, 1936, 1954; McAtee, 1932). It is a simple step then to accept perfect mimicry as a valid concept, but to deny any other kind of resemblance as mimetic: for example, Glumac (1962) and Drees (1997) thought there were no mimics among syrphids except for those with elongated antennae (even bumblebee mimics, which they attributed to convergent requirements of thermoregulation). Waldbauer and his colleagues (see Maier, 1978; Waldbauer, 1988) simply ignored poor mimics altogether because it was impossible to determine whether they were really mimics or not.

However, denial of the reality of mimicry in these syrphids has not been credible since Mostler's (1935) detailed and large-scale experiments on learning in naïve and experienced insectivorous birds. Single, either hand-reared or experienced (wild-caught) birds were free to fly in a large windowed room with naturalistic perches, and models and mimics were released alive into the back of the room, usually to fly directly to the window: Mostler recorded the subsequent behaviour of the birds. He divided his study into two parts: the first was concerned with the beginning and the end points of the learning of birds, i.e. the responses of young naïve birds and of old experienced birds to models and mimics; the second part investigated the learning process itself. Although he did not plot or hardly even analyse his data, the great value of his work is not just that it was the first well-designed large-scale experimental approach to testing the theory of mimicry, but also that he obtained comparable data on all three hymenopteran models and their hoverfly mimics. These established beyond doubt that the colour patterns of all the syrphids he used did give substantial protection from predation, and that the protective effect was proportional to mimetic similarity (Figs 9.1 and 9.2). Honeybees and their mimics were a less successful mimetic system than the wasp system, while the bumblebee system was the most successful of all. When mimics were offered soon after their models (within 50 min) in the wasp mimicry system, the wasp mimics were strongly protected (Fig. 9.2), fading with time, but this protection vanished when they were offered before models, and in fact the wasps suffered, since more wasps than normal were attacked. Assessment was easier when the birds could compare models and mimics at the same time, but when this could not be done, they were much more cautious, and the protective effect lasted much longer. Mostler also conducted some trials of wasps and wasp mimics where he gave insects in the sequence model – mimic – model, ensuring in each case that the mimic was eaten before the second model was presented. The proportion of models attacked after experiencing the mimic was much higher than those attacked beforehand, demonstrating very clearly the negative effect that mimics have on models.

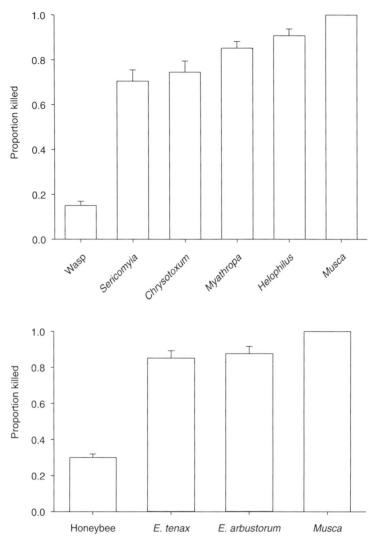

Fig. 9.1. The mean proportion (±SE) of offered model and mimetic insects that were attacked and killed by a set of insectivorous birds (*n* = 48) of various species. Above: wasps and their mimics; below: honeybees and their mimics (two *Eristalis* species). In each case, mimics are ordered along a gradient of mimetic quality as assessed visually by the observer. Data from Mostler (1935).

A further and powerful argument about whether the colour patterns are mimetic is the observation that in New Zealand there are no native bumblebees or yellow-and-black noxious wasps, and uniquely there are no wasp or bumblebee mimics either among native New Zealand syrphids (S.D. Wratten, personal communication).

The only field-based experiments on the protection afforded to syrphid mimics were done by Dlusskii (1984) in a forest close to Moscow. He undertook

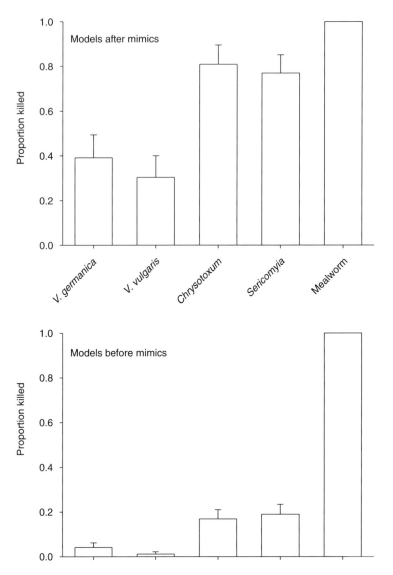

Fig. 9.2. The average proportion (±SE) of models (*Vespula germanica, V. vulgaris*) and mimics (*Sericomyia borealis, Chrysotoxum arcuatum*) from the wasp model–mimic complex, and mealworms (control) killed and eaten by birds when models were offered after mimics (above) or before mimics (below). Data from Mostler (1935).

a series of choice tests under natural conditions, by placing near nests a table on which were offered live insects tethered in pairs, able to move and even to start up flying. Dlusskii paired a series of different insect species with *Eristalis nemorum*, which he knew from earlier experiments to be acceptable, and he determined which of the two insects was taken first. If one insect appeared to be palatable and the other unpalatable, then the palatable insect was taken at least

90% of the time, and even if the unpalatable insect was taken, it was never delivered to the nest. If both seemed unpalatable because of prior experience, then the bird sat and looked at them, but flew away again without sampling either. If both seemed palatable, then the bird ate them both, but the most attractive one was eaten first, and usually this was the larger of the two alternatives. Thus it was possible to test whether insects were considered to be palatable or not to the birds being tested. The Hymenoptera were always considered to be unpalatable, even *Eucera longicornis*, which was rare in the area and therefore (Dlusskii thought) unlikely to have been encountered before by these birds. These observations (and pigeon experiments by W. Dittrich (unpublished data)) suggest that the birds had a polymorphous concept (see Lea and Harrison, 1978) of what a hymenopterous insect was, and could apply it effectively to identify the insects on offer even if they had never encountered them before. On the other hand, all the syrphids were considered to be palatable, and even the superb wasp mimic *Temnostoma vespiforme* was eaten by spotted flycatchers despite the fact that its model was rejected. Dlusskii concluded that these experienced birds usually distinguished between models and mimics, even the good ones, and thus mimicry was ineffective here. There were cases where prior experiences caused the birds to reject mimics, however, and thus on occasion even weak similarity to a model could protect mimics.

Apart from Mostler's and Dlusskii's work, there are really only fragments of information in the literature about the protective effects of syrphid mimicry (see Pocock, 1911; Steiniger, 1937a,b; Lane, 1957; Liepelt, 1963; Davies and Green, 1976; Evans and Waldbauer, 1982; Heal, 1982, 1995; Evans, 1984; Grewcock, 1992). I can summarize the available data no more clearly than Steiniger (1937b) did many years ago:

- Syrphids form part of the normal dipteran diet of many insectivorous birds such as robins (*Erithacus rubecula*), redstarts, and *Sylvia* and *Phylloscopus* warblers.
- Wasps are not normal food for these birds, which do not have any innate avoidance of wasps, and eventually come to try one; once tried, they are not eaten any more.
- Syrphids are also removed from the normal diet of these insectivores as soon as they become acquainted with wasps, since wasps and syrphids are apparently confused.
- The protective effects of this process are related to the similarity between model and mimic, but even poor-quality mimics benefit to some degree.

6.2 The nature of predator perception: 'poor mimics' appear perfect to their predators

It is possible that poor mimicry is an artefact of human perception, and that to the predators that generate the selection on the pattern, these mimics appear to be just as perfect as any other mimic. We have already seen that Dittrich *et al.* (1993) have suggested that the apparent imperfections of two very common syrphid mimics were only so for human eyes, because they appeared to be categorized as extremely good mimics by pigeons. Against this interpretation is the overall pattern of mimicry obtained from the pigeon experiment, which matched the ordering of mimetic quality of the human eye. It may well be true that certain species have managed to exploit some feature of bird perception in order to appear more perfect than they do to us, but in general this is not the case, and hence it is not a solution to the problem of imperfect mimicry. I have already shown above that the fact that birds can see UV does not seem to be an important feature of syrphid predation, and therefore is an invalid interpretation of the two anomalous patterns (cf. Cuthill and Bennett, 1993; Church *et al.*, 2004).

Alternatively, the fact that bumblebee mimics are on average substantially bigger than wasp mimics might cause them to have *different* predators, and this might underlie the observed differences in mimetic quality and relative abundance (Edmunds, 2000). According to current thinking, however, this requirement for a visually selective agent restricts the potential candidates basically to birds, because it seems to rule out virtually all invertebrate predators. But is this true?

Kassarov (2003) tries to argue that not even birds are suitable candidates because even their visual abilities are inadequate for perceiving adequately the largest insect patterns (butterflies). He thinks that birds only see the flight movements of potential prey, relying on these to make decisions about whether to attack or not. He rejects the idea that birds can be the selective agents generating mimetic colour patterns. If true, then the patterns must have arisen by magic!

Dragonflies are perhaps the most obvious of insect predators that hunt visually, although their eyes seem to be adapted to detecting potential prey moving against the sky, rather than forming an image that would include the colour pattern (Corbet, 1999: 341). Most insect eyes appear to be primarily movement detectors, and the conventional wisdom is that they probably do not form an image sufficiently detailed to be able to generate selection for high quality mimicry. Some large aeshnids do take many bees (Corbet, 1999: 354, 379), but no-one has recorded them taking bee mimics. While they may sometimes avoid certain prey types, such as wasps (Alonso-Meija and Marquez, 1994; O'Donnell, 1996; Howarth, 1998: 16), unless this avoidance is visually based it is hard to imagine this contributing to the evolution of mimicry. Recent work (Kauppinen and Mappes, 2003; T. Sherratt, personal communication), however, shows that dragonflies are able to select between wasps and flies, and that this discrimination is largely visual: black-and-yellow stripes alone reduce rates of attack. Thus dragonflies may well have contributed to the evolution of mimicry in insect colour patterns.

The beewolves of the genus *Philanthus* (Sphecidae) are well-known bee predators (Osten-Sacken, 1894: 11). The European *P. triangulum* takes almost exclusively honeybees, mostly from flowers but also from the hive entrance; they never take any mimetic Diptera (Iwata, 1976: 150). Fabre (1913) tried to deceive one by offering it an *Eristalis tenax*, which it 'rejected with supreme contempt'! In the USA, *P. bicinctus* preys on bumblebees in Yellowstone

National Park, but is apparently never deceived by mimetic flies; it makes an interesting contrast with the sphecid *Bembix pruinosa*, a fly predator that feeds very often on *E. tenax* (Evans, 1966: 131). In neither case is there any evidence of any protection gained from the resemblance. This is not surprising, since *Philanthus* is well known to respond visually first to motion, and then when close odour becomes the crucial cue, stimulating the final pounce. This odour-directed prey capture explains why honeybee mimics are never captured. Similarly predators specializing on flies, such as *Bembix*, often take mimetic syrphids but do not take wasps, for the same reason (Evans and Eberhard, 1970: 52).

Robberflies (Diptera: Asilidae) are also voracious predators, but there is only a single study that suggests they can generate selection based on vision: a study of tiger beetles by Shelly and Pearson (1978) suggested that a robberfly may have been responsible for the evolution of both chemical and aposematic defences. However, the red pattern involved is just a block of colour, very crude in comparison to hoverflies. Brues (1946) suggested that asilids had a 'fondness' for worker honeybees, but this seems unlikely, given the evidence of all the prey records (listed at www.geller-grimm.de/catalog/lavigne.htm). He thought that because they frequently catch *E. tenax*, this meant that 'to the insect eye *Eristalis* really looks like a bee'. However, the appropriate null hypothesis of no ability to discriminate, that asilids merely catch both because they co-occur in the same habitats, has never been tested.

Sphecidae (Hymenoptera) are well-known insect predators, but members of only two of the subfamilies (Nyssoninae and Crabroninae) take syrphids (Bohart and Menke 1976; Iwata, 1976). *Bembix* (Nyssoninae) are large, very fast-flying wasps that deliberately target flies on flowers, or swarming males, and hence often take a large number of muscids, tabanids and syrphids. In Tsuneki's (1956) Japanese study, for example, *B. nipponica* took prey belonging to 11 families of flies overall, but in some sites and years syrphids formed almost half the prey (47%), mainly *Eristalis cerealis*. A large number of the hoverflies caught were mimetic, such as the wasp-like *Takaomyia* and *Chrysotoxum*, and the bee-like eristalines. Large flies took disproportionately longer to find, capture, subdue and bring back to the nest: although it took on average 0.7 s longer to catch a syrphid rather than a non-syrphid, this difference was not significant and there was no evidence that syrphids were harder to catch than other flies. The other subfamily, the Crabroninae, are virtually all dipteran specialists, and the Crabronini are especially significant as hoverfly predators. A number of genera are important, especially *Ectemnius*. At least one species, *E. cavifrons*, seems to be a syrphid specialist; it is the commonest species in the UK (Pickard, 1975). The great majority of the prey in Pickard's study consisted of poor mimics (*Syrphus* spp. and *Episyrphus balteatus*). There was some selectivity involved because small dark species were greatly underrepresented, and no *Eristalis* were taken at all: Pickard thought their resemblance to honeybees might have protected them. Thus discrimination on the basis of colour patterns is possible, with a preference for yellow-and-black species. Some other studies have suggested that visual mimicry may protect against solitary wasps (e.g. predation on salticid spiders: Edmunds, 1993).

Social wasps (Spradberry, 1973: 141) take a huge variety of prey, but they concentrate on adult Diptera: some colonies have been recorded as taking up to 84% flies. Hornets can certainly take a lot of honeybees. It is usually thought very unlikely that these predators identify their prey visually, but recently (Tibbetts, 2002) *Polistes* wasps have been shown to identify individual colony members via variation in facial markings. Such a sophisticated ability indicates the capability for social wasps to generate natural selection for visual mimicry: this needs testing.

It is also possible that spiders make the sort of visual mistakes that would select for mimetic colour patterns, even though Bristowe (1941, vol. 2: 319) states that 'the yellow and black wasp-like appearance of certain syrphids is of no avail against spiders'. In fact spiders treat social wasps and bees with great caution, and only the largest species can tackle them successfully. Pocock (cited in Osten-Sacken, 1894: 11) noticed that *Agelena labyrinthica* used special precautions before overpowering a honeybee enmeshed in the web, whilst they pounced immediately on normal prey: when offered *Eristalis*, spiders approached and finally killed them, but used the same precautions as for honeybees.

Spider webs are in general not very good at catching hoverflies (Nentwig, 1982) since these flies are too large, strong and active. Insects with kinetic energies of about 150 µJ are able to fly straight through a web, and those with energies greater than about 500 µJ always do. The weight and flight speeds of syrphids indicate kinetic energies between 25 and 500 µJ, and therefore the smaller species should generally get caught, while the larger species (*Eristalis* spp., for example) should be able to ignore webs altogether. After becoming entangled in a web, insects differ considerably in their behaviour, and these differences determine whether they escape. Insects such as syrphids that react to web entanglement by continuous vigorous activity are able to escape in the few seconds available before the spider attacks, and syrphids weighing as little as 9 mg escape rather easily: most syrphids are larger and more powerful than this. Orb-web spiders (*Araneus diadematus*) studied by Myers (1935) seized non-mimetic Diptera such as *Calliphora* immediately with no precautions, and never wrapped them in silk. When wasps were the victims, the spiders would carefully rotate the prey, showing great skill and alacrity in avoiding both the mouthparts and the apex of the abdomen, and swathe it in silk until completely helpless before biting near the centre of the dorsal surface – the safest position. Honeybees and *Eristalis* were, if tackled at all, treated with great caution and were nearly always swathed in silk. Thus spiders treated their victims differently, not according to size and vigour, but according to the perceived risk; *Eristalis* was treated like a bee rather than a fly of the same size.

Flower spiders (Thomisidae), especially *Misumena*, have evolved to catch flower-visiting insects, and most of their prey are syrphids, honeybees and bumblebees (see Morse, 1986). Only a single study has addressed whether colour patterns might be protective. About half the individuals studied by Tyshchenko (1961) would avoid both wasp models and their mimics, and the other half would eat them; the reluctance of the former group to attack mimics was in proportion to their visual similarity to the model. Thus visually based predation by spiders needs systematic study, since it too seems perfectly capable of generating selection for mimicry.

Which birds might be candidates for agents of selection for mimetic colour patterns? Reviewing Palaearctic bird diets (using Cramp, 1977–1994) does not get us very far since really we need information about whether birds undergo the process of learning to avoid models, and whether they confuse models with mimics, rather than data about the endpoint of the learning process where the birds never take either (i.e. the adult and nestling diets that are normally reported). Bee-eaters (*Merops apiaster*) feed on both models and mimics, but they hawk in the open savanna on hot days; this is quite different from the habitat of most hoverfly mimics, which are overwhelmingly forest dwellers (Speight *et al.*, 1975; Maier, 1978; Speight, 1983) and avoid the hot midday (Gilbert, 1985). No protective effect occurs with this bird, since it specializes on wasps and bees, preferring them to all other prey. Hirundines such as the swallow (*Hirundo rustica*) also take syrphids and honeybees, but with their high-speed aerial scooping feeding method it is unlikely that they perceive the colour patterns before or after capture. Spotted flycatchers also take both wasps and hoverflies, but they are not deceived by the resemblance (Davies, 1977), and have an effective method of dealing with the venom of wasps. Thus for these birds, wasps and bees may not be noxious but instead may form part of their normal diet; it is possible that syrphids have longer handling times and are therefore unprofitable.

More likely candidates are birds such as *Phylloscopus*, *Sylvia* and *Hippolais* warblers, and others such as stonechats (*Saxicola torquata*). All these feed on syrphids, but we know virtually nothing about their selectivity among syrphid species. What we would be looking for would be evidence that: (i) birds had contact with noxious models; (ii) they also took syrphids; and (iii) the spectrum of syrphids upon which they fed was biased towards non-mimetic species. This sort of evidence is amazingly sparse in the literature. For example, Greig-Smith and Quicke (1983) noted that stonechats fed many warningly coloured ichneumonids and large numbers of syrphids to their nestlings, but we do not know what kinds of syrphids these were, and hence whether they might have been mimics. Similarly, we know that wheatears (*Oenanthe oenanthe*) feed bees and 'large Diptera' to their older nestlings (Cramp, vol. 5: 779), but were these large Diptera bee mimics? Then there are other birds such as *Ficedula* flycatchers, *Acrocephalus* warblers, and small passerines such as wagtails (Motacillidae), redstarts, robins and titmice (Paridae). The diet evidence suggests that these birds are minor or insignificant as hoverfly predators, but this is based only on samples of prey caught by experienced adults. Their reluctance to feed on syrphids may be entirely learned behaviour, taught to them by disastrous experiences they had when fledglings.

It is thus certainly possible that spiders and wasps are the main agents of selection for the smaller wasp mimics, whilst birds are the selective agents for bumblebee mimics. Superficially this is an attractive explanation because the much cruder visual abilities of the invertebrates would mean that imperfection of the colour pattern would not matter. However, until we know more about visual aspects of their predation, current knowledge really rules them out. Thus at the moment we can only conclude that inexperienced fledgling birds must be the selective agents responsible for the evolution and maintenance of mimicry in syrphids across the mimicry spectrum, but very little is known about their foraging behaviour. Birds that swoop down on flower-visiting insects from perches are probably the major candidates (see below). Since very high mortalities occur between fledging and recruitment into the adult population, the numbers of such young birds are probably very high relative to those of breeding pairs of adults (the usual density estimates), and hence their selective impact on syrphids might be very large.

6.3 The nature of the mimicry: are poor mimics Müllerian rather than Batesian mimics?

For a long time various people have wondered whether poor mimics are actually Müllerian mimics: for example, Jacobi (1913) labelled wasp mimicry a specially striking form of mimicry ('sphecoidy'), but could not decide whether it was Batesian or Müllerian. Is the gradient from good to poor mimics therefore a Batesian–Müllerian spectrum? Since Müllerian mimics are not thought to evolve to resemble one another particularly closely, but Batesian mimics are, this is an attractive explanation. The basis of the 'noxiousness' of a model need not be unpalatability or stings, despite the fact that most discussions about mimicry have focused upon these elements. As Holling (1965) noted, a considerable number of other features generally related to defence can affect the acceptability of prey: stings, noxious sprays, sticky exudates, colonial defence, tough and spiny integuments, and effective escape behaviours. Thus some syrphids may be benefiting from their unprofitability, rather than hiding under the cloak of noxious models. What sort of unprofitability might be involved? There are a number of related, but subtly different possibilities of the ways in which this kind of mechanism might account for the poor-to-good spectrum of mimetic quality.

- *Is there a trade-off between flight agility and mimetic resemblance? Are poor mimics particularly agile, and therefore do they not have to be accurate visually?* Aside from their conspicuous coloration, syrphids are very agile fliers and may therefore be particularly difficult for birds to catch. It would be no surprise if these two notable features were connected in some way. One obvious hypothesis is that these are alternative strategies (Grewcock, 1992). Species with relatively slow unaccomplished flight may be placed under strong selection for high-quality mimicry if their mimetic strategy is to be successful. More agile species may achieve a similar degree of protection with a less close resemblance because their agility reduces the predator's opportunity for assessing the pattern. However, cause and effect are difficult to disentangle since where selection acts to perfect resemblance, this may include mimicry of the typically slow meandering and weaving hymenopteran flight pattern, in sharp contrast to the direct, darting flight of most syrphids. Slow flight might be an integral part of high-quality mimicry, rather than a factor that promotes its evolution: high-quality mimics such as *Callicera* and *Temnostoma* are well known for this kind of behaviour (Glumac, 1962; Haeseler, 1976; Morgan and Heinrich, 1987; Speight and Lucas, 1992; Nickol, 1994; Gilbert, 2001).

- *Is flight agility an alternative to mimetic resemblance? Is accurate mimicry only required in particular circumstances?* Flight agility and mimicry might be substitutes rather than alternatives, working at different times or places. For example, it could be argued (Grewcock, 1992) that the agility of hoverflies is such that they can rely entirely on escape as a means of protection (see below), and that the colour patterns, if they represent a protective strategy at all, confer protection under more particular circumstances, such as during the pre- and post-active flight periods of the day, or at a time when flight is hampered (e.g. when pairs fly around *in copula*). Thus Hartley and Quicke (1994) were amazed to find half of the nestling diet of corn buntings (*Miliaria calandra*) consisting of syrphids (*Helophilus* and *Rhingia*), and assumed that they must have been caught in the early morning before they had had a chance to get warm enough to fly. Although syrphids have their own endothermic warming mechanisms, remarkable for such small flies, which shorten this vulnerable time window relative to other similar-sized flies, it is certain that endothermic predators will be able to remain active for a considerable period of the day during which hoverflies will be unable to use flight as an escape response. If this were a vital component of the evolution of mimicry in syrphids, then we might predict differences in the thermoregulatory abilities between mimics and non-mimics. The only study of this question (Morgan and Heinrich, 1987) found no such differences. Alternatively, perhaps mating is the vulnerable period. Allen (1964) caught a mating pair of the rare *Pocota personata* in the early afternoon and tried in vain to persuade the couple to separate by gently pulling them apart. Most mated pairs of syrphids break apart immediately upon capture, but this pair remained *in copula* in the jar until the following morning, more than 20 h later, the male in a 'cataleptic state'. Allen thought that the mated state must therefore be a very vulnerable one, 'protected solely by the remarkable bumblebee mimicry' of this species. There are other possibilities for especially critical places and times: Speight (2000) suggested that the vulnerable period when mimicry becomes effective is when females are immobilized while ovipositing, or on hot days when models and mimics co-occur on the ground drinking at streams.
- *Escape mimicry: do syrphids constitute a Müllerian complex advertising their unprofitability?* The syrphid colour patterns may have arisen to advertise the unprofitability of great flight agility to predators (Grewcock, 1992; Dittrich *et al.*, 1993; Edmunds, 2000), and hence these syrphids would then constitute a Müllerian complex. This complex could then be mimicked by other non-agile insects, and hence such syrphids could be models rather than mimics. This would imply that predators try to catch syrphids in flight rather than on flowers, and as we shall see below, the latter seems more likely to be a significant selective force. If they were advertising their agility, then it would imply that the evolution of conspicuous coloration represents a low-cost strategy, at least in syrphids. If hoverflies are so difficult to catch, what is the point of advertising this fact? The widely accepted explanation is that, providing that the cost of advertisement is low, it can reduce an already low

risk of attack to near-zero at very little cost (Grewcock, 1992). The ability to escape from bird predators is a well-established alternative to unpalatability in butterflies (Marden and Chai, 1991; Srygley, 1994), and therefore it would be reasonable to expect it in syrphids. Thus the prediction is that syrphids with wasp-like colour patterns (the 'poor' mimics) are particularly agile, and advertise this fact to potential predators; these 'poor' Müllerian mimics should then be more agile than either the true Batesian mimics (which should mimic the unconcerned flight of their models) or non-mimics. Using phylogenetically independent contrasts, Azmeh (1999) performed a preliminary test of this hypothesis by measuring the centre-of-body-mass of 14 species (correlated with flight agility: Srygley and Dudley, 1993; Srygley, 1994, 1999) and relating it to similarity to the model: there was no relationship, but the power of the test was low because of the small sample size. Davies (1977) is the only person I know actually to have some idea of the relative difficulty of capture of different Diptera for any bird, derived from his studies of spotted flycatchers. Interestingly, he thought that syrphids were not noticeably more difficult to catch than other large Diptera, and actually seemed to be easier than muscids, whose tricky erratic flight is harder for the birds to follow.

- *Poor syrphid mimics normally have aphid-feeding larvae. Are aphidophagous hoverflies unpalatable?* Malcolm (1976, 1981) put forward the idea that there is a fundamental difference in the nature of the mimicry between the good (Batesian) and the poor (Müllerian) mimics. The good mimics usually have non-predatory larvae, and he suggested that these were true Batesian mimics. The poor wasp mimics almost always have aphidophagous larvae, and he thought some or all the individuals of a population of hoverflies might be sequestering plant poisons via their aphid prey to make themselves noxious to predators. Thus the poor mimics of the aphidophagous Syrphinae would form a Müllerian complex based on unpalatability. Statements in the literature that syrphids might be unpalatable are usually derived from very doubtful interpretations of data by Pocock (1911), Carrick (1936), Parmenter (1953) and Lane (1957): we need some strong quantitative evidence of toxin sequestering and subsequent rejections by birds. Malcolm studied the South African species *Ischiodon aegyptius*, whose bright green larvae often feed on the bright yellow *Aphis nerii* on *Asclepiadaceae*. Plants of the *Asclepiadaceae* often contain large quantities of cardenolides as a defence against herbivores. *Aphis nerii* is restricted to these plants and at least some populations on some plants sequester the host-plant cardenolides. Using thin-layer chromatography, many glycosides were detected in the plants, and some were also detected in both aphids and syrphids. Extracts of *Ischiodon* reared on *A. nerii*, however, produced four completely different spots which could not be reliably identified in any of the plant extracts. Furthermore, extracts from *Ischiodon* reared on two other aphid–plant combinations produced the same spots. Thus the experiment failed to show any evidence of transfer of cardiac glycosides from the host-plant through the aphid to the predator. There did seem to be cardiac glycosides in *Ischiodon* adults, but these were apparently not sequestered from the prey. All extracts from *Ischiodon* fed on *A. nerii* produced a huge impact on exposed toad and chameleon hearts, particularly severe in chameleon hearts where there was a dramatic and sudden drop in heart rate, followed by a slow recovery. A similar but weaker response to *Ischiodon* reared on non-aposematic aphids was seen in the frog heart. Thus cardiac glycoside activity seemed to be present in *Ischiodon* irrespective of the plant–aphid association on which it was reared. The gut lining of vertebrates is relatively impermeable to highly polar glycosides, and hence *Ischiodon* is likely to produce an emetic response rather than cardiac arrest; low polarity glycosides are readily absorbed and could result in serious cardiac toxicity and death. These hoverflies should therefore be noxious to potential predators, and their yellow-and-black colour pattern could be aposematic rather than mimetic. Malcolm (1981) tried to repeat this work in greater detail in Oxford, but rearing problems kept sample sizes very low, and *Ischiodon* showed no sign of any cardenolides after feeding on *A. nerii*. This is perhaps the reason why this remarkable work remains unpublished.

In contrast to the paucity of evidence for unpalatability, many papers show that hoverflies are extremely palatable (e.g. Mostler, 1935; Steiniger, 1937a; Heal, 1979, 1982; Evans and Waldbauer, 1982; Dlusskii, 1984). These data are very compelling, and the conclusion must be that most individual syrphids are probably palatable most of the time to most, if not all, predators. However, the relationships among plant, aphid and syrphid defences is both fascinating and almost unexplored, and deserve further study, taking into account the predicted variability in toxin levels among different individual aphid–host-plant rearings (cf. Vanhaelen *et al.*, 2001, 2002). Thus, in principle it is not impossible for some individual syrphids to be unpalatable and others to be wholly palatable, but it would be more convincing were there many more systematic observations of captive predators displaying behaviours indicating that apparently innocuous aphidophagous hoverflies were unpalatable.

6.4 The nature of the model: wasp models of poor mimics are exceptionally noxious, so mimics do not need to be perfect

This hypothesis certainly appears to explain at least some of the mimicry spectrum in syrphids, since wasps are particularly noxious (and hence their mimics need not be very accurate), whereas bumblebees are apparently only mildly noxious (and hence require accurate copying). Whether this is also the explanation for the differences in relative abundance is not clear, but it is certainly possible, since a very noxious model should in theory be able to support a greater relative abundance of mimics. Thus, although this hypothesis seems certain to be part of the explanation, it still leaves unexplained the force counteracting the constant selection for improved mimicry, even though this force is almost certainly very weak when mimics evolve a reasonably close similarity to the model (see below).

6.5 The type of mimicry involved: poor mimics use behaviour or other factors to compensate for their visual discriminability

Dlusskii (1984) suggested that for poor mimics, mimicry is only one of a range of strategies designed by natural selection to reduce the impact of predators; for good mimics, in contrast, mimicry is *the* mechanism, and low abundance is not the consequence but the reason for their high mimetic fidelity. Mimicry may be strongly promoted by flower-visiting behaviour, since insects are usually fairly conspicuous on flowers whatever their pattern, negating the main disadvantage of aposematic coloration, i.e. the increased attack rate caused by greater visibility (Guilford, 1990). Quantitative comparisons between the flight behaviour of models and hoverfly mimics have only just begun to be carried out, on honeybees and their *Eristalis* mimics (Golding and Edmunds, 2000; Golding *et al.*, 2001), and demonstrate some behavioural convergence in flight characteristics and visitation rates. Nickol (1994) made the interesting further suggestion that mimicry may be associated specifically with interspecific competition on flowers. Direct interactions on flowers occur very frequently, and usually the winner is the larger of the two insects (see Kikuchi, 1963). If the winner remains on the flower despite interruption by smaller insects, then the former should either be extremely good at discriminating the approaching shapes of predators from those of other insects, or be able to deter the attacks of predators via mimicry. This might explain the association between large body size and mimetic quality in syrphids.

The exact details of prey capture by birds can be very important, but we have hardly any accounts of the behaviour of wild birds when hunting for and catching bumblebees, wasps or hoverflies, especially those birds identified above as probable candidates as selective agents for mimicry. It would be particularly useful to have data on the foraging behaviour of juvenile birds, and any differences from adults. However, predation consists of a set of relatively rare behavioural events, often occurring in widely separated parts of the environment, and is therefore extremely difficult to study systematically, especially in juveniles learning to forage for the first time. This lack gives Dlusskii's (1984) work special significance. He and his students watched four bird species taking both models and mimics from flowers during the nesting period in the forests of Russia. The birds were two specialist flycatchers (pied and spotted) and two generalists (redstart and pied wagtail, *Motacilla alba*). When hunting for flies, redstarts and spotted flycatchers, in particular, concentrated their searching at flowers, where models and mimics were also found. Redstarts, and especially wagtails, foraged on the ground amongst low vegetation, but all four bird species used the same strategy to take insects from flowers, catching them either before or just after they tried to escape. From a perch often more than 10 m away, an individual bird noticed its prey and swooped down at speeds of between 3.5 m/s (redstart) to 6.6 m/s (pied flycatcher). Experienced birds dived down with folded wings and steered only with the tail, taking the prey without any noticeable reduction in speed. Davies (1977) described a very similar hunting technique in spotted flycatchers in an Oxford garden, and in the USA similar behaviour has been noted in the painted redstart (*Myioborus pictus*). Called 'flush pursuit', the birds may be exploiting escape responses built into the neural circuits of the flies: particular body movements, patterns of contrast on the wings and tail, and the looming image trigger these primitive escape responses and increase the success of the hunt (Jablonski and Strausfeld, 2000). While learning, young birds used a similar strategy, but usually after approaching the flower they braked, slowing down and practically coming to a standstill, and only then lunged for the insect. Probably the long-range fast dive requires some experience to bring off successfully.

By filming *Eristalis arbustorum* and *Lucilia* (Calliphoridae) startled by a lifelike model of a bird moved with a piece of string, Dlusskii discovered that the flies could perceive the bird and start to fly up only at a maximum distance of about 30 cm (and usually at smaller distances). This gave the fly only about 0.02 s (maximum 0.07 s) to respond before the bird was in a position to take it. Dlusskii measured the speed of *Eristalis* in level flight at about 1.1 (maximum 1.6) m/s (also possibly 10 m/s over short distances: see Golding *et al.*, 2001), but from a standing start this speed could only be reached after about 0.1–0.15 s. Thus the fly only had time to move about 0.5 cm to a maximum of 3 cm before being attacked. Since the bird could manoeuvre as well, the insect had almost no chance of escape. This was just as well, since despite flying four times faster than the insect, if the bird got there too late, or missed, it was practically impossible to take the fly in free flight. However, these calculations are only correct if the bird dived directly and did not reduce its speed, i.e. like experienced adults but not like naïve juvenile birds. They also effectively mean that birds must select their target before making a move, making long-range cues crucial. In butterfly predation by jacamars (Galbulidae), Chai (1990) inferred a particular behavioural sequence of responses to signals from the prey, exploited by butterflies in their defence: a similar sequence for hoverfly predation is likely. Vitally, by far the most frequent responses by jacamars to butterflies were the categories of 'sight-rejected' and 'eaten'; thus long-range visual cues that encourage rejection on sight can make a huge difference to survival of insect prey. Only a small minority of butterflies were rejected after having been tasted, presumably because the jacamars were experienced adult birds. The most important factor leading to butterflies being rejected on sight was their colour pattern, with locally common mimetic or conspicuous patterns being rejected much more frequently than cryptic or intermediate ones. Other components contributing to being rejected on sight were a regular slow flight pattern and a long slender body.

In Dlusskii's study, flies never flew directly forwards in front of the bird, but instead always went perpendicular to the bird's path, either sideways or downwards. Thus to take the prey, the bird needed to turn quickly, and this reduced its speed, providing the fly with the possibility of escape. Of course, often or even usually there are many insects on a flower head, a complex mixture of models, mimics and non-mimetic flies. A crucial finding of Dlusskii's experiments was that upon attack by a bird, most of the flies took off in their escape response (although 40% did not, or only started up after the bird had passed – assuming they had not been snatched). However, the social hymenopterans (wasps, honeybees and bumblebees) practically never reacted to the fake bird, or even to a light touching of the flower. This has very important consequences for the ratio of models to mimics as perceived by the attacker. Not only could a young bird not discriminate between model and mimic from a distance, but it also slowed down at the crucial moment, allowing the flies to escape and leaving only the models for the bird to experience. Thus observed model–mimic ratios by a human observer may not constitute good estimates of the ratio encountered by young birds, which can have a greatly elevated encounter frequency with models. It is therefore very likely indeed that a young bird will take a bee or a wasp during its first few foraging bouts, even if their numbers are many times lower than those of the mimics. Since the first experiences are remembered longest and recalled best (Speed, 2000), they are important in developing the long-term protective effects of mimicry. At first the stricken bird will probably generalize to all insects that even vaguely resemble the model, and imperfect mimicry will be protective. Later on, with more experience, the bird may learn how to distinguish between models and their mimics by selecting particular components of their appearance; it may even come to discriminate very good mimics from models, and then even near-perfect mimicry will not be protective. Dlusskii considered this scenario as a powerful explanation for why poor mimics appeared in such numbers only in the second half of summer, when the number of insectivorous predators was augmented several times over by the recruitment of naïve young birds.

Insectivorous birds use a variety of feeding techniques to gather prey, the main ones being aerial pursuit, swooping down from a perch, gleaning from vegetation, and picking from the ground. It seems clear from the comments above that gleaning and picking are not important methods of obtaining adult hoverflies, except perhaps early in the morning when the flies are immobilized by cold (cf. Hövemeyer, 1995). Surprisingly, aerial pursuit also does not seem to be an important foraging technique for syrphids: even the most agile of birds cannot easily catch them on the wing. Future studies of bird predation in nature should probably concentrate on those species that swoop down from a perch on to flower-visiting flies, just as Dlusskii (1984) described. On current information, this seems to be a critical forum where the hunting behaviour of birds could generate selection for mimicry.

There are many components of the signal produced by an insect aimed at predators, transmitted in various different modalities (visual, sound, vibration, odour). Experimentally each modality on its own can evoke the same or similar responses in a predator because prey categorization by the predator is almost certainly multimodal, i.e. it uses information from all modalities, and these can interact in surprising ways (see Rowe, 1999). For example, in Brower and Brower's (1965) experiments with toads feeding on honeybees and their *Palpada* mimics, producing a buzz with the wings caused a 38% drop in predation, whereas the use of the sting caused only a 21% decrease in the mortality of the mimic. Thus sound seems to be a very important component of the signal that toads associate with noxiousness. The gradient of mimic quality evident in pigeon responses to whole images (Dittrich et al., 1993) could also be evoked by images just of the abdomen, or just the head+thorax. In addition, the birds could also make the same categorizations (creating the same gradient of mimetic quality) using pictures of the insects taken under natural conditions, with various different orientations and relative sizes: a quite extraordinary feat. They obviously have sophisticated polymorphous concepts of these prey categories. Each of the signal modalities may trade-off against one another, or act synergistically, in evoking predator responses, and thus one could easily imagine some mimics emphasizing one modality rather than another. The visual impact of the colour pattern is merely one component of the overall signal, more or less important to different species: these differences may result in the 'poor' mimicry of some syrphids.

6.6 The effect on the predator: poor mimics confuse rather than deceive

It is possible that the effects of good and poor mimics on a predator are different. For example, little is known about the impact of different nasty experiences on birds, but there are indications that emetic experiences are learned in a fundamentally different way from ones that merely make birds feel ill (Testa and Ternes, 1970), and have longer lasting effects that are relatively resistant to subsequent modification (Cowan and Reynolds, 2000).

Rothschild (1984) introduced the idea of 'aide-mémoire mimicry' as a way of explaining the often poor resemblance between models and mimics. In this, predators are induced to remember an unpleasant experience at the hands of a model by features of a mimic that reproduce some but not all the characteristics of the model. Howse and Allen (1994) invented the idea of satyric mimicry to account for poor mimics among the Syrphidae; in this hypothesis, imperfection is regarded as a true ambiguity in the signal, where the black-and-yellow sign of noxiousness is placed in the 'wrong' context, on to a fly shape. Predators are thereby presented with two conflicting signals, and are confused, allowing more time for the insect to escape. There is no evidence that either of these explanations represent significant factors, as far as I am aware.

6.7 The speed of evolution: poor mimics are still evolving their mimetic resemblance

Poor mimics could still be in the process of evolving to be perfect (Edmunds, 2000). For example, Glumac (1962) reasoned that the low abundance of mimics relative to models was a precondition (caused by greater pre-imaginal mortality) that speeded up selection for mimetic colour patterns. The basic components of the mimetic pattern he regarded as having evolved for other non-mimicry-related reasons (parallel evolution with models); thus species that are common relative to potential models have much slower evolution of mimicry, and hence are still poor mimics. I have already discussed (see Section 6.5) the view that mimicry might form a less important part of the lifestyle of some species, and hence selection pressures for perfection of mimetic resemblance might be lower. With lower selection pressure and slower evolutionary rates, such mimics might be less likely to catch up in the coevolutionary chase, and hence be imperfect.

I have also already discussed (see Section 1.2) why I think this non-equilibrial view of mimicry evolution is an unlikely scenario. Although it might sound reasonable, no mathematical theory or computer simulation predicts it, and there is no evidence at all of mimetic patterns changing through time, despite there being lots of spatial discrete variation in colour pattern morphs. On the contrary, as noted above, even simulations of the coevolutionary chase predict stasis (Franks and Noble, 2004). In my view, it is better to assume that poor mimetic patterns have evolved to an equilibrium state, rather than being in the process of being perfected by constant directional selection.

6.8. Disturbance by man: poor mimics have recently become abundant, causing mimetic degradation

A related idea also involving a non-equilibrial situation concerns changes in the habitats where syrphids live. Rather than the continual process of the coevolutionary chase, this supposes that habitat change is only a very recent phenomenon that has impacted on relative abundances only.

Drawing a distinction between good mimics ('specialized Batesian mimics') and what he called 'non-mimetic syrphids' (i.e. including the poor mimics), Maier (1978) suggested that the former only occur in forested areas. The overwhelming number of such good mimics are non-predators as larvae, requiring decaying wood habitats for their development. Maier thought they had evolved their excellent mimicry because they spent a higher proportion of their time than non-mimetic syrphids in forests where potential avian predators are abundant; they have conspicuous foraging and mating behaviour at flowers that increase the chance that they will be noticed by birds; and they share foraging behaviour on flowers with the models. This habitat specificity is difficult to maintain because virtually all syrphids are native to forests and glades within forests (Speight et al., 1975; Speight, 1983), including aphidophagous species (and hence many poor mimics).

The relationship between the original habitat of syrphids and modern disturbed habitats is an interesting one, which has undoubtedly affected syrphid communities and the relative abundances of mimetic species (the 'disturbed ecology' hypothesis: Grewcock, 1992). While classical mimicry theory predicts that there will be a limit to the abundance of Batesian mimics relative to their models, this is only true of undisturbed habitats. In the very disturbed habitats created by man, relative abundances of models and mimics may have greatly changed. As Maier noted, many high-fidelity mimics are restricted to relatively undisturbed forests which provide suitable larval habitats, and deforestation and human agriculture may have caused a severe reduction in the availability of such larval sites. However, this activity has also created huge areas of new habitats, allowing the expansion of many plants and insects with appropriate life-history traits. Aphids in particular are extremely common in open or ecotone habitats with a well-developed herbaceous layer (Dixon, 1998), and increases in aphid availability may boost the abundance of aphidophagous syrphids: this may account for the fact that poor syrphid mimics greatly outnumber their models. Since most research involves habitats that are relatively disturbed as compared to truly pristine areas, we may have an unrealistic view of these relative abundances. Furthermore, it is possible that these syrphids were originally good mimics, but their vastly increased abundance over the last 1000 years or so may have caused a breakdown in mimetic fidelity (cf. Turner, 1984: 336; Carpenter and Ford, 1933: 112–114). Whether or not the mimetic pattern has degraded, it is certainly the case that syrphid communities are sensitive to human disturbance (Bankowska, 1980), and the proportion of good mimics increases in the more pristine habitats (Azmeh *et al.*, 1998).

6.9 Resistance to evolving mimetic perfection

Imperfect mimicry would be more understandable if we could identify selective forces that oppose the putative constant directional selection (in favour of more and more perfect resemblance), allowing an equilibrium state of the pattern. Virtually all considerations of mimicry, theoretical or empirical, show that there should always be an advantage, however small, in becoming more like the model. Hence without any opposing forces all mimics should either be perfect, or in the process of becoming perfect. What could the opposing forces be? There are several possibilities, as described below.

6.9.1 There are costs of producing a perfectly mimetic pattern

Various types of costs of the patterns are imaginable. The most frequently invoked is thermoregulation: syrphids are extremely good thermoregulators for their size (Morgan and Heinrich, 1987; Heinrich, 1993), and the colour pattern may play some role in this function. The development of perfect mimicry might compromise thermoregulatory abilities, placing constraints on the evolution of colour patterns: Heal (1981, 1989) put this forward as an explanation of *Eristalis* colour patterns, subsequently followed by others. In both sexes, individual and particularly seasonal variation in the pattern (generated by the way the pattern responds to rearing temperature) has usually also been interpreted as adaptations to thermal balance (e.g. in Holloway, 1993; Ottenheim *et al.*, 1998), since darker insects are active in cooler weather.

The cost is unlikely to involve conspicuousness. Although there may be a relationship between similarity to the model and the probability of detection, it is probably non-linear. Models may represent an optimum signal for detection and as the appearance of the prey converges on that of the model, prey are likely to suffer similar probabilities of detection under any given circumstances. However, almost any arrangement of bright pattern features will be much more conspicuous than their lack, and hence even very poor mimics are likely to be almost as detectable as their models. Thus there will probably be an asymptotic curved relationship between similarity and detectability (Grewcock, 1992), with rapid increases as the initially poor mimicry originates, followed by small or non-existent changes as mimicry is perfected.

Other types of costs are certainly possible. Heal (1995) had the novel idea that it might be much more costly to produce high-fidelity mimicry, particularly in terms of pupal duration, when the adult body is formed and the colour pattern laid down. Poor syrphid mimics are generally fast-developing aphid predators with a very short pupal duration of about a week, but it might take a long time for good mimetic patterns to develop during the pupal stage: good mimics tend to be ones with a univoltine life cycle and long pupal phases. Thus one could imagine there being two alternative strategies involved here. It may be difficult in resource terms to make the appropriate pigments for creating mimetic patterns. The black colours of syrphids are presumably created from eumelanin, a nitrogen-containing compound present as granules in the exocuticle (Chapman, 1998: 660). The yellows are probably xanthopterins (heavily nitrogenized compounds made from the nucleotide guanosine) synthesized in the epidermis, as in the wasp models. For there to be a cost of producing a mimetic pattern, it must be more costly to produce xanthopterins than melanins (since an all-black hoverfly is not mimetic); while a single molecule of xanthopterin has similar numbers of carbon, hydrogen and oxygen atoms to a single molecule of the quinone monomer of melanin, it has five times the number of nitrogen atoms. Thus it is possible that the yellow colours of hoverfly mimics are costly to the nitrogen budget; according to White (1993) nitrogen is the limiting factor for most animals. However, it is not the case that good-quality mimics have more yellow than poor-quality ones – it is the distribution of the yellow that matters.

There is only one piece of information in the literature that might indicate some possible costs; Lyon (1973) mentioned almost in passing that varieties of *Merodon equestris* differ in their ovarian development, and possibly the mechanism involves different balances between larval nutrient carryover and adult feeding. The varieties *narcissi*, *equestris* and *flavicans* showed normal development of the ovaries, whereas *nobilis* and *transversalis* had much more poorly developed gonads, more pupal fat and required less food in the adult stage. According to Conn (1972), the *transversalis* morph does not seem to be a worse or better mimic of UK bumblebees than the other morphs (the *nobilis* and *flavicans* morphs do not occur in the UK). It is therefore possible that different mimetic morphs have different costs and benefits associated with them, and hence have evolved different life histories.

6.9.2 Kin selection opposes individual selection for perfect mimicry

Johnstone (2002) used an analytical theory of predator perception and signal detection to suggest that kin selection can stabilize imperfect mimicry. Signal detection theory addresses the question of where a predator should set its visual attack threshold in order to optimize the benefit (of attacking a mimic) to cost (of attacking a model) ratio, for model and mimic populations with differing dissimilarities (see Fig. 9.3). Just as Oaten et al. (1975) showed some time ago, it predicts that inaccurate mimics (and their models) will suffer a lower overall attack rate than perfect mimics when the mimics are relatively common. This arises because perfect mimics and their models are all attacked, since there is no threshold that gives a higher benefit-to-cost ratio than simply

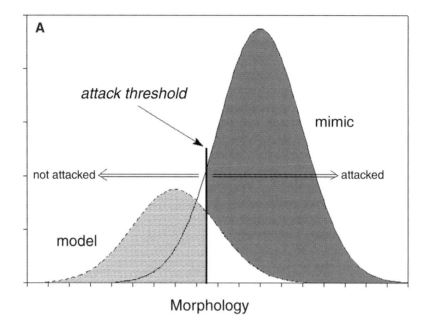

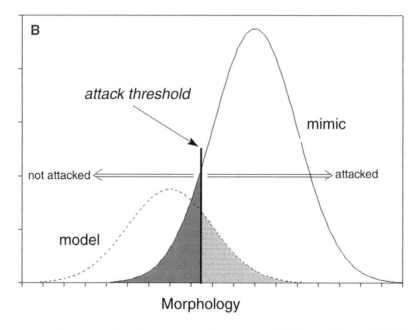

Fig. 9.3. Basis of the signal detection models of Johnstone (2002) and Sherratt (2002). A model and its Batesian imperfect mimic overlap in the distributions of their appearance (indexed by the *x*-axis, 'morphology'). A predator sets a threshold appearance above which it attacks, and below which it does not attack. The optimal attack threshold maximizes the ratio between **A** the benefits (i.e. mimics attacked and models not attacked); and **B** the costs (mimics not attacked and models attacked in error).

attacking everything: models are uncommon (low costs) and mimics are very common (high benefits). When mimicry is inaccurate it will always be worthwhile for the predator to set some threshold of acceptance, and hence a proportion of mimics (the more accurate ones) will escape predation. However, even here mortality is biased against the less perfect mimics in the population, creating selection for more perfect mimicry. This is where kin selection comes in. It is assumed that directional selection for perfect mimicry will not occur under the above conditions if localized groups of mimics have high relatedness: inaccurate mimetic kin effectively pay the costs for the survival of their more accurate relatives. The main prediction is that when models are common and/or strongly aversive, and hence the incentive to attack is low, mimics should evolve ever more accurate resemblance; when models are rare and/or weakly aversive, and the incentive to attack is high, kin selection can oppose individual selection sufficiently strongly for imperfect mimics to evolve. Higher levels of local relatedness and greater incentives to attack therefore both favour greater dissimilarity to the model. Thus his theory suggests that: (i) kin selection will be more important in imperfect mimics; (ii) more noxious models will result in more perfect mimics; and (iii) high mimic abundance will favour imperfect mimicry.

On the basis of model noxiousness, this model predicts that social-wasp mimics should be the most and bumblebee mimics the least accurate of hoverfly mimics. On relative abundance grounds, however, the opposite predictions are made. Relative abundances in pristine habitats where mimicry evolved are probably less different among the mimicry complexes than they are currently in the degraded habitats of Europe. This suggests that noxiousness was the prime cause of the evolved patterns, and hence the first prediction is more appropriate; however, the predicted pattern is exactly opposite to observed situation. Furthermore, the requirement for localized relatedness is vanishingly unlikely among the very common imperfect wasp-mimicking hoverflies, since virtually all of these species migrate southwards in enormous numbers in autumn (Gatter and Schmid, 1990). In addition, the high mobility of syrphids is notorious in mark–release–recapture studies (e.g. Holloway and McCaffery, 1990). Kin selection thus forms a very poor basis upon which to develop a theory of imperfect mimicry for hoverflies.

6.9.3 The existence of multiple models means the optimum is a 'jack-of-all-trades' mimetic pattern

To explain imperfect mimicry in hoverflies, Edmunds (2000) suggested that a perfect hoverfly mimic only achieves protection when it lives within the joint ranges of its model and appropriate predators, whereas a poorer mimic with some degree of resemblance to several models obtains a lower degree of protection but can be distributed over the combined ranges of all models and predators, and thus can occupy a much greater distributional range. Edmunds' argument was limited to non-overlapping model distributions, but even if models were sympatric, it is possible that a mimic evolves to resemble the average of all the models rather than any one model; Barnard (1984) put forward the same idea in a different, behavioural, context a 'jack-of-all-trades' mimic. Just such a verbal model has been proposed as an explanation for the colour pattern of a European burnet moth (Sbordoni et al., 1979). Thus the observed mimetic pattern of a widely distributed mimic is suggested to be an optimal compromise among a set of models, each with a slightly different colour pattern and (possibly) with different ranges. Glumac (1962) noted that a 'considerable number' of mimics are more widespread in their distributions than their models, but a quantitative analysis remains to be done.

This idea was modelled by Sherratt (2002) using a signal-detection approach (like Johnstone: see Fig. 9.3) and receiver operating characteristic (ROC) curves (see Shettleworth, 1998) to set an optimal attack threshold that maximizes the benefit (of avoiding models and attacking mimics) to cost (of attacking models and avoiding mimics) ratio, for model and mimic populations with differing dissimilarities. The relative frequency of mimics to models that can be supported turns out to be directly proportional to the cost–benefit ratio with no upper limit (unlike in traditional ideas of mimicry). In addition, the greater the relative frequency of mimics, and the lower the cost–benefit ratio, the closer the resemblance of the mimic needs to be to gain complete protection. As Dittrich et al. (1993) found experimentally with pigeon pecking rates, the relationship between attack rates and resemblance is highly non-linear, with very low selection for improvement near the model phenotype: as one might predict, the width of this region of weak selection is determined by relative frequencies and the cost–benefit ratio (Fig. 9.4A). Thus an imperfect resemblance is often sufficient, with further improvements being very close to being selectively neutral with respect to predation. The increased variability of hoverfly colour patterns with increasing similarity to their wasp model (Holloway et al., 2002) is consistent with this prediction of relaxed selection.

When there are several visually different, sympatric aposematic models encountered by predators at the same time, mimics either evolve to resemble one of them, or if the models are similar, to some intermediate 'imperfect' phenotype. When the models differ in space or time, there should be selection on a mimic that spans both areas or times to develop an intermediate phenotype. The optimal intermediate phenotype should resemble more closely the model with which the mimic spends most time, or (if this is equal) the less noxious and less numerous model.

This theory is very attractive as an explanation for the patterns of mimicry in hoverflies, because it is the only one consistent with all the evidence, capable of explaining the nature of all three of the major mimicry complexes. It predicts that imperfect mimicry will arise whenever overlapping regions of protection exist in the morphological space among models. Since the sizes of these regions of protection depend on model densities and the costs of attack for a predator, the prediction is that models that are at low density and/or are not very noxious (such as bumblebees) would be least likely to overlap in their regions of protection, and therefore would be the most likely to produce discrete Müllerian mimicry rings, least likely to generate imperfect mimics, and the most likely to result in polymorphic mimics (see Fig. 9.4B). Highly noxious models (such as wasps) are much more likely to have overlapping regions of protection, and

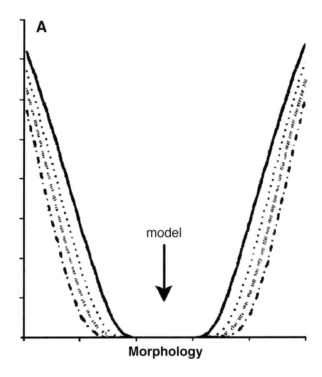

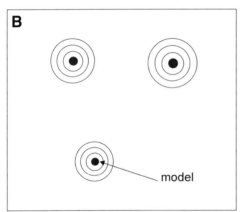
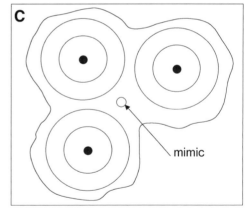

Fig. 9.4. Sherratt's (2002) multiple-model theory. **A** Mortality of a Batesian mimic in relation to its appearance ('morphology') relative to a noxious model, whose position on the morphology axis is indicated. The different lines are for gradually increasing model noxiousness. Note that there is a zone of protection where mortality is effectively zero, and hence where selection for improved resemblance is more or less neutral. **B** Three slightly noxious models in a two-dimensional morphological space, each surrounded by three isoclines of mortality (the 'zones of protection' they afford their mimics). Since the zones do not overlap, selection draws mimetic resemblance to one or other of the models, producing very accurate resemblance. **C** Three very noxious models in the same two-dimensional morphological space. Note that the 'zones of protection' now overlap because of the increased noxiousness, and mimetic resemblance is now at a selection balance of imperfect mimicry.

hence they should themselves form a large and diffuse Müllerian mimicry ring (since by definition they are already protecting one another); there should be more imperfect mimics of them, but they are not likely to promote the evolution of mimetic polymorphism (see Fig. 9.4C). It may even explain why mimicry is such a feature of the Syrphidae: since flight agility will reduce profitability, the cost–benefit ratio increases. This allows a greater density of mimics at equilibrium and a broader range of selectively neutral phenotypes, thus making the evolution of any mimetic pattern easier.

7. Conclusions

Theories of mimicry have become more sophisticated and realistic over the 20th century as they began to incorporate more of the features known to be important in the evolution of mimics. Holling (1965) identified a set of key elements of predators: rate of search, area of detection, the time exposed to prey, handling time of prey, hunger levels and their effects on attack thresholds, impact of encounters with prey on attack thresholds (i.e. learning, forgetting and extinction) which interact with prey characteristics (relative densities, unprofitabilities, similarity to models, presence of alternative prey). To this we can now add the signal-detection process, and memory limitations which make memorizing several categories of prey much harder than keeping just one or two in mind (Bernays, 2001; cf. Schuler, 1980, in the context of mimicry).

The major conclusion to be drawn from this assessment of the ability of theory to account for the spectrum of mimicry in hoverflies is that we do have one candidate theory that potentially can explain why imperfect mimicry is so common. Sherratt's (2002) hypothesis of the evolution of imperfect mimicry under the influence of multiple models is the only one that accounts for all the evidence. The theory and its elaborations (Sherratt, 2003) now require detailed testing. With this approach we are starting to reach a level at which real advances can be made in a more general understanding of mimicry complexes.

There are many other more restricted conclusions to be drawn from my review. Most if not all of the hoverflies labelled as mimetic actually are mimics. The apparently poor nature of their resemblance does not prevent them from obtaining at least some protection from suitably experienced birds. Mimicry is a dominant theme of this very large family of Diptera, with at least a quarter of all species in Europe being mimetic.

Hoverfly mimics fall into three major groups according to their models, involving bumblebees, honeybees and social wasps. There are striking differences in the general levels of mimetic fidelity and relative abundances of the three groups, with accurate mimicry, low abundance and polymorphism characterizing the bumblebee mimics: more than half of all the species of bumblebee mimics are polymorphic. Mimics of social wasps tend to be poor mimics, have high relative abundance, and polymorphism is completely absent. At least some apparently 'poor' mimetic resemblances may be much closer in birds' perception than we imagine, and more work needs to be done on this. Thermoregulatory constraints on the evolution of colour patterns also need clarifying.

Bumblebee models fall into a small number of Müllerian mimicry rings which are very different between the Palaearctic and Nearctic regions. Social wasps and associated models form one large Müllerian complex. Together with honeybees, these complexes probably form real clusters of forms as perceived by many birds. Bumblebees are the least noxious and wasps the most noxious of the three main model groups. The basis of noxiousness seems to be different, with bumblebees being classified as non-food, whereas honeybees and wasps are nasty tasting and (rarely) sting. The distribution of mimicry is exactly what would be expected from this ordering, with polymorphic and accurate forms being a key feature of mimics of the least noxious models, while highly noxious models have poor-quality mimicry.

Mimics may not have to occur at the same season as their models, but usually do. Different model groups have different phenologies, and this may account in part for the different phenologies of their mimics. Waldbauer's phenological hypothesis that mimics can be separated in time from their models needs much more thorough testing, as does the idea that mimics avoid the early summer when fledgling birds are common.

Not enough is known about bird predation on syrphids, yet in principle this should not be too difficult to study with captive birds in large cages, or even in the field. The rates at which naïve birds encounter models and mimics are likely to be very different from their relative abundances measured by human researchers, as Dlusskii's (1984) study clearly shows. We need to know more about the relationships between the noxiousness of the model, the abundances of models and mimics, and the impact of the abundance of alternative prey on the decisions that birds make about whether to attack model–mimic complexes. Similarly, we need to know much more about the psychology of birds as predators. There are at least four processes that need elucidating: (i) learning about the noxiousness of models; (ii) the erasing of that learning through contact with mimics ('extinction', or learned forgetting); (iii) forgetting; and (iv) deliberate risk-taking and the physiological states that promote it.

Acknowledgements

It is a pleasure to acknowledge all who have collaborated in this research and discussed these ideas, especially David Grewcock, who started it all off, but also: Salma Azmeh, Chris Barnard, Sarah Collins, Winand Dittrich, Malcolm Edmunds, Pat Green, Graham Holloway, Brigitte Howarth, Pavel Laska, Steven Malcolm, Peter McGregor, Valeri Mutin, Graham Rotheray, Candy Rowe and Tom Sherratt. Several of these friends and an anonymous referee made valuable comments on earlier drafts of this contribution, but I thank Malcolm Edmunds, Bryan Clarke and Jim Mallet in particular for their careful and extremely thorough reviews.

References

Alcock, J. (1970) Punishment levels and the response of white-throated sparrows (*Zonotrichia albicollis*) to three kinds of artificial models and mimics. *Animal Behaviour* 18, 733–739.

Alford, D.V. (1975) *Bumblebees*. Davis-Poynter, London.

Allen, A.A. (1964) *Pocota personata* Harr. (Diptera, Syrphidae) in SE London; with a note on the copula, etc. *Entomologists Monthly Magazine* 100, 163–164.

Alonso-Meija, J. and Marquez, M. (1994) Dragonfly predation on butterflies in a tropical dry forest. *Biotropica* 26, 341–344.

Archer, M.E. (1988) The aculeate wasp and bee assemblage (Hymenoptera: Aculeata) of a woodland: Bernwood Forest in the English Midlands. *Entomologist* 107, 24–33.

Azmeh, S. (1999) Mimicry and the hoverflies. PhD thesis, Nottingham University, Nottingham, UK.

Azmeh, S., Owen, J., Sorensen, K., Grewcock, D. and Gilbert, F. (1998) Mimicry profiles are affected by human-induced habitat changes. *Proceedings of the Royal Society of London, Series B* 265, 2285–2290.

Bankowska, R. (1980) Fly communities of the family Syrphidae in natural and anthropogenic habitats of Poland. *Memorabilia Zoologica* 33, 3–93.

Barnard, C.J. (1984) When cheats may prosper. In: Barnard, C.J. (ed.) *Producers and Scroungers: Strategies of Exploitation and Parasitism*. Chapman and Hall, London, pp. 6–33.

Benton, F. (1903) Bees working on chrysanthemums. *Proceedings of the Entomological Society of Washington* 6, 102–103.

Bernays, E.A. (2001) Neural limitations in phytophagous insects: implications for diet breadth and evolution of host affiliation. *Annual Review of Entomology* 46, 703–727.

Birkhead, T.R. (1974) Predation by birds on social wasps. *British Birds* 67, 221–229.

Bohart, R.M. and Menke, A.S. (1976) *Sphecid Wasps of the World: A Generic Revision*. University of California Press, Berkeley, California.

Bombosch, S. (1963) Untersuchungen zur Vermehrung von *Aphis fabae* Scop. in Samenrübenbeständen unter besonderer Berücksichtigung der Schwebfliegen (Diptera, Syrphidae). [Studies on the increase of *Aphis fabae* Scop. in seed-beet crops with special reference to hoverflies (Diptera, Syrphidae)]. *Zeitschrift für Angewandte Entomologie* 52, 105–141.

Bristowe, W.S. (1941) *The Comity of Spiders*, 2 vols. Ray Society, London.

Brower, J. van Z. (1960) Experimental studies of mimicry. IV. The reactions of starlings to different proportions of models and mimics. *American Naturalist* 94, 271–282.

Brower, J. van Z. and Brower, L.P. (1962) Experimental studies of mimicry. 6. The reaction of toads (*Bufo terrestris*) to honeybees (*Apis mellifera*) and their dronefly mimics (*Eristalis vinetorum*). *American Naturalist* 96, 297–307.

Brower, J. van Z. and Brower, L.P. (1965) Experimental studies of mimicry. 8. Further investigations of honeybees (*Apis mellifera*) and their dronefly mimics (*Eristalis* spp.). *American Naturalist* 99, 173–187.

Brower, L.P. (1988) Avian predation on the Monarch butterfly and its implications for mimicry theory. *American Naturalist* 131, S4–S6.

Brower, L.P. and Brower, J. van Z. (1972) Parallelism, convergence, divergence, and the new concept of advergence in the evolution of mimicry. In: Deevey, E.A. (ed.) *Ecological Essays in Honour of G. Evelyn Hutchinson. Transactions of the Connecticut Academy of Science* 44, 59–67.

Brower, L.P., Brower, J. van Z. and Westcott, P.W. (1960) Experimental studies of mimicry. 5. The reactions of toads (*Bufo terrestris*) to bumblebees (*Bombus americanorum*) and their robberfly mimics (*Mallophora bomboides*), with a discussion of aggressive mimicry. *American Naturalist* 94, 343–356.

Brown, E.S. (1951) Mimicry as illustrated by the British fauna. *New Biology* 10, 72–94.

Brues, C.T. (1946) *Insect Diet: An Account of the Food Habits of Insects*. Harvard University Press, Cambridge, Massachusetts.

Carpenter, G.D.H. and Ford, E.B. (1933) *Mimicry*. Methuen, London.

Carrick, R. (1936) Experiments to test the efficiency of protective adaptations in insects. *Transactions of the Royal Entomological Society of London* 85, 131–140.

Chai, P. (1990) Relationships between visual characteristics of rainforest butterflies and responses of a specialized insectivorous bird. In: Wickstein, M. (ed.) *Adaptive Coloration in Invertebrates*. Proceedings of a Symposium sponsored by the American Society of Zoologists. Texas A&M University Sea Grant College Program, Galveston, Texas, pp. 31–60.

Chapman, R.F. (1998) *The Insects: Structure and Function*, 4th edn. Cambridge University Press, Cambridge, UK.

Church, S.C., Bennett, A.T.D., Cuthill, I.C. and Partridge, J.C. (2004) Avian ultraviolet vision and its implications for insect protective coloration. In: van Emden, H. and Rothschild, M. (eds) *Insect and Bird Interactions*. Intercept, Andover, UK, pp. 165–184.

Conn, D.L.T. (1972) The genetics of mimetic colour polymorphism in the large narcissus bulb fly, *Merodon equestris* Fab. (Diptera, Syrphidae). *Proceedings of the Royal Society of London, Series B* 264, 353–402.

Corbet, P.S. (1999) *Dragonflies: Behaviour and Ecology of Odonata*. Harley Books, Essex, UK.

Cott, H.B. (1940) *Adaptive Coloration in Animals*. Methuen, London.

Cowan, D.P., Reynolds, J.C. and Gill, E.L. (2000) Reducing predation through conditioned taste aversion. In: Gosling, L.M. and Sutherland, W.J. (eds) *Behaviour and Conservation*. Cambridge University Press, Cambridge, UK, pp. 281–299.

Cramp, S. (ed.) (1977–1994) *Birds of the Western Palaearctic*, 9 vols. Oxford University Press, Oxford, UK.

Cuthill, I.C. and Bennett, A.T.D. (1993) Mimicry and the eye of the beholder. *Proceedings of the Royal Society of London, Series B* 253, 203–204.

Davies, N.B. (1977) Prey selection and the search strategy of the spotted flycatcher (*Muscicapa striata*): a field study on optimal foraging. *Animal Behaviour* 25, 1016–1033.

Davies, N.B. and Green, R.E. (1976) The development and ecological significance of feeding techniques in the Reed Warbler (*Acrocephalus scirpaceus*). *Animal Behaviour* 24, 213–229.

Dill, L. (1975) Calculated risk-taking by predators as a factor in Batesian mimicry. *Canadian Journal of Zoology* 53, 1614–1621.

Dittrich, W., Gilbert, F., Green, P., McGregor, P. and Grewcock, D. (1993) Imperfect mimicry: a pigeon's perspective. *Proceedings of the Royal Society of London, Series B* 251, 195–200.

Dixon, A.F.G. (1998) *Aphid Ecology*, 2nd edn. Chapman and Hall, London.

Dlusskii, G.M. (1984) Are dipterous insects protected by their similarity to stinging hymenopterans? (in Russian). *Byulleten' Moskovskogo Obshchestva Ispytatelei Prirody, Otdel Biologicheskii* 89(5), 25–40 (translation available from http://www.nottingham.ac.uk/~plzfg/syrphweb/Dlusski1984.doc).

Drees, M. (1997) Zur Hautflüglermimikry bei Schwebfliegen (Diptera, Syrphidae) [On hymenopteran mimicry by hoverflies (Diptera, Syrphidae)]. *Entomologische Zeitschrift* 107, 498–503.

Duncan, C.J. and Sheppard, P.M. (1965) Sensory discrimination and its role in the evolution of Batesian mimicry. *Behaviour* 24, 269–282.

Edmunds, M. (1974) *Defence in Animals: A Survey of Anti-Predator Defences.* Longman, Harlow, Essex, UK.

Edmunds, M. (1993) Does mimicry of ants reduce predation by wasps on salticid spiders? *Memoirs of the Queensland Museum* 11, 507–512.

Edmunds, M. (2000) Why are there good and poor mimics? *Biological Journal of the Linnean Society* 70, 459–466.

Evans, D.L. (1984) Reactions of some adult passerines to *Bombus pennsylvanicus* and its mimic *Mallota bautias. Ibis* 126, 50–58.

Evans, D.L. and Waldbauer, G.P. (1982) Behavior of adult and naive birds when presented with a bumblebee and its mimic. *Zeitschrift für Tierpsychologie* 59, 247–259.

Evans, H.E. (1966) *Life on a Little-Known Planet: A Journey into the Insect World.* Andre Deutsch, London.

Evans, H.E. and Eberhard, M.J.W. (1970) *The Wasps.* University of Michigan Press, Ann Arbor, Michigan.

Fabre, J.-H. (1913) *The Life of the Fly* (translated by A. Teixera de Mattos). Hodder and Stoughton, London.

Fincher, F. (1951) Mimicry in *Volucella bombylans* (L.) (Diptera, Syrphidae). *Entomologists Monthly Magazine* 87, 31.

Ford, H.A. (1971) The degree of mimetic protection gained by new partial mimics. *Heredity* 27, 227–236.

Franks, D.W. and Noble, J. (2004) Batesian mimics influence mimicry ring evolution. *Proceedings of the Royal Society of London, Series B* 271, 191–196.

Gabritschevsky, E. (1924) Farbenpolymorphismus und Vererbung mimetischer Varietäten der Fliege *Volucella bombylans* und anderer 'hummelähnlicher' Zweiflügler. [Colour polymorphism and genetics of mimetic varieties of the fly *Volucella bombylans* and other 'bee-like' Diptera]. *Zeitschrift für Induktive Abstamms- und Vererbungslehre, Berlin* 32, 321–353.

Gabritschevsky, E. (1926) Convergence of coloration between American pilose flies and bumblebees (*Bombus*). *Biological Bulletin* 51, 269–287.

Gatter, W. and Schmid, U. (1990) Die Wanderungen der Schwebfliegen (Diptera: Syrphidae) am Randecker Maar. [The migration of hoverflies (Diptera: Syrphidae) at Randecker Maar]. *Spixiana* 15(S), 1–100.

Getty, T. (1985) Discriminability and the sigmoid functional response: how optimal foragers could stabilize model–mimic complexes. *American Naturalist* 125, 239–256.

Gilbert, F. (1985) Diurnal activity patterns in hoverflies (Diptera, Syrphidae). *Ecological Entomology* 10, 385–392.

Gilbert, J.D.J. (2001) *Callicera aurata* in Surrey. *Entomologists Monthly Magazine* 137, 153.

Glumac, S. (1962) The problem of mimicry in Syrphoidea. *Archiv Bioloskih Nauka, Beogradu* 14, 61–67.

Golding, Y.C. and Edmunds, M. (2000) Behavioural mimicry of honeybees (*Apis mellifera*) by droneflies (*Eristalis* spp., Diptera, Syrphidae). *Proceedings of the Royal Society of London, Series B* 267, 903–909.

Golding, Y.C., Ennos, A.R. and Edmunds, M. (2001) Similarity in flight behaviour between the honeybee *Apis mellifera* (Hymenoptera: Apidae) and its presumed mimic, the dronefly *Eristalis tenax* (Diptera: Syrphidae). *Journal of Experimental Biology* 204, 139–145.

Goodale, M.A. and Sneddon, I. (1977) The effect of distastefulness of the model on the predation of artificial Batesian mimics. *Animal Behaviour* 25, 660–665.

Goulson, D. (2003) *Bumblebees: Their Behaviour and Ecology*. Cambridge University Press, Cambridge, UK.

Green, P.R., Gentle, L., Peake, T.M., Scudamore, R.E., McGregor, P.K., Gilbert, F. and Dittrich, W. (1999) Conditioning pigeons to discriminate naturally lit insect specimens. *Behavioural Processes* 46, 97–102.

Greenwood, J.J.D. (1986) Crypsis, mimicry, and switching by optimal foragers. *American Naturalist* 128, 294–300.

Greig-Smith, P.W. and Quicke, D.L.J. (1983) The diet of nestling Stonechats. *Bird Study* 30, 47–50.

Grewcock, D. (1992) The Hoverflies: A Case of 'Poor' Mimicry? PhD thesis, Nottingham University, Nottingham, UK.

Guilford, T. (1990) The evolution of aposematism. In: Evans, D.L. and Schmidt, J.O. (eds) *Insect Defenses: Adaptive Mechanisms and Strategies of Prey and Predators*. State University of New York Press, Albany, New York, pp. 23–61.

Haeseler, V. (1976) *Ceriana conopsoides* (L.) near Oldenburg (Diptera, Syrphidae). *Drosera* 1976, 19–21.

Hartley, I.R. and Quicke, D.L.J. (1994) The diet of nestling corn buntings of North Uist: insects not grain. *Scottish Birds* 17(3), 169–170.

Heal, J.R. (1979) Colour patterns of Syrphidae. II. *Eristalis intricarius*. *Heredity* 43, 229–238.

Heal, J.R. (1981) Colour patterns of Syrphidae. III. Sexual dimorphism in *Eristalis arbustorum*. *Ecological Entomology* 6, 119–127.

Heal, J.R. (1982) Colour patterns of Syrphidae. 4. Mimicry and variation in natural populations of *Eristalis tenax*. *Heredity* 49, 95–110.

Heal, J.R. (1989) Variation and seasonal changes in hoverfly species: interactions between temperature, age and genotype. *Biological Journal of the Linnean Society* 36, 251–269.

Heal, J.R. (1995) Of what use are the bright colours of hoverflies? *Dipterists Digest* 2(1), 1–4.

Heikertinger, F. (1918) Die Bienenmimikry von *Eristalis*: eine kritische Untersuchung. [The bee mimicry of *Eristalis*: a critical study]. *Zeitschrift für Wissenschaftliche Insektenbiologie* 14, 1–5, 73–79.

Heikertinger, F. (1936) Noch ein Wort über Wespenmimikry. [A further note on wasp mimicry]. *Zeitschrift für Morphologie und Ökologie der Tiere* 29, 140.

Heikertinger, F. (1954) *Das Rätsel der Mimikry und seine Lösung: eine kritische Darstellung des Werdens, des Wesens und der Widerlegung der Tiertrachthypothesen*. [The enigma of mimicry and its solution: a critical description of the development, nature and refutation of the hypothesis of animal colour patterns]. Gustav Fischer Verlag, Jena, Germany.

Heinrich, B. (1993) *The Hot-Blooded Insects: Strategies and Mechanisms of Thermoregulation*. Harvard University Press, Cambridge, Massachusetts.

Hespenheide, H.A. (1973) A novel mimicry complex: beetles and flies. *Journal of Entomology* (A) 48, 49–56.

Hetz, M. and Slobodchikoff, C.N. (1988) Predation pressure on an imperfect Batesian mimicry complex in the presence of alternative prey. *Oecologia* 76, 570–573.

Holling, C.S. (1965) The functional response of predators to prey density and its role in mimicry and population regulation. *Memoirs of the Entomological Society of Canada* 45, 1–60.

Holloway, B.A. (1976) Pollen feeding in hoverflies (Diptera, Syrphidae). *New Zealand Journal of Zoology* 3, 339–350.

Holloway, G.J. (1993) Phenotypic variation in colour pattern and seasonal plasticity in *Eristalis* hoverflies (Diptera: Syrphidae). *Ecological Entomology* 18(3), 209–217.

Holloway, G.J. and McCaffery, A.R. (1990) Habitat utilization and dispersion in *Eristalis pertinax* (Diptera, Syrphidae). *Entomologist* 109, 116–124.

Holloway, G.J., Gilbert, F. and Brandt, A. (2002) The relationship between mimetic imperfection and phenotypic variation in insect colour patterns. *Proceedings of the Royal Society of London, Series B* 269, 411–416.

Hövemeyer, K. (1995) Seasonal and diurnal activity patterns in the hoverfly species *Cheilosia fasciata* (Diptera: Syrphidae). *Entomologia Generalis* 20, 87–102.

Howarth, B. (1998) An ecological study of Batesian mimicry in the British Syrphidae (Diptera). PhD thesis, University of Central Lancashire, Preston, UK.

Howarth, B. and Edmunds, M. (2000) The phenology of Syrphidae (Diptera): are they Batesian mimics of Hymenoptera? *Biological Journal of the Linnean Society* 71, 437–457.

Howarth, B., Clee, C. and Edmunds, M. (2000) The mimicry between British Syrphidae (Diptera) and Aculeate Hymenoptera. *British Journal of Entomology and Natural History* 13, 1–40.

Howarth, B., Edmunds, M. and Gilbert, F. (2004) Does the abundance of hoverfly mimics (Diptera: Syrphidae) depend on the numbers of their hymenopteran models? *Evolution* 58(2), 367–375.

Howse, P.E. and Allen, J.A. (1994) Satyric mimicry: the evolution of apparent imperfection. *Proceedings of the Royal Society of London, Series B* 257, 111–114.

Iwata, K. (1976) *Evolution of Instinct: Comparative Ethology of Hymenoptera.* Amerind Publishing, New York.

Jacobi, A. (1913) *Mimikry und verwandte Erscheinungen.* [*Mimicry and related phenomena*]. Vieweg, Braunschweig, Germany.

Jablonski, P.G. and Strausfeld, N.J. (2000) Exploitation of an ancient escape circuit by an avian predator: prey sensitivity to model predator display in the field. *Brain Behavior and Evolution* 56(2), 94–106.

Johnstone, R.A. (2002) The evolution of inaccurate mimics. *Nature* 418, 524–526.

Joron, M. and Mallet, J.L.B. (1998) Diversity in mimicry: paradox or paradigm? *Trends in Ecology and Evolution* 13, 461–466.

Kassarov, L. (2003) Are birds the primary selective force leading to evolution of mimicry and aposematism in butterflies? An opposing point of view. *Behaviour* 140, 433–451.

Kauppinen, J. and Mappes, J. (2003) Why are wasps so intimidating: field experiments on hunting dragonflies (Odonata: *Aeschna grandis*). *Animal Behaviour* 66, 505–511.

Kearns, C.A. and Thomson, J.D. (2001) *The Natural History of Bumblebees: A Sourcebook for Investigations.* University Press of Colorado, Boulder, Colorado.

Kikuchi, T. (1963) Studies on the coaction among insects visiting flowers. III. Dominance relationships among flower-visiting flies, bees and butterflies. *Science Reports of the Tohoku (Imperial) University, 4th Series, Biology* 29, 1–8.

Komarek, S. (1998) *Mimicry, Aposematism and Related Phenomena in Animals and Plants: A Bibliography 1800–1990.* Vesmir, Prague.

Kratochwil, A. (1983) Zur Phänologie von Pflanzen und blütenbesuchenden Insekten eines versäumten Halbtrockenrasens im Kaiserstuhl: ein Beitrag zur Erhaltung brachliegender Wiesen als Lizenz-Biotope gefährdeter Tierarten. [Phenological phenomena of plants and flower-visiting insects in a fallow limestone grassland in the Kaiserstuhl: remarks about the conservation of fallow grassland as 'licence biotopes' for endangered invertebrate animals]. *Beihefte der Veröffentliche Naturschutz Landschaftpflege Bad-Würtenbergs* 34, 57–108.

Lane, C. (1957) Preliminary note on insects eaten and rejected by a tame Shama (*Kittacincla malabarica* Gm.), with the suggestion that in certain species of butterflies and moths females are less palatable than males. *Entomologists Monthly Magazine* 93, 172–179.

Lea, S.E.G. and Harrison, S.N. (1978) Discrimination of polymorphous stimulus sets by pigeons. *Quarterly Journal of Experimental Psychology* 30, 521–537.

Liepelt, W. (1963) Zur Schutzwirkung des Stachelgiftes von Bienen und Wespen gegenüber Trauerfliegenschnäpper und Gartenrotschwanz. [On the defensive action of the poison stings of bees and wasps against flycatchers and robins]. *Zoologisches Jahrbuch, Abteilung Zoologie und Physiologie* 70, 167–176.

Lindström, L., Alatalo, R. and Mappes, J. (1997) Imperfect Batesian mimicry: the effects of the frequency and the distastefulness of the model. *Proceedings of the Royal Society of London, Series B* 264, 149–153.

Lyon, J.-P. (1973) La mouche des Narcisses (*Merodon equestris* F., Diptère, Syrphidae). I. Identification de l'insecte et de ses dégats et biologie dans le Sud-Est de la France. [The Narcissus fly (*Merodon equestris* F., Diptera, Syrphidae). I. Identification of the insect, and its damage and biology in the south-east of France]. *Revue de Zoologie Agricole et de Pathologie Végétale* 72(3), 65–92.

Maier, C.T. (1978) Evolution of Batesian mimicry in the Syrphidae (Diptera). *Journal of the New York Entomological Society* 86, 307.

Malcolm, S.B. (1976) An investigation of plant-derived cardiac glycosides as a possible basis for aposematism in the aphidophagous hoverfly *Ischiodon aegyptius* (Wiedemann) (Diptera, Syrphidae). MSc thesis, Rhodes University, Grahamstown, South Africa.

Malcolm, S.B. (1981) Defensive use of plant-derived cardenolides by *Aphis nerii* Boyer de Fonscolombe against predation. DPhil thesis, Oxford University, Oxford, UK.

Malcolm, S.B. (1990) Mimicry: status of a classical evolutionary paradigm. *Trends in Ecology and Evolution* 5, 57–62.

Mallet, J. (1999) Causes and consequences of a lack of coevolution in Müllerian mimicry. *Evolutionary Ecology* 13, 777–806.

Mallet, J. and Gilbert, L.E. (1995) Why are there so many mimicry rings? Correlations between habitat, behaviour and mimicry in *Heliconius* butterflies. *Biological Journal of the Linnean Society* 55, 159–180.

Mallet, J. and Joron, M. (1999) Evolution of diversity in warning color and mimicry: polymorphisms, shifting balance, and speciation. *Annual Review of Ecology and Systematics* 30, 201–233.

Mappes, J. and Alatalo, R.V. (1997) Batesian mimicry and signal accuracy. *Evolution* 51(6), 2050–2053.

Marden, J.H. and Chai, P. (1991) Aerial predation and butterfly design: how palatability, mimicry, and the need for evasive flight constrain mass allocation. *American Naturalist* 138, 15–36.

McAtee, W.L. (1932) Effectiveness in nature of the so-called protective adaptations in the animal kingdom, as illustrated by the food habits of Nearctic birds. *Smithsonian Miscellaneous Collections* 87(7), 1–201.

Morgan, K.R. and Heinrich, B. (1987) Temperature regulation in bee- and wasp-mimicking syrphid flies. *Journal of Experimental Biology* 133, 59–71.

Morse, D.H. (1986) Predatory risk to insects foraging at flowers. *Oikos* 46, 223–228.

Mostler, G. (1935) Beobachtungen zur Frage der Wespenmimikry. [Studies on the question of wasp mimicry]. *Zeitschrift für Morphologie und Ökologie der Tiere* 29, 381–454.

Myers, J.G. (1935) Experiments with spiders and the bee-like *Eristalis tenax* Linn. *Proceedings of the Royal Entomological Society of London, Series A* 9, 93–95.

Nentwig, W. (1982) Why do only certain insects escape from a spider's web? *Oecologia* 53, 412–417.

Nicholson, A.J. (1927) A new theory of mimicry in insects. *Australian Zoologist* 5, 10–104.

Nickol, M. (1994) *Volucella zonaria* (Diptera: Syrphidae) in Rheinland-Pfalz: Nachweise nebst Bemerkungen über Blütenbesuch, Verhalten, Färbung und Ökologie sowie andere Gattungsvertreterinnen. [*Volucella zonaria* (Diptera: Syrphidae) in Rheinland-Pfalz: records and remarks on flower visits, behaviour, colour and ecology, as well as for other members of the genus]. *Mitteilungen der Pollichia der Pfälzischen Vereins für Naturkunde und Naturschutz* 81, 383–405.

Nur, U. (1970) Evolutionary rates of models and mimics in Batesian mimicry. *American Naturalist* 104, 477–486.

Oaten, A., Pearce, C.E.M. and Smyth, B. (1975) Batesian mimicry and signal-detection theory. *Bulletin of Mathematical Biology* 37, 367–387.

O'Donnell, S. (1996) Dragonflies (*Gynacantha nervosa* Rambur) avoid wasps (*Polybia aequatorialis* Zavattari and *Mischocyttarus* sp) as prey. *Journal of Insect Behavior* 9, 159–162.

Osten-Sacken, C.R. (1894) *On the Oxen-Borne Bees of the Ancients (Bugonia), and their Relation to* Eristalis Tenax, *a Two-winged Insect*. Heidelberg, Germany.

Ottenheim, M.M., Henseler, A. and Brakefield, P.M. (1998) Geographic variation in plasticity in *Eristalis arbustorum*. *Biological Journal of the Linnean Society* 65, 215–229.

Owen, J. (1991) *Ecology of a Garden*. Cambridge University Press, Cambridge, UK.

Parmenter, L. (1953) The hoverflies (Syrphidae). *Entomologists Record and Journal of Variation* 65, 122–238.

Pickard, R.S. (1975) Relative abundance of syrphid [sic] species in a nest of the wasp *Ectemnius cavifrons* compared with that of the surrounding habitat. *Entomophaga* 20, 143–151.

Pilecki, C. and O'Donald, P. (1971) The effects of predation on artificial mimetic polymorphisms with perfect and imperfect mimics at varying frequencies. *Evolution* 25, 365–370.

Plowright, R.C. and Owen, R.E. (1980) The evolutionary significance of bumble bee color patterns: a mimetic interpretation. *Evolution* 34, 622–637.

Pocock, R.I. (1911) On the palatability of some British insects. *Proceedings of the Zoological Society of London* 1911, 809–868.

Poulton, E.B. (1890) *The Colours of Animals*. Trübner, London.

Prŷs-Jones, O.E. and Corbet, S.A. (1991) *Bumblebees. Naturalists Handbook 6*. Richmond Publishing, Slough, UK.

Rettenmeyer, C.W. (1970) Insect mimicry. *Annual Review of Entomology* 15, 43–74.

Ritland, D.B. and Brower, L.P. (1991) The viceroy butterfly is not a Batesian mimic. *Nature* 350, 497–498.

Röder, G. (1990) *Biologie der Schwebfliegen Deutschlands (Diptera: Syrphidae)*. [Biology of the hoverflies of Germany (Diptera: Syrphidae)]. Erna Bauer Verlag, Keltern-Weiler, Germany.

Rotheray, G.E. (1993) Colour guide to hoverfly larvae. *Dipterists Digest* 9, 1–155.

Rotheray, G.E. and Gilbert, F. (1999) Phylogeny of Palaearctic Syrphidae (Diptera): evidence from larval stages. *Zoological Journal of the Linnean Society* 127, 1–112.

Rothschild, M. (1984) Aide-mémoire mimicry. *Ecological Entomology* 9, 311–319.

Rowe, C. (1999) Receiver psychology and the evolution of multicomponent signals. *Animal Behaviour* 58, 921–931.

Rowe, C., Lindström, L. and Lyttinen, A. (2004) The importance of pattern similarity between Müllerian mimics in predator avoidance learning. *Proceedings of the Royal Society of London, Series B* 271, 407–413.

Rupp, L. (1989) Die mitteleuropäischen Arten der Gattung *Volucella* (Diptera, Syrphidae) als Kommensalen und Parasitoide in den Nestern von Hummeln und sozialen Wespen: Untersuchungen zur Wirtsfindung, Larvalbiologie und Mimikry. [The central European species of the genus *Volucella* (Diptera, Syrphidae) as commensals and parasitoids in the nests of bees and social wasps: studies on host finding, larval biology and mimicry]. PhD thesis, Albert-Ludwigs-Universität, Freiburg-im-Breisgau, Germany.

Sbordoni, V., Bullini, L., Scarpelli, G., Forestiero, S. and Rampini, M. (1979) Mimicry in the burnet moth *Zygaena ephialtes*: population studies and evidence of a Batesian–Müllerian situation. *Ecological Entomology* 4, 83–93.

Schmidt, R.S. (1960) Predator behaviour and the perfection of incipient mimetic resemblances. *Behaviour* 16, 149–158.

Schuler, W. (1980) Zum Meidenlernen ungeneissbarer Beute bei Vögeln: der Einfluss der Faktoren Umlernen, neue Alternativbeute und Ähnlichkeit der Alternativbeute. [Factors influencing learning to avoid unpalatable prey in birds: relearning, new alternative prey, and similarity of appearance of alternative prey]. *Zeitschrift für Tierpsychologie* 54, 105–143.

Seeley, T.D. (1985) *Honeybee Ecology: A Study of Adaptation in Social Life*. Princeton University Press, Princeton, New Jersey.

Shelly, T.D. and Pearson, D.L. (1978) Size and colour discrimination of the robberfly *Efferia tricella* (Diptera: Asilidae) as a predator on tiger beetles (Coleoptera: Cincidelidae). *Environmental Entomology* 7, 790–793.

Sheppard, P.M. (1959) The evolution of mimicry: a problem in ecology and genetics. *Cold Spring Harbor Symposia in Quantitative Biology* 24, 131–140.

Sherratt, T.N. (2002) The evolution of imperfect mimicry. *Behavioral Ecology* 13(6), 821–826.

Sherratt, T.N. (2003) State-dependent risk-taking by predators in systems with defended prey. *Oikos* 103, 93–100.

Shettleworth, S.J. (1998) *Cognition, Evolution and Behaviour*. Oxford University Press, Oxford, UK.

Southwood, T.R.E. and Henderson, P.A. (2000) *Ecological Methods*, 3rd edn. Blackwell Scientific, Oxford, UK.

Speed, M.P. (1993) Müllerian mimicry and the psychology of predation. *Animal Behaviour* 45, 571–580.

Speed, M.P. (2000) Warning signals, receiver psychology and predator memory. *Animal Behaviour* 60, 269–278.

Speed, M.P. (2001) Batesian, quasi-Batesian or Müllerian mimicry? Theory and data in mimicry research. *Evolutionary Ecology* 13, 755–776.

Speed, M.P. and Turner J.R.G. (1999) Learning and memory in mimicry. II. Do we understand the mimicry spectrum? *Biological Journal of the Linnean Society* 67, 281–312.

Speight, M.C.D. (1983) Flies: Diptera. In: Feehan, J. (ed.) *Laois: An Environmental History*. Ballykilcavan Press, Ireland, pp. 172–178, 502–507.

Speight, M.C.D. (2000) Hoverflies (Dip.: Syrphidae) with a drinking habit. *Entomologists Record and Journal of Variation* 112, 107–113.

Speight, M.C.D. and Lucas, J.A.W. (1992) Liechtenstein Syrphidae (Diptera). *Bericht der Botanische-Zoologische Gesellschaft für Liechtenstein-Sargans-Werdenberg* 19, 327–463.

Speight, M.C.D., Chandler, P.J. and Nash, R. (1975) Irish Syrphidae (Diptera): notes on the species and an account of their known distribution. *Proceedings of the Royal Irish Academy B* 75, 1–80.

Spradberry, J.P. (1973) *Wasps*. Sidgwick and Jackson, London.

Srygley, R.B. (1994) Locomotor mimicry in butterflies? The associations of positions of centres of mass among groups of mimetic, unprofitable prey. *Philosophical Transactions of the Royal Society of London, Series B* 343, 145–155.

Srygley, R.B. (1999) Locomotor mimicry in *Heliconius* butterflies: contrast analysis of flight morphology and kinematics. *Philosophical Transactions of the Royal Society of London, Series B* 354, 203–214.

Srygley, R.B. (2004) The aerodynamic costs of warning signals in palatable mimetic butterflies and their distasteful models. *Proceedings of the Royal Society of London, Series B* 271, 589–594.

Srygley, R.B. and Dudley, R. (1993) Correlations of the position of the centre of body mass with butterfly escape tactics. *Journal of Experimental Biology* 174, 155–166.

Steiniger, F. (1937a) 'Ekelgeschmack' und visuelle Anpassung einiger Insekten (Fütterungsversuche an Vögeln. ['Disgusting taste' and visual recognition of some insects (feeding experiments with birds)]. *Zeitschrift für Wissenschaftliche Zoologie* 149, 221–229.

Steiniger, F. (1937b) Beobachtungen und Bemerkungen zur Mimikryfrage. [Observations and remarks on the mimicry question]. *Biologisches Zentralblatt* 57, 47–58.

Stiles, E.W. (1979) Evolution of color pattern and pubescence characteristics in male bumblebees: automimicry vs. thermoregulation. *Evolution* 33(3), 941–957.

Stubbs, A.E. and Falk, S. (1983) *British Hoverflies: An Illustrated Identification Guide*. British Entomological and Natural History Society, London.

Testa, T.J. and Ternes, J.W. (1970) Specificity of conditioning mechanisms in the modification of food preferences. In: Barker, L.M., Best, M.R. and Domjan, M. (eds) *Learning Mechanisms in Food Selection*. Baylor University Press, Texas, pp. 229–253.

Tibbetts, E.A. (2002) Visual signals of individual identity in the wasp *Polistes fuscatus*. *Proceedings of the Royal Society of London, Series B* 269, 1423–1428.

Tsuneki, K. (1956) Ethological studies on *Bembix niponica* Smith, with emphasis on the psychobiological analysis of behaviour inside the nest (Hymenoptera: Sphecidae). 1. Biological part. *Memoirs of the Faculty of Liberal Arts, Fukui University, Series II: Natural Sciences* 6, 77–172.

Turner, J.R.G. (1984) Darwin's coffin and Doctor Pangloss: do adaptationist models explain mimicry? In: Shorrocks, B. (ed.) *Evolutionary Ecology*. Blackwell Scientific, Oxford, UK, pp. 313–361.

Turner, J.R.G. (1987) The evolutionary dynamics of Batesian and Müllerian mimicry: similarities and differences. *Ecological Entomology* 12, 81–95.

Tyshchenko, V.P. (1961) On the relationship between flower spiders of the family Thomisidae and mimetic Diptera and their models (in Russian). *Vestnik Leningradskogo Gosudarstvennogo Universiteta. Ser. Biologiya* 1961(3), 133–139.

van Someren, V.G.L. and Jackson, T.H.E. (1959) Some comments on the protective resemblance amongst African Lepidoptera (Rhopalocera). *Journal of the Lepidopterists Society* 13(3), 121–150.

Vane-Wright, R.I. (1978) Mimicry. In: Stubbs, A. and Chandler, P. (eds) *A Dipterists Handbook. Amateur Entomologist* 15, 239–241.

Vanhaelen, N., Haubruge, E., Lognay, G. and Francis, F. (2001) Hoverfly glutathione S-transferases and effect of Brassicaceae secondary metabolites. *Pesticide Biochemistry and Physiology* 71(3), 170–177.

Vanhaelen, N., Gaspar, C. and Francis, F. (2002) Influence of prey host plant on a generalist aphidophagous predator, *Episyrphus balteatus* (Diptera: Syrphidae). *European Journal of Entomology* 99(4), 561–564.

Verrall, G.H. (1901) *British Flies*, Vol. VIII. Taylor and Francis, London.

von Hagen, E. and Aichhorn, A. (2003) *Hummeln*. [*Bumblebees*]. Fauna Verlag, Nottuln, Germany.

Waldbauer, G.P. (1970) Mimicry of hymenopteran antennae by Syrphidae. *Psyche* 77, 45–49.

Waldbauer, G.P. (1988) Asynchrony between Batesian mimics and their models. *American Naturalist* 131, S103–S121.

Waldbauer, G.P. and LaBerge, W.E. (1985) Phenological relationships of wasps, bumblebees, their mimics and insectivorous birds in northern Michigan. *Ecological Entomology* 10, 99–110.

Waldbauer, G.P. and Sheldon, J.K. (1971) Phenological relationships of some aculeate Hymenoptera, their dipteran mimics, and insectivorous birds. *Evolution* 25, 371–382.

Waldbauer, G.P., Sternberg, J.G. and Maier, C.T. (1977) Phenological relationships of wasps, bumblebees, their mimics, and insectivorous birds in an Illinois sand area. *Ecology* 58, 583–591.

Watts, J.O. (1983) A comparative study of the Syrphidae (Diptera) from different habitats within Bernwood Forest. PhD thesis, Oxford Polytechnic, Oxford, UK.

White, T.C.R. (1993) *The Inadequate Environment: Nitrogen and the Abundance of Animals*. Springer Verlag, Berlin, Germany.

Williams, P.H. (1991) The bumble bees of the Kashmir Himalaya (Hymenoptera: Apidae, Bombini). *Bulletin of the British Museum of Natural History* (*Entomology*) 60, 1–204.

Williams, P.H. (1998) An annotated checklist of bumble bees with an analysis of patterns of description (Hymenoptera: Apidae, Bombini). *Bulletin of the Natural History Museum, London* (*Entomology*) 67, 79–152.

10 Evolutionary Ecology of Insect Host–Parasite Interactions: an Ecological Immunology Perspective

KENNETH WILSON

Department of Biological Sciences, Lancaster Environment Centre, Lancaster University, Lancaster, UK

1. Introduction

Parasites and pathogens pose a ubiquitous threat to all organisms. However, until relatively recently, the impact of parasites on the ecology and evolution of their hosts had been largely ignored by biologists. Now, of course, parasites are recognized as an important selective force on their hosts and a key factor influencing their population dynamics (Grenfell and Dobson, 1995; Hudson *et al.*, 2002). The majority of studies examining the evolutionary ecology of host–parasite interactions have been conducted on vertebrate hosts, despite the fact that most animals are insects and other invertebrates. In recent years, though, this taxonomic bias has been challenged, as biologists have exploited the logistical advantages that insects and their parasites often provide.

   The aim of this chapter is to review recent advances in our understanding of insect host–parasite interactions. The theoretical framework underpinning this work has largely come from the field of 'ecological immunology' (Sheldon and Verhulst, 1996; Norris and Evans, 2000; Rolff and Siva-Jothy, 2003; Schmid-Hempel, 2003), a new and growing field concerned with understanding the evolutionary ecology of parasite resistance mechanisms. Ecological immunology is about the ecological and evolutionary causes and consequences of variation in immunity, defined in its widest sense (see below), to include any mechanism that improves the capacity of an organism to resist a parasite or pathogen. Essentially, ecological immunology is examining the proximate and ultimate causes of variation in disease resistance, and it takes an evolutionary ecology approach. In the past, this field has been dominated by biologists working on vertebrate systems, particularly birds (Norris and Evans, 2000). However, there is a growing belief that because the insect immune system is relatively simple in comparison with that of vertebrates, in that it lacks a conventional acquired immune system, ecological immunology studies are likely to be most successful when applied to insects.

This chapter is divided into three main parts. In Section 2 we explore the diverse range of behavioural, physical and immunological defence mechanisms employed by insects to combat the threat that parasites pose. A fundamental assumption of life-history theory in general, and ecological immunology in particular, is that life history and related traits (including immunity) are costly and result in physiological or genotypic trade-offs (Stearns, 1989, 1992; Roff, 2002). Therefore, in Section 3 we review the evidence for costs of resistance in insects. Finally, in Section 4, we examine three case studies from our own work in which the ecological immunology approach has been used to potentially gain new insights into the evolutionary ecology of insect host–parasite interactions.

2. A Cascade of Defence Components

Over evolutionary time, organisms have evolved a suite of mechanisms that combine and interact to reduce the probability and impact of parasitic infections. Schmid-Hempel and Ebert (2003) refer to a 'cascade of defence components', to reflect the fact that the different resistance mechanisms act at different levels, in different sequences, and with different specificities. For example, there are behavioural mechanisms which generally reduce the probability of an animal encountering parasites in the first place; there are physical or physiological mechanisms which often reduce the probability that a given parasite will become established in its host upon contact; and there are immunological mechanisms which increase the probability that the parasite will be killed or its pathological effects will be reduced (Fig. 10.1). Thus, behavioural mechanisms are often the first line of defence against parasites and pathogens (but see below), whereas immunological mechanisms, *sensu stricto*, are usually the last.

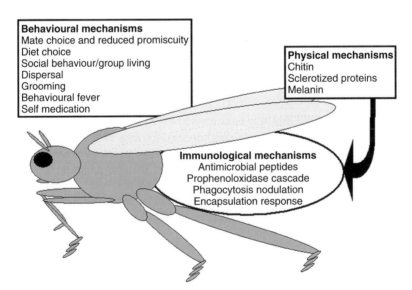

Behavioural mechanisms
Mate choice and reduced promiscuity
Diet choice
Social behaviour/group living
Dispersal
Grooming
Behavioural fever
Self medication

Physical mechanisms
Chitin
Sclerotized proteins
Melanin

Immunological mechanisms
Antimicrobial peptides
Prophenoloxidase cascade
Phagocytosis nodulation
Encapsulation response

Fig. 10.1. A cascade of defence components.

2.1 Behavioural mechanisms

There are a number of behavioural mechanisms that insects could employ to reduce their probability of becoming exposed to parasites and pathogens, and to ameliorate their effects should they become infected. These include: mate selection (to avoid parasites that are transmitted by potential sexual partners); selective diet choice (to avoid ingesting parasites); avoiding or dispersing from areas of high infection risk; social behaviours and group-living (to take advantage of 'dilution effects'); self-grooming and reciprocal allogrooming (e.g. to remove fungal spores or ectoparasites); behavioural fever and chills (i.e. altering body temperature to reduce parasite fitness and/or to increase immune function efficacy); and oral self-medication (i.e. ingesting nutrients that enhance physiological/immunological resistance mechanisms). Some of these behavioural mechanisms are reviewed here.

2.1.1 Mate choice and promiscuity

For parasites that are transmitted between sexual partners, there is an obvious benefit to being choosy or reducing levels of promiscuity. By avoiding mates with parasitic infections, an individual may reduce its probability of becoming infected. Whilst intuitively appealing, there is little evidence for mate choice in relation to infection status in insects. In a recent review of insect sexually transmitted diseases (STDs), Knell and Webberley (2004) reported finding only two examples where mate choice in relation to STD infection has been examined. Both studies involved beetles infected with mites: the leaf beetle, *Labidomera clivicollis*, infected with *Chrysomelobia labidomera* (Abbot and Dill, 2001), and the two-spotted ladybird, *Adalia bipunctata*, infected with the mite *Coccipolipus hippodamiae* (Webberley *et al.*, 2002). In neither case was there any evidence that females discriminated between males with or without the sexually transmitted mite, suggesting that there was no mate choice in relation to the potential mate's infection status. While at first this might seem puzzling, Knell (1999) has argued that avoidance of STDs is unlikely to be important in female choice, since there would be strong selection on STDs to become cryptic in such circumstances. Indeed, Graves and Duvall (1995) argued that STDs could select against mate choice in females because popular males might be more likely to infect them during mating (but see Boots and Knell, 2002; Kokko *et al.*, 2002).

So, if selection favours STDs that avoid revealing themselves to potential new hosts, an alternative parasite-avoidance strategy might be simply to reduce the number of mates, so minimizing contact with potentially infective hosts (Hamilton, 1990; Sheldon, 1993; Loehle, 1995; Lockhart *et al.*, 1996). Modelling studies suggest that STDs may select for reduced promiscuity, but if there is a fitness benefit to being promiscuous then this can result in a polymorphism, with some individuals being promiscuous and others being less so (Boots and Knell, 2002). At present, however, there is little evidence with which to test this prediction (Knell and Webberley, 2004).

Multiple mating might not only result in an increased risk of encountering an infected individual, but might also result in increased susceptibility to infection.

This is because a recent study on the mealworm beetle (*Tenebrio molitor*) has shown that, in both sexes, mating reduces the activity levels of phenoloxidase, a key enzyme in the insect immune system (Rolff and Siva-Jothy, 2002) (see below). This mating-induced downregulation of immunity lasts for at least 24 h and appears to be mediated by juvenile hormone. Thus, by mating frequently, individuals may be compromising their immune systems. Similar conclusions were reached in an earlier study by McKean and Nunney (2001) using the fruitfly *Drosophila melanogaster* (Fig. 10.2). They found that as the number of females housed with each male was increased (from 0 to 4), so there was a highly significant decrease in the male's ability to immunologically clear an experimental injection of *E. coli* bacteria. This reduced immunocompetence was not due to food shortage, since it was prevalent also when food was in excess. Nor was it simply due to crowding effects, since males housed with four males cleared the bacteria significantly faster than those housed with four females. Since both courtship behaviour and mating activity of males increase with the number of females available, it seems likely that one or both of these activities results in male immunosuppression. Thus, decreased immunocompetence may be a significant 'cost of reproduction' in insects.

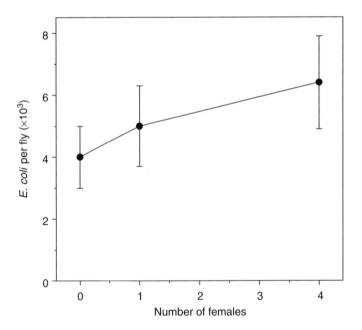

Fig. 10.2. Relationship between mating activity and immunity in the fruitfly, *Drosophila melanogaster*. The symbols (and bars) show the number of *E. coli* bacteria (±95% confidence interval) recovered from males kept in vials alone, or with either one or four virgin females that were replaced daily. As the number of females each male has access to increases, so his ability to clear an experimental infection of *E. coli* declines. Redrawn after McKean and Nunney (2001). Copyright (2001) National Academy of Sciences, USA.

2.1.2 Parasite avoidance behaviours

It is not just sexually transmitted parasites that need to be avoided, other types of parasites do also. One strategy used by many vertebrates to reduce contact rates with directly transmitted macroparasites (e.g. nematodes) is to spatially segregate feeding and defecating areas (often by using specialized 'latrines'), thereby avoiding faeces contaminated with infective eggs or larvae (Hart, 1994; Hutchings et al., 1998, 1999). While we are not aware of any examples of such behaviour in insects, there are instances in which insects are known to avoid parasites directly (e.g. Evans, 1982; Hajek and St Leger, 1994) or avoid dead conspecifics which might harbour such parasites (e.g. Kramm et al., 1982).

Other avoidance patterns may be more subtle. For example, it appears that the foraging behaviour of the leaf-cutter ant, Atta cephalotes, is influenced by the presence of the diurnal parasitoid Neodohrniphora curvinervis (Orr, 1992). Ants foraging above ground for leaves during the day are frequently parasitized by this day-flying phorid fly and so most ants forage at night instead. That this behavioural shift is a consequence of parasite pressure is indicated by the fact that when artificial lighting was supplied to allow phorids to hunt past dusk, ants foraged less than when light was provided but flies were removed (Orr, 1992). In this example, it appears that the ants were responding directly to the physical presence of the parasites. But in other instances, this may not be necessary to evoke avoidance behaviour. For example, in laboratory studies, it has been shown that gravid Aedes aegypti mosquitoes oviposit fewer eggs in water that contains conspecific larvae parasitized by the digenean parasite Plagiorchis elegans, than in water that contains non-parasitized larvae (Lowenberger and Rau, 1994). More extraordinary still, they also discriminate against water that was previously home to parasitized larvae, even if this water was subsequently boiled, treated with antibiotics, or filter-sterilized. This suggests that A. aegypti females are deterred from ovipositing in response to chemical cues produced by parasitized larvae.

2.1.3 Self-grooming and allogrooming

In many social insects, self-grooming and allogrooming are extremely prevalent and may be an effective strategy for removing the spores of entomopathogenic fungi or the infective stages of entomopathogenic nematodes (Schmid-Hempel, 1998). In the dampwood termite, Zootermopsis angusticollis, allogrooming increases in frequency during and after exposure to the spores of the fungus Metarhizium anisopliae (Rosengaus et al., 1998b). This appears to be effective in removing potentially infectious spores from the cuticle, so increasing termite survival. Thus, allogrooming plays a crucial role in the control of disease in termites, and Rosengaus and colleagues speculate that this advantage of group living may have been significant in the evolution of social behaviour in the Isoptera. Across ant species, the frequency of allogrooming tends to increase with colony size (Schmid-Hempel, 1998), whilst the frequency of self-grooming tends to do the opposite (Schmid-Hempel, 1990). In some ways, these results appear counterintuitive. This is because, if parasite pressure is greater in larger colonies (as seems likely; Schmid-Hempel, 1998), then we might expect the rate of self-grooming to increase (to maximize hygienic behaviour) and the rate of allogrooming to decline (to minimize risk of exposure to pathogens). However, the relationship between group size and infection risk is not straightforward and predictions regarding grooming behaviour are not easily made (e.g. Coté and Poulin, 1995; Wilson et al., 2003a; see below).

In vertebrates, it has been shown that grooming is a costly activity (Moore, 2002) and may be traded-off against vigilance behaviour (antelopes; Hart, 1992; Hart et al., 1992), resting (bats; Giorgi et al., 2001), thermoregulatory capacity (moose; Samuel, 1991), energy expenditure (bats; Giorgi et al., 2001) or even saliva production (rats; Ritter and Epstein, 1974). For example, in bats, the experimental inoculation of 20 or 40 ectoparasitic mites resulted in an increase in grooming activity and a concomitant decrease in resting activity, resulting in a drastic increase in overall metabolism (oxygen consumption) and weight loss (Giorgi et al., 2001). However, as far as we are aware, the costs of grooming have not yet been quantified in any insect species.

2.1.4 Feeding behaviour and self-medication

It is now well established that the impact of parasites on their hosts is often dependent on the quality and quantity of the host's diet (e.g. Duffey et al., 1995) and thus by choosing what (and how much) to eat, an animal is in a position to influence its exposure and susceptibility to parasitism. For example, the susceptibility of gypsy moth (Lymantria dispar) larvae to nucleopolyhedrovirus (GMNPV) varied significantly between four host-plant species, with the LD_{50} ranging from just 8000 viral occlusion bodies (OB) per larva when they were fed GMNPV on bigtooth aspen (Populus grandidentata) to more than 500,000 OB per larva when they were fed on black oak (Quercus velutina); this amounts to a 60-fold difference in susceptibility. In this particular example, decreased viral pathogenicity (increased larval survival) was correlated with increased acidity and hydrolysable tannin content of the leaves (Keating et al., 1988).

Differences in parasite virulence in relation to diet may be: because the structural, physical or chemical properties of a particular food (e.g. host plant) alter parasite transmission; because the diet influences the host's parasite resistance mechanisms, and hence susceptibility to the parasite; or because of an interaction between these two factors. Determining the precise nature of the host–parasite–diet interaction is difficult, because these different chemical and physical components interact in ways that are often complex and non-addictive (Duffey and Stout, 1996). In the case of baculoviruses of insect herbivores like gypsy-moth larvae, it appears that leaf phenolics may present a chemical barrier to viral infection (e.g. Young et al., 1995). The precise mechanism by which this occurs remains to be established, but phenolic extracts from leaves that inhibit NPV activity cause the viral OBs to form large aggregations (though this does not appear to reduce their infectivity; Keating et al., 1990).

It is not just herbivorous insects that are affected by the properties of their diet. The malarial vector, Anopheles stephensi, requires both a recent blood meal and a sugar source before it can successfully develop an effective melanization immune response (Koella and Sørensen, 2002). Thus, there is a significant interaction between dietary components. Similarly, a recent study of the Egyptian leafworm, *Spodoptera littoralis*, showed that resistance to NPV was critically dependent on the carbohydrate–protein balance of the diet, with insects that were fed on a high-protein, low-carbohydrate diet showing the highest survival following NPV challenge (Lee, 2002). Moreover, when insects were given *ad libitum* choice of two nutritionally complementary foods, those that survived virus challenge had selected a diet containing a higher protein content than those that had succumbed (or larvae that had not been challenged with NPV). It remains to be established, however, whether the larvae were engaging in nutritional 'self-medication' (i.e. actively seeking out a protein-rich diet in response to infection) or if there is a distribution of intake ratios within a population of caterpillars, and infection simply culled those larvae that had chosen a low-protein diet. Further studies are clearly required to distinguish between these possibilities.

2.1.5 Behavioural fever and chills

In mammals and other endotherms, parasitic infection is often associated with an endogenous fever response, in which the host's body temperature is increased above the norm by several degrees. It is usually argued that this is an adaptive response that helps the body to fight infection by stimulating natural defence mechanisms, though the issue remains controversial (Kluger *et al.*, 1998). Whilst it is true that under many circumstances the body's elevated temperature can help fight off an infection, this is not always the case, and excessive fever can cause problems such as dehydration. Demonstrating the adaptive value of fever in endotherms has proved problematic because reducing fever (e.g. by the administration of aspirin) usually has other major physiological effects such as pain reduction, reduced inflammation and other responses that could help the animal to defend itself against parasites. Thus, isolating the effects of fever from these other responses is extremely difficult in endotherms.

Many ectotherms, including insects, also exhibit a fever response to infection. However, this response is behavioural rather than physiological, as found by Kluger *et al.* (1975), working on the desert iguana (*Dipsosaurus dorsalis*). Since then, similar experiments have been conducted on a range of insects and their parasites. For example, the North American grasshopper *Melanoplus sanguinipes*, exhibits behavioural fever in response to infection by the lethal protozoan *Nosema acridophagus*. Grasshoppers kept at these fever temperatures survive longer and gain weight more rapidly than when kept at temperatures (6°C cooler) preferred by uninfected conspecifics (Boorstein and Ewald, 1987). However, fever is not shown in this same grasshopper species in response to infection by a closely related parasite, *Nosema locustae* (Hanley, 1989; cited in Ewald, 1994). Indeed, even when fever is elicited by a parasite, its effectiveness may depend critically on the susceptibility of the parasite to extreme temperatures. For example, when the clearwinged grasshopper, *Camnula pellucida*, is infected with the fungus *Entomopgaga grylli*, it generates a fever in excess of 40°C. This fever response is capable of killing the US strain of the fungus, but not an Australian strain, which is resistant to high temperatures (Carruthers et al., 1992; Ewald, 1994).

Behavioural fever is a taxonomically widespread phenomenon and examples have been found in annelids, arthropods (including insects), fish, amphibians and reptiles. Insect examples include: hissing Madagascar cockroaches (Gromphadorhina portentosa) infected with an E. coli suspension or an endotoxin (Bronstein and Conner, 1984); field crickets (Gryllus bimaculatus) infected with the lethal intracellular parasite Rickettsiella grylli (Louis et al., 1986); the house cricket, Acheta domesticus, when infected with the same parasite, but not when infected with the bacterium Serratia marascens (Adamo, 1998); and the desert locust Schistocerca gregaria, infected with the entomopathogenic fungus Metarhizium anisopliae (Blanford and Thomas, 1999; Wilson et al., 2002). However, behavioural fever is not a ubiquitous phenomenon (e.g. Ballabeni et al., 1995). In particular, there is little experimental evidence at present that macroparasites (nematodes, tapeworms, parasitoids, etc.) induce behavioural fever in insect hosts (but see Karban, 1998).

Changes in thermal response in relation to parasitism are not restricted to behavioural fever; they may also be manifested in behavioural chills. For example, although foraging bumblebee workers (Bombus terrestris) normally return to the nest at night, those that have become parasitized by a conopid fly (Conopidae, Diptera) remain outside the nest at night, where the temperatures are cooler (Müller and Schmid-Hempel, 1993). Moreover, when given a choice of temperatures, parasitized worker bees spend more time in cold areas than do non-parasitized ones, and these colder temperatures have been shown to increase the survival rate of infected workers, and fewer parasitoids complete their development (Müller and Schmid-Hempel, 1993).

Of course, these changes in thermal behaviour in response to parasitism may not always be adaptive from the host's point of view. In some cases they may be selectively neutral (a simple by-product of the pathology of infection), whereas in others they may be adaptive manipulations of host behaviour by the parasites themselves (e.g. Maitland, 1994), although conclusively demonstrating parasite-manipulation of host behaviour is notoriously difficult (Moore and Gotelli, 1990; Poulin, 2000; Moore, 2002).

In the vast majority of instances in which changes in host thermal behaviour have been observed in parasitized hosts, it remains to be established to what extent the survival benefits of the behavioural change result: from direct negative effects of temperature on the parasite (e.g. Starks et al., 2000); from indirect effects via changes in the host's immune response (e.g. Ouedraogo et al., 2003); or from interactions between these two effects. However, a recent study of the migratory locust Locusta migratoria, suggests that fever temperatures may enhance haemocyte production and thus host immunity via phagocytosis (Ouedraogo et al., 2003). Moreover, an earlier study of a refractory strain of the malarial vector, Anopheles gambiae, showed that the ability of the mosquito to melanize a Sephadex bead (a response similar to that which results in the death of ingested Plasmodium) declined significantly as the environmental temperature increased from 24°C to 30°C

(Suwanchaichinda and Paskewitz, 1998). Thus, determining the relative importance of direct and indirect effects of temperature on the host–parasite interaction is a challenge for future studies of this phenomenon (Thomas and Blanford, 2003). Understanding the relationship between temperature and pathogen resistance is complicated still further by the fact that the response may also vary between host genotypes infected with the same pathogen. For example, Stacey et al. (2003) found that although two clones of the aphid Acyrthosiphon pisum showed changes in susceptibility to the fungal entomopathogen Erynia neoaphidis, which mirrored changes in the pathogen's in vitro vegetative growth rate at different temperatures, two other clones exhibited responses that differed considerably from this expected response. Such interactions between genotype and temperature may have important implications for our understanding of disease dynamics in natural populations.

The extreme sensitivity of insects and their pathogens to temperature (and the insects' thermoregulatory response to infection) may have important implications for the use of entomopathogens in biocontrol (Ewald, 1994; Thomas and Blanford, 2003). Typically, biocontrol agents are chosen on the basis of their virulence to a chosen pest species, based on laboratory bioassays undertaken at one or a few constant temperatures. Yet, it is now known that for many of the fungi used in biocontrol (such as Beauveria bassiana and Metarhizium anisopliae, for example), the speed and magnitude of kill is critically dependent on the host's body temperature and that of its environment. As a consequence, a pathogen may cause rapid and extensive mortality to its host under some thermal conditions, but may be virtually benign under others (Thomas and Blanford, 2003). This is true not just for locusts and grasshoppers and their fungal pathogens, but also for a range of other host species (including flies, moths, leafhoppers and aphids) and many different types of parasite (including microsporidia, bacteria, viruses, rickettsia and nematodes). Thus, it is becoming increasingly apparent that future biocontrol studies will have to factor into their analyses the relative thermal sensitivity profiles (Thomas and Blanford, 2003) of both the insect they are trying to control and the potential biocontrol agents. It is only by understanding the potentially complex interactions between the host, its parasite and their shared environment that effective biocontrol strategies will be developed (Thomas and Blanford, 2003).

2.2 Physiological mechanisms

The first physical line of defence against most parasites and pathogens is the integument (the outer layer of the insect). It comprises the epidermis and the cuticle. The epidermis is one cell thick; the cuticle is a secretion of the epidermis and covers the whole of the outside of the body, as well as lining ectodermal invaginations, such as the trachea (Chapman, 1998). The inner region of the cuticle contains chitin, a polysaccharide. Hardening of the cuticle is primarily a consequence of cross-links between cuticular protein molecules so that they form a rigid matrix – this is a process known as tanning or sclerotization

(Chapman, 1998). The greater the proportion of proteins that become cross-linked, the greater the degree of sclerotization, and the more rigid the cuticle becomes. Hard, heavily sclerotized cuticles may be difficult for parasites to penetrate (Chapman, 1998). As the cuticle hardens, it also darkens. This darkening may be a consequence of quinone tanning, but it may also involve the polymerization of quinones to form melanin.

2.2.1 Melanism and parasite resistance

Melanin is a nitrogen-containing polymer. In the integument, it is either incorporated into granules or scattered throughout the cuticle (Hiruma and Riddiford, 1988). It has at least two properties that are likely to impact on potential pathogens. First, because it is a polymer, melanin is likely to strengthen the cuticle and so improve its ability to act as a physical barrier to the penetration of parasites and pathogens that enter via the cuticle, such as fungi, bacteria and even parasitoids (St Leger et al., 1988). Second, melanin is highly toxic to microorganisms and has potent antimicrobial activity (e.g. Montefiori and Zhou, 1991; Ourth and Renis, 1993; Sidibe et al., 1996; Ishikawa et al., 2000), probably because it binds to a range of proteins (e.g. Doering et al., 1999) and inhibits many of the lytic enzymes produced by microorganisms, including proteases and chitinases (Kuo and Alexander, 1967; Bull, 1970).

The antifungal properties of melanin are demonstrated nicely by two in vitro experiments conducted in the 1980s. Söderhäll and Ajaxon (1982) showed that when the crayfish-parasitic fungus Aphanomyces astaci was grown on agar, there was significant inhibition of fungal growth when the growth-medium contained melanin or any of its precursor quinones (e.g. 5,6-dihydroxyindole), but not L-dopa. Subsequently, St Leger et al. (1988) showed that when larval cuticles of Manduca sexta were induced to melanize by suspending them overnight in L-dopa, they resisted fungal penetration nearly twice as long as non-melanized cuticles (dopa-melanized cuticles 72 h, non-melanized cuticles 40 h). Thus, melanin may enhance disease resistance in insects not only by improving the physical properties of the cuticle, but also by enhancing its chemical properties.

Recently, the association between cuticular melanization and resistance to parasites and pathogens that enter their hosts percutaneously has been examined more explicitly. Wilson et al. (2001) examined variation in resistance to an entomopathogenic fungus (Beauveria bassiana) and an ectoparasitic wasp (Euplectrus laphygmae) in two phase-polyphenic lepidopteran species (Spodoptera exempta and S. littoralis). In both of these species, the colour of the isolated larval cuticle varies from near-transparent to nearly black, due to the deposition of melanin granules in the cuticle (Fig. 10.3A–C). Wilson and colleagues found that, relative to non-melanic conspecifics, melanic S. exempta larvae melanized a greater proportion of the ectoparasitoid's eggs (Fig. 10.3D), and that melanic S. littoralis were more resistant to the entomopathogenic fungus (in S. exempta, the association between melanism and fungal resistance was non-significant, possibly because the fungal dose was too high; see Wilson et al., 2001, for details). Similarly, when larvae of the Oriental armyworm,

Mythimna separata, were percutaneously infected with the entomopathogenic fungus *Nomuraea rileyi*, non-melanic larvae were substantially more susceptible than melanic ones (LC$_{50}$s: non-melanic = 4.9×10^7 OBs, melanic = 27.2×10^7 OBs; Mitsui and Kunimi, 1988).

In the mealworm beetle, *Tenebrio molitor*, cuticular colour varies from red-brown ('tan') to black, due to the combined effects of sclerotization and melanization of the cuticle (Thompson *et al.*, 2002; and references therein). As with the lepidopteran examples above, mealworm beetles are more likely to develop darker cuticles under crowded conditions (see below), although the phenotypic response to population density is not as strong (Barnes and Siva-Jothy, 2000). Barnes and Siva-Jothy (2000) examined variation in susceptibility

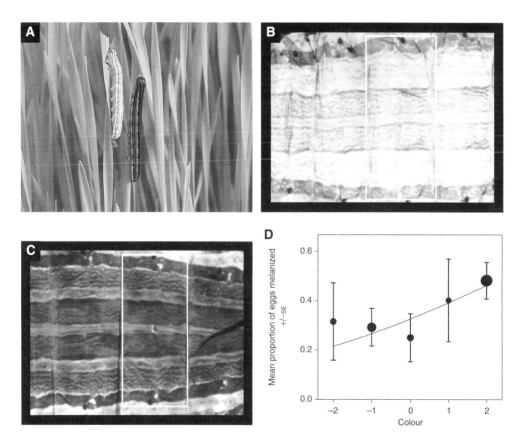

Fig. 10.3. Density-dependent cuticular melanization in *Spodoptera exempta*. **A** Live larvae, showing the pale, low-density phenotype on the left and the dark, high-density phenotype on the right; **B** the dorsal cuticle of the pale phenotype; **C** the dorsal cuticle of the dark phenotype; **D** relationship between cuticle colour and resistance to the ectoparasitoid, *Euplectrus laphygmae*. The vertical axis shows the proportion of melanized eggs (±SE) as a function of degree of cuticular melanization, scored on a scale from −2 (very pale) to +2 (very dark). Symbol size reflects sample size. The line is the fitted logistic regression to the raw data. Reprinted from Wilson *et al.* (2001) with permission from Blackwell Publishing Ltd.

to a generalist fungal pathogen (*Metarhizium anisopliae*) in relation to the cuticular colour of mealworm beetles and found that the highest mortality was observed in the palest (tan) beetles (91%) and the lowest in the darkest (black) beetles (29%).

Thus, the available evidence suggests that melanization and sclerotization of the cuticle is associated with greater resistance to parasites that enter their hosts via the cuticle. However, it is unclear, at present, how much of this effect is due to the physical and chemical properties of the cuticle and how much is due to correlated responses to aspects of the insect immune system (see below). Either way, the association between insect melanism and parasite resistance suggests that we may need to review our current understanding of the evolution of colour in insects. In fact, an association between melanism and disease resistance may not be restricted to insects (Owens and Wilson, 1999): recently, it has been suggested that similar correlations may also exist in vertebrate taxa including birds (Møller *et al.*, 1996; Gonzalez *et al.*, 1999; Evans *et al.*, 2000; Jawor and Breitwisch, 2003) and humans (Mackintosh, 2001).

2.3 Immunological mechanisms

One of the advantages of examining evolutionary questions about immunity using insects is that the insect immune system is simpler than its vertebrate equivalent, in that it lacks an adaptive or acquired immune system (in the conventional sense; but see below). There are many similarities between the vertebrate and invertebrate innate immune systems, and this makes insects potent models for understanding innate immunity (Vilmos and Kurucz, 1998). Even so, the insect innate immune system is still not fully understood and there remain gaps in our knowledge concerning the relative importance of its different components and how they interact. The immune defences of insects include constituitive and inducible defences, and both cellular and humoral components figure prominently. They include: cell-mediated responses, such as nodulation, phagocytosis, and cellular encapsulation; the enzymes of the prophenoloxidase cascade; and inducible antimicrobial peptides (including lysozymes, attacins, cecropins and insect defensins). The description of the insect immune system given here draws heavily on that proposed by Gupta (2001a,b).

2.3.1 Cellular immune system

Insects produce several different types of haemocytes that can be distinguished to a large degree on morphological grounds, though these classifications have been the subject of intense debate. Gupta (2001a) identifies six cell types: granulocytes, plasmatocytes, spherulocytes, oenoctoids, adipohaemocytes and prohaemocytes (a stem cell which differentiates into the other morphotypes). Of these, only the granulocytes and plasmatocytes appear to participate in defence reactions, and are hence sometimes referred to as immunocytes (Gupta, 1991). These immunocytes are responsible for a number of cell-mediated defence reactions, including phagocytosis, cellular encapsulation and nodulation, as well as the production and storage of the enzyme prophenoloxidase (see below). In the Diptera, including *Drosophila melanogaster*, the lamellocytes take on these phagocytic and encapsulation roles, and prophenoloxidase is produced in the crystal cells.

Phagocytosis is a dynamic and energy-requiring defence reaction (Gupta, 2001a), in which bacteria and other small (<1 μm diameter) particles are engulfed by the host haemocytes. This is followed by endocytosis and digestion of the foreign antigen in membrane-bound vesicles. Phagocytosis is usually associated with oxidative or respiratory burst and intracellular oxygen radical activity (Gupta, 2001a).

Encapsulation by immunocytes occurs when foreign biotic (e.g. bacteria, fungi, protozoa, nematodes or insect eggs and larvae) or abiotic (e.g. latex or Sephadex beads, nylon thread, etc.) antigens are too large to be phagocytosed (Gupta, 2001a). Humoral (cell-free) encapsulation also occurs in some insects, especially in those insects that have a very low number of haemocytes (e.g. Diptera). Cellular encapsulation begins when, following random contact between immunocytes (granulocytes) and foreign antigens, the antigens are recognized by surface receptors on the immunocytes, which trigger exocytosis and the release of coagulation-, recognition-, growth- and opsonin-like factors, and/or phenoloxidase (see below). This is followed by the formation of 20–70 more layers of immunocytes (plasmatocytes) and the melanization of one or more layers of the cellular capsule (Gupta, 2001a). Melanization of the capsule appears to be the main mechanism for the killing or sequestering of most biotic and abiotic antigens via 'disinfectants' (Taylor, 1969). These probably include the highly toxic quinones produced during melanin production, but may also include other toxic molecules, such as nitric oxide and superoxide, released by cells localized in the innermost layers of the cellular capsule.

Although encapsulation is a highly effective defence mechanism against many parasites and pathogens, it is ineffective against some bacteria and parasitoids, which have evolved a range of strategies to inactivate or evade encapsulation. These strategies include: secretion of defensive membranes, exhibiting particular surface charge properties, molecular mimicry of host antigens, secretion or injection of anti-immune factors or particles, causing haemocytopaenia, and stimulating acquired immune tolerance in the host (see Gupta, 2001a, for details). For example, many hymenopteran endoparasitoids belonging to the Braconidae and Ichneumonidae inject polydnaviruses into their lepidopteran hosts during parasitization (see Beckage, 2003). These viruses are integrated into the genomic DNA of the wasp and undergo replication only in the female's ovary. When they are injected into the host during oviposition, they rapidly enter the host's haemocytes and the viral genes are expressed. The haemocytes then either alter their behaviour and fail to spread (so inhibiting the encapsulation response) or undergo fragmentation and programmed cell death (depending on the host and parasitoid species involved). Thus, it appears that the living parasitoid larvae either escape being detected as foreign by mechanisms that may involve host antigen mimicry or masking, or by the presence of specific, and as yet unidentified, surface molecules that prevent their recognition as 'non-self' by immunocytes.

Nodule (or granuloma) formation involves both phagocytosis and encapsulation, and occurs in response to both inanimate and animate particulates (e.g. masses of bacteria) that are too numerous to be phagocytosed effectively. It is often difficult to distinguish between nodulation and encapsulation, as nodule formation also involves multicellular sheaths being formed by the immunocytes, but it appears that some of the encapsulating immunocytes detach from the nodule and enter the particulate mass (e.g. bacteria) to phagocytose them. Melanization of the nodule occurs in much the same way as in encapsulation. Nodulation can effect the production and release of antibacterial molecules from the fat body (the insect's functional analogue of the liver and the main source of circulating immune-related components; Christophides et al., 2002).

Another important cellular process is coagulation and wound healing, which prevents any further loss of haemolymph following injury, and so maintains haemostasis. Wound healing is performed by granulocytes (sometimes referred to as coagulocytes). Recently it has been suggested that the frequency of wounding may be an important selective pressure influencing an organism's optimal investment in immune defences (Plaistow et al., 2003). This is because using and maintaining an efficient immune system is costly (see below), and wound-healing and immunity share mechanisms and substrates in common. Thus, if the frequency of wounding is high, then many of the resources (haemocytes, phenoloxidase substrates, etc.) that are utilized in healing the wound may not be available for immune function.

2.3.2 Humoral immune system

Whilst the immunocytes (granulocytes and plasmatocytes) provide the first line of defence against parasites and pathogens, via phagocytosis, encapsulation and nodule formation, during a prolonged microbial insult the cellular defences may become impaired due to immunocyte depletion. Under these circumstances, antimicrobial proteins secreted by the host's immunocytes and/or fat body provide a second line of defence (reviewed by Boman and Hultmark, 1987; Dunn, 1991; Gupta, 1991, 2001b; Hetru et al., 1998). These include inducible antimicrobial peptides and polypeptides, agglutinins (lectins), cytokines ('factors'), neuropeptides and opioids, complement-like molecules, and the prophenoloxidase cascade (Gupta, 2001b).

Innate immunity in insects (as in mammals) is provided by various inducible antibacterial, antifungal and (presumably) antiviral proteins, which are produced in the haemocytes, fat body and epithelial tissues and are usually secreted into the plasma, where they act on the parasite either directly or via altering the behaviour of the immunocytes to enhance the immune response. Several hundred inducible antimicrobial peptides have now been described and these have been grouped into a number of types, including cecropins, attacins, insect defensins and lysozymes (for a review, see Faye and Hultmark, 1993). In the tobacco hornworm, Manduca sexta, injection of bacteria induced more than 25 different proteins, including cecropins, attacins and lysozymes (Hurlbert et al., 1985).

The upregulation of antimicrobial peptide production takes time. For example, in the two giant silkworm moths *Hyalophora cecropia* and *Antheraea pernyi*, the production of inducible antimicrobial peptides following bacterial challenge peaked at around 7–8 days post-infection, whereas in the smaller *Samia cynthia* it peaked in about half this time (Boman and Hultmark, 1987). After the peak, antimicrobial activity gradually declines, and disappears in about the same time as it took to reach the peak. In these particular examples, the antimicrobial peptides produced included attacins and cecropins.

Cecropins are a group of basic peptides (35–39 amino acid residues long) that affect the integrity of the cell membrane of most Gram-positive and especially Gram-negative bacteria, causing leakage of K^+ ions, interfering with ATP generation and ultimately causing cell lysis (Faye and Hultmark, 1993). Cecropins from species other than *Cecropia* have often been given new names, such as lepidopterin, bactericidin and sarcotoxin. However, given their obvious homology most are now referred to simply as cecropins. In each insect species, a family of cecropin forms is generally produced, and it is likely that each of these is adapted to the different needs of different compartments or developmental stages. For example, in *Drosophila*, cecropin A is mainly expressed in the fat body of larvae and adults, whereas the B and C forms appear to be preferentially expressed in the haemocytes of metamorphosing pupae (Samakovlis et al., 1990; Tryselius et al., 1992).

Insect defensins (such as sapecin) are cysteine-rich predominantly antibacterial peptides, named because of their apparent sequence similarity to the defensins produced from mammalian macrophages and neutrophils (Lehrer et al., 1991). In the meat fly *Sarcophaga peregrina*, the defensins appear to be produced only by the haemocytes, whereas in the flesh fly *Phormia terranovae*, they are synthesized also in the fat body (Faye and Hultmark, 1993). Like the cecropins and mammalian defensins, the insect defensins are believed to attack the bacterial cell membrane. Since insect defensins are most effective against Gram-positive bacteria, their antibacterial properties complement those of dipteran cecropins, which preferentially kill Gram-negative bacteria.

Recently, a number of peptides with defensin-like structures have been found to have potent antifungal properties, but no antibacterial activity. These include drosomycin (from *Drosophila melanogaster*; Fehlbaum et al., 1994; Michaut et al., 1996), heliomycin (from the tobacco budworm *Heliothis virescens*; Lamberty et al., 1999), termicin (from the termite *Pseudacanthotermes spiniger*; Da Silva et al., 2003) and gallerimycin (from the greater waxmoth *Galleria mellonella*; Schuhmann et al., 2003). These antifungal peptides will provide the first line of defence against fungal pathogens that infect via the cuticle.

Attacins are a family of bactericidal proteins that affect only growing cells, by interfering with cell division. Their antibacterial spectra are rather narrow and only a few Gram-negative bacteria are killed with purified attacin (Hetru et al., 1998). The target of the attacins appears to be the biosynthesis of the outer membrane, which becomes permeabilized, and so may allow easier access for lysozymes and cecropins. As with the cecropins, it appears that several different forms of these proteins may coexist within the same insect.

Lysozymes were the first antibacterial factors purified from insect haemolymph (Powning and Davidson, 1973). They are enzymes that degrade peptidoglycan in the bacterial cell wall. They are ubiquitous and are found not only in animals (including insects) but also plants, fungi and bacteriophages (see review by Jollès and Jollès, 1984). In insects, lysozymes are an important part of the immune defence. As with the other immune proteins, the main source of lysozyme in the haemolymph appears to be the fat body, although low levels are also found in the haemocytes. For example, in *Locusta*, lysozyme is stored in granulocytes and released during haemocyte coagulation. As well as the haemocoel, lysozyme is also found in the intestinal tract and, in *Manduca sexta*, it is specifically secreted there during metamorphosis (Russell and Dunn, 1991). In lepidopteran species (*Galleria*, *Bombyx*, *Spodoptera*, etc.) lysozyme is induced in the haemolymph, along with other antibacterial factors, but it is also sometimes present at significant constituitive levels in non-induced insects. In *Drosophila*, and indeed most other non-lepidopteran insects, lysozyme does not seem to be induced at all by exposure to bacterial antigens (Faye and Hultmark, 1993). The lysozyme from *Cecropia* is bactericidal to only a few Gram-positive bacteria, and it has been suggested that its main function might not be to kill sensitive bacteria, but to clear up the debris left behind after the action of cecropins and attacins (Boman and Hultmark, 1987). In addition, it is likely that lysozyme works in synergy with cecropins and attacins (Engström *et al.*, 1985).

Agglutinins (lectins) are carbohydrate-binding proteins, produced in both the immunocytes and fat body, which agglutinate bacteria, protozoa and metazoan parasites, because of the presence of particular polysaccharides on their cell surface (Hetru *et al.*, 1998). Agglutinins have many functions: they act as surface receptors, opsonins, can cause the recruitment and proliferation of immunocytes during encapsulation, and are also involved in histogenesis and wound healing.

Specific antiviral (interferon-like) molecules have not yet been found in insects, despite antiviral resistance having been well-characterized in a number of field and laboratory populations (e.g. Boots and Begon, 1993; Abot *et al.*, 1995). It remains to be established whether specific antiviral molecules are waiting to be discovered.

2.3.3 Prophenoloxidase cascade

The enzymes of the prophenoloxidase cascade oxidize tyrosine and its derivatives to their corresponding quinones and their polymerization product, melanin (Mason, 1955; Hiruma and Riddiford, 1988; Nappi and Vass, 1993). These enzymes are involved not only in cuticular melanization (see above), but also in the various immune responses directed against parasites and pathogens, including cellular encapsulation, humoral (cell-free) encapsulation and nodule formation (Poinar, 1974; Gotz, 1986; Paskewitz *et al.*, 1988; Hung and Boucias, 1992; Beckage *et al.*, 1993; Washburn *et al.*, 1996, 2000).

The highly reactive enzyme phenoloxidase (PO) is synthesized and stored as an inactive zymogen called prophenoloxidase (proPO). Sequence data from a range of insect prophenoloxidase enzymes clearly indicate that insect proPO is homologous to arthropod haemocyanin, with the overall amino acid sequence homology being between 30% and 40% (Ashida and Brey, 1998). The exact localization of proPO and PO in insect haemolymph is controversial. Leonard *et al.* (1985) argued that proPO is stored in granulocytes and released into the haemolymph by degranulation (exocytosis). However, Ashida and Yamazaki (1990) have argued that because the proPO genes lack a signal peptide signature, the proPOs are not released into the haemolymph by secretion but by rupture of haemocytes. Once in the haemolymph, the proPO is activated by a specific serine protease cascade (known as the prophenoloxidase cascade or the prophenoloxidase activating system). This cascade is triggered by minute amounts of microbial cell wall components, such as β-1,3 glucans (found in yeast and fungal membranes), and lipopolysaccharides and peptidoglycans (found in bacterial cell walls).

The insect cuticle also contains phenoloxidase enzymes (Lai-Fook, 1966; Hiruma and Riddiford, 1988), which have previously been referred to as wound phenoloxidase and granular phenoloxidase (Ashida and Yamazaki, 1990). It appears that wound (or cuticular) PO is synthesized in the haemolymph and transported to the cuticle, suggesting that the polypeptide backbone of the enzyme is likely to be identical to that of haemolymph PO (Ashida and Brey, 1995; Asano and Ashida, 2001). The relationship between haemolymph PO and granular PO is less clear, but Hiruma and Riddiford (1988) found that polyclonal antibody raised against granular PO did not cross-react with haemolymph proPO.

The midgut is an important site for resisting pathogens that enter the host orally, such as baculoviruses, protozoa and many bacteria, and phenoloxidase has often been implicated in midgut defence reactions. For example, the spread of baculovirus in non-permissive lepidopteran hosts appears to be blocked by aggregations of haemocytes that form melanotic capsules around infected cells in the midgut trachea (Washburn *et al.*, 1996, 2000). In tsetse flies (*Glossina* spp.), there was a significant positive association between PO activity and refractoriness to the protozoan *Trypanosoma brucei rhodesiense*, both within species (male versus female *G. morsitans morsitans*) and among species (*G. m. morsitans* versus *G. palpalis palpalis*) (Nigam *et al.*, 1997). And, in *Anopheles gambiae* mosquitoes selected for resistance to malaria, parasite ookinetes were melanized between the midgut epithelial cells and the basal laminae, completely blocking parasite transmission (Collins *et al.*, 1986). These resistant insects also showed higher phenoloxidase activity in the midgut following exposure to the parasite, indicating that the melanotic encapsulation was under phenoloxidase control (Paskewitz *et al.*, 1989). Recent studies on another *Anopheles* species (*A. stephensi*) indicate that the phenoloxidase isozyme isolated from the haemolymph differs from that extracted from the midgut (Sidjanski *et al.*, 1997). However, as yet, the relative importance of the two isozymes in the melanization of malarial ookinetes remains to be established.

It is becoming increasingly apparent that most, if not all, insects possess not a single prophenoloxidase enzyme, but a number of functionally related isoenzymes that may be activated in different tissues. For example, recent studies indicate that the genome of the mosquito *Anopheles gambiae* encodes for nine different proPO isoenzymes (Christophides *et al.*, 2002), and both *Drosophila melanogaster* (De Gregorio *et al.*, 2001) and the silkworm *Bombyx mori* (Yamamoto *et al.*, 2003) produce three isoforms of proPO. However, regardless of how many different forms of the enzyme there are, all of the proPO enzymes are structurally similar, generate the same end-product (melanin), and are activated by similar combinations of substrates and triggers. It seems, therefore, that similar selection pressures are likely to prevail on all forms of the enzyme.

2.3.4 Molecular genetics of immune function

Recent advances in molecular genetics and the sequencing of the entire genomes of two important model insect species – *Drosophila melanogaster* (Adams *et al.*, 2000) and *Anopheles gambiae* (Holt *et al.*, 2002) – has allowed progress to be made in trying to understand the molecular genetic basis of immune function in insects (e.g. De Gregorio *et al.*, 2001; Christophides *et al.*, 2002).

Using high-density oligonucleotide microarrays encompassing almost the entire *D. melanogaster* genome, De Gregorio *et al.* (2001) examined the gene expression profile of adult flies in response to microbial infection, due to either septic injury with a mixture of Gram-negative (*Escherichia coli*) and Gram-positive (*Micrococcus luteus*) bacteria, or to a natural infection with an entomopathogenic fungus (*Beauvaria bassiana*). Out of 13,197 genes tested, they identified 230 genes that were induced following microbial infection and a further 170 that were repressed. Septic injury caused gene-expression changes in nearly all of the 400 *Drosophila* immune-related genes, whereas fungal infection regulated just 157, including one gene that responded only to *B. bassiana* infection, suggesting that it could be specific to this fungus (see below).

Much is now known about immune-related genes of *Drosophila*, and this can inform future studies of the molecular genetic basis of immunity in other insects of greater economic or health importance. Recently, Christophides *et al.* (2002) used this approach to conduct a comparative analysis of immune-related genes in *Anopheles* mosquitoes. They found that, relative to the genome as a whole, the immune systems of both species, and especially *Anopheles*, shows a deficit of well-conserved 1:1 orthologues and an overabundance of specific gene expansions with species-specific functions. In *Anopheles*, for instance, there are four genes that code for cecropins (effective against Gram-negative bacteria) and another four genes that encode insect defensins (most effective against Gram-positive bacteria). Thus, these important antimicrobial peptides are more numerous and more diverged than in *Drosophila* (Christophides *et al.*, 2002). Similarly, as indicated above, there are nine genes coding for prophenoloxidase in *Anopheles*, but just three in *Drosophila*. Christophides and colleagues interpret these patterns as indicating that in many immune gene families, orthologues are under selection pressure to diversify, or are lost, whereas certain immune genes reduplicate and then diversify. In other words, there is strong selection to adjust and expand the innate immune repertoire in response to new ecological and physiological conditions. In the case of *Anopheles*, these challenges include blood-borne infectious agents such as *Plasmodium*.

Of course, both *Drosophila* and *Anopheles* are dipterans and, as such, their immune systems differ in a number of important ways from many other insect groups (e.g. they have a weak cellular encapsulation response, due to low haemocyte numbers). Therefore, one of the major challenges for future studies will be to determine the similarities and differences between the molecular genetics of immune function in dipterans and non-dipterans, especially lepidopterans, which include many of the most economically-important pest species. Another challenge will be to integrate whole-genome studies of insects with those of their parasites (e.g. *Autographa californica* and AcNPV) to gain new insights into insect host–parasite coevolution (Taylor *et al.*, 2000).

2.3.5 Adaptive immunity and specific memory in innate immune responses

Acquired or adaptive immunity in vertebrates is characterized by immunological memory and specificity. Because invertebrates lack the T-cell receptors and immunoglobulins that mediate vertebrate adaptive immunity, it is generally assumed that their immune responses also lack specificity and memory (e.g. Lemaitre *et al.*, 1997). With the advent of new molecular methods, however, these assumptions are being challenged. In the fruitfly *D. melanogaster*, it has been shown that the genes encoding antibacterial and antifungal peptides are differentially expressed depending on whether the insect is challenged by Gram-negative bacteria, Gram-positive bacteria, or fungal pathogens (De Gregorio *et al.*, 2001). Indeed, following natural infection by the entomopathogenic fungus *Beauvaria bassiana*, only peptides with antifungal activity were expressed, indicating that the antimicrobial response shows some degree of specificity (Lemaitre *et al.*, 1997; De Gregorio *et al.*, 2001). However, the exact degree of specificity remains to be established. Certainly, there are a number of invertebrate studies suggesting that infection patterns in the wild are dependent on the match between host and parasite genotypes (e.g. Lively and Dybdahl, 2000; Carius *et al.*, 2001), and there is a growing number of studies suggesting specific memory in invertebrate host–parasite interactions, arguing for specificity in host immune responses.

Kurtz and Franz (2003) examined the specificity of immunological memory in an invertebrate, the copepod *Macrocyclops albidus*, infected with one of its natural parasites, the tapeworm *Schistocephalus solidus*. Their experimental design involved exposing each copepod to three tapeworm larvae and then, 3 days later, exposing them to either three sibling parasites or to three unrelated parasites from a different sibling group. In both cases, the tapeworms used in the second challenge were fluorescently labelled so that they could be distinguished from the parasites used in the primary challenge. They found that when the copepods had prior exposure to related parasites, re-infection success was approximately 48%, whereas when they had previous experience of unrelated parasites, the success rate was nearly 60%, indicating that this invertebrate is exhibiting specific immunological memory to its parasites.

Similar results were gained by Little *et al.* (2003), using two strains of the pathogenic bacterium *Pasteuria ramosa*, infecting the water flea *Daphnia magna*. They observed that there appears to be maternal transfer of strain-specific immunity in this crustacean following exposure to the bacterium. When *D. magna* females were exposed to one strain of the bacterium and their offspring were exposed to either the same strain (i.e. homologous challenge) or a different strain (i.e. heterologous challenge), the proportion of offspring that became infected was greater for the heterologous than the homologous challenge. This indicates that *D. magna* is exhibiting strain-specific immunity to this parasite. Little and colleagues also examined the overall fitness benefit of this maternal transfer of immunity, by determining the timing and magnitude of differences in the rate of offspring production. They found that when exposed to a homologous challenge, *D. magna* females started reproducing earlier and produced up to 21% more offspring than when they were exposed to a heterologous challenge. Moreover, these fecundity effects translate into a significant fitness benefit of maternally transferred immunity, such that populations exhibiting this maternal effect would double in size, relative to populations lacking such an effect, in just two to three generations. The mechanisms underlying this phenomenon remain to be determined.

So far, there is evidence for transgenerational transfer of immunity in just one insect, the bumblebee, *Bombus terrestris*. Moret and Schmid-Hempel (2001) used a split-colony design in which most of the worker bees in the treated half of the colony were injected weekly with LPS in Ringer's solution to activate the immune system, and control workers in the other half of the colony were treated with Ringer's solution alone. As expected, they found that workers in the treated half of the colony showed higher antibacterial activity than those in the control half. However, their phenoloxidase activity levels were lower, suggesting a possible phenotypic trade-off between the two immune responses (see below). They also found that immune-challenged groups produced fewer queens and had lower reproductive output overall, indicating a possible cost of deploying the immune system (see below). Most significantly, perhaps, they found that male offspring from challenged groups showed higher phenoloxidase activity (and encapsulation response) than controls, although both haemocyte counts and antibacterial activity were comparable. Again, the mechanism by which this transgenerational transfer of immunity occurs remains to be determined.

All of these studies indicate that there may be greater levels of specificity and memory in the invertebrate immune response than had previously been assumed. The challenge for future studies is to determine the mechanisms by which these are achieved and their costs (Schmid-Hempel and Ebert, 2003).

3. Costs of Resistance

An important assumption of the life-history theory approach to examining immune defence is that immunity is costly and is traded off against other fitness traits (Sheldon and Verhulst, 1996; Owens and Wilson, 1999; Lochmiller and Deerenberg, 2000). This assumption is based on the idea that if there were no costs then there would be no penalty associated with maintaining or activating the immune system at its maximal level. The fact that we observe considerable genetic and phenotypic variation in immune function implies that there must be costs. There are three different kinds of costs that have been identified and are discussed below: deployment costs, maintenance costs and evolutionary costs (Table 10.1).

3.1 Deployment costs

The most obvious costs associated with parasite resistance are those that are incurred when the immune system (or other resistance mechanism, see above) is deployed in response to a parasitological challenge. In both vertebrates and invertebrates, there is now good evidence that using the immune system is costly (see review by Schmid-Hempel, 2003). For example, in *Drosophila melanogaster*, those flies that have successfully encapsulated an egg of the parasitoid *Asobara tabida* have reduced fitness as adults (Fellowes *et al.*, 1999c): capsule-bearing adults of both sexes are smaller than control flies, females produce significantly fewer eggs, and males allowed to copulate just once produce fewer offspring (though capsule-bearing males allowed repeated copulations with females do not show a reduction in fecundity). Thus, it appears that there is a trade-off between using resources in defence against parasitism and using those same resources in processes promoting fecundity and mating success. Similarly, when larvae of *Culex pipiens* mosquitoes were infected with the microsporidian *Vavraia culicis*, infected females pupated significantly earlier than uninfected females and tended to emerge as smaller adults, indicating a cost to their fecundity. However, the age and size at maturity of infected male mosquitoes was no different from uninfected males, indicating possible gender differences in the costs of resisting parasitic infection (Agnew *et al.*, 1999; see below).

Whilst both of these studies suggest that immune defence against parasites is costly, it is unclear whether the observed costs are due: to the host initiating an encapsulation response; to the pathological damage caused by the parasite; or to a combination of the two. Thus, the only way to reliably determine the magnitude of the costs associated with deploying the immune system is to measure the costs in the absence of the parasite. For example, in insect studies this has often involved triggering an immune response with Sephadex beads, a nylon implant, or a microbial cell wall component (e.g. lipopolysaccharide; LPS), and comparing the subsequent fitness of these individuals with control insects whose immune systems have not been activated. Because the immune elicitors are non-pathogenic and non-living, yet trigger the invertebrate immune system, the costs of deploying the immune system are isolated from the pathological costs associated with the parasites themselves.

One of the best examples of this approach is a recent study by Moret and Schmid-Hempel (2000) working with the bumblebee, *Bombus terrestris*. They

Table 10.1. Costs of resistance in insects. Adapted after Schmid-Hempel (2003).

Type of cost	Species	Protocol	Effect	Reference
Deployment costs (effects of immune challenge on fitness traits)	Captive bumblebee (*Bombus terrestris*)	Antigenic challenge by injection of LPS and Sephadex beads	Reduced survival only when starved	Moret and Schmid-Hempel, 2000
	Cabbage white butterfly (*Pieris brassicae*)	Pupae challenged with nylon implant	Standard metabolic rate of challenged pupae raised by nearly 8%	Freitak et al., 2003
	Mosquito (*Armigeres subalbatus*)	Experimental infection with microfilariae taken from mammalian host	Infection reduces egg development	Ferdig et al., 1993
	Mosquito (*Culex pipiens*)	Experimental infection with a microsporidian (*Vavraia culicis*)	Infected females pupate earlier and emerge as smaller adults, but no effect on males	Agnew et al., 1999
	Fruitfly (*Drosophila melanogaster*)	Experimental infection by larval parasitoid (*Asobara tabida*)	Infected females smaller as adults and less fecund; adult males smaller and less fecund if allowed to mate just once. Puparial walls thinner	Fellowes et al., 1998b, 1999c
Deployment costs (effects of work, mating or resource limitation on immune function)	Captive bumblebee (*Bombus terrestris*)	Restricted access to food	Reduces reproductive success but has no effect on encapsulation response	Schmid-Hempel and Schmid-Hempel, 1998
	Free-flying bumblebee (*Bombus terrestris*)	Clipping wings to prevent foraging and flying	Foraging bees show reduced encapsulation response	König and Schmid-Hempel, 1995 Doums and Schmid-Hempel, 2000
	Wild damselfly (*Matrona basilaris*)	Observation of activity	After copulation or oviposition, encapsulation response decreased	Siva-Jothy et al., 1998
	Mosquito (*Anopheles gambiae*)	Restricted larval access to food	Encapsulation response weaker in adults that had been nutritionally deprived	Suwanchaichinda and Paskewitz, 1998

	Fruitfly (*Drosophila melanogaster*)	Variation in access to receptive females	Ability to clear experimental injection of *E. coli* negatively related to mating activity	McKean and Nunney, 2001
	Mealworm beetle (*Tenebrio molitor*)	Short-term starvation	Phenoloxidase activity reduced (but not cellular encapsulation response)	Siva-Jothy and Thompson, 2002
	Mealworm beetle (*Tenebrio molitor*)	Comparison of experimentally mated and non-mated beetles	Mating reduces PO activity through juvenile hormone	Rolff and Siva-Jothy, 2002
Maintenance costs	Leaf-cutting ant (*Acromyrmex octospinosus*)	Experimental closure of paired exocrine metapleural glands	Reduced production of antimicrobial peptides leads to reduced energy expenditure	Poulsen *et al.*, 2002
	Mealworm beetle (*Tenebrio molitor*)	Observation of relationship between cuticular melanism and patterns of sperm precedence	Melanic males showed reduced sperm precedence but probably with low fitness consequences	Drnevich *et al.*, 2002
	Egyptian leafworm (*Spodoptera littoralis*)	Observation of relationship between cuticular melanism and larval fitness	Melanic larvae smaller, have lower haemolymph protein levels and reduced lifespan	Cotter *et al.*, 2004a
	African armyworm (*Spodoptera exempta*)	Observation of relationship between cuticular melanism and larval and adult fitness	Melanic moths smaller, but produce more eggs when fed on sucrose solution	Mensah and Gatehouse, 1998
Evolutionary costs	Mosquito (*Aedes aegypti*)	Selection on age at pupation (and hence reproduction)	Earlier reproduction correlated with lower encapsulation response	Koella and Boëte, 2002
	Mosquito (*Aedes aegypti*)	Selection for increased resistance to nematode infection	Increased resistance correlated with reduced reproductive success	Ferdig *et al.*, 1993
	Fruitfly (*Drosophila melanogaster*)	Selection for increased encapsulation response to common larval parasitoid (*Asobara tabida*)	Increased resistance correlated with reduced larval competitive ability	Kraaijeveld and Godfray, 1997

Continued

Table 10.1. *continued.*

Type of cost	Species	Protocol	Effect	Reference
Evolutionary costs *continued*	Fruitfly (*Drosophila melanogaster*)	Selection for increased encapsulation response to virulent larval parasitoid (*Leptopilina boulardi*)	Increased resistance correlated with reduced larval competitive ability	Fellowes *et al.*, 1998a
	Honeybee (*Apis mellifera*)	Selection for increased resistance to bacterial disease	Increased resistance correlated with higher larval mortality	Sutter *et al.*, 1968
	Honeybee (*Apis mellifera*)	Selection for increased resistance to bacterial disease	Increased resistance correlated with higher larval mortality	Rothenbuhler and Thompson, 1956
	Indian meal moth (*Plodia interpunctella*)	Selection for increased resistance to granulosis virus	Increased resistance correlated with slower development, lower egg viability, but increased pupal mass	Boots and Begon, 1993
	House cricket (*Acheta domesticus*)	Quantitative genetics of immunity and body size	Positive genetic correlations between haemocyte load, encapsulation response and body size in males; no evidence for costs	Ryder and Siva-Jothy, 2001
	Egyptian leafworm (*Spodoptera littoralis*)	Quantitative genetics of immune function and life history traits	Negative genetic correlations between antibacterial (lysozyme-like) activity and haemocyte density, haemolymph phenoloxidase activity and cuticular melanization	Cotter *et al.*, 2004b
	Velvetbean caterpillar (*Anticarsia gemmatalis*)	Selection for increased resistance to nucleopolyhedrovirus	Increased resistance correlated with lower larval survival rates, lower pupal weights, longer lifespans, and reduced fertility	Fuxa and Richter, 1998

found that when they activated the bumblebee immune system by injecting individuals with LPS and/or small, sterile micro-latex beads (4.5 μm diameter), the immune system was activated (as indicated by an increase in antibacterial activity, measured by a zone-inhibition assay). Moreover, the survival time of worker bees that had been injected with LPS and/or beads was reduced by factor of 1.5–1.7 (odds ratios) relative to control bees that were injected with Ringer solution alone (Fig. 10.4). Importantly, however, this cost of deployment was revealed only in worker bees that were starved post-challenge; activation of the immune system had no measurable cost to worker bees that were allowed to feed on sugar-water *ad libitum*. This result demonstrates that deploying immune defence mechanisms is costly in this species, but that these costs may be ameliorated by compensatory feeding to replenish the limiting resources lost to the immune system. An important implication of this observation is that, provided sufficient resources are available in the wild, the costs of immune activation may be insignificant under field conditions. Alternatively, given that it is probable that resources are more constrained under field conditions than in the laboratory, it suggests that the fitness costs of deploying the immune system may be extremely important in wild insect populations. It is therefore extremely important to determine the magnitude of nutrient constraints in the wild.

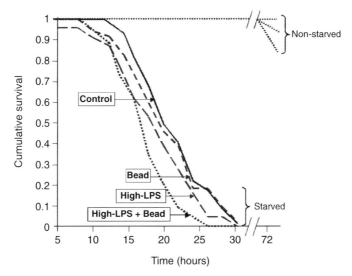

Fig. 10.4. Effect of diet and immune challenge on survival of the bumblebee, *Bombus terrestris*. Worker bees were either starved or fed sugar-water *ad libitum*, and were challenged with lipopolysaccharides (LPS), Sephadex beads, both, or neither. The top, nearly horizontal, lines refer to non-starved animals (no difference was found among treatments). High-LPS + beads refer to the combined injection of a high dose of lipopolysaccharides and beads. Cox regression analysis for the starved animals (sloping lines) showed that worker survival rate was reduced by the injection of Sephadex beads (odds ratio, OR = 1.56) or LPS (high dose, OR = 1.75). The addition of beads to LPS had an additive survival cost compared with LPS and beads alone. Redrawn with permission from Moret and Schmid-Hempel (2000). Copyright (2000) AAAS.

Another take-home message from this study is that resistance to parasites is likely to be condition-dependent. In other words, hosts with higher levels of the key limiting resources should be better able to fight off parasitic infections. In fact, there is good evidence for this prediction from another study of *B. terrestris* (Brown et al., 2000). This showed that when worker bees had *ad libitum* access to pollen and sugar-water and were infected with an intestinal trypanosome (*Crithidia bombi*), the bees suffered negligible mortality rates (<6%), which did not differ from those of conspecifics in a control group that were not infected with the parasite (<5%). However, when workers were starved, the mortality rate of bees in the infected treatment group was approximately 1.5 times higher than that of bees in the control group. Thus, the virulence of the parasite was enhanced under conditions of host food stress.

Whilst these results suggest that parasite resistance is condition-dependent, the experimental protocol employed in both of these experiments is rather extreme (though they may mimic the situation faced by many insects when foraging is prevented for long periods by rain and/or cold weather; Brown et al., 2000). It remains to be seen whether such effects are also observed under more realistic conditions, in which resources are constrained but not withdrawn completely. However, two studies indicate that such effects will prevail: Suwanchaichinda and Paskewitz (1998) found that when a *Plasmodium*-refractory strain of the mosquito *Anopheles gambiae*, was well nourished during the larval stages, they produced stronger encapsulation responses as adults than those that were nutritionally deprived. Similarly, Siva-Jothy and Thompson (2002) observed that in the mealworm beetle, *Tenebrio molitor*, phenoloxidase activity (but not cellular encapsulation) was downregulated in individuals that had experienced short-term starvation relative to conspecifics that were given *ad libitum* access to food.

All of the studies outlined above measured the costs of immune deployment by activating the immune system and measuring the fitness costs. Another way of demonstrating that immunity is costly is to increase the demands of other fitness components (e.g. by increasing workload) and to measure the impact on immune function. For example, when worker bumblebees (*Bombus terrestris*) were prevented from engaging in energetically demanding foraging activity by cutting their wings, their encapsulation response against a novel antigen (a nylon implant) was significantly stronger than that of conspecifics that were allowed to forage (König and Schmid-Hempel, 1995; Doums and Schmid-Hempel, 2000). Reproductive activity may also be costly. For example, the encapsulation response of Japanese calopterygid damselflies (*Matrona basilaris japonica*) was significantly reduced following copulation or oviposition (Siva-Jothy et al., 1998); the ability of male fruitflies (*Drosophila melanogaster*) to clear an experimental injection of *E. coli* was significantly negatively related to the number of receptive females they had access to (McKean and Nunney, 2001; see above); and phenoloxidase activity (but not haemocyte load) was significantly reduced in mealworm beetles (*Tenebrio molitor*) following mating (Rolff and Siva-Jothy, 2002).

A key question that needs to be addressed in future studies is what are the key resources limiting immune expression? Although it is often assumed that immunity is *energetically* costly, only a few studies on vertebrates have detected a change in energy budget as a consequence of immune stimulation (Lochmiller and Deerenberg, 2000). These suggest that when the vertebrate host mounts an immune response following vaccination or sepsis, the metabolic rate typically increases by around 15–30% (relative to resting metabolic rate). In the leaf-cutting ant *Acromyrmex octospinosus*, experimental closure of paired exocrine metapleural glands prevented the production of a highly effective cocktail of antibacterial and antifungal peptides (Poulsen *et al.*, 2002). Consequently, there was a significant reduction in the respiration rate of treated ants, suggesting that metapleural gland secretion incurs a substantial cost, though these costs may best be referred to as maintenance rather than deployment costs, because they are presumably paid even in the absence of the pathogen (see below).

As far as we are aware, only one study has examined the metabolic costs of mounting an immune response in an insect. Freitak *et al.* (2003) showed that when diapausing cabbage white butterfly (*Pieris brassicae*) pupae were challenged with a nylon implant, they raised their standard metabolic rate by nearly 8% relative to controls. In contrast, Alleyne *et al.* (1997) found that when tobacco hornworm (*Manduca sexta*) larvae were infected by the braconid wasp *Cotesia congregata*, their metabolic rates decreased 1 day following parasitization. However, as Freitak and colleagues point out, the polydnaviruses (or venom) that accompany wasp eggs are known to temporarily suppress the host's cellular immune response (Strand and Pech, 1995; Shelby and Webb, 1999; Shelby *et al.*, 2000), and so the lower metabolic rate of parasitized larvae may be a result of their reduced investment in immune defence. These two experiments demonstrate the importance of distinguishing between the costs of deploying the immune system from the costs associated with parasitism per se. Clearly, more studies are required before we can establish whether the metabolic cost of immune activation demonstrated in cabbage white butterfly pupae is typical of insects as a whole.

In insects, available nitrogen is often considered to be the major limiting resource (McNeill and Southwood, 1978; Crawley, 1983), and certain key nutrients, such as salt, may be essential for normal growth and survival of some species (Trumper and Simpson, 1993). However, the nutritional requirements for growth, reproduction and survival may be very different from those for an effective immune response. There is therefore a pressing need to identify the key macro- and micronutrients constraining insect immune function (but see Lee, 2002).

Given that the dietary precursors of melanin are phenylalanine (an essential amino acid) and tyrosine (a non-essential amino acid), it seems possible that these may constrain aspects of insect immune function (Johnson *et al.*, 2003). Experimental evidence in support of this idea comes from a study of *Armigeres subalbatus*, the vector for the filarial worm *Brugia malayi*, which is responsible for elephantiasis (Ferdig *et al.*, 1993). This mosquito has a very effective encapsulation response, which can kill more than 80% of the microfilarial infective stages of its parasite within 36 h of ingestion. Tyrosine is required both for melanotic encapsulation and for egg-chorion tanning (amongst other things). Thus, because the blood-meal both initiates egg development and is the source of the parasite, Ferdig and colleagues examined the possibility that the process of encapsulating microfilaria in an infected blood-meal would impose a reproductive cost on the host. As predicted, relative to control females, insects that had fed on an infected blood-meal took significantly longer to develop their ovaries and to start oviposition. Moreover, normal processes of egg development, including vitelline accumulation, were drastically altered, and the tyrosine and total protein levels in their ovaries were less than half those of the controls. Another possible example comes from the *D. melanogaster–A. tabida* system, where the pupae of larvae that have successfully encapsulated a parasitoid have relatively thinner puparial walls (Fellowes *et al.*, 1998b). Since both encapsulation and puparium formation utilize the substrates and enzymes of the prophenoloxidase cascade, this is a likely mechanism linking these two phenomena. Regardless of the precise mechanism, the thinner walls of the puparium appear to impose a significant fitness cost, since the pupal parasitoid *Pachycrepoideus vindemiae* preferentially attacks *Drosophila* pupae that have previously survived attack by the larval parasitoid, presumably because the thinner puparium reduces the handling time required for successful attack (Fellowes *et al.*, 1998b). Clearly, further studies are required in which important nutritional components are manipulated directly and their impacts on immune function in insects examined.

Another possibility is that the costs of resistance are mediated not by limiting resources but via physiological constraints. In the case of the downregulation of phenoloxidase activity exhibited by mealworm beetles (*T. molitor*) following mating, it appears that the costs of mating are mediated by juvenile hormone (JH), secreted from the *corpora allata* (Rolff and Siva-Jothy, 2002). Mating-induced JH secretion functions to switch on physiological processes associated with gametogenesis and spermatophore production (Wigglesworth, 1965), processes vital to both female and male fitness. It appears that a side-effect of this is that phenoloxidase activity is reduced, though the mechanism underlying this physiological trade-off remains to be elucidated. The immunological costs of mating detected in earlier studies on damselflies (Siva-Jothy *et al.*, 1998) and fruitflies (McKean and Nunney, 2001) may also be a consequence of similar physiological antagonisms.

A hitherto underrated potential cost of activating an immune response is the risk of self-reactivity or auto-immunity. This seems particularly likely in invertebrates, with their open circulatory system which exposes all of the host's vital organs to the immune effector systems it switches on in response to the recognition of non-self. When the prophenoloxidase cascade is activated, during melanotic and cellular encapsulation, a number of cytotoxic by-products are produced, including quinones and reactive oxygen species (see above). Whilst these may be extremely effective in combating the parasite, because of the open haemocoel, they may also cause significant damage to the host tissues if not rapidly detoxified (Nappi and Vass, 1993; Nappi *et al.*, 1995; Sugumaran *et al.*, 2000). The magnitude and frequency of this autoreactive threat remains to be determined. However, it has been argued that immune responses may be actively suppressed by animals under conditions when their risk of self-reactivity is high (Råberg *et al.*, 1998; Westneat and Birkhead, 1998).

3.2 Maintenance costs

Even when the immune system is not being used to combat a parasite or pathogen, insects that maintain high levels of immunological 'readiness' are still likely to pay significant costs. The distinction between deployment and maintenance costs has been likened to the respective penalties of fighting a war (deployment costs) and maintaining a standing army (maintenance costs) (Fellowes and Godfray, 2000; Wilson, 2001). Some of the costs associated with maintaining an effective immune system will be evolutionary costs (see below), with different genotypes expressing different constitutive levels of immune investment. However, within the constraints imposed by their genetic make-up, individuals will also exhibit phenotypic plasticity with respect to levels of prophylactic immune investment (Wilson and Reeson, 1998), and so are likely to incur phenotypic costs associated with physiological trade-offs.

In a number of recent studies of insects, it has been shown that immune function is upregulated in response to cues associated with high population density (e.g. Reeson et al., 1998, 2000; Barnes and Siva-Jothy, 2000; see below for details). Similarly, in rodents, it has been shown that immune function is seasonally upregulated at the start of winter (Nelson and Demas, 1996; Nelson et al., 1998) and, in humans, it appears that immune function is downregulated in high-performance athletes relative to non-athletes (Kumae et al., 1994). Measuring the costs of immune maintenance in these and similar studies is inherently difficult because of the way in which the immune system is integrated with other physiological systems (Lochmiller and Deerenberg, 2000; Schmid-Hempel, 2003).

The costs associated with investment in prophylactic resistance have been measured in several insect species. In adult mealworm (T. molitor) beetles, individuals with cuticles that are dark and melanized are better able to survive exposure to fungal pathogens relative to individuals with pale, less melanized cuticles (Barnes and Siva-Jothy, 2000; see above). Thus, it is assumed that melanism incurs a marginal cost in the absence of the pathogen (otherwise selection would favour individuals always possessing a fungus-resistant, melanized cuticle). Drnevich et al. (2002) examined the relationship between male colour and sperm competitive ability and found that although dark males did lose sperm precedence over time relative to light males, this was unlikely to result in lower fitness for darker males under normal female remating frequencies. It remains to be seen whether there are other costs associated with being melanic in this species.

In the Egyptian leafworm, Spodoptera littoralis, cuticular melanism is also associated with resistance to an entomogenous fungus (Wilson et al., 2001; see above). In this species, the melanic phenotype is smaller, has lower haemolymph protein levels and dies at a significantly earlier age (Cotter et al., 2004a). Thus, there do appear to be small, but detectable, costs of maintaining high levels of investment in immune function. In contrast, in a related species, the African armyworm Spodoptera exempta, the costs are less obvious. In this species, resistance to a range of entomopathogens (including nucleopoly-hedrovirus and an ectoparasitoid) is positively associated with larval density and melanism, and the melanic, high-density phenotype has significantly higher levels of phenoloxidase activity than the non-melanic, low-density phenotype (Reeson *et al.*, 1998, 2000; Wilson and Reeson, 1998; see below). Contrary to expectation, females raised under high-density conditions laid approximately 26% more eggs than those reared at low density when they were fed only water as adults (there was no phase difference in fecundity when the adult moths were fed on sucrose solution) (Mensah and Gatehouse, 1998). This is despite the fact that low-density females were significantly smaller, and fecundity tends to increase with body weight in this species (B.A. Mensah, unpublished). Thus, under laboratory conditions at least, high-density, melanic females do not appear to incur a fecundity cost to investing in immune function. However, in the wild, their lower body mass may impose a survival cost that is not evident in the laboratory. Moreover, the costs of melanism may not be physiological, but ecological, since melanic larvae may be more conspicuous and so suffer greater predation rates.

Thus, it appears that the phenotypic costs of maintaining an efficient immune system are sometimes evident, but they may not be large. In their review of the costs of immunity in mammals, Lochmiller and Deerenberg (2000) argued that the immune system may be analogous to the female reproductive system, where the costs of maintenance are minor in comparison with the costs of actually using it. Further studies on insects will be required before we are in a position to be able to say whether the same applies to insects.

3.3 Evolutionary costs

Evolutionary costs of resistance arise when the alleles associated with enhanced resistance have negative pleiotropic effects on other important fitness traits, such as growth or reproduction (Stearns, 1989, 1992; Roff, 2002), such that high levels of immunity can be expressed only at the expense of other life-history traits. As a result of these negative genetic covariances, individuals are constrained in the decisions they can make.

A number of methods have been used to characterize possible evolutionary costs of immunity in insects. These include micro-evolutionary experiments (e.g. Boots and Begon, 1993), artificial selection experiments (e.g. Kraaijeveld and Godfray, 1997; Fellowes *et al.*, 1998a) and breeding studies (e.g. Ryder and Siva-Jothy, 2001; Cotter *et al.*, 2004b).

Boots and Begon (1993) took a novel approach to examine the genotypic costs of resistance in the Indian meal moth (*Plodia interpunctella*). They set up six replicate populations of the moth in small boxes. In three of these, they added a granulosis virus (GV) infection and, after 2 years, compared these populations against the three virus-free control populations. Based on LD_{50} bioassays with the GV, they found that moths from the virus-selected populations were nearly twice as resistant to infection as moths derived from the virus-free control populations, indicating that selection for increased viral resistance had indeed occurred. More significantly, they found that this increase in resistance was correlated, at the population level, with a lengthening of the larval development time, a reduction in egg viability, and an increase in pupal weight. These changes in life-history traits resulted in a fitness cost of resistance of around 15%. Subsequently, it was argued that the reduced fitness of moths from the selected populations may have been due to fitness costs associated with maintaining a covert infection of the virus, rather than the costs of evolving virus resistance (Goulson and Hauxwell, 1995). Although it is now recognized that covert baculovirus infections may be highly prevalent in field and laboratory populations of Lepidoptera (e.g. Burden et al., 2003), it seems unlikely that the magnitude of their effects on fitness are large enough to account for the results from this micro-evolutionary experiment (Begon and Boots, 1995).

Conventional artificial selection experiments using insects have shown repeatedly that there is often considerable additive genetic variation for resistance to parasites and pathogens (for a summary of selection experiments involving entomopathogenic viruses, see Fuxa and Richter, 1998). For example, laboratory experiments using colonies of velvetbean caterpillars, *Anticarsia gemmatalis*, collected from the USA and Brazil showed that following intense artificial selection for increased resistance to its nucleopolyhedrovirus (AgNPV) over four and 13 generations, respectively, resistance levels had increased by 5× and >1000× relative to insects in respective control colonies (Abot et al., 1996). Two lines of evidence from subsequent studies on the US population suggest that virus resistance was costly. First, when artificial selection was discontinued, the resistant insects returned to their original level of susceptibility within just three generations (Fuxa and Richter, 1998). Moreover, two additional cycles of resistance selection and reversion were repeated in the same insect population with similar results, except that the responses were even quicker. This indicates that resistance was costly and that resistant genotypes were rapidly replaced by susceptible ones in the absence of the pathogen. Second, when insects from the resistant colony were compared with those from the susceptible colony, the former exhibited reduced fitness. Resistant females produced significantly fewer eggs and these had significantly lower hatch rates, resulting in 61% fewer viable offspring being produced by resistant females than by susceptible ones. Resistant insects also had shorter lifespans, a lower rate of larval survival, and lower pupal weights than susceptible insects (Fuxa and Richter, 1998). Thus, the available evidence indicates that resistance to NPV in this moth population is 'bought' at the expense of costly reductions in reproduction and survival in the absence of the pathogen. However, these results must be treated with a certain degree of caution because the experiment was not replicated and so the robustness of the conclusions is unclear.

Two replicated artificial selection experiments using the fruitfly *Drosophila melanogaster* and its parasitoids provide convincing evidence for a cost of resistance. Four genetic lines were selected for resistance to attack by the braconid parasitoid wasp *Asobara tabida* and compared to four control lines that were not exposed to the parasitoid (Kraaijeveld and Godfray, 1997). Selected lines rapidly increased their cellular encapsulation response from 5%

at the start of the experiment to greater than 60% after five generations of selection (Fig. 10.5A). Initial investigations revealed little indication for a cost of resistance. However, after examining various life-history and other traits in the selected and control lines under a range of conditions, it became apparent that the main cost of resistance was a decline in the competitive ability of larvae in the selected lines relative to controls when food was in short supply (Fig. 10.5B). Similar experiments with the eucoilid wasp *Leptopilina boulardi* showed a similarly rapid response to selection, from less than 1% encapsulation at the start of selection to 45% after just five generations (Fellowes *et al.*, 1998a). Significantly, increased resistance to parasitism by *L. boulardi* was also achieved at the expense of competitive ability when food was in limited supply. Subsequent studies indicated that this was because larvae from the selected lines had a lower feeding rate, an important determinant of larval competitive ability (Fellowes *et al.*, 1999a). Larvae in the lines selected for resistance to *A. tabida* also had approximately twice the number of circulating haemocytes as those in the control lines (Kraaijeveld *et al.*, 2001; Fig. 10.5C), and so it seems likely that their enhanced resistance was because they were better able to encapsulate the parasitoid larvae. This interpretation is consistent with the positive correlation observed across *Drosophila* species between encapsulation ability and both haemocyte counts (Eslin and Prévost, 1996, 1998; Fig. 10.5D), and metabolic rate (Fellowes and Godfray, 2000).

Thus, it appears that in the *D. melanogaster–A. tabida* system, there is a trade-off between feeding rate/competitive ability and haemocyte number/encapsulation response. Why there should be a negative correlation between haemocyte count and feeding rate is unclear at present. One possibility is that there is a switch in the general energy budget of the larvae away from investment in trophic function to investment in the immune system. Alternatively, since both the head musculature and the haemopoietic organ (where the haemocytes are produced) both originate from the same part of the embryo, perhaps increased allocation of tissue to the future haemopoietic organ is at the expense of future muscle tissue. A third possibility is that a doubling of the number of circulating haemocytes increases the viscosity of the haemolymph, leading to lower rates of resource supply (e.g. glucose) to the muscles (Kraaijeveld *et al.*, 2001, 2002).

A reverse approach to looking at the evolutionary costs of immunity was taken by Koella and Boëte (2002), who selected six lines of the mosquito *Aedes aegypti*, for either early or late pupation and measured the extent to which this selection procedure changed the mosquito's ability to encapsulate and melanize a Sephadex bead. They found that after ten generations of selection, the age at pupation in the two selection regimes differed by about 0.7 days, showing that the selection procedure had worked. More significantly, they also found that whereas 32% of individuals from the lines selected for late pupation melanized the Sephadex bead, only 6% of mosquitoes from the lines selected for early pupation did so. Thus, there appears to be a genetic trade-off between the rate of larval development and the efficacy of the melanization response.

A third approach to examining the evolutionary costs of resistance has been to employ breeding studies (i.e. sib-analyses, offspring–parent regressions and

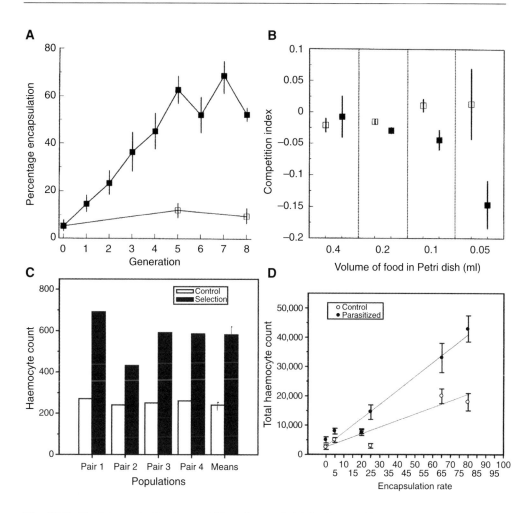

Fig. 10.5. Resistance to the braconid larval parasitoid *Asobara tabida* in the fruitfly *Drosophila melanogaster*. **A** The frequency of encapsulation in control lines of fruitflies and lines selected for resistance to *A. tabida*, showing means and standard errors of the four selected (filled symbols) and control lines (open symbols). **B** The competitive ability of experimental flies relative to a tester strain, showing means and standard errors of the four selected or control lines at four levels of competition for larval food. **C** Haemocyte counts in the four pairs of selected and control lines plus the overall mean (±SE) for both sets of lines. **D** Relationship between the encapsulation rate and the total haemocyte count recorded in both parasitized larvae (closed symbols), and control larvae (open symbols). Mean values (±SE) are given for six *Drosophila* species of the *melanogaster* subgroup: *D. sechellia, D. melanogaster, D. mauritiana, D. yakuba, D. teissieri* and *D. simulans*. Figures **A** and **B** reprinted from Kraaijeveld and Godfray (1997) with permission from Nature Publishing Group (http://www.nature.com); **C** reprinted from Kraaijeveld *et al.* (2001) with permission from the Royal Society, and **D** from Eslin and Prévost (1998) copyright (1998) with permission from Elsevier.

pedigree analyses; Falconer and Mackay, 1996; Roff, 1997). For example, Ryder and Siva-Jothy (2001) applied a sib-analysis approach to house crickets (*Acheta domesticus*) to examine the genetic correlations between two immune function traits (haemocyte load and encapsulation response) and a sexually-selected trait (male body size, which is strongly correlated with calling rate and the ability of a male to attract a mate). They found that all three traits were heritable and positively genetically correlated. Thus, females choosing large males with particular call characteristics will not only tend to produce larger offspring, but those offspring will have a greater ability to produce an encapsulation response. The costs associated with genetically large body size and encapsulation response have yet to be established. Recently, Cotter *et al.* (2004b) have applied a similar approach to the Egyptian leafworm (*Spodoptera littoralis*) to examine the genetic correlations between a suite of immune function traits (total haemocyte count, haemolymph phenoloxidase activity, lysozyme-like antibacterial activity, encapsulation response and degree of cuticular melanization) and a number of life-history traits (larval and pupal development rate, pupal weight, and adult longevity). They found that there was a complex mixture of positive and negative genetic correlations between immune function traits and life-history traits. But, most interestingly perhaps, they also observed a potential genetic trade-off within the immune system itself, with lysozyme-like antibacterial activity exhibiting negative genetic correlations with haemocyte density, haemolymph phenoloxidase activity and cuticular melanization. This result is consistent with those from other studies showing negative phenotypic correlations between antibacterial activity and either phenoloxidase activity (in the bumblebee, *Bombus terrestris*; Moret and Schmid-Hempel, 2001) or encapsulation response (in the Mediterranean field cricket, *Gryllus bimaculatus*; Rantala and Kortet, 2003). Although conflicts between responses tailored to specific parasites have been predicted (Hamilton and Zuk, 1982), this is perhaps less surprising than the possibility of a trade-off between general components of the innate immune system. Further studies are clearly required before a trade-off between the humoral and cellular arms of the insect immune system can be verified, but it is worth noting that there is some evidence for these sorts of negative genetic correlations from vertebrate studies (e.g. Gross *et al.*, 1980; Grencis, 1997; Gehad *et al.*, 1999; Ibanez *et al.*, 1999; Johnsen and Zuk, 1999; Gill *et al.*, 2000).

A likely consequence of negative genetic correlations between different components of the immune system is that we can also expect negative correlations between levels of resistance to different parasite types. So far, few studies have examined this possibility. Fellowes *et al.* (1999b) found no evidence for the existence of trade-offs between resistance to different parasitoid species in *D. melanogaster*. In fact, selection for increased resistance to *A. tabida* and *L. boulardi* both resulted in concomitant increases in resistance to a third parasitoid, *Leptopolina heterotoma*, suggesting that in both sets of selection lines some attribute of general utility in resisting parasitoids was being selected for. However, some more specific attributes were also selected for in the lines resistant to *L. boulardi*. This is indicated by the fact that although selection for increased resistance to *L. boulardi* led to levels of resistance against *A. tabida* that were similar to those in the *A. tabida* selection lines, selection for increased resistance to *A. tabida* did not result in increased levels of resistance to the more specialized parasitoid, *L. boulardi* (which produces immunodepressive virus-like particles, unlike *A. tabida*, which evades the *Drosophila* immune system by hiding within host tissues). It remains to be established whether there are trade-offs between resistance to parasitoids and resistance to more distantly related parasites (e.g. bacteria, fungi, etc.) that may be combated by different components of the immune system.

Understanding the costs of resistance potentially has a number of applied ramifications. For example, there are hopes that it will one day be possible to genetically manipulate the immune response of *Anopheles* mosquitoes in order to make them refractory towards *Plasmodium* parasites, and so control the spread of malaria (Collins, 1994; Collins and Paskewitz, 1995). These hopes have been raised recently by the observation that the immune response has a genetic basis (e.g. Collins *et al.*, 1986; Dimopoulos *et al.*, 2002; Thomasova *et al.*, 2002), and genes involved in the immune response are being located and mapped in the mosquito's genome (e.g. Gorman *et al.*, 1997). However, the efficacy of such an approach may depend on inherent genetic trade-offs, as well as on the extent to which the insect's immune response depends on non-genetic factors, such as the host's diet, age, sex and reproductive status (Koella and Sørensen, 2002).

4. Case Studies

In this section, we apply the ecological immunology approach to examine three areas of interest to evolutionary ecologists: (i) density-dependent prophylaxis; (ii) group-living and disease risk; and (iii) sex-biased parasitism. Each of these examples comes from our own work, but hopefully illustrate general principles that apply not only to insects, but also to other animal groups, including vertebrates.

4.1 Density-dependent prophylaxis

Most pest species of insects, almost by definition, exhibit wide fluctuations in population density from one generation to the next: when conditions are favourable, population densities are high, and when they are less favourable they are low. For example, population densities of the larch budmoth (*Zeiraphera diniana*), a lepidopteran forest pest, may vary by more than 20,000-fold over five generations (Speight *et al.*, 1999). Low- and high-density environments differ in a number of qualitative and quantitative ways. Most obviously, of course, the levels of competition for food will vary, and this may impact on the insect's capacity to 'feed' its immune system (see above). Another important aspect of the environment that differs between low- and high-density populations is the degree of exposure to parasites and pathogens (Wilson and Reeson, 1998). This is because most pathogens tend to be transmitted in a positively density-dependent manner (Anderson and May, 1979; McCallum *et* al., 2001) – as population density increases, so the probability of infectious and susceptible hosts coming into contact increases. As a consequence, insects in crowded populations will generally experience greater risk of exposure to pathogens than those in low-density populations (though the relationship between infection risk and population density may not be linear; e.g. Hochberg, 1991a; McCallum et al., 2001). Thus, we can expect animals to tailor their investment in disease resistance mechanisms to match the perceived risk of infection, using population density as a cue to their infection risk.

This 'density-dependent prophylaxis (DDP) hypothesis' (Wilson and Reeson, 1998) rests on two important assumptions. The first is that investment in prophylactic resistance mechanisms is costly – if resistance was cost-free, then these resistance mechanisms would always be expressed. As we saw in the previous section, there are good theoretical grounds for assuming that resistance will be costly, though the exact costs of maintaining prophylactic resistance are not well characterized (see above). The second assumption is that insects have reliable mechanisms for perceiving local population density and adjusting their phenotype accordingly. It has long been established that such phenotypic plasticity may be manifested in insects, and the phenomenon has been termed 'density-dependent phase polyphenism'. It is typified by the desert locust, *Schistocerca gregaria*, which, in the nymphal stages, exhibits a green, cryptic phenotype adapted to low-density conditions, and a conspicuous yellow-and-black phenotype adapted to high-density conditions. Switching between these two phenotypes is largely determined in response to tactile cues, especially those perceived by the hind-femurs, during nymphal development (though visual and olfactory cues are also important; Simpson et al., 2001).

Given that at least some insects are capable of adjusting their phenotype in response to cues reflecting local population density (and infection risk), and that maintaining disease resistance mechanisms exact some cost, is there any evidence that insects alter investment in disease resistance mechanisms to reflect their perceived risk of infection? A number of studies over the past few years have provided evidence in support of this notion. The first direct test of the DDP hypothesis was conducted using the African armyworm, *Spodoptera exempta* (Reeson et al., 1998, 2000). The larval stage of this moth is an economically important pest of pasture grasses and graminaceous crops in eastern Africa, and densities of this 2–3-cm-long caterpillar can reach anything up to $1000/m^2$, and occasionally more (Rose et al., 2000). Population densities vary considerably between years: in Tanzanian light traps, a survey conducted over a 30-year period indicated that the number of moths caught varied between just 150 to more than 300,000; a 2000-fold difference (Harvey and Mallya, 1995). Moreover, there is little correlation between the numbers of moths caught in successive years. For example, in 1976 more than 60,000 moths were caught in Tanzanian light traps, compared with just 800 moths in the following year. In response to this unpredictable variation in population density, just like the desert locust, *S. exempta* has evolved density-dependent phase polyphenism: at low population densities, the larvae are usually green or pale brown, whereas at high densities, they are invariably jet black, due to the deposition of melanin in the cuticle (see Fig. 10.3; Wilson et al., 2001).

To determine whether disease resistance was related to larval phenotype, Reeson *et al.* (1998) reared *S. exempta* larvae under either high- or low-density conditions in the laboratory and then orally challenged early fourth-instar larvae with one of five known doses of *S. exempta* NPV via the diet-plug method. This baculovirus is probably the main pathogen impacting on the fitness of African armyworms and has been known to cause greater than 90% mortality in some larval outbreaks (Rose *et al.*, 2000). Its main mode of transmission is probably horizontal, via the ingestion of virus-contaminated vegetation (but vertical transmission is also known to occur; see Swaine, 1966), and so the *per capita* risk of infection is likely to increase in a positive density-dependent manner.

In this experiment, approximately 15% of larvae reared under low-density conditions expressed a phenotype that was more like that of larvae reared under high-density conditions (reflecting genetic variation in the larval density required to trigger the switch into the high-density phenotype). Thus, Reeson and colleagues determined the LD_{50}s for the typical green-brown low-density phenotype, the typical melanic high-density phenotype, and the atypical melanic phenotype produced under low-density conditions (which, in fact, is less melanic than the typical crowded phenotype; Wilson *et al.*, 2001). They found that, as predicted, the LD_{50} increased as the larval phenotype switched from the typical solitary to the typical crowded forms (Fig. 10.6A). In fact, approximately ten times more viral occlusion bodies are required to kill a typical crowded larva as a typical solitary one (Reeson *et al.*, 1998). Moreover, similar results were gained when the viral infection was gained via natural exposure to the virus in a field situation (Reeson *et al.*, 2000). Significantly, this difference in viral susceptibility was mirrored by similar variation in constituitive levels of phenoloxidase activity (Fig. 10.6B), suggesting a possible involvement of this enzyme cascade in resistance to baculoviruses (see also Washburn *et al.*, 1996).

Similar experiments on other Lepidoptera (Mitsui and Kunimi, 1988; Kunimi and Yamada, 1990; Goulson and Cory, 1995; Wilson *et al.*, 2001; Cotter *et al.*, 2004a), tenebrionid beetles (Barnes and Siva-Jothy, 2000) and the archetypal phase polyphenic species, the desert locust (Wilson *et al.*, 2002), generally show similar density-dependent increases in disease resistance and/or immune function. Future studies on this subject need to identify the precise costs associated with the density-dependent increase in immune function. Attempts to do this so far have generally proved unsatisfactory (see above), due mainly to the fact that a whole suite of coordinated immunological and non-immunological traits are simultaneously altered in relation to population density, and so dissecting the relative costs and benefits of any particular phenotypic change is difficult. This is particularly so because there will also be compensatory changes in diet and metabolism to ameliorate any costs. The problem associated with identifying the costs of density-dependent prophylaxis is analogous to that associated with quantifying the costs of migration (Dingle, 1991, 1996; Wilson, 1995). The emergence of 'migration syndromes' (Dingle, 1991, 1996) to minimize the costs of migration has thwarted efforts to accurately quantify the costs involved. Ecological immunology might benefit from the lessons learned by migration biologists.

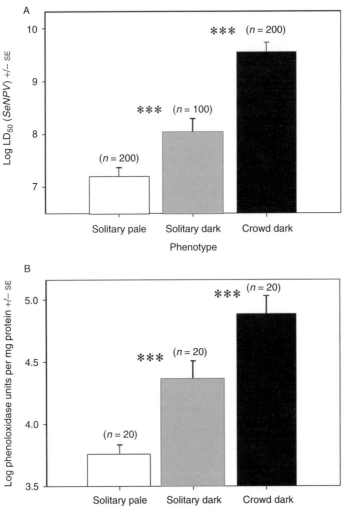

Fig. 10.6. Relationship between *Spodoptera exempta* larval phenotype and **A** resistance to *S. exempta* nucleopolyhedrovirus and **B** phenoloxidase (PO) activity. In **A**, resistance is measured via mean LD_{50} (±SE) determined using five different doses of the baculovirus (redrawn after Reeson *et al.*, 1998 with permission from The Royal Society). In **B**, mean PO activity (±SE) is expressed as PO units per mg protein (redrawn after Wilson *et al.*, 2001 with permission from Blackwell Publishing Ltd). The three larval phenotypes are typical (non-melanic) larvae produced under solitary rearing conditions, typical (melanic) larvae produced under crowded rearing conditions, and (atypical) melanic larvae produced under solitary rearing conditions (see Fig. 10.3).

Future studies also need to establish the speed and specificity of density-dependent changes in prophylactic immune function. Desert locust nymphs start exhibiting gregarious behaviour within just 4 h of receiving tactile stimuli characteristic of high-density, crowded conditions (Simpson *et al.*, 2001). This leads to the obvious questions of how quickly prophylactic disease resistance mechanisms can be upregulated in response to the appropriate stimuli; whether downregulation subsequently occurs in the absence of such cues; and whether the cues used for immunological transformation are the same as those used for behavioural transformation.

Although we have focused on density-dependent changes in investment in prophylactic immune function, the same principle applies to any cue that reliably predicts infection risk. For example, if there are reliable seasonal changes in infection risk, then we might expect insects to use seasonal cues (such as daylength, temperature and humidity) when making decisions about relative investment in immune function. While we are aware of no insect examples of this phenomenon, it has been hypothesized that adaptive seasonal variation in immune function occurs in rodents (Nelson *et al.*, 1996a,b, 1998). Similarly, geographical or climatic variation in prophylactic immune function might be expected to evolve under some scenarios.

Recently, Moret and Siva-Jothy (2003) have added another example to this list. They argue that one of the best cues predicting future exposure to parasites is previous exposure to them and, therefore, we can expect 'responsive-mode prophylaxis' to evolve. They tested this idea using larvae of the mealworm beetle, *Tenebrio molitor*. They found that when larvae were pre-challenged with LPS, they produced antibacterial responses that lasted at least 7 days. Moreover, relative to control insects, the LPS-treated larvae exhibited greater resistance to infection by the entomopathogenic fungus *Metarhizium anisopliae*, when they were exposed 4 or 7 days after the pre-challenge. Thus, Moret and Siva-Jothy argue that the long-lasting antimicrobial responses of invertebrates may serve a similar function to the adaptive immune responses of vertebrates in providing better protection against repeated parasitic infections, though the specific memory of these prophylactic responses is clearly not comparable (see above). An earlier study of the dampwood termite, *Zootermopsis angusticollis*, also found that prior exposure to parasite antigens can offer enhanced protection against parasites in insects. Rosengaus *et al.* (1999b) found that nymphs immunized with an injection of glutaraldehyde-killed bacteria (*Pseudomonas aeruginosa*) had significantly higher survival than controls following a challenge with a lethal concentration of live bacteria. Similarly, nymphs exposed to a low dose of an entomopathogenic fungus (*Metarhizium anisopliae*) survived better than control insects after a challenge with a lethal concentration of live spores. Thus, prior exposure to a pathogen conferred upon termites a degree of protection during a subsequent encounter with the same pathogen. Unfortunately, Rosengaus and colleagues did not determine whether exposure to dead *P. aeruginosa* afforded protection to *M. anisopliae*, and so it is not possible to determine the specificity of this response.

4.2 Group-living and disease risk

An intuitive extrapolation of the DDP hypothesis outlined above is that, across species, we might expect group-living insects to invest more in prophylactic disease resistance than solitary-living insects. This is because group-living insects will typically experience higher local densities than solitary-living ones and hence, presumably, higher *per capita* infection risk. Increased parasitism as a cost of group-living has long been assumed, but the evidence for it is equivocal at best (Freeland, 1979; Davies *et al.*, 1991; Coté and Poulin, 1995). The idea was first tested in insects by Hochberg (1991b), who used data extracted from the literature to show that, in laboratory bioassays using baculoviruses, larvae of gregariously feeding species exhibited an age-related increase in virus resistance that was not observed in solitary-feeding species. Although these results are consistent with the idea that gregariously feeding insects are investing more in disease resistance mechanisms, this analysis was potentially confounded by host and pathogen phylogenies, as well as several other variables such as host diet, rearing conditions and coevolved resistance mechanisms.

To test this idea further, Wilson *et al.* (2003a) took a different approach to the problem. Instead of measuring pathogen resistance directly, they measured several aspects of immune function. To minimize the influence of potentially confounding variables, they assayed immune function in 12 lepidopteran species arranged in six phylogenetically matched pairs of species, assayed at similar times and raised on the same diet and in the same environmental conditions. They found that, although there was considerable variation across species in the density of haemocytes in the haemolymph (i.e. total haemocyte count; THC) across the six species-pairs, the THC of the solitary species was, on average, 40% higher than that of the gregarious species. Moreover, all six solitary species had THC values that were higher than their paired gregarious species (Fig. 10.7A). Similar results were apparent in the phenoloxidase activity, except that levels were higher in the solitary species in only five of the six species-pairs. Thus, contrary to the initial expectation, it appears that solitary species are investing more in immune function than gregarious species.

There are at least two possible explanations for this counterintuitive result. The first is that high densities of haemocytes in the haemolymph and high phenoloxidase activity levels do not, in fact, reflect differences in disease resistance. However, there is good evidence from a number of studies that high phenoloxidase activity enhances parasite resistance (see above), and that high THC enhances both the capacity to phagocytose novel antigens (e.g. Kurtz, 2002) and the capacity to encapsulate at least some parasites (Eslin and Prévost, 1996, 1998; Rantala *et al.*, 2000; Kraaijeveld *et al.*, 2001; Fig. 10.5D). Moreover, Wilson and colleagues showed that, for the species in their analyses, there was a strong positive correlation between THC (and PO) and the magnitude of the encapsulation response (Fig. 10.7B). This suggests that the difference in THC and PO related to larval feeding habit genuinely reflects investment in immune function.

A second possible explanation for the observed results is that the relationship between group-living and infection risk is not as previously believed. To investigate this possibility further, Wilson *et al.* (2003a) developed a dynamic, susceptible/infected spatially explicit model in which different degrees of host-clustering were created by allowing different proportions of local (nearest-neighbour) and distant (random) reproduction. In the model, there was a regular network of sites, each taking one of three possible states: empty, occupied by a

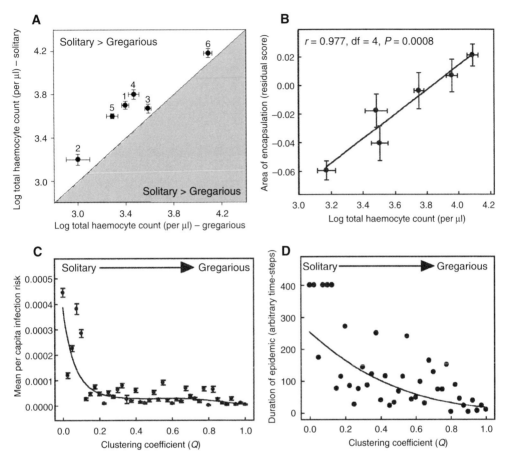

Fig. 10.7. Group-living and risk of disease in lepidopteran larvae. **A** Relationship between feeding style and total haemocyte count (THC); gregariously feeding species have significantly lower THC than solitary-feeding species (means ± SE shown). **B** Relationship between THC and magnitude of encapsulation response directed against a nylon implant by six species of lepidopteran larvae (means ± SE shown). Relationship between the degree of host-clustering and **C** mean *per capita* infection risk (±SE) and **D** duration of epidemic. Data are output from a dynamic, susceptible/infected spatially-explicit model. Reprinted from Wilson *et al.* (2003a) with permission from Blackwell Publishing Ltd.

susceptible host, or occupied by an infected host (see Boots and Sasaki, 2000, for a detailed explanation of this modelling approach). Infection occurs when there is contact between neighbouring infected and susceptible individuals. A site becomes empty when an individual dies, and it is then available to be re-occupied by the progeny of other individuals. The key parameter in this model, in terms of the present discussion, is the clustering coefficient, *Q*, which determines the proportion of offspring that are born into neighbouring sites, as opposed to randomly across the whole lattice. Changes in this clustering coefficient therefore produce populations with different average local clustering.

Using this model, Wilson and colleagues showed that, for a significant proportion of parameter space, host clustering can reduce individual infection risk in both endemic and epidemic host–pathogen interactions. Thus, in the epidemic scenario, in which a disease epidemic spreads through the population, as the clustering coefficient increases (and individuals become increasingly aggregated in groups), so the mean *per capita* risk of becoming infected declines, as does the duration of the epidemic (Fig. 10.7C,D).

So, it appears that aggregation in clusters might actually reduce the probability of becoming infected by a disease agent. But how does this happen? The mechanism behind this phenomenon is fairly simple. If pathogen transmission requires close proximity between potential hosts, then any process that increases the distance between infected and susceptible hosts will lead to reduced pathogen transmission. By increasing the variance in nearest-neighbour distance, host clustering increases the probability that the pathogen will fail to breach the gap between the host it is infecting and the nearest susceptible hosts. Therefore, the model indicates that part of the advantage of group living is attributable to the fact that any disease epidemics will tend to fade out faster within populations of group-living animals than within populations of solitary hosts (Wilson *et al.*, 2003a). Similar conclusions were drawn from a simpler epidemiological model by Watve and Jog (1997). Of course, group living will fail to be advantageous in this context when the parasite is highly mobile (or transmitted by a mobile vector) and so not constrained by the spatial distribution of its host, or when hosts are highly mobile or at such low densities that the infection risk is low for all hosts.

Thus, both the models and the immune function assays suggest that group living may lead to reduced, rather than increased, risk of becoming infected by parasites under some circumstances. What are needed now are replicated experiments in which insect group size is manipulated and the spread of an introduced pathogen is monitored. Only then can we be sure whether group living enhances or reduces disease risk in Lepidoptera–pathogen interactions. Regardless of the results of such experiments, the take-home message from this study is that, in combination with other relevant methods, ecological immunology studies can both challenge and generate novel predictions regarding the evolutionary ecology of insect host–pathogen interactions.

It seems likely that the relationship between sociality and disease risk will depend critically on the interaction between the insect's social system and the nature of the pathogenic challenge (Wilson *et al.*, 2003a). For example, it has recently been suggested that 'social transfer' of infection resistance might yield a survivorship bonus in social insects (Traniello *et al.*, 2002). In the termite *Zootermopsis angusticollis*, survival following infection with the entomopathogenic fungus *M. anisopliae* was significantly enhanced if infected individuals were subsequently reared in groups rather than in isolation. Moreover, termite survival following exposure to a lethal dose of fungal spores was significantly enhanced if naïve individuals were allowed to associate with previously immunized nestmates before they were infected (Traniello *et al.*, 2002). These 'socially immunized' termites were approximately 34% less susceptible to a lethal infection than termites that were not allowed contact with immunized siblings. Yet again, the precise mechanisms underpinning these patterns remain to be established, but other mechanisms used by termites to minimize infection risk include: mutual grooming, scaled in frequency to pathogen prevalence (Rosengaus *et al.*, 1998b, 2000b; see above); the production of antibiotic secretions in exocrine glands and other exudates (Rosengaus *et al.*, 1998a, 2000a); and the communication of information about the presence of pathogens in the nest (Rosengaus *et al.*, 1999a).

4.3 Sex-biased parasitism

Male and female life histories are usually very different, especially in species with highly polygynous mating systems, in which females maximize fitness by living for a long time and producing many young, and males maximize fitness by mating with many reproductively active females (Bateman, 1948; Trivers, 1972). These disparate selection pressures often lead to the production of weapons and ornaments in males, as well as large body size, especially in mammals (Andersson, 1994). These traits are envisaged to be costly to produce and maintain and, in mammals, it has been shown that in species in which there is strong sexual selection (as measured by mating system and sexual size dimorphism), males appear to suffer viability costs (i.e. there is male-biased mortality). Recently, it has been suggested that one mechanism by which this viability cost might be exerted is via increased susceptibility or exposure to parasites (Moore and Wilson, 2002). Indeed, across a range of mammal species, males suffer significantly greater prevalence of parasitism (via a range of parasites and pathogens) than do females, and this sex-bias in parasitism (SBP) is positively correlated with the strength of sexual selection: SBP is greater in polygynous than monogamous species, and is positively correlated with the extent of sexual size dimorphism (Moore and Wilson, 2002), but not with sex differences in home-range size (Wilson *et al.*, 2003b). Moreover, there was a positive correlation between SBP and sex-biased mortality, with species in which there was male-biased parasitism also exhibiting male-biased mortality. These analyses are consistent with the idea that sexual selection imposes a viability cost on males and that this cost is mediated, in part at least, by sex differences in parasitism, although the precise mechanism causing differential parasitism of the sexes remains to be identified (Moore and Wilson, 2002).

If sex-biased parasitism is a general cost of sexual selection, then we should also observe SBP in other taxonomic groups. Sex differences in parasitism have been observed in humans (e.g. Owens, 2002; Wilson *et al.*, 2003b) and birds (Poulin, 1996), but were not found in fish (Poulin, 1996) or in invertebrates, including insects (Sheridan *et al.*, 2000). However, a recent study of insects, using a much larger dataset than that used by Sheridan and colleagues, observed a small, but highly significant, male-bias in the prevalence of parasitism by a range of entomopathogens (S.L. Moore and K. Wilson, unpublished). Moreover, as with the mammal analysis, the male bias was statistically significant for polygynous species, but not for non-polygynous (monogamous and polyandrous) ones, and was significantly positively correlated with the degree of sexual size dimorphism.

Interestingly, a meta-analysis of published information on sex differences in traits associated with immune function in insects, namely haemocyte number, antibacterial activity and phenoloxidase activity, indicated that females have consistently higher levels than males (S.L. Moore and K. Wilson, unpublished). Thus, sex differences in immune function may explain why males are generally more heavily parasitized than females, though clearly more detailed species-specific studies are required. This result is particularly interesting because it shows that in both vertebrates and invertebrates females have better immunity and lower parasite loads than males, suggesting a common explanation. This is interesting because it suggests that we can probably dismiss the immuno-depressive effects of testosterone as an important explanation for sex differences in parasitism in vertebrates (Folstad and Karter, 1992), since we get the same pattern in insects which, of course, lack testosterone or any sex-specific hormones (Nijhout, 1994). However, recent studies have shown that mating may have immunodepressive side effects in insects (e.g. McKean and Nunney, 2001; Rolff and Siva-Jothy, 2002), and that these effects may be mediated by juvenile hormone (Rolff and Siva-Jothy, 2002). So, if some males mate more frequently, on average, than females, then they may spend relatively more time in an immunodepressed state and this might explain their greater parasitism.

5. Concluding remarks

These three case studies illustrate the sort of approach taken by ecological immunologists to understand the evolutionary ecology of insect host–parasite interactions. They show that there are consistent and predictable differences between individuals, sexes and species in terms of their investment strategies in parasite defence mechanisms, and that these can be understood by considering details of the ecology and evolutionary background of the species concerned. For example, within species, individuals respond to the perceived risk of parasitism by investing in mechanisms that reduce their susceptibility to pathogens. Across species, group-living insects tend to exhibit reduced haemocyte production and reduced phenoloxidase activity, suggesting that they may experience a reduced risk of infection from at least some types of parasite relative to solitary-feeding species. And finally, within species, males tend to have weaker immune responses than females and, as a result, tend to have a higher prevalence of parasitism. The prospects for ecological immunology making further significant basic and applied contributions to the study of insect host–parasite interactions look promising.

Acknowledgements

Drs Mike Boots, Sheena Cotter, Rob Knell, Kwang Lee and Sarah Moore are gratefully acknowledged for stimulating discussions; Sarah Moore for access to unpublished work, Yannick Moret for Figure 10.4, and the editors of this volume for their patience.

References

Abbot, P. and Dill, L.M. (2001) Sexually transmitted parasites and sexual selection in the milkweed leaf beetle, *Labidomera clivicollis*. *Oikos* 92, 91–100.

Abot, A.R., Moscardi, F., Fuxa, J.R., Sosagomez, D.R. and Richter, A.R. (1995) Susceptibility of populations of *Anticarsia gemmatalis* (Lepidoptera, Noctuidae) from Brazil and the United States to nuclear polyhedrosis virus. *Journal of Entomological Science* 30, 62–69.

Abot, A.R., Moscardi, F., Fuxa, J.R., SosaGomez, D.R. and Richter, A.R. (1996) Development of resistance by *Anticarsia gemmatalis* from Brazil and the United States to a nuclear polyhedrosis virus under laboratory selection pressure. *Biological Control* 7, 126–130.

Adamo, S.A. (1998) The specificity of behavioral fever in the cricket *Acheta domesticus* *Journal of Parasitology* 84, 529–533.

Adams, M.D., Celniker, S.E., Holt, R.A., Evans, C.A., Gocayne, J.D., Amanatides, P.G., Scherer, S.E., Li, P.W., Hoskins, R.A., Galle, R.F., George, R.A., Lewis, S.E., Richards, S., Ashburner, M., Henderson, S.N., Sutton, G.G., Wortman, J.R., Yandell, M.D., Zhang, Q., Chen, L.X., Brandon, R.C., Rogers, Y.H.C., Blazej, R.G., Champe, M., Pfeiffer, B.D., Wan, K.H., Doyle, C., Baxter, E.G., Helt, G., Nelson, C.R., Miklos, G.L.G., Abril, J.F., Agbayani, A., An, H.J., Andrews-Pfannkoch, C., Baldwin, D., Ballew, R.M., Basu, A., Baxendale, J., Bayraktaroglu, L., Beasley, E.M., Beeson, K.Y., Benos, P.V., Berman, B.P., Bhandari, D., Bolshakov, S., Borkova, D., Botchan, M.R., Bouck, J., Brokstein, P., Brottier, P., Burtis, K.C., Busam, D.A., Butler, H., Cadieu, E., Center, A., Chandra, I., Cherry, J.M., Cawley, S., Dahlke, C., Davenport, L.B., Davies, A., de Pablos, B., Delcher, A., Deng, Z.M., Mays, A.D., Dew, I., Dietz, S.M., Dodson, K., Doup, L.E., Downes, M., Dugan-Rocha, S., Dunkov, B.C., Dunn, P., Durbin, K.J., Evangelista, C.C., Ferraz, C., Ferriera, S., Fleischmann, W., Fosler, C., Gabrielian, A.E., Garg, N.S., Gelbart, W.M., Glasser, K., Glodek, A., Gong, F.C., Gorrell, J.H., Gu, Z.P., Guan, P., Harris, M., Harris, N.L., Harvey, D., Heiman, T.J., Hernandez, J.R., Houck, J., Hostin, D., Houston, D.A., Howland, T.J., Wei, M.H., Ibegwam, C., *et al.* (2000) The genome sequence of *Drosophila melanogaster*. *Science* 287, 2185–2195.

Agnew, P., Bedhomme, S., Haussy, C. and Michalakis, Y. (1999) Age and size at maturity of the mosquito *Culex pipiens* infected by the microsporidian parasite *Vavraia culicis*. *Proceedings of the Royal Society of London, Series B Biological Sciences* 266, 947–952.

Alleyne, M., Chappell, M.A., Gelman, D.B. and Beckage, N.E. (1997) Effects of parasitism by the braconid wasp *Cotesia congregata* on metabolic rate in host larvae of the tobacco hornworm, *Manduca sexta*. *Journal of Insect Physiology* 43, 143–154.

Anderson, R.M. and May, R.M. (1979) Population biology of infectious diseases: Part I. *Nature* 280, 361–367.

Andersson, M. (1994) *Sexual Selection*. Princeton University Press, Princeton, New Jersey.

Asano, T. and Ashida, M. (2001) Cuticular pro-phenoloxidase of the silkworm, *Bombyx mori* – Purification and demonstration of its transport from haemolymph. *Journal of Biological Chemistry* 276 (14), 11100–11112.

Ashida, M. and Brey, P.T. (1995) Role of the integument in insect defense – pro-phenol oxidase cascade in the cuticular matrix *Proceedings of the National Academy of Sciences USA* 92, 10698–10702.

Ashida, M. and Brey, P. (1998) Recent advances in research on the insect prophenoloxidase cascade. In: Brey, P. and Hultmark, D. (eds) *Molecular*

Mechanisms of Immune Responses in Insects. Chapman & Hall, London, pp. 135–172.

Ashida, M. and Yamazaki, H.I. (1990) Biochemistry of the phenoloxidase system in insects: with special reference to its activation. In: Ohnishi, E. and Ishizaki, H. (eds) *Molting and Metamorphosis.* Springer-Verlag, Berlin, pp. 239–265.

Ballabeni, P., Benway, H. and Jaenike, J. (1995) Lack of behavioral fever in nematode-parasitized *Drosophila. Journal of Parasitology* 81, 670–674.

Barnes, A.I. and Siva-Jothy, M.T. (2000) Density-dependent prophylaxis in the mealwork beetle *Tenebrio molitor* L. (Coleoptera:Tenebrionidae): cuticular melanization is an indicator of investment in immunity. *Proceedings of the Royal Society of London, Series B Biological Sciences* 267, 177–182.

Bateman, A.J. (1948) Intra-sexual selection in *Drosophila. Heredity* 2, 349–368.

Beckage, N.E. (2003) Immunology. In: Resh, V.H. (ed.) *Encyclopedia of Insects.* Academic Press, Orlando, Florida, pp. 555–560.

Beckage, N., Thompson, S. and Federici, B. (eds) (1993) *Parasites and Pathogens of Insects.* Academic Press, San Diego, California.

Begon, M. and Boots, M. (1995) Covert infection or, simply, resistance. *Functional Ecology* 9, 549–550.

Blanford, S. and Thomas, M.B. (1999) Host thermal biology: the key to understanding insect-pathogen interactions and microbial pest control? *Agricultural and Forest Entomology* 1, 195–202.

Boman, H.G. and Hultmark, D. (1987) Cell-free immunity in insects. *Annual Review of Microbiology* 41, 103–126.

Boorstein, S.M. and Ewald, P.W. (1987) Costs and benefits of behavioral fever in *Melanoplus sanguinipes* infected by *Nosema acridophagus. Physiological Zoology* 60, 586–595.

Boots, M. and Begon, M. (1993) Trade-offs with resistance to a granulosis virus in the Indian meal moth, examined by a laboratory evolution experiment. *Functional Ecology* 7, 528–534.

Boots, M. and Knell, R.J. (2002) The evolution of risky behaviour in presence of a sexually transmitted disease. *Proceedings of the Royal Society of London, Series B Biological Sciences* 269, 585–589.

Boots, M. and Sasaki, A. (2000) The evolutionary dynamics of local infection and global reproduction in host-parasite interactions. *Ecology Letters* 3, 181–185.

Bronstein, S.M. and Conner, W.E. (1984) Endotoxin-induced behavioral fever in the Madagascar cockroach, *Gromphadorhina portentosa. Journal of Insect Physiology* 30, 327–330.

Brown, M.J.F., Loosli, R. and Schmid-Hempel, P. (2000) Condition-dependent expression of virulence in a trypanosome infecting bumblebees. *Oikos* 91, 421–427.

Bull, A.T. (1970) Inhibition of polysaccharases by melanin: enzyme inhibition in relation to mycolysis. *Archives of Biochemistry and Biophysics* 137, 345–356.

Burden, J.P., Nixon, C.P., Hodgkinson, A.E., Possee, R.D., Sait, S.M., King, L.A. and Hails, R.S. (2003) Covert infections as a mechanism for long-term persistence of baculoviruses. *Ecology Letters* 6, 524–531.

Carius, H.J., Little, T.J. and Ebert, D. (2001) Genetic variation in a host-parasite association: potential for coevolution and frequency-dependent selection. *Evolution* 55, 1136–1145.

Carruthers, R.I., Larkin, T.S., Firstencel, H. and Feng, Z.D. (1992) Influence of thermal ecology on the mycosis of a rangeland grasshopper. *Ecology* 73, 190–204.

Chapman, R.F. (1998) *The Insects: Structure and Function.* Ed. 4. Cambridge University Press, Cambridge, UK.

Christophides, G.K., Zdobnov, E., Barillas-Mury, C., Birney, E., Blandin, S., Blass, C., Brey, P.T., Collins, F.H., Danielli, A., Dimopoulos, G., Hetru, C., Hoa, N., Hoffmann, J.A., Kanzok, S.M., Letunic, I., Levashina, E.A., Loukeris, T.G., Lycett, G., Meister, S., Michel, K., Muller, H.M., Osta, M.A., Paskewitz, S.M., Reichhart, J.M., Rzhetsky, A., Troxler, L., Vernick, K.D., Vlachou, D., Volz, J., von Mering, C., Xu, J.N., Zheng, L.B., Bork, P. and Kafatos, F.C. (2002) Immunity-related genes and gene families in *Anopheles gambiae. Science* 298, 159–165.

Collins, F.H. (1994) Prospects for malaria control through the genetic manipulation of its vectors. *Parasitology Today* 10, 370–371.

Collins, F.H. and Paskewitz, S.M. (1995) Malaria – current and future – prospects for control. *Annual Review of Entomology* 40, 195–219.

Collins, F.H., Sakai, R.K., Vernick, K.D., Paskewitz, S., Seeley, D.C., Miller, L.H., Collins, W.E., Campbell, C.C. and Gwadz, R.W. (1986) Genetic selection of a *Plasmodium*-refractory strain of the malaria vector *Anopheles gambiae. Science* 234, 607–610.

Coté, I.M. and Poulin, R. (1995) Parasitism and group-size in social animals – a metaanalysis. *Behavioral Ecology* 6, 159–165.

Cotter, S.C., Hails, R.S., Cory, J.S. and Wilson, K. (2004a) Density-dependent prophylaxis and condition-dependent immune function in lepidopteran larvae: a multivariate approach. *Journal of Animal Ecology* 73, 283–293.

Cotter, S.C., Kruuk, L.E.B. and Wilson, K. (2004b) Costs of resistance: genetic correlations and potential trade-offs in an insect immune system. *Journal of Evolutionary Biology* 17, 421–429.

Crawley, M.J. (1983) *Herbivory: The Dynamics of Animal–Plant Interactions.* Blackwell Scientific, Oxford, UK.

Da Silva, P., Jouvensal, L., Lamberty, M., Bulet, P., Caille, A. and Vovelle, F. (2003) Solution structure of termicin, an antimicrobial peptide from the termite *Pseudacanthotermes spiniger. Protein Science* 12, 438–446.

Davies, C.R., Ayres, J.M., Dye, C. and Deane, L.M. (1991) Malaria infection-rate of Amazonian primates increases with body-weight and group-size. *Functional Ecology* 5, 655–662.

De Gregorio, E., Spellman, P.T., Rubin, G.M. and Lemaitre, B. (2001) Genome-wide analysis of the *Drosophila* immune response by using oligonucleotide microarrays. *Proceedings of the National Academy of Sciences USA* 98, 12590–12595.

Dimopoulos, G., Christophides, G.K., Meister, S., Schultz, J., White, K.P., Barillas-Mury, C. and Kafatos, F.C. (2002) Genome expression analysis of *Anopheles gambiae*: responses to injury, bacterial challenge, and malaria infection. *Proceedings of the National Academy of Sciences USA* 99, 8814–8819.

Dingle, H. (1991) Evolutionary genetics of animal migration. *American Zoologist* 31, 253–264.

Dingle, H. (1996) *Migration: The Biology of Life on the Move.* Oxford University Press, Oxford, UK.

Doering, T.L., Nosanchuk, J.D., Roberts, W.K. and Casadevall, A. (1999) Melanin as a potential cryptococcal defence against microbicidal proteins. *Medical Mycology* 37, 175–181.

Doums, C. and Schmid-Hempel, P. (2000) Immunocompetence in workers of a social insect, *Bombus terrestris* L., in relation to foraging activity and parasitic infection. *Canadian Journal of Zoology–Revue Canadienne De Zoologie* 78, 1060–1066.

Drnevich, J.M., Barnes, A.I. and Siva-Jothy, M.T. (2002) Immune investment and sperm competition in a beetle. *Physiological Entomology* 27, 228–234.

Duffey, S.S. and Stout, M.J. (1996) Antinutritive and toxic components of plant defense against insects. *Archives of Insect Biochemistry and Physiology* 32, 3–37.

Duffey, S.S., Hoover, K., Bonning, B. and Hammock, B.D. (1995) The impact of host plant on the efficacy of baculoviruses. *Reviews in Pesticide Toxicology* 3, 137–275.

Dunn, P.E. (1991) Insect antibacterial proteins. In: Warr, G.W. and Cohen, M. (eds) *Phylogenesis of Immune Functions*. CRC Press, Boca Raton, Florida, pp. 19–44.

Engström, A., Xanthopoulos, K.G., Boman, H.G. and Bennich, H. (1985) Amino acid and cDNA sequences of lysozyme from *Hyalophora cecropia*. *EMBO Journal* 4, 2119–2122.

Eslin, P. and Prévost, G. (1996) Variation in *Drosophila* concentration of haemocytes associated with different ability to encapsulate *Asobara tabida* larval parasitoid. *Journal of Insect Physiology* 42, 549–555.

Eslin, P. and Prévost, G. (1998) Hemocyte load and immune resistance to *Asobara tabida* are correlated in species of the *Drosophila melanogaster* subgroup. *Journal of Insect Physiology* 44, 807–816.

Evans, H.C. (1982) Entomogenous fungi in tropical forest ecosystems: an appraisal. *Ecological Entomology* 7, 47–60.

Evans, M.R., Goldsmith, A.R. and Norris, S.R.A. (2000) The effects of testosterone on antibody production and plumage coloration in male house sparrows (*Passer domesticus*). *Behavioral Ecology and Sociobiology* 47, 156–163.

Ewald, P.W. (1994) *Evolution of Infectious Disease*. Oxford University Press, New York.

Falconer, D.S. and Mackay, T.F.C. (1996) *Introduction to Quantitative Genetics*. Ed. 4. Longman, Essex.

Faye, I. and Hultmark, D. (1993) The insect immune proteins and the regulation of their genes. In: Beckage, N.E., Thompson, S.N. and Federici, B.A. (eds) *Parasites and Pathogens of Insects*, Vol.2. Academic Press, San Diego, California, pp. 25–53.

Fehlbaum, P., Bulet, P., Michaut, L., Lagueux, M., Broekaert, W.F., Hetru, C. and Hoffmann, J.A. (1994) Insect immunity: septic injury of *Drosophila* induces the synthesis of a potent antifungal peptide with sequence homology to plant antifungal peptides. *Journal of Biological Chemistry* 269, 33159–33163.

Fellowes, M.D.E. and Godfray, H.C.J. (2000) The evolutionary ecology of resistance to parasitoids by *Drosophila*. *Heredity* 84, 1–8.

Fellowes, M.D.E., Kraaijeveld, A.R. and Godfray, H.C.J. (1998a) Trade-off associated with selection for increased ability to resist parasitoid attack in *Drosophila melanogaster*. *Proceedings of the Royal Society of London Series B Biological Sciences* 265, 1553–1558.

Fellowes, M.D.E., Masnatta, P., Kraaijeveld, A.R. and Godfray, H.C.J. (1998b) Pupal parasitoid attack influences the relative fitness of *Drosophila* that have encapsulated larval parasitoids. *Ecological Entomology* 23, 281–284.

Fellowes, M.D.E., Kraaijeveld, A.R. and Godfray, H.C.J. (1999a) Association between feeding rate and parasitoid resistance in *Drosophila melanogaster*. *Evolution* 53, 1302–1305.

Fellowes, M.D.E., Kraaijeveld, A.R. and Godfray, H.C.J. (1999b) Cross-resistance following artificial selection for increased defense against parasitoids in *Drosophila melanogaster*. *Evolution* 53, 966–972.

Fellowes, M.D.E., Kraaijeveld, A.R. and Godfray, H.C.J. (1999c) The relative fitness of *Drosophila melanogaster* (Diptera, Drosophilidae) that have successfully defended themselves against the parasitoid *Asobara tabida* (Hymenoptera, Braconidae). *Journal of Evolutionary Biology* 12, 123–128.

Ferdig, M.T., Beerntsen, B.T., Spray, F.J., Li, J.Y. and Christensen, B.M. (1993) Reproductive costs associated with resistance in a mosquito-filarial work system. *American Journal of Tropical Medicine and Hygiene* 49, 756–762.

Folstad, I. and Karter, A.J. (1992) Parasites, bright males, and the immunocompetence handicap. *American Naturalist* 139, 603–622.

Freeland, W.J. (1979) Primate social groups as biological islands. *Ecology* 60, 719–728.

Freitak, D., Ots, I., Vanatoa, A. and Hõrak, P. (2003) Immune response is energetically costly in white cabbage butterfly pupae. *Proceedings of the Royal Society of London, Series B Biological Sciences* 270, 220–222.

Fuxa, J.R. and Richter, A.R. (1998) Repeated reversion of resistance to nucleopolyhedrovirus by *Anticarsia gemmatalis*. *Journal of Invertebrate Pathology* 71, 159–164.

Gehad, A.E., Mashaly, M.M., Siegel, H.S., Dunnington, E.A. and Siegel, P.B. (1999) Effect of genetic selection and MHC haplotypes on lymphocyte proliferation and Interleukin-2 like activity in chicken lines selected for high and low antibody production against sheep red blood cells. *Veterinary Immunology and Immunopathology* 68, 13–24.

Gill, H.S., Altmann, K., Cross, M.L. and Husband, A.J. (2000) Induction of T helper 1- and T helper 2-type immune response during *Haemonchus contortus* infection in sheep. *Immunology* 99, 458–463.

Giorgi, M.S., Arlettaz, R., Christe, P. and Vogel, P. (2001) The energetic grooming costs imposed by a parasitic mite (*Spinturnix myoti*) upon its bat host (*Myotis myotis*). *Proceedings of the Royal Society of London, Series B Biological Sciences* 268, 2071–2075.

Gonzalez, G., Sorci, G. and de Lope, F. (1999) Seasonal variation in the relationship between cellular immune response and badge size in male house sparrows (*Passer domesticus*). *Behavioral Ecology and Sociobiology* 46, 117–122.

Gorman, M.J., Severson, D.W., Cornel, A.J., Collins, F.H. and Paskewitz, S.M. (1997) Mapping a quantitative trait locus involved in melanotic encapsulation of foreign bodies in the malaria vector *Anopheles gambiae*. *Genetics* 146, 965–971.

Gotz, P. (1986) Encapsulation in arthropods. In: Brehelin, M. (ed.) *Immunity in Invertebrates: Cells, Molecules and Defense Reactions*. Springer, Berlin.

Goulson, D. and Cory, J.S. (1995) Responses of *Mamestra brassicae* (Lepodoptera, Noctuidae) to crowding: interactions with disease resistance, color phase and growth. *Oecologia* 104, 416–423.

Goulson, D. and Hauxwell, C. (1995) Resistance or covert infection: baculovirus studies reexamined. *Functional Ecology* 9, 548–549.

Graves, B.M. and Duvall, D. (1995) Effects of sexually-transmitted diseases on heritable variation in sexually selected systems. *Animal Behaviour* 50, 1129–1131.

Grencis, R.K. (1997) Th2-mediated host protective immunity to intestinal nematode infections. *Philosophical Transactions of the Royal Society of London, Series B Biological Sciences* 352, 1377–1384.

Grenfell, B.T. and Dobson, A.P. (eds) (1995) *Ecology of Infectious Diseases in Natural Populations*. Cambridge University Press, Cambridge, UK.

Gross, W.G., Siegel, P.B., Hall, W., Domersmuth, C.H. and DuBoise, R.T. (1980) Production and persistence of antibodies in chickens to sheep erythrocytes. 2. Resistance to infectious disease. *Poultry Science* 59, 205–210.

Gupta, A.P. (ed.) (1991) *Immunology of Insects and Other Arthropods*. CRC Press, Boca Raton, Florida.

Gupta, A.P. (2001a) Immunology of invertebrates: cellular. In: *Nature Encyclopedia of Life Sciences*, Vol. 10. Nature Publishing Group, London, pp. 72–81.

Gupta, A.P. (2001b) Immunology of invertebrates: humoral. In: *Nature Encyclopedia of Life Sciences*, Vol. 10. Nature Publishing Group, London, pp. 81–86.

Hajek, A.E. and St Leger, R.J. (1994) Interactions between fungal pathogens and insect hosts. *Annual Review of Entomology* 39, 293–322.

Hamilton, W.D. (1990) Mate choice near or far. *American Zoologist* 30, 341–352.

Hamilton, W.D. and Zuk, M. (1982) Heritable true fitness and bright birds: a role for parasites. *Science* 218, 384–387.

Hanley, K.A. (1989) Pathogenic threat and variation in febrile responses. Bachelor's honours, Amherst College, Amherst, Massachusetts.

Hart, B.L. (1992) Behavioral adaptations to parasites – an ethological approach. *Journal of Parasitology* 78, 256–265.

Hart, B.L. (1994) Behavioral defense against parasites – interaction with parasite invasiveness. *Parasitology* 109, S139–S151.

Hart, B.L., Hart, L.A., Mooring, M.S. and Olubayo, R. (1992) Biological basis of grooming behavior in antelope: the body size, vigilance and habitat principles. *Animal Behaviour* 44, 615–631.

Harvey, A.W. and Mallya, G.A. (1995) Predicting the severity of *Spodoptera exempta* (Lepidoptera: L. Noctuidae) outbreak seasons in Tanzania. *Bulletin of Entomological Research* 85, 479–487.

Hetru, C., Hoffmann, D. and Bulet, P. (1998) Antimicrobial peptides from insects. In: Brey, P.T. and Hultmark, D. (eds) *Molecular Mechanisms of Immune Responses in Insects*. Chapman and Hall, London, pp. 40–66.

Hiruma, K. and Riddiford, L.M. (1988) Granular phenoloxidase involved in cuticular melanization in the tobacco hornworm: regulation of its synthesis in the epidermis by juvenile hormone. *Developmental Biology* 130, 87–97.

Hochberg, M.E. (1991a) Nonlinear transmission rates and the dynamics of infectious diseases. *Journal of Theoretical Biology* 153, 301–321.

Hochberg, M.E. (1991b) Viruses as costs to gregarious feeding behavior in the Lepidoptera. *Oikos* 61, 291–296.

Holt, R.A., Subramanian, G.M., Halpern, A., Sutton, G.G., Charlab, R., Nusskern, D.R., Wincker, P., Clark, A.G., Ribeiro, J.M.C., Wides, R., Salzberg, S.L., Loftus, B., Yandell, M., Majoros, W.H., Rusch, D.B., Lai, Z.W., Kraft, C.L., Abril, J.F., Anthouard, V., Arensburger, P., Atkinson, P.W., Baden, H., de Berardinis, V., Baldwin, D., Benes, V., Biedler, J., Blass, C., Bolanos, R., Boscus, D., Barnstead, M., Cai, S., Center, A., Chatuverdi, K., Christophides, G.K., Chrystal, M.A., Clamp, M., Cravchik, A., Curwen, V., Dana, A., Delcher, A., Dew, I., Evans, C.A., Flanigan, M., Grundschober-Freimoser, A., Friedli, L., Gu, Z.P., Guan, P., Guigo, R., Hillenmeyer, M.E., Hladun, S.L., Hogan, J.R., Hong, Y.S., Hoover, J., Jaillon, O., Ke, Z.X., Kodira, C., Kokoza, E., Koutsos, A., Letunic, I., Levitsky, A., Liang, Y., Lin, J.J., Lobo, N.F., Lopez, J.R., Malek, J.A., McIntosh, T.C., Meister, S., Miller, J., Mobarry, C., Mongin, E., Murphy, S.D., O'Brochta, D.A., Pfannkoch, C., Qi, R., Regier, M.A., Remington, K., Shao, H.G., Sharakhova, M.V., Sitter, C.D., Shetty, J., Smith, T.J., Strong, R., Sun, J.T., Thomasova, D., Ton, L.Q., Topalis, P., Tu, Z.J., Unger, M.F., Walenz, B., Wang, A.H., Wang, J., Wang, M., Wang, X.L., Woodford, K.J., Wortman, J.R., Wu, M., Yao, A., Zdobnov, E.M., Zhang, H.Y., Zhao, Q., *et al.* (2002) The genome sequence of the malaria mosquito *Anopheles gambiae*. *Science* 298, 129–149.

Hudson, P.J., Rizzoli, A., Grenfell, B.T., Heesterbeek, H. and Dobson, A.P. (eds) (2002) *The Ecology of Wildlife Diseases*. Oxford University Press, Oxford, UK.

Hung, S.Y. and Boucias, D.G. (1992) Influence of *Beauveria bassiana* on the cellular defense response of the beet armyworm, *Spodoptera exigua*. *Journal of Invertebrate Pathology* 60, 152–158.

Hurlbert, R.E., Karlinsey, J.E. and Spence, K.D. (1985) Differential synthesis of bacteria-induced proteins of *Manduca sexta* larvae and pupae. *Journal of Insect Physiology* 31, 205–215.

Hutchings, M.R., Kyriazakis, I., Anderson, D.H., Gordon, I.J. and Coop, R.L. (1998) Behavioural strategies used by parasitized and non-parasitized sheep to avoid ingestion of gastro-intestinal nematodes associated with faeces. *Animal Science* 67, 97–106.

Hutchings, M.R., Kyriazakis, I., Gordon, I.J. and Jackson, F. (1999) Trade-offs between nutrient intake and faecal avoidance in herbivore foraging decisions: the effect of animal parasitic status, level of feeding motivation and sward nitrogen content. *Journal of Animal Ecology* 68, 310–323.

Ibanez, O.M., Mouton, D., Ribeiro, O.G., Bouthillier, Y., De Franco, M., Cabrera, W.H.K., Siqueira, M. and Biozzi, G. (1999) Low antibody responsiveness is found to be associated with resistance to chemical skin tumorigenesis in several lines of Biozzi mice. *Cancer Letters* 136, 153–158.

Ishikawa, H., Mitsui, Y., Yoshitomi, T., Mashimo, K., Aoki, S., Mukuno, K. and Shimizu, K. (2000) Presynaptic effects of botulinum toxin type A on the neuronally evoked response of albino and pigmented rabbit iris sphincter and dilator muscles. *Japanese Journal of Ophthalmology* 44, 106–109.

Jawor, J.M. and Breitwisch, R. (2003) Melanin ornaments, honesty, and sexual selection. *Auk* 120, 249–265.

Johnsen, T.S. and Zuk, M. (1999) Parasites and tradeoffs in the immune response of female red jungle fowl. *Oikos* 86, 487–492.

Johnson, J.K., Rocheleau, T.A., Hillyer, J.F., Chen, C.C., Li, J. and Christensen, B.M. (2003) A potential role for phenylalanine hydroxylase in mosquito immune responses. *Insect Biochemistry and Molecular Biology* 33, 345–354.

Jollès, P. and Jollès, J. (1984) What's new in lysozyme research: always a model system, today as yesterday. *Molecular and Cellular Biochemistry* 63, 165–189.

Karban, R. (1998) Caterpillar basking behavior and nonlethal parasitism by tachinid flies. *Journal of Insect Behavior* 11, 713–723.

Keating, S.T., Yendol, W.G. and Schultz, J.C. (1988) Relationship between susceptibility of gypsy moth larvae (Lepidoptera, Lymantriidae) to a baculovirus and host plant foliage constituents. *Environmental Entomology* 17, 952–958.

Keating, S.T., Hunter, M.D. and Schultz, J.C. (1990) Leaf phenolic inhibition of gypsy moth nuclear polyhedrosis virus: role of polyhedral inclusion body aggregation. *Journal of Chemical Ecology* 16, 1445–1457.

Kluger, M.J., Ringler, D.J. and Anver, M.R. (1975) Fever and survival. *Science* 188, 166–168.

Kluger, M.J., Kozak, W., Conn, C.A., Leon, L.R. and Soszynski, D. (1998) Role of fever in disease. In: *Molecular Mechanisms of Fever*, Vol. 856, pp. 224–233.

Knell, R.J. (1999) Sexually transmitted disease and parasite-mediated sexual selection. *Evolution* 53, 957–961.

Knell, R.J. and Webberley, K.M. (2004) Sexually transmitted diseases of insects: distribution, evolution ecology and host behavior. *Biological Reviews of the Cambridge Philosophical Society* 79, 557–581.

Koella, J.C. and Boëte, C. (2002) A genetic correlation between age at pupation and melanization immune response of the yellow fever mosquito *Aedes aegypti*. *Evolution* 56, 1074–1079.

Koella, J.C. and Sørensen, F.L. (2002) Effect of adult nutrition on the melanization immune response of the malaria vector *Anopheles stephensi*. *Medical and Veterinary Entomology* 16, 316–320.

Kokko, H., Ranta, E., Ruxton, G. and Lundberg, P. (2002) Sexually transmitted disease and the evolution of mating systems. *Evolution* 56, 1091–1100.

König, C. and Schmid-Hempel, P. (1995) Foraging activity and immunocompetence in workers of the bumble bee, *Bombus terrestris*. *Proceedings of the Royal Society of London, Series B Biological Sciences* 260, 225–227.

Kraaijeveld, A.R. and Godfray, H.C.J. (1997) Trade-off between parasitoid resistance and larval competitive ability in Drosophila melanogaster. Nature 389, 278–280.

Kraaijeveld, A.R., Limentani, E.C. and Godfray, H.C.J. (2001) Basis of the trade-off between parasitoid resistance and larval competitive ability in Drosophila melanogaster. Proceedings of the Royal Society of London, Series B Biological Sciences 268, 259–261.

Kraaijeveld, A.R., Ferrari, J. and Godfray, H.C.J. (2002) Costs of resistance in insect-parasite and insect-parasitoid interactions. Parasitology 125, S71–S82.

Kramm, K.R., West, D.F. and Rockenbach, P.G. (1982) Termite pathogens: transfer of the entomopathogen Metarhizium anisopliae between Reticulitermes sp. termites. Journal of Invertebrate Pathology 40, 1–6.

Kumae, T., Kurakake, S., Machida, K. and Sugawar, K. (1994) Effect of training on physical exercise-induced changes in non-specific humoral immunity. Japanese Journal of Physical Fitness and Sports Medicine 43, 75–83.

Kunimi, Y. and Yamada, E. (1990) Relationship of larval phase and susceptibility of the armyworm, Pseudaletia separata Walker (Lepidoptera, Noctuidae) to a nuclear polyhedrosis virus and a granulosi virus. Applied Entomology and Zoology 25, 289–297.

Kuo, M.J. and Alexander, M. (1967) Inhibition of the lysis of fungi by melanins. Journal of Bacteriology 94, 624–629.

Kurtz, J. (2002) Phagocytosis by invertebrate hemocytes: causes of individual variation in Panorpa vulgaris scorpionflies. Microscopy Research and Technique 57, 456–468.

Kurtz, J. and Franz, K. (2003) Evidence for memory in invertebrate immunity. Nature 425, 37–38.

Lai-Fook, J. (1966) The repair of wounds in the integument of insects. Journal of Insect Physiology 12, 195–226.

Lamberty, M., Ades, S., Uttenweiler-Joseph, S., Brookhart, G., Bushey, D., Hoffmann, J.A. and Bulet, P. (1999) Insect immunity: isolation from the lepidopteran Heliothis virescens of a novel insect defensin with potent antifungal activity. Journal of Biological Chemistry 274, 9320–9326.

Lee, K.P. (2002) Ecological factors impacting on the nutritional biology of a generalist and specialist caterpillar: effects of pathogens and plant structural compounds on macro-nutrient balancing. Zoology Department and Wolfson College. Oxford University, Oxford, UK.

Lehrer, R.I., Ganz, T. and Selsted, M.E. (1991) Defensins: endogenous antibiotic peptides of animal cells. Cell 64, 229–230.

Lemaitre, B., Reichhart, J.M. and Hoffmann, J.A. (1997) Drosophila host defense: differential induction of antimicrobial peptide genes after infection by various classes of microorganisms. Proceedings of the National Academy of Sciences USA 94, 14614–14619.

Leonard, C., Söderhall, K. and Ratcliffe, N.A. (1985) Studies on prophenoloxidase and protease activity of Blaberus craniifer hemocytes. Insect Biochemistry 15, 803–810.

Little, T.J., O'Connor, B., Colegrave, N., Watt, K. and Read, A.F. (2003) Maternal transfer of strain-specific immunity in an invertebrate. Current Biology 13, 489–492.

Lively, C.M. and Dybdahl, M.F. (2000) Parasite adaptations to locally common host genotypes. Nature 405, 679–681.

Lochmiller, R.L. and Deerenberg, C. (2000) Trade-offs in evolutionary immunology: just what is the cost of immunity? Oikos 88, 87–98.

Lockhart, A.B., Thrall, P.H. and Antonovics, J. (1996) Sexually transmitted diseases in animals: ecological and evolutionary implications. Biological Reviews of the Cambridge Philosophical Society 71, 415–471.

Loehle, C. (1995) Social barriers to pathogen transmission in wild animal populations. *Ecology* 76, 326–335.

Louis, C., Jourdan, M. and Cabanac, M. (1986) Behavioral fever and therapy in a Rickettsia-infected Orthoptera. *American Journal of Physiology* 250, R991–R995.

Lowenberger, C.A. and Rau M.E. (1994) Selective oviposition by *Aedes aegypti* (Diptera: Culicidae) in response to a larval parasite, *Plagiorchis elegans* (Trematoda: Plagiorchiidae). *Environmental Entomology* 23, 1269–1276.

Mackintosh, J.A. (2001) The antimicrobial properties of melanocytes, melanosomes and melanin and the evolution of black skin. *Journal of Theoretical Biology* 211, 101–113.

Maitland, D.P. (1994) A parasitic fungus infecting yellow dungflies manipulates host perching behavior. *Proceedings of the Royal Society of London, Series B Biological Sciences* 258, 187–193.

Mason, H. (1955) Comparative biochemistry of the phenolase complex. *Advances in Enzymology* 16, 105–184.

McCallum, H., Barlow, N. and Hone, J. (2001) How should pathogen transmission be modelled? *Trends in Ecology and Evolution* 16, 295–300.

McKean, K.A. and Nunney, L. (2001) Increased sexual activity reduces male immune function in *Drosophila melanogster*. *Proceedings of the National Academy of Sciences USA* 98, 7904–7909.

McNeill, S. and Southwood, T.R.E. (1978) Role of nitrogen in the development of insect-plant relations. In: Harborne, J. (ed.) *Biochemical Aspects of Plant and Animal Coevolution*. Academic Press, New York, pp. 77–98.

Mensah, B.A. and Gatehouse, A.G. (1998) Effect of larval phase and adult diet on fecundity and related traits in *Spodptera exempta*. *Entomologia Experimentalis et Applicata* 86, 331–336.

Michaut, L., Fehlbaum, P., Moniatte, M., VanDorsselaer, A., Reichhart, J.M. and Bulet, P. (1996) Determination of the disulfide array of the first inducible antifungal peptide from insects: drosomycin from *Drosophila melanogaster*. *FEBS Letters* 395, 6–10.

Mitsui, J. and Kunimi, Y. (1988) Effect of larval phase on susceptibility of the armyworm, *Pseudaletia separata* Walker (Lepidoptera, Noctuidae) to an entomogeneous deuteromycete, *Nomuraea rileyi*. *Japanese Journal of Applied Entomology and Zoology* 32, 129–134.

Møller, A.P., Kimball, R.T. and Erritzoe, J. (1996) Sexual ornamentation, condition, and immune defence in the house sparrow *Passer domesticus*. *Behavioral Ecology and Sociobiology* 39, 317–322.

Montefiori, D.C. and Zhou, J.Y. (1991) Selective antiviral activity of synthetic soluble L-tyrosine and L-dopa melanins against human-immunodeficiency-virus *in vitro*. *Antiviral Research* 15, 11–26.

Moore, J. (2002) *Parasites and the Behavior of Animals*. Oxford University Press, New York.

Moore, J. and Gotelli, N.J. (1990) A phylogenetic perspective on the evolution of altered host behaviours. In: Barnard, C.J. and Behnke, J.M. (eds) *Parasitism and Host Behaviour*. Taylor & Francis, London, pp. 193–233.

Moore, S.L. and Wilson, K. (2002) Parasites as a viability cost of sexual selection in natural populations of mammals. *Science* 297, 2015–2018.

Moret, Y. and Schmid-Hempel, P. (2000) Survival for immunity: the price of immune system activation for bumblebee workers. *Science* 290, 1166–1168.

Moret, Y. and Schmid-Hempel, P. (2001) Entomology – immune defence in bumble-bee offspring. *Nature* 414, 506–506.

Moret, Y. and Siva-Jothy, M.T. (2003) Adaptive innate immunity? Responsive-mode prophylaxis in the mealworm beetles, *Tenebrio molitor*. *Proceedings of the Royal Society of London, Series B Biological Sciences* 270, 2475–2480.

Müller, C.B. and Schmid-Hempel, P. (1993) Exploitation of cold temperature as defense against parasitoids in bumblebees. *Nature* 363, 65–67.

Nappi, A.J. and Vass, E. (1993) Melanogenesis and the generation of cytotoxic molecules during insect cellular immune-reactions. *Pigment Cell Research* 6, 117–126.

Nappi, A.J., Vass, E., Frey, F. and Carton, Y. (1995) Superoxide anion generation in *Drosophila* during melanotic encapsulation of parasites. *European Journal of Cell Biology* 68, 450–456.

Nelson, R.J. and Demas, G.E. (1996) Seasonal changes in immune function. *Quarterly Review of Biology* 71, 511–548.

Nelson, R.J., Asfaw, B., DeVries, A.C. and Demas, G.E. (1996a) Reproductive response to photoperiod affects corticosterone and immunoglobulin G concentrations in prairie voles (*Microtus ochrogaster*). *Canadian Journal of Zoology–Revue Canadienne De Zoologie* 74, 576–581.

Nelson, R.J., Fine, J.B., Demas, G.E. and Moffatt, C.A. (1996b) Photoperiod and population density interact to affect reproductive and immune function in male prairie voles. *American Journal of Physiology: Regulatory Integrative and Comparative Physiology* 39, R571–R577.

Nelson, R.J., Demas, G.E. and Klein, S.L. (1998) Photoperiodic mediation of seasonal breeding and immune function in rodents: a multifactorial approach. *American Zoologist* 38, 226–237.

Nigam, Y., Maudlin, I., Welburn, S. and Ratcliffe, N.A. (1997) Detection of phenoloxidase activity in the hemolymph of tsetse flies, refractory and susceptible to infection with *Trypanosoma brucei rhodesiense*. *Journal of Invertebrate Pathology* 69, 279–281.

Nijhout, H.F. (1994) *Insect Hormones*. Princeton University Press, Princeton, New Jersey.

Norris, K. and Evans, M.R. (2000) Ecological immunology: life history trade-offs and immune defense in birds. *Behavioral Ecology* 11, 19–26.

Orr, M.R. (1992) Parasitic flies (Diptera, Phoridae) influence foraging rhythms and caste division of labor in the leaf-cutter ant, *Atta cephalotes* (Hymenoptera, Formicidae). *Behavioral Ecology and Sociobiology* 30, 395–402.

Ouedraogo, R.M., Cusson, M., Goettel, M.S. and Brodeur, J. (2003) Inhibition of fungal growth in thermoregulating locusts, *Locusta migratoria*, infected by the fungus *Metarhizium anisopliae* var *acridum*. *Journal of Invertebrate Pathology* 82, 103–109.

Ourth, D.D. and Renis, H.E. (1993) Antiviral melanization reaction of *Heliothis virescens* hemolymph against DNA and RNA viruses *in vitro*. *Comparative Biochemistry and Physiology B Biochemistry and Molecular Biology* 105, 719–723.

Owens, I.P.F. (2002) Sex differences in mortality rate. *Science* 297, 2008–2009.

Owens, I.P.F. and Wilson, K. (1999) Immunocompetence: a neglected life history trait or conspicuous red herring? *Trends in Ecology and Evolution* 14, 170–172.

Paskewitz, S.M., Brown, M.R., Lea, A.O. and Collins, F.H. (1988) Ultrastructure of the encapsulation of *Plasmodium cynomolgi* (b-strain) on the midgut of a refractory strain of *Anopheles gambiae*. *Journal of Parasitology* 74, 432–439.

Paskewitz, S.M., Brown, M.R., Collins, F.H. and Lea, A.O. (1989) Ultrastructural-localization of phenoloxidase in the midgut of refractory *Anopheles-gambiae* and association of the enzyme with encapsulated *Plasmodium-cynomolgi*. *Journal of Parasitology* 75, 594–600.

Plaistow, S.J., Outreman, Y., Moret, Y. and Rigaud, T. (2003) Variation in the risk of being wounded: an overlooked factor in studies of invertebrate immune function? *Ecology Letters* 6, 489–494.

Poinar, G. (1974) Insect immunity to parasite nematodes. In: Cooper, E. (ed.) *Contemporary Topics in Immunobiology*, Vol. 4. Plenum, New York, pp. 167–178.

Poulin, R. (1996) Sexual inequalities in helminth infections: a cost of being a male? *American Naturalist* 147, 287–295.

Poulin, R. (2000) Manipulation of host behaviour by parasites: a weakening pardigm? *Proceedings of the Royal Society of London, Series B Biological Sciences* 267, 787–792.

Poulsen, M., Bot, A.N.M., Nielsen, M.G. and Boomsma, J.J. (2002) Experimental evidence for the costs and hygienic significance of the antibiotic metapleural gland secretion in leaf-cutting ants. *Behavioral Ecology and Sociobiology* 52, 151–157.

Powning, R. and Davidson, W.J. (1973) Studies on insect bacteriolytic enzymes. I. Lysozyme in haemolymph of *Galleria mellonella* and *Bombyx mori*. *Comparative Biochemistry and Physiology* 45, 669–681.

Råberg, L., Grahn, M., Hasselquist, D. and Svensson, E. (1998) On the adaptive significance of stress-induced immunosuppression. *Proceedings of the Royal Society of London, Series B Biological Sciences* 265, 1637–1641.

Rantala, M.J. and Kortet, R. (2003) Courtship song and immune function in the field cricket *Gryllus bimaculatus*. *Biological Journal of the Linnean Society* 79, 503–510.

Rantala, M.J., Koskimaki, J., Taskinen, J., Tynkkynen, K. and Suhonen, J. (2000) Immunocompetence, developmental stability ad wingspot size in the damselfly *Calopteryx splendens* L. *Proceedings of the Royal Society of London, Series B Biological Sciences* 267, 2453–2457.

Reeson, A.F., Wilson, K., Gunn, A., Hails, R.S. and Goulson, D. (1998) Baculovirus resistance in the noctuid *Spodoptera exempta* is phenotypically plastic and responds to population density. *Proceedings of the Royal Society of London, Series B Biological Sciences* 265, 1787–1791.

Reeson, A.F., Wilson, K., Cory, J.S., Hankard, P., Weeks, J.M., Goulson, D. and Hails, R.S. (2000) Effects of phenotypic plasticity on pathogen transmission in the field in a Lepidoptera-NPV system. *Oecologia* 124, 373–380.

Ritter, R.C. and Epstein, A.N. (1974) Saliva lost by grooming: a major item in the rat's water economy. *Behavioral Biology* 11, 581–585.

Roff, D.A. (1997) *Evolutionary Quantitative Genetics*. Chapman & Hall, New York.

Roff, D.A. (2002) *Life History Evolution*. Sinauer Associates, Sunderland, Massachusetts.

Rolff, J. and Siva-Jothy, M.T. (2002) Copulation corrupts immunity: a mechanism for a cost of mating in insects. *Proceedings of the National Academy of Sciences USA* 99, 9916–9918.

Rolff, J. and Siva-Jothy, M.T. (2003) Invertebrate ecological immunology. *Science* 301, 472–475.

Rose, D.J.W., Dewhurst, C.F. and Page, W.W. (2000) *The African Armyworm Handbook: The Status Biology Ecology Epidemiology and Management of Spodoptera exempta (Lepidoptera: Noctuidae)*. Natural Resources Institute, Chatham, UK.

Rosengaus, R.B., Guldin, M.R. and Traniello, J.F.A. (1998a) Inhibitory effect of termite fecal pellets on fungal spore germination. *Journal of Chemical Ecology* 24, 1697–1706.

Rosengaus, R.B., Maxmen, A.B., Coates, L.E. and Traniello, J.F.A. (1998b) Disease resistance: a benefit of sociality in the dampwood termite *Zootermopsis angusticollis* (Isoptera:Termopsidae). *Behavioral Ecology and Sociobiology* 44, 125–134.

Rosengaus, R.B., Jordan, C., Lefebvre, M.L. and Traniello, J.F.A. (1999a) Pathogen alarm behavior in a termite: a new form of communication in social insects. *Naturwissenschaften* 86, 544–548.

Rosengaus, R.B., Traniello, J.F.A., Chen, T., Brown, J.J. and Karp, R.D. (1999b) Immunity in a social insect. *Naturwissenschaften* 86, 588–591.

Rosengaus, R.B., Lefebvre, M.L. and Traniello, J.F.A. (2000a) Inhibition of fungal spore germination by *Nasutitermes*: evidence for a possible antiseptic role of soldier defensive secretions. *Journal of Chemical Ecology* 26, 21–39.

Rosengaus, R.B., Traniello, J.F.A., Lefebvre, M.L. and Carlock, D.M. (2000b) The social transmission of disease between adult male and female reproductives of the dampwood termite *Zootermopsis angusticollis. Ethology, Ecology and Evolution* 12, 419–433.

Rothenbuhler, W.C. and Thompson, V.C. (1956) Resistance to American foulbrood in honeybees. I. Differential survival of larvae of different genetic lines. *Journal of Economic Entomology* 49, 470–475.

Russell, V.W. and Dunn, P.E. (1991) Lysozyme in the midgut of *Manduca sexta* during metamorphosis. *Archives of Insect Biochemistry and Physiology* 17, 67–80.

Ryder, J.J. and Siva-Jothy, M.T. (2001) Quantitative genetics of immune function and body size in the house cricket, *Acheta domesticus. Journal of Evolutionary Biology* 14, 646–653.

Samakovlis, C., Kimbrell, D.A., Kylsten, P., Engstrom, A. and Hultmark, D. (1990) The immune response in *Drosophila*: pattern of cecropin expression and biological activity. *EMBO Journal* 9, 2969–2976.

Samuel, W.M. (1991) Grooming by moose (*Alces alces*) infested with the winter tick,*Dermacentor albipictus* (Acari) – a mechanism for premature loss of winter hair. *Canadian Journal of Zoology–Revue Canadienne De Zoologie* 69, 1255–1260.

Schmid-Hempel, P. (1990) Reproductive competition and the evolution of work load in social insects. *American Naturalist* 135, 501–526.

Schmid-Hempel, P. (1998) *Parasites in Social Insects.* Princeton University Press, Princeton, New Jersey.

Schmid-Hempel, P. (2003) Variation in immune defence as a question of evolutionary ecology. *Proceedings of the Royal Society of London, Series B Biological Sciences* 270, 357–366.

Schmid-Hempel, P. and Ebert, D. (2003) On the evolutionary ecology of specific immune defence. *Trends in Ecology and Evolution* 18, 27–32.

Schmid-Hempel, R. and Schmid-Hempel, P. (1998) Colony performance and immunocompetence of a social insect, *Bombus terrestris*, in poor and variable environments. *Functional Ecology* 12, 22–30.

Schuhmann, B., Seitz, V., Vilcinskas, A. and Podsiadlowski, L. (2003) Cloning and expression of gallerimycin, an antifungal peptide expressed in immune response of greater wax moth larvae, *Galleria mellonella. Archives of Insect Biochemistry and Physiology* 53, 125–133.

Shelby, K.S. and Webb, B.A. (1999) Polydnavirus-mediated suppression of insect immunity. *Journal of Insect Physiology* 45, 507–514.

Shelby, K.S., Adeyeye, O.A., Okot-Kotber, B.M. and Webb, B.A. (2000) Parasitism-linked block of host plasma melanization. *Journal of Invertebrate Pathology* 75, 218–225.

Sheldon, B.C. (1993) Sexually-transmitted disease in birds – occurrence and evolutionary significance. *Philosophical Transactions of the Royal Society of London, Series B Biological Sciences* 339, 491–497.

Sheldon, B.C. and Verhulst, S. (1996) Ecological immunology: costly parasite defences and trade-offs in evolutionary ecology. *Trends in Ecology and Evolution* 11, 317–321.

Sheridan, L.A.D., Poulin, R., Ward, D.F. and Zuk, M. (2000) Sex differences in parasitic infections among arthropod hosts: is there a male bias? *Oikos* 88, 327–334.

Sidibe, S., Saal, F., RhodesFeuillette, A., Lagaye, S., Pelicano, L., Canivet, M., Peries, J. and Dianoux, L. (1996) Effects of serotonin and melanin on *in vitro* HIV-1 infection. *Journal of Biological Regulators and Homeostatic Agents* 10, 19–24.

Sidjanski, S., Mathews, G.V. and Vanderberg, J.P. (1997) Electrophoretic separation and indentification of phenoloxidases in hemolymph and midgut of adult *Anopheles stephensi* mosquitoes. *Journal of Parasitology* 83, 686–691.

Simpson, S.J., Despland, E., Hagele, B.F. and Dodgson, T. (2001) Gregarious behavior in desert locusts is evoked by touching their back legs. *Proceedings of the National Academy of Sciences USA* 98, 3895–3897.

Siva-Jothy, M.T. and Thompson, J.J.W. (2002) Short-term nutrient deprivation affects immune function. *Physiological Entomology* 27, 206–212.

Siva-Jothy, M.T., Tsubaki, Y. and Hooper, R.E. (1998) Decreased immune response as a proximate cost of copulation and oviposition in a damselfly. *Physiological Entomology* 23, 274–277.

Söderhäll, K. and Ajaxon, R. (1982) Effect of quinones and melanin on mycelial growth of *Aphanomyces* spp. and extracellular protease of *Aphanomyces astaci* a parasite on crayfish. *Journal of Invertebrate Pathology* 39, 105–109.

Speight, M.R., Hunter, M.D. and Watt, A.D. (1999) *Ecology of Insects: Concepts and Applications*. Blackwell Science, Oxford, UK.

St Leger, R.J., Cooper, R.M. and Charnley, A.K. (1988)The effect of melanization of *Manduca sexta* cuticle on growth and infection by *Metarhizium anisopliae*. *Journal of Invertebrate Pathology* 52, 459–470.

Stacey, D.A., Thomas, M.B., Blanford, S., Pell, J.K., Pugh, C. and Fellowes, M.D.E. (2003) Genotype and temperature influence pea aphid resistance to a fungal entomopathogen. *Physiological Entomology* 28, 75–81.

Starks, P.T., Blackie, C.A. and Seeley, T.D. (2000) Fever in honeybee colonies. *Naturwissenschaften* 87, 229–231.

Stearns, S.C. (1989) Trade-offs in life-history. *Functional Ecology* 3, 259–268.

Stearns, S.C. (1992) *The Evolution of Life Histories*. Oxford University Press, Oxford, UK.

Strand, M.R. and Pech, L.L. (1995) *Microplitis demolitor* polydnavirus induces apoptosis of a specific hemocyte morphotype in *Pseudoplusia includens*. *Journal of General Virology* 76, 283–291.

Sugumaran, M., Nellaiappan, K. and Valivittan, K. (2000) A new mechanism for the control of phenoloxidase activity: inhibition and complex formation with quinone isomerase. *Archives of Biochemistry and Biophysics* 379, 252–260.

Sutter, G.R., Rothenbuhler, W.C. and Raun, E.S. (1968) Resistance to American foulbrood in honey bees. VII. Growth of resistant and susceptible larvae. *Journal of Invertebrate Pathology* 12, 25–28.

Suwanchaichinda, C. and Paskewitz, S.M. (1998) Effects of larval nutrition, adult body size, and adult temperature on the ability of *Anopheles gambiae* (Diptera: Culicidae) to melanize Sephadex beads. *Journal of Medical Entomology* 35, 157–161.

Taylor, J.G., Ferdig, M.T., Su, X.Z. and Wellems, T.E. (2000) Toward quantitative genetic analysis of host and parasite traits in the manifestations of *Plasmodium falciparum* malaria. *Current Opinion in Genetics and Development* 10, 314–319.

Taylor, R.L. (1969) A suggested role for the polyphenol-phenoloxidase system in invertebrate immunity. *Journal of Invertebrate Pathology* 14, 427–428.

Thomas, M.B. and Blanford, S. (2003) Thermal biology in insect-parasite interactions. *Trends in Ecology and Evolution* 18, 344–350.

Thomasova, D., Ton, L.Q., Copley, R.R., Zdobnov, E.M., Wang, X.L., Hong, Y.S., Sim, C., Bork, P., Kafatos, F.C. and Collins, F.H. (2002) Comparative genomic analysis in the region of a major *Plasmodium* refractoriness locus of *Anopheles gambiaed*. *Proceedings of the National Academy of Sciences USA* 99, 8179–8184.

Thompson, J.J.W., Armitage, S.A.O. and Siva-Jothy, M.T. (2002) Cuticular colour change after imaginal eclosion is time-constrained: blacker beetles darken faster. *Physiological Entomology* 27, 136–141.

Traniello, J.F.A., Rosengaus, R.B. and Savoie, K. (2002) The development of immunity in a social insect: evidence for the group facilitation of disease resistance. *Proceedings of the National Academy of Sciences USA* 99, 6838–6842.

Trivers, R.L. (1972) Parental investment and sexual selection. In: Campbell, B. (ed.) *Sexual Selection and the Descent of Man 1871–1971*. Aldine Press, Chicago, Illinois, pp. 136–175.

Trumper, S. and Simpson, S.J. (1993) Regulation of salt intake by nymphs of *Locusta migratoria*. *Journal of Insect Physiology* 39, 857–864.

Tryselius, Y., Samakovlis, C., Kimbrell, D.A. and Hultmark, D. (1992) Cecc, a cecropin gene expressed during metamorphosis in *Drosophila* pupae. *European Journal of Biochemistry* 204, 395–399.

Vilmos, P. and Kurucz, E. (1998) Insect immunity: evolutionary roots of the mammalian innate immune system. *Immunology Letters* 62, 59–66.

Washburn, J.O., Kirkpatrick, B.A. and Volkman, L.E. (1996) Co-infection of *Manduca sexta* larvae with polydnavirus from *Cotesia congregata* increases susceptibility to fatal infection by *Autographa californica* M. Nucleopolyhedrovirus. *Nature* 383, 767–767.

Washburn, J.O., Haas-Stapleton, E.J., Tan, F.F., Beckage, N.E. and Volkman, L.E. (2000) Insect protection against virus. *Journal of Insect Physiology* 46, 179–190.

Watve, M.G. and Jog, M.M. (1997) Epidemic diseases and host clustering: an optimum cluster size ensures maximum survival. *Journal of Theoretical Biology* 184, 167–171.

Webberley, K.M., Hurst, G.D.D., Buszko, J. and Majerus, M.E.N. (2002) Lack of parasite-mediated sexual selection in a ladybird/sexually transmitted disease system. *Animal Behaviour* 63, 131–141.

Westneat, D.F. and Birkhead, T.R. (1998) Alternative hypotheses linking the immune system and mate choice for good genes. *Proceedings of the Royal Society of London, Series B Biological Sciences* 265, 1065–1073.

Wigglesworth, V.B. (1965) *The Principles of Insect Physiology*. Methuen, London.

Wilson, K. (1995) Insect migration in heterogeneous environments. In: Drake, V.A. and Gatehouse, A.G. (eds) *Insect Migration: Tracking Resources Through Space and Time*. Cambridge University Press, Cambridge, UK, pp. 243–264.

Wilson, K. (2001) The costs of resistance in *Drosophila*: blood cells count. *Trends in Ecology and Evolution* 16, 72–73.

Wilson, K. and Reeson, A.F. (1998) Density-dependent prophylaxis: evidence from Lepidoptera-baculovirus interactions. *Ecological Entomology* 23, 100–101.

Wilson, K., Cotter, S.C., Reeson, A.F. and Pell, J.K. (2001) Melanism and disease resistance in insects. *Ecology Letters* 4, 637–649.

Wilson, K., Thomas, M.B., Blanford, S., Doggett, M., Simpson, S.J. and Moore, S.L. (2002) Coping with crowds: density-dependent disease resistance in desert locusts. *Proceedings of the National Academy of Sciences USA* 99, 5471–5475.

Wilson, K., Knell, R., Boots, M. and Koch-Osborne, J. (2003a) Group living and investment in immune defence: an interspecific analysis. *Journal of Animal Ecology* 72, 133–143.

Wilson, K., Moore, S.L. and Owens, I.P.F. (2003b) Response to comment on 'Parasites as a viability cost of sexual selection in natural populations of mammals'. *Science* 300, 55.

Yamamoto, K., Fuji, H., Banno, Y., Aso, Y. and Ishiguro, M. (2003) Polymorphism of prophenoloxidase in the silkworm, *Bombyx mori*. *Journal of the Faculty of Agriculture, Kyushu University* 47, 319–324.

Young, S.Y., Yang, J.G. and Felton, G.W. (1995) Inhibitory effects of dietary tannins on the infectivity of a nuclear polyhedrosis virus to *Helicoverpa zea* (Noctuidae, Lepidoptera). *Biological Control* 5, 145–150.

11 Adaptive Plasticity in Response to Predators in Dragonfly Larvae and Other Aquatic Insects

FRANK JOHANSSON[1] AND ROBBY STOKS[2]

[1]Department of Ecology and Environmental Science, Umeå University, Umeå, Sweden; [2]Laboratory of Aquatic Ecology, University of Leuven, Leuven, Belgium

1. Introduction

Predators are one of the strongest selection agents for prey organisms because they can immediately reduce a prey's fitness to zero. Not surprisingly, prey organisms have developed a bewildering array of traits to avoid predation. Antipredator mechanisms can be divided into fixed and flexible types. Fixed traits are present irrespective of whether predators are present or absent. Since many traits are costly to produce and maintain, fixed traits may come with an unavoidable cost if they occur in the wrong environment. Similarly, predation risk may vary spatially and temporally within the same environment, and for that reason antipredator mechanisms are not always necessary. Flexible traits avoid such unnecessary production costs because they are expressed only in the presence of predators. They represent a specific case of phenotypic plasticity: the ability to express different trait values in response to distinct environmental conditions. In the case of predation, these conditions are different levels of predation risk. Four prerequisites are required for circumstances favouring plasticity over fixed responses (Harvell, 1990):

1. The presence of predators has to be variable and sometimes strong.
2. The induced defence should be effective against the predator.
3. A reliable cue must be present for detecting the predator.
4. The induced defence should be cost-saving; that is, the benefits must be greater than the costs.

If plasticity is beneficial and maintained by selection, it is adaptive (Gotthard and Nylin, 1995). A useful tool for studying the adaptiveness of phenotypic plasticity is the concept of reaction norms. We here refer to a reaction norm as the set of phenotypes expressed by a single genetic category (e.g. family, population, species). The adaptiveness of plasticity can be demonstrated by showing that the reaction norm agrees with predictions of verbal or mathematical optimality models (Newman, 1992) or by performing an experiment where the plastic genotype is demonstrated to have higher fitness than a non-plastic or less plastic genotype (Gotthard and Nylin, 1995). In the case of plasticity to predators the latter can be done, for example in reciprocal experiments where the induced phenotype has higher fitness in the predator environment and the non-induced phenotype has higher fitness in the non-predator environment.

    Aquatic insects show plasticity in many different types of traits in order to avoid predation: predators may induce changes in behaviour, life history and morphology. Behavioural plasticity to predators, including reduced activity and increased refuge use, has been widely documented in dragonfly larvae (Pierce et al., 1985; Jeffries, 1990; McPeek, 1990a; Johansson, 1993; Ryazanova and Mazokhinporshnyakov, 1993; Wiseman et al., 1993; Claus-Walker et al., 1997; Koperski, 1997; Dixon and Baker, 1998; Schaffner and Anholt, 1998; Stoks, 1998, 1999a; Baker et al., 1999; Stoks and Johansson, 2000; Steiner et al., 2000; Hopper, 2001; McPeek et al., 2001; Suhling and Lepkojus, 2001; Brodin and Johansson, 2002; Stoks et al., 2003; Stoks and McPeek, 2003b) and other aquatic insect taxa (e.g. Sih, 1980; Kohler and McPeek, 1989; Peckarsky et al., 1993; McIntosh and Townsend, 1994; Peckarsky, 1996; Juliano and Gravel, 2002). We will not elaborate on behavioural plasticity to predators as it has been the subject of several excellent reviews (Lima and Dill, 1990; Lima, 1998). The focus in our chapter is on plasticity response to predators with regard to life history and morphology. Life-history responses to predators have been less frequently documented in dragonfly larvae. Generally, when food is kept constant, growth rate and development rate of dragonfly larvae were reduced in the presence of a predator (McPeek, 1990b, 1998; Schaffner and Anholt, 1998; Stoks et al., 1999a,b, 2001; Johansson et al., 2001; McPeek et al., 2001; Stoks and McPeek, 2003a). Similar life-history responses are also found in other aquatic insects (e.g. Peckarsky et al., 1993; Ball and Baker, 1996; Hechtel and Juliano, 1997; Peckarsky et al., 2001, 2002). Finally, predator-induced morphological plasticity has only rarely been documented in dragonfly larvae (but see the work of Johansson) or in other aquatic insects (Dahl and Peckarsky, 2002).

    In this chapter we give examples of plasticity in morphology and life-history traits as a way of avoiding predators in dragonfly larvae and other aquatic insects, discuss its adaptiveness and the four prerequisites for the evolution of plasticity, and suggest avenues for further research.

2. Predation Risk and Morphological Defence

Predator-induced morphological defence is a phenotypic change in a morphological character that is triggered by cues associated with a potential predator. The cues are generally believed to be mediated by chemical stimuli. The phenomenon has been described in many organisms including plants, protozoa, gastropods, crustaceans, amphibians, rotifers, bryozoans and fish. In animals, the morphological changes consist of production or modification of spines, and variation in shape or thickness of various body characters (Tollrian and Harvell, 1999). At present, only two examples are known in aquatic insects and we will review those two.

2.1 Morphological defence in the dragonfly *Leucorrhinia dubia*

Many dragonfly larvae show prominent abdominal spines. The lengths of these spines vary both within and among species (Askew, 1988; Johansson and Samuelsson, 1994; Needham *et al.*, 2000; Westman *et al.*, 2000; Hovmöller and Johansson, 2004). Johansson and Samuelsson (1994) showed that a great deal of the variation within *Leucorrhinia dubia* (Odonata: Libelullidae) larvae was accounted for by the presence or absence of fish predators. Larvae from lakes with fish had significantly longer dorsal and lateral spines on their abdomen than larvae from lakes without fish (Fig. 11.1). Similarly, *L. dubia* larvae raised in tanks with fish had longer spines than larvae raised in tanks

No fish lake

Fish lake

Fig. 11.1. *Leucorrhinia dubia* exuviae (cast exoskeleton) showing the difference in spine length between larvae from a lake without and with fish. Note the long dorsal and lateral spines on the lower exuviae from a lake with fish.

without fish (Fig. 11.2) (Arnqvist and Johansson, 1998; Johansson, 2002). Hence, the longer spines of larvae from fish lakes are probably a result of the presence of fish, as supported by the laboratory experiment. Two other interesting results emerged from these laboratory-rearing experiments. First, for some of the dorsal spines, families differed in spine length. Such variation between genetic categories explains part of the variation in spine length seen within a population. Second, for some of the spines there was a significant interaction between genetic category (family) and environment (fish absent/present). This interaction suggests a potential for the evolution of plasticity of spine length. Hence, if the spatial variation in a region with regard to the presence or absence of fish becomes larger we should expect a higher plasticity in spine length in the future and vice versa.

Johansson and Wahlström (2002) also manipulated whole lakes to observe whether the phenotypic plasticity in larval spine length of *L. dubia* could be

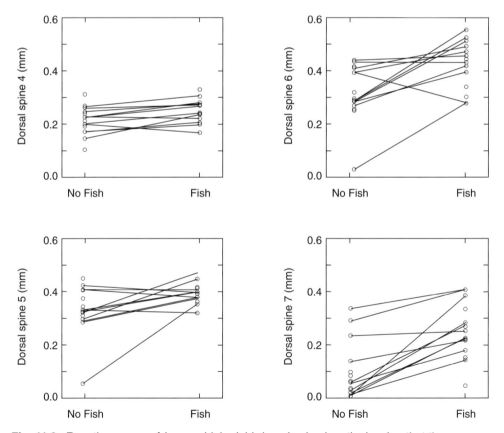

Fig. 11.2. Reaction norms of *Leucorrhinia dubia* larval spine length showing that the presence of fish induces longer spines. Each dot represents mean spine length of larvae of the same family raised without or with fish. Lines connect larvae belonging to the same family (same genotype). When no line is present between dots a full family replicate is missing. Modified from Johansson (2002).

observed in natural systems that have been manipulated. These whole-lake experiments supported the laboratory experiments because mean spine length of larvae decreased when fish were removed from the manipulated lakes. The strength of performing such whole-lake manipulation experiments is that the experiment is performed under natural conditions, and hence many other environmental variables are accounted for.

2.2 Morphological defence in the mayfly *Drunella coloradensis*

A second example of predator-induced morphological defence in aquatic insects has been demonstrated by Dahl and Peckarsky (2002). They showed that *Drunella coloradensis* (Ephemeroptera: Ephemerellidae) larvae had longer caudal filaments and a heavier exoskeleton in stream sites that contained brook trout (Fig. 11.3). Rearing experiments showed that chemical cues from brook

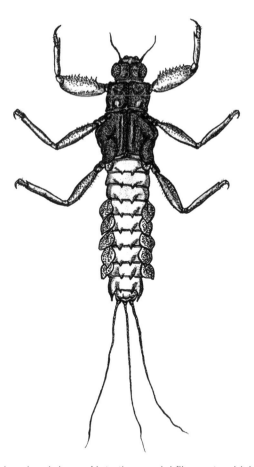

Fig. 11.3. *Drunella coloradensis* larva. Note the caudal filaments which grow longer in the presence of fish. Illustration by Ming Huang.

trout induced longer caudal filaments in these larvae. Interestingly, chemical cues from fish only induced longer caudal filaments in larvae originating from fishless sites. If larvae originated from fish sites, chemical cues from brook trout had no effect on the length of the caudal filaments. The authors do not discuss this result but our interpretation is that longer caudal filaments are induced early during development, since larvae from fish sites had already experienced the fish cues.

2.3 Are the four prerequisites for the evolution of plasticity supported?

Temporal and spatial variation in predation risk is probably common in the two systems studied. Both *L. dubia* adults and *D. coloradensis* were studied in regions where fish and fishless waterbodies are intermixed and because these insects disperse as adults they should have a high probability of encountering a fish or fishless site when females are ovipositing. Similarly, since these insects occur in small waterbodies, severe physical conditions such as floods, drought and freezing can affect fish populations dramatically and therefore create temporal variations in fish abundance (Seegrist and Gard, 1972; Tonn and Magnusson, 1982). Therefore, selection pressures exerted by predators should be variable and hence this prerequisite is fulfilled. If females could distinguish between fish and fishless sites this could reduce the probability of eggs being deposited at high predation risk sites, and then spatial variation in predation risk would be very low. Currently we do not know whether females of these species can distinguish between fish and fishless sites, but studies on other odonates and mayflies showed that females did not distinguish between both types of site (McPeek 1989; Caudill, 2003).

The induced defence is effective in both species. For the *L. dubia* system, Johansson and Samuelsson (1994) showed that yellow perch had a longer handling time when eating long-spined larvae, suggesting that spines do provide some protection. In an among-species study on the protective value of spines, Mikolajewski and Johansson (2004) showed that long-spined *Leucorrhinia* species were ejected more often than short-spined *Leucorrhinia* species when attacked by yellow perch. *D. coloradensis* larvae from fish streams had lower mortality in predation trials compared with larvae from fishless sites (Dahl and Peckarsky, 2002). However, this result might be flawed because larvae from fish sites are pre-adapted to the presence of fish. Therefore the effect of caudal filaments might interact with other behavioural or morphological traits that were not measured. In the worst case it could be these other traits that caused the difference in mortality. However, larvae with artificially shortened filaments survived less well than larvae with intact filaments, suggesting that filament length is important for survival from attacking fish. Support for the idea that caudal filaments of aquatic insects provide protection against fish predation has also been found by Otto and Sjöström (1983). Hence, evidence that induced morphological characters are adaptive is starting to be gathered, but more data are certainly needed.

Induced defences of aquatic organisms are often triggered by chemical cues released by the predator or the injured prey (Adler and Harvell, 1990), and many aquatic organisms, including dragonflies and mayflies, can recognize predators through chemical cues (Huryn and Chivers, 1999; Hopper, 2001). In the *L. dubia* experiments, dragonfly larvae received both visible and chemical cues. The induced defence in *D. coloradensis* was induced by chemical cues because fish were not visible in this experiment. Since chemical cues travel further than visual cues in water, we should expect chemical cues to be detected before visible cues, and it might therefore be advantageous to use chemical cues for predator detection.

Both studies estimated character production costs, which are coupled with the production of the defensive trait. These are not true costs of plasticity, since individuals with a fixed response also have these costs. That is, producing a character requires energy, tissue etc. no matter whether it is a fixed response or a flexible response. Nevertheless, production costs are important because they assess the cost of producing a character when it is not necessary. Johansson (2002) measured time to last instar and size at last instar and found no evidence of production costs, since no difference in larvae that had been raised with and without fish was evident in the two traits. Similarly, spine length of *L. dubia* is not correlated with larval size in larvae from natural lakes (Johansson and Samuelsson, 1994). Other variables related to fitness might still have been affected through production costs, and more work is certainly needed to explore the cost issue. *D. coloradensis* females emerging from fish sites were smaller than females emerging from fishless sites, which may suggest production costs or behavioural effects influencing growth and development. In contrast to this field pattern, no such difference in size of females at emergence was found in the experiments where fish cues had been manipulated. Thus, this result suggests that other environmental cues or indirect behavioural traits affected the size of emerging females in the fish sites rather than a production cost. In the absence of any plasticity costs we would expect to find organisms perfectly adapted by plasticity to all environments. We do not find such organisms and therefore plasticity costs should exist, but so far very little evidence of plasticity costs has been amassed and reported (Pigliucci, 2001).

In summary, these two studies on insects show adaptive plasticity in morphology in response to the presence of potential predators. Although our focus is limited to aquatic insects, we would like to point at a third example of predator-induced morphological shift in insects. Weisser *et al.* (1999) showed that the presence of predatory ladybirds increased the proportion of winged morphs in the pea aphid. The morphological shift in the aphids does not provide a direct defence against the predators, but enables the prey to leave patches with high predation risk. As far as we know these three examples of predator-induced morphological shifts are the only three described in insects. Since predator-induced morphological defence is common in other organisms, we see no reason for it to not be common in insects as well, and we suggest that the future will show more examples from the insect world.

3. Predation Risk, Time Constraints and Life History

Organisms differ enormously in life-history traits such as age and size at maturity, mortality rate, and fecundity (Roff, 1992). Since life-history traits are important for fitness and environments are heterogeneous from a spatial and temporal viewpoint, we should expect phenotypic plasticity in life-history traits (Sultan and Spencer, 2002). Many insects have complex life cycles with metamorphosis and where natural variation in hatching and emergence is common. Size at, and time to, metamorphosis influences fitness in insects. Large females are usually more fecund and large males often gain more matings (Thornhill and Alcock, 1983). Similarly, early-metamorphosing males may gain access to high-quality resources and early-metamorphosing females may minimize the pre-reproductive period (Wiklund and Fagerström, 1977; Fagerström and Wiklund, 1982). However, achieving a large size and early emergence should come at the cost of higher predation risk (Rowe and Ludwig, 1991). Therefore, insects are expected to balance the trade-off between growth benefits and mortality risk during the larval stage in an adaptive way.

In seasonal environments the reproductive period is restricted to a certain time and therefore insects are time-constrained with respect to growth, development and time at metamorphosis. Accelerating development in order to have a long mating season comes at the cost of reduced growth opportunities in the larval stage. On the other hand, spending more time in the larval stage for growth comes at the cost of less time available for mating (and potentially more time exposed to larval predators). Hence a trade-off with respect to early metamorphosis and further growth exists. Rowe and Ludwig (1991) modelled this scenario and their model shows that late individuals should trade-off further growth against earlier maturation. In the simplest case, when no significant growth occurs in the adult stage, the model predicts that size at metamorphosis should decline as the season progresses. Such a pattern is indeed common in insects (Vannote and Sweeney, 1980; Falck and Johansson, 2000).

One of the strongest mortality factors during the larval stage is predation risk. Many models have suggested how animals should respond in life history if predation risk is present (Roff, 1992; Abrams and Rowe, 1996). The relevant theory as far as insects are concerned is based on stage-dependent models, and the response to increased predation risk depends on two assumptions. First, whether age and size at maturity is flexible or fixed and, second, whether increased growth effort results in high or low increase in predation risk. The most general result of the theory is that size at metamorphosis will decrease with increased predation risk in the larval stage. Several empirical studies in insects have found such results (Peckarsky et al., 1993; Ball and Baker, 1996; Hechtel and Juliano, 1997). The decrease in size can result from two effects: (i) a direct effect through changes in life-history traits such as an increase in development rate without a balancing increase in growth rate; or (ii) through an indirect effect mediated through behaviour. Foraging behaviour affects intake rate of food and therefore changes in behaviour such as a decreased foraging activity result in slower growth and a smaller size at emergence. Few studies have tried to disentangle the direct and indirect effects, but see Ball and Baker (1996), Johansson and Rowe (1999) and Johansson *et al.* (2001).

Both time constraints and predation risk are ubiquitous in nature and insects are therefore expected to make adaptive trade-offs with respect to the conflicting goal these two factors present. Few studies have considered the joint effect of time constraints and predation risk. Rowe and Ludwig (1991) developed a model exploring these conflicting goals. In their model they assumed a natural variation in body size of individuals due to variation in hatching date, temperature etc. By allowing individuals to follow growth trajectories they focused on the conflict between large size at emergence and early emergence and predicted the optimal time and size of emergence given a certain starting date and size. The outcome of their model depends on growth and mortality rates in the larval and adult stages. Figure 11.4 shows a case derived from their model, where growth occurs in both the larval and the adult stage and reproduction is constrained to a fixed period. If mortality is higher in the adult stage then the model predicts a descending optimal switch curve of

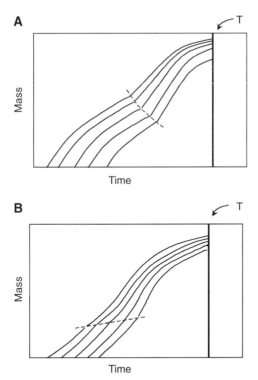

Fig. 11.4. A Optimal size and time at metamorphosis for growth and mortality rates as in Fig. 11.2A in Rowe and Ludwig (1991). Mortality rate is higher in habitat 2, the adult stage. Each line represents the growth trajectory of an individual starting at a certain size and time in habitat 1 and switching to habitat 2 above the dotted line. The dotted line represents the optimal time and size for the switch from the juvenile to the adult stage. T indicates payoff time where reproduction occurs. **B** as in **A**, but with mortality rate reversed so that it is higher in habitat 1. Simplified and modified from Rowe and Ludwig (1991).

time and size for metamorphosis (Fig. 11.4A). The descending switch curve arises because the high predation rate in the adult stage discourages individuals from metamorphosing early. The individuals that are still small late in the season do not have the opportunity to delay shifting because a delay results in a short reproductive period. If mortality is higher in the larval stage, an ascending switch curve is expected (Fig. 11.4B). The reason is that the smallest individuals must maximize growth or they will not make it to the minimum size of reproduction. In contrast the large individuals will shift early in order to avoid predation.

3.1 Test of the model

Johansson et al. (2001) tested the Rowe and Ludwig (1991) model by estimating behaviour, growth, development rate and size at emergence in the damselfly Lestes sponsa. The species has a 1-year life cycle and diapauses over the winter in the egg stage. The eggs hatch in spring and larvae develop rapidly during spring and early summer, after which an adult period follows during the summer. The model was tested by manipulating time constraints and predation risk simultaneously in a laboratory experiment. Eggs from L. sponsa were brought to the laboratory where they were hatched at two different photoperiods. The starting time for the time-constrained larvae was set to 1 June and the starting time for the non-time-constrained larvae to 15 March. Photoperiod was thereafter adjusted weekly until the larvae emerged. Half of the larvae in each time-constraint treatment received a non-lethal threat of a fish predator.

In accordance with theory, larvae that were time-constrained increased their foraging activity (Fig. 11.5). Such increases in foraging behaviour of time-constrained insects have been found in other studies as well (Johansson and Rowe, 1999). The presence of fish predators reduced foraging activity (Fig. 11.5), which is in accordance with many other studies on insects (e.g. Peckarsky et al., 1993; Ball and Baker, 1996). Time-constrained larvae with no predators had the highest foraging activity, while non-time-constrained larvae with predators had the lowest foraging activity. The behavioural responses to time constraints and predation risk were thus additive.

The time-constrained larvae accelerated their development as predicted by theory. Several other studies have found this response in insects subjected to time constraints (Nylin et al., 1989; Blanckenhorn, 1998). This acceleration of development comes at the cost of a reduced size at emergence. The reduction in size occurred independently of any associations with behaviour that mediated changes in growth rate because the increased foraging activity did not result in an increased growth rate (Fig. 11.6). Similar results showing that development rate may change independently of behaviour have been found by Johansson and Rowe (1999) and De Block and Stoks (2003). Hence the change in development rate was a direct effect mediated by physiological processes rather than increased development mediated by behaviour.

Larvae subjected to a predation threat had a lower growth rate, which resulted in a longer development time (Fig. 11.6). The longer development and

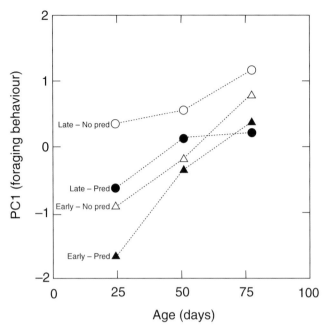

Fig. 11.5. Foraging activity extracted from a principal components analysis against development day of *Lestes sponsa* larvae. Time-constrained larvae (late larvae) with no predator present had the highest foraging activity while non-time-constrained (early) larvae without predators had the lowest foraging activity. Modified from Johansson *et al.* (2001).

slower growth were in accordance with the lower foraging activity in the presence of predators. In contrast to the effect of time constraint, where behaviour did not seem to affect growth and development, a reduction in foraging affected development and growth under predation risk, and hence an indirect effect mediated through behaviour on life history was observed.

Interestingly, mass at emergence did not differ between the predator and non-predator treatments under time constraints but did differ in the absence of time constraints (Fig. 11.6). This indicates that there is a critical size for emergence. Accordingly, the time-constrained larvae did not reduce their foraging activity in the presence of predators as much as the larvae in the absence of predators, which further supports the notion that there might be a critical size at emergence. Hence larvae seem to adjust their behaviour adaptively in response to predators and time constraints.

The model by Rowe and Ludwig (1991) predicted a shift from descending to ascending optimal switch curves from the larval stage to the adult stage as mortality rate shifted from being higher at the adult stage to being higher at the larval stage (Fig. 11.4A,B). Under the assumption that predation risk is higher for dragonflies in the adult stage, the results of Johansson *et al.* (2001) support this prediction. By subjecting larvae to predation in the larval stage we expected the switch curve to shift from a descending to an ascending curve

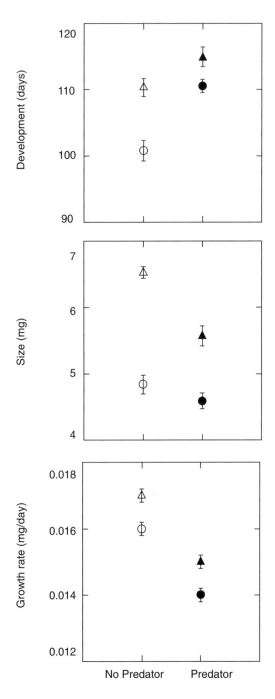

Fig. 11.6. The effect of time constraints and predation risk on life-history variables in *Lestes sponsa* larvae. Triangles denote non-time-constrained larvae and circles denote time-constrained larvae. Filled symbols indicate the presence of predators and unfilled symbols indicate the absence of predators. Modified from Johansson *et al.* (2001).

because mortality rate was increased in the larval stage. The result showed that the slope of the switch curve was less steep in the presence of predators; hence support for the theory was found (Fig. 11.7). One possibility for the absence of a complete switch could be that the increase in predation rate in the larval stage relative to predation rate in the adult stage was not high enough to cause a shift to an ascending switch curve. The model by Rowe and Ludwig (1991) assumes a fixed growth rate. Since changes in growth rate did occur as a consequence of predation risk, some of the response in development might be mediated through changes in growth rate.

Also, for the life-history case, an evaluation of the four prerequisites for the evolution of plasticity (Harvell, 1990) can be made. As discussed for the *Leucorrhinia* system, and also for the *Lestes* system, fish predation pressure varies considerably, both on a spatial and on a temporal scale because *Lestes sponsa* occurs both in temporary ponds and in permanent fish lakes (R. Stoks, personal observation). Similarly, *L. sponsa* larvae probably use the same combination of visual and chemical cues to detect fish predation risk (Johansson *et al.*, 2001). With regard to the observed life-history plasticity, the reduced size at emergence is a clear cost, as it will directly reduce survival until maturation in the adult stage (Stoks, 1999b; De Block and Stoks, 2005). The effectiveness of the life-history response to avoid fish predation is indirect. The response is a consequence of an adaptive lowering of foraging activity in the

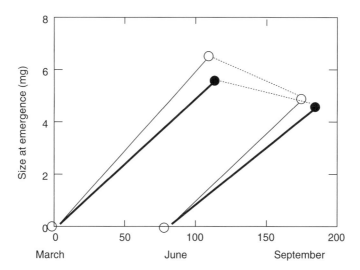

Fig. 11.7. The effect of time constraints and predation risk on size and time at metamorphosis of *Lestes sponsa* larvae from a laboratory experiment. Non-time-constrained larvae were started on 15 March and time-constrained larvae were started on 1 June. At the start of the experiment weight was assumed to be 0 mg. Filled dots and solid lines represent larvae raised with predators and unfilled dots and thin lines represent larvae raised without predators. Note the decline in size of time-constrained larvae, which results in a descending switch curve (dotted line). The switch curve for larvae with predators is less steep because larvae must reach a certain size before metamorphosis.

presence of fish, and a low foraging activity has been shown to be adaptive in terms of survival in the presence of fish (Stoks and Johansson, 2000).

The results of several studies, including the Johansson et al. (2001) study, suggest that there is no direct (physiological) effect of predators on development, although theory predicts such a response. However, negative physiological effects on growth rates induced by predators have been shown recently in damselfly larvae (McPeek et al., 2001; Stoks, 2001; Stoks and McPeek, 2003b). More studies are certainly needed to disentangle the direct effects from the indirect behavioural effects under predation risk.

4. General Conclusions and Future Directions

The case studies presented here clearly went a long way towards demonstrating that the observed plasticity in response to predators is adaptive and likely to evolve. The adaptiveness of the predator-induced plasticity in life history and morphology has been demonstrated using two very different methods (Newman, 1992; Gotthard and Nylin, 1995). For life history, the response agreed with the predictions of the mathematical optimality model of Rowe and Ludwig (1991). For morphology, experiments suggested that animals with induced spines had higher fitness in the predator environment compared with non-plastic or less plastic animals. However, our present knowledge of our selected model system is not complete. Showing the adaptiveness of the life-history response to predators and time stress will be stronger when it also incorporates fitness effects in a longitudinal study, i.e. including the adult stage (see e.g. Altwegg, 2002; Altwegg and Reyer, 2003, for a set of studies on anurans). Moreover, for several of the model systems we have no definite answer on the validity of all four prerequisites for the evolution of plasticity. Needless to say, there is a field of opportunity not only for deepening our understanding of plasticity in these model systems but also in exploring other systems in aquatic insects. One of the greatest challenges at the moment in plasticity issues is the cost of plasticity, and insects may provide a good system for exploring this issue since we have extensive knowledge about the behaviour, ecology and life history of many insect systems. Moreover, they are relatively easy to manipulate and their short generation time may be crucial in the necessary longitudinal studies. We now go on to suggest some directions we think may be promising to explore in order to broaden our understanding of the current model systems, which at the same time may be useful for the selection of new model systems.

4.1 Long-term experiments

We would like to emphasize the importance of doing long-term experiments, ideally bridging the life-history transition, when assessing the costs of plasticity. Costs and benefits may be small and hard to detect and compare when measured on a short timescale. However, such relatively small changes in daily growth rate or survival associated with plastic responses to predators may have considerable fitness consequences when summed across the entire larval lifespan. For example, a plastic response to a predator that increases daily survival from 0.95 to 0.96 will almost double survivorship after a larval period of 60 days: from 0.046 (= 0.95×60) to 0.086 (= 0.96×60) (B. Anholt, personal communication). However, even including the entire larval period may not be enough to detect costs. Despite the fact that most aquatic insects, including dragonfly larvae, have a complex life cycle, studies of plasticity to predators have largely been limited to the larval stage. Including the adult stage in such studies may broaden our understanding of the adaptiveness and costs of predator-induced plasticity. As shown, for example, by the model of Rowe and Ludwig (1991), knowledge of the relative mortality rates on larval and adult stages may be crucial in predicting the response of prey in the larval stage. More importantly, costs may only become detectable in the adult stage. For example, a recent study on predator-induced defences in larval tadpoles showed that costs could only be found in the adult stage (Benard and Fordyce, 2003; but see Van Buskirk and Saxer, 2001). This begs the more general question to what extent larval stresses (such as predation risk and low food level) are captured by our traditional life-history end points (age and size at emergence) included in optimality models and empirical studies. A disturbing finding is that damselfly larvae reared under time and nutritional stress showed clear responses in age and size at emergence, but survival to emergence was best explained by the larval stress conditions and not by age and size at emergence (De Block and Stoks, 2005). This may explain the existence of 'hidden costs' of plasticity that only become detectable when they translate into the variables (like survival) that we typically measure. In most cases such 'delayed' costs were already present in the larval stage but were not strongly related to traditional fitness proxies like age and size at emergence. This cries out for studies trying to find the direct targets of natural and sexual selection, because maybe our traditional assumed targets are simply indirectly selected for.

4.2 Connection between larval and adult stages

Another complicating factor with prey which have a complex life cycle is that larval decisions may be offset by compensatory mechanisms in the adult stage or that larval decisions themselves are not free to evolve because of constraints in the adult stage. Using a simple model, McPeek and Peckarsky (1998) showed that predator-induced life-history plasticity in the larval stage is much more important for the population dynamics of mayflies than for damselflies. One important difference between these taxa is the extent to which the adult stage can compensate for costs paid in the larval stage. In aquatic insects such as mayflies with a non-feeding adult stage, fecundity is determined in the larval stage and any compensation in the adult stage seems unlikely. In damselflies, however, fecundity is largely independent of the larval life-history characteristics and is mainly determined by adult feeding (Richardson and Baker, 1997). This does not mean that size and mass at emergence are not relevant for adult fitness. A minimal mass is needed in order to emerge and a higher mass at emergence is positively correlated with survival and the probability of sexual maturation (Stoks, 1999b). The extent to which larval and adult stages of aquatic insects are also genetically decoupled will prove crucial in evaluating the possibility of independent evolution of plasticity in both stages (Moran, 1994). Preliminary evidence suggests that genetic coupling within the larval stage is much more important than genetic coupling between larval and adult stages in the damselfly *Ischnura elegans* (R. Stoks *et al.*, unpublished results).

4.3 Costs of plasticity

None of the reviewed plasticity studies in insects have tried to estimate costs of plasticity, and studies on other taxa have failed to show convincing evidence for costs of plasticity (Pigliucci, 2001). Actual plasticity costs may indeed be very small and hard to demonstrate. This demands more precise measurements. The common approach to detecting costs of plasticity in these studies has been to use a statistical approach based on a model by Van Tienderen (1991). A more effective approach is probably an actual understanding of the developmental and physiological mechanisms underlying phenotypic plasticity. This approach has been successfully used by, for example, Krebs and Feder (1997). They compared lines of *Drosophila melanogaster* with and without heat shock proteins (Hsp 70) turned on. Their results showed that lines with Hsp 70 turned on survived better under heat stress. In contrast, flies with Hsp 70 turned on produced fewer offspring compared with females with the Hsp turned off in the absence of heat stress, hence indicating a cost of plasticity. Also, instead of the traditional life-history end points in measuring costs, other more refined physiological variables may be scored.

4.4 Gender differences

Because the adaptiveness of plasticity in general depends on the costs and benefits associated with it, we expect that plasticity may be sex-specific. Sex differences have largely been ignored so far, and have only recently been explicitly considered in optimality models (Crowley, 2000; Crowley and Johansson, 2002). In dragonflies, males seem to optimize their date of emergence while females seem to optimize mass at emergence (De Block and Stoks, 2003). Similar differences have been found in other aquatic insects including chironomids (Ball and Baker, 1996) and may affect the balance between the costs and benefits of responses to predators and ultimately shape the predator-induced plasticity. Furthermore, also in the adult stage, optimal traits may differ between the sexes, with maximizing lifespan being especially important for females (Rolff, 2002). No sex differences in behavioural responses to predators have been found so far in larval dragonflies (e.g. Johansson *et al.*, 2001).

4.5 Graded responses

Very little is known about the specificity of the predator-induced plasticity in behaviour, life history and morphology in dragonfly larvae or other aquatic insects. Because syntopic predators may differ in predation risk, plasticity may be predator-specific as shown, for example, in mayfly larvae (Kohler and McPeek, 1989; McIntosh and Peckarsky, 1999). Moreover, there may be graded responses in function of the actual predation risk imposed by the same predator. Tadpoles, for example, seem to have continuous dosage response curves for behavioural and morphological antipredator traits (Van Buskirk and Arioli, 2002). Such sensitivities are important for testing the adaptiveness of plasticity because a difference caused by dosage provides strong evidence for adaptiveness. Unfortunately, even for the well-studied behavioural plasticity responses to predators, these responses are largely unknown in insects. A recent optimality model of Lima and Bednekoff (1999) predicts the optimal allocation of foraging behaviour in situations that differ in predation risk. However, a critical assumption is that prey organisms are able to discriminate between situations that differ in their grade of predation risk (e.g. one or two fish predators), and the temporal variation in the time spent in these situations. Evidence is being gathered that some insects are able to sense differences in predation threat. Drift rate of *Baetis* larvae was affected by fish odour concentrations (McIntosh *et al.*, 1999). Similarly, a recent study on the antipredator behaviour and life history of the damselfly *Ischnura elegans* reared under different combinations of predation risk (none, one or two fish) and frequency at high risk (never, 50% or 100% of the time) suggest they are able to perceive such differences to some extent (S. Slos and R. Stoks, unpublished). Obviously, more studies are needed that establish a range of predation risk and test the dosage response in antipredator traits. An elegant approach may be skipping the predator exposure part entirely and directly manipulating the prey animals with a linear gradient of chemicals known to play an intermediate role in the expression of antipredator traits. Recent progress on other aquatic taxa (e.g. Barry, 2002) suggests that this kind of phenotypic engineering may not be so futuristic after all.

4.6 The genetic approach

Too often, researchers ignore the fact that prey individuals have a particular genetic background, which may be crucial to explain what at first sight appear to be erratic findings. Little effort has been done to look at variation among genetic categories (families, populations, species etc.) in predator-induced phenotypic plasticity. Such a comparative approach may prove rewarding in understanding the evolution of plasticity and is an important step in identifying plasticity as an adaptation (Gotthard and Nylin, 1995). Expression of flexible antipredator traits may differ among families and positive genetic correlations of, for example, foraging activity between situations with and without a predator may limit the independent evolution of such traits in both situations, hence the evolution of phenotypic plasticity to predators (Sih *et al.*, 2003). Recent studies show such strong positive coupling of foraging activity between situations with and without fish in the damselflies *Coenagrion hastulatum* (Brodin and Johansson, 2004) and *L. sponsa* (R. Stoks and M. De Block, unpublished data). This may explain among other things the persistence of prey with risky foraging activity in fish populations. Such behaviours might at a first sight seem to be maladaptive but can be understood in the context of genetic constraints, i.e. limiting genetic correlations (Sih *et al.*, 2003). Hence, antipredator behaviours going in opposite directions as predicted by optimality theory might not always have to be an effect of low statistical power or experimental noise. Plasticity may also differ among populations, as suggested by the results of McIntosh and Townsend (1994), McIntosh and Peckarsky (1996) and Dahl and Peckarsky (2002). In general, plasticity is expected to be larger in more heterogeneous environments (Sultan and Spencer, 2002), which in our case means populations with temporally variable predation pressures (Harvell, 1990). We have, however, no data to compare the degree of plasticity among populations that differ in such variability. For example, populations of *L. dubia* with and without fish are largely intermixed in the landscape in Sweden, and gene flow is probably high, so that both kinds of populations are expected to show the discussed morphological plasticity. However, in the southern parts of its range, *L. dubia* almost exclusively occupies fishless water bodies and plasticity may be much smaller. Such patterns may explain striking differences where some studies show strong predator-induced plasticity and others not, within the same species. Also, very few among-species comparisons exist. If available for congeneric species with known phylogeny this may provide an opportunity to learn about the evolution of antipredator traits. For example, phylogenetic reconstructions showed that *Enallagma* damselfly species that shifted to fishless waterbodies lost their behavioural plasticity to fish predators (Stoks *et al.*, 2003). Moreover, as we have seen for genetic correlations, phylogenetic correlations of traits across predator environments may also limit the adaptive evolution of antipredator mechanisms and explain the absence of plasticity where it would be adaptive (e.g. Richardson, 2001, for an example in anurans). Knowledge of the ancestral history may be crucial in understanding cases where no plasticity is seen although it would be adaptive. Caudill and Peckarsky (2003) report the absence of appropriate behavioural and developmental responses to trout by the mayfly *Callibaetis ferrugineus* living in ponds with trout. The most plausible reason for this pattern was phylogenetic inertia whereby the adaptations of the species to its ancestral temporary fishless habitat persist, leading to maladaptive responses to trout. Adaptations to trout may not develop because coexistence with trout is too recent or because of ongoing dispersal from fishless ponds to fish ponds that act as sinks for the species. So, although it may appear to be paradoxical, we feel that the study of adaptive predator-induced plasticity may benefit enormously from the study of situations where it is lacking.

5. Conclusion

It can be concluded that we have basic information on adaptive plasticity in insects and that this information in many cases conforms to theory. Nevertheless, future exciting areas in the study of adaptive plasticity in response to predators will probably include genomic and molecular techniques. This approach has the strength that it focuses directly on the mechanism responsible for the traits and allows the controlled manipulation of plastic responses (phenotypic engineering; Feder *et al.*, 2000). However, at present these methods are only available for certain model organisms and in many cases are limited to laboratory conditions. An alternative approach is to use an among-population comparison, where populations or species known to differ in predator environment are compared. This could be done on an ecological time-scale or on an evolutionary time-scale using a phylogenetic approach. A third approach that might be fruitful is to use experimental evolution by using similar approaches as those taken to study, for example, temperature acclimation in *E. coli* (Leroi *et al.*, 1994). This approach requires insects with rapid life cycles, but a recent study of Juliano and Gravel (2002) on mosquitoes showed that it is possible. We are entering an exciting era where the integration of information from such hitherto widely dispersed fields as modelling, evolutionary ecology, phylogeography, physiology and molecular biology is becoming possible. We now have plenty of studies showing plasticity of prey to predators, and it is time to switch our limited time and energy towards the daunting task of combining all these fields if we really want to understand adaptive plasticity to predators.

Acknowledgements

Thanks to Brad Anholt, Mark McPeek and an anonymous referee for valuable comments on this manuscript. This work was partly supported by a grant from the Swedish Natural Research Council to F.J., and by a postdoctoral fellowship and research grants of the Fund for Scientific Research Flanders (FWO-Flanders) to R.S.

References

Abrams, P.A. and Rowe, L. (1996) The effects of predation on the age and size of maturity of prey. *Evolution* 50, 1052–1061.

Adler, F.R. and Harvell, C.D. (1990) Inducible defences, phenotypic variability and biotic environments. *Trends in Ecology and Evolution* 5, 407–410.

Altwegg, R. (2002) Predator-induced life-history plasticity under time constraints in pool frogs. *Ecology* 83, 2542–2551.

Altwegg, R. and Reyer, H.U. (2003) Patterns of natural selection on size at metamorphosis in water frogs. *Evolution* 57, 872–882.

Arnqvist, G. and Johansson, F. (1998) Ontogenetic reaction norms of predator induced defensive morphology in dragonfly larvae. *Ecology* 79, 1847–1858.

Askew, R.K. (1988) *The Dragonflies of Europe*. Harley Books, Colchester, UK.

Baker, R.L., Elkin, C.M. and Brennan, H.A. (1999) Aggressive interactions and risk of fish predation for larval damselflies. *Journal of Insect Behavior* 12, 213–223.

Ball, S.L. and Baker, R.L. (1996) Predator-induced life history changes: antipredator behavior costs or facultative life history shifts? *Ecology* 77, 1116–1124.

Barry, M.J. (2002) Progress toward understanding the neurophysiological basis of predator-induced morphology in *Daphnia pulex*. *Physiological and Biochemical Zoology* 75, 179–186.

Benard, M.F. and Fordyce, J.A. (2003) Are induced defenses costly? Consequences of predator-induced defenses in western toads, *Bufo boreas*. *Ecology* 84, 68–78.

Blanckenhorn, W.U. (1998) Adaptive phenotypic plasticity in growth, development, and body size in the yellow dung fly. *Evolution* 52, 1394–1407.

Brodin, T. and Johansson, F. (2002) Effects of predator-induced thinning and activity changes on life history in a damselfly. *Oecologia* 132, 316–322.

Brodin, T. and Johansson, F. (2004) Conflicting selection pressures on the growth/predation-risk trade-off in a damselfly. *Ecology* 85, 2927–2932.

Caudill, C.C. (2003) Empirical evidence for nonselective recruitment and a source–sink dynamic in a mayfly metapopulation. *Ecology* 84, 2119–2132.

Caudill, C.C. and Peckarsky, B.L. (2003) Lack of appropriate behavioral or developmental responses by mayfly larvae to trout predators. *Ecology* 84, 2133–2144.

Claus-Walker, D.B., Crowley, P.H. and Johansson, F. (1997) Fish predation, cannibalism, and larval development in the dragonfly *Epitheca cynosura*. *Canadian Journal of Zoology* 75, 687–696.

Crowley, P.H. (2000) Sexual dimorphism with female demographic dominance: age, size, and sex ratio at maturation. *Ecology* 81, 2592–2605.

Crowley, P.H. and Johansson, F. (2002) Sexual dimorphism in Odonata: age, size, and sex ratio at emergence. *Oikos* 96, 364–378.

Dahl, J. and Peckarsky, B.L. (2002) Induced morphological defenses in the wild: predator effects on a mayfly, *Drunella coloradensis*. *Ecology* 1620–1634.

De Block, M. and Stoks, R. (2003) Adaptive sex-specific life history plasticity to temperature and photoperiod in a damselfly. *Journal of Evolutionary Biology* 16, 986–995.

De Block, M. and Stoks, R. (2005) Fitness effects from egg to reproduction: bridging the life history transition. *Ecology* 86, 185–197.

Dixon, S.M. and Baker, R.L. (1988) Effects of predation risk, behavioral response to fish and cost of reduced feeding in larvae *Ischnura verticalis* (Coenagrionidae, Odonata). *Oecologia* 76, 200–205.

Fagerström, T. and Wiklund, C. (1982) Why do males emerge before females? Protandry as a mating strategy in male and female butterflies. *Oecologia* 52, 164–166.

Falck, J. and Johansson, F. (2000) Patterns in size, sex ratio and time at emergence in a south Swedish population of *Sympetrum sanguineum* (Odonata). *Aquatic Insects* 22, 311–317.

Feder, M.E., Bennett, A.F. and Huey, R.B. (2000) Evolutionary physiology. *Annual Review of Ecology and Systematics* 31, 315–341.

Gotthard, K. and Nylin, S. (1995) Adaptive plasticity and plasticity as an adaptation: a selective review of plasticity in animal morphology and life history. *Oikos* 74, 3–17.

Harvell, C.D. (1990) The ecology and evolution of inducable defenses. *Quarterly Review of Biology* 65, 323–340.

Hechtel, L.J. and Juliano, S.A. (1997) Effects of a predator on prey metamorphosis: plastic responses by prey or selective mortality. *Ecology* 78, 838–851.

Hopper, K.R. (2001) Flexible antipredator behaviour in a dragonfly species that coexists with different predator types. *Oikos* 93, 470–476.

Hovmöller, R. and Johansson, F. (2004) A phylogenetic perspective on larval spine morphology in *Leucorrhinia* (Odonata: Libellulidae) based on ITS1, 5.8S and ITS2 rDNA sequences. *Molecular Phylogenetics and Evolution* 30, 653–662.

Huryn, A.D. and Chivers, D.P. (1999) Contrasting behavioral responses by detritivorous and predatory mayflies to chemicals released by injured conspecifics and their predators. *Journal of Chemical Ecology* 25, 2729–2740.

Jeffries, M. (1990) Interspecific differences in movement and hunting success in damselfly larvae (Zygoptera, Insecta): responses to prey availability and predation threat. *Freshwater Biology* 23, 191–196.

Johansson, F. (1993) Effects of prey type, prey density and predator presence on behavior and predation risk in a larval damselfly. *Oikos* 68, 481–489.

Johansson, F. (2002) Reaction norms and production costs of predator-induced morphological defence in a larval dragonfly (*Leucorrhinia dubia*: Odonata). *Canadian Journal of Zoology* 80, 944–950.

Johansson, F. and Rowe, L. (1999) Life history and behavioral responses to time constraints in a damselfly. *Ecology* 80, 1242–1252.

Johansson, F. and Samuelsson, L. (1994) Fish-induced variation in abdominal spine-length of *Leucorrhinia dubia* (Odonata) larvae? *Oecologia* 100, 74–79.

Johansson, F. and Wahlström, E. (2002) Induced morphological defence: evidence from whole-lake manipulation experiments. *Canadian Journal of Zoology* 80, 199–206.

Johansson, F., Stoks, R., Rowe, L. and De Block, M. (2001) Life history plasticity in a damselfly: effects of combined time and biotic constraints. *Ecology* 82, 1857–1869.

Juliano, S.A. and Gravel, M.E. (2002) Predation and the evolution of prey behavior: an experiment with tree hole mosquitoes. *Behavioral Ecology* 13, 301–311.

Kohler, S. and McPeek, M.A. (1989) Predation risk and the foraging behaviour of competing stream insects. *Ecology* 70, 1811–1825.

Koperski, P. (1997) Changes in feeding behaviour of the larvae of the damselfly *Enallagma cyathigerum* in response to stimuli from predators. *Ecological Entomology* 22, 167–175.

Krebs, R.A. and Feder, M.E. (1997) Natural variation in the expression of the heat-shock protein Hsp70 in a natural population of *Drosophila melanogaster* and its correlation with tolerance of ecologically relevant thermal stress. *Evolution* 51, 173–179.

Leroi, A.M., Bennett, A.F. and Lenski, R.E. (1994) Temperature acclimation and competitive fitness: an experimental test of the beneficial acclimation assumption. *Proceedings of the National Academy of Sciences USA* 91, 1917–1921.

Lima, S.L. (1998) Nonlethal effects in the ecology of predator-prey interactions – what are the ecological effects of anti-predator decision-making? *BioScience* 48, 25–34.

Lima, S.L. and Bednekoff, P.A. (1999) Temporal variation in danger drives antipredator behavior: the predation risk allocation hypothesis. *American Naturalist* 153, 649–659.

Lima, S.L. and Dill, L.M. (1990) Behavioral decisions made under the risk of predation: a review and prospectus. *Canadian Journal of Zoology* 68, 619–640.

Ludwig, D. and Rowe, L. (1990) Life history strategies for energy gain and predator avoidance under time constraints. *American Naturalist* 135, 686–707.

McIntosh, A.R. and Peckarsky, B.L. (1996) Differential behavioural responses of mayflies from streams with and without fish to trout odour. *Freshwater Biology* 35, 141–148.

McIntosh, A.R. and Peckarsky, B.L. (1999) Criteria determining behavioural responses to multiple predators by a stream mayfly. *Oikos* 85, 554–564.

McIntosh, A.R. and Townsend, C.R. (1994) Interpopulation variation in mayfly anti-predator tactics: differential effects of contrasting predatory fish. *Ecology* 75, 2078–2090.

McIntosh, A.R., Peckarsky, B.L. and Taylor, B.W. (1999) Rapid size-specific changes in the drift of *Baetis bicaudatus* (Ephemeroptera) caused by alterations in fish odour concentration. *Oecologia* 118, 256–264.

McPeek, M.A. (1989) Differential dispersal tendencies among *Enallagma* damselflies (Odonata) inhibiting different habitats. *Oikos* 56, 187–195.

McPeek, M.A. (1990a) Behavioral differences between *Enallagma* species (Odonata) influencing differential vulnerability to predators. *Ecology* 71, 1714–1726.

McPeek, M.A. (1990b) Determination of species composition in the *Enallagma* damselfly assemblages of permanent lakes. *Ecology* 71, 83–98.

McPeek, M.A. (1998) The consequences of changing the top predator in a food web: a comparative experimental approach. *Ecological Monographs* 68, 1–23.

McPeek, M.A. and Peckarsky, B.L. (1998) Life histories and the strengths of species interactions: combining mortality, growth, and fecundity effects. *Ecology* 79, 867–879.

McPeek, M.A., Grace, M. and Richardson, J.M.L. (2001) Physiological and behavioral responses to predators shape the growth/predation risk trade-off in damselflies. *Ecology* 82, 1535–1545.

Mikolajewski, D.J. and Johansson, F. (2004) Morphological and behavioral defences in dragonfly larvae: trait compensation and cospecialization. *Behavioral Ecology* 15, 614–620.

Moran, N.A. (1994) Adaptation and constraints in the complex life cycles of animals. *Annual Reviw of Ecology and Systematics* 25, 573–600.

Needham, J.G., Westfall M.J., Jr and May, M.L. (2000) *Dragonflies of North America.* Scientific Publishers, Gainesville, Florida.

Newman, R.A. (1992) Adaptive plasticity in amphibian metamorphosis. *Bioscience* 42, 671–678.

Nylin, S., Wickman, P.-O. and Wiklund, C. (1989) Seasonal plasticity in growth and development of the speckled wood butterfly, *Pararge aegeria* (Satyrinae). *Biological Journal of the Linnean Society* 38, 155–171.

Otto, C. and Sjöström, P. (1983) Cerci as antipredatory attributes in stonefly nymphs. *Oikos* 41, 200–204.

Peckarsky, B.L. (1996) Alternative predator avoidance syndromes of stream-dwelling mayfly larvae. *Ecology* 77, 1888–1905.

Peckarsky, B.L. and McIntosh, A.R. (1998) Fitness and community consequences of avoiding multiple predators. *Oecologia* 113, 565–576.

Peckarsky, B.L., Cowan, C.A., Penton, M.A. and Anderson, C. (1993) Sublethal consequences of stream dwelling predatory stoneflies on mayfly growth and fecundity. *Ecology* 74, 1836–1846.

Peckarsky, B.L., Taylor, B.W., McIntosh, A.R., McPeek, M.A. and Lytle, D.A. (2001) Variation in mayfly size at metamorphosis as a developmental response to risk of predation. *Ecology* 82, 740–757.

Peckarsky, B.L., McIntosh, A.R., Taylor, B.W. and Dahl, J. (2002) Predator chemicals induce changes in mayfly life history traits: a whole-stream manipulation. *Ecology* 83, 612–618.

Pierce, C.L., Crowley, P.H. and Johnson, D.M. (1985) Behavior and ecological interactions of larval Odonata. *Ecology* 66, 1504–1512.

Pigliucci, M. (2001) *Phenotypic Plasticity: Beyond Nature and Nurture.* Johns Hopkins University Press, Baltimore, Maryland.

Richardson, J.M.L. (2001) A comparative study of activity levels in larval anurans and response to the presence of different predators. *Behavioral Ecology* 12, 51–58.

Richardson, J.M.L. and Baker, R.L. (1997) Effect of body size and feeding on fecundity in the damselfly *Ischnura verticalis* (Odonata: Coenagrionidae). *Oikos* 79, 477–483.

Roff, D.A. (1992) *The Evolution of Life Histories: Theory and Analysis*. Chapman and Hall, New York.

Rolff, J. (2002) Bateman's principle and immunity. *Proceedings of the Royal Society of London, Series B* 269, 867–872.

Rowe, L. and Ludwig, D. (1991) Size and timing of metamorphosis in complex life cycles: time constraints and variation. *Ecology* 72, 413–427.

Ryazanova, G.I. and Mazokhinporshnyakov, G.A. (1993) Effect of the presence of fish on spatial distribution of dragonfly larvae *Calopteryx splendens* (Odonata). *Zoologichesky Zhurnal* 72, 68–75.

Schaffner, A.K. and Anholt, B.R. (1998) Influence of predator presence and prey density on behavior and growth of damselfly larvae (*Ischnura elegans*) (Odonata: Zygoptera). *Journal of Insect Behavior* 11, 793–809.

Seegrist, D.W. and Gard, R. (1972) Effects of floods on trout in Saghen Creek, California. *Transactions of the American Fisheries Society* 101, 478–482.

Sih, A. (1980) Optimal behavior: can foragers balance two conflicting demands? *Science* 210, 1041–1043.

Sih, A., Kats, L.B. and Maurer, E.F. (2003) Behavioural correlations across situations and the evolution of antipredator behaviour in a sunfish–salamander system. *Animal Behaviour* 65, 29–44.

Steiner, C., Siegert, B., Schulz, S. and Suhling, F. (2000) Habitat selection in the larvae of two species of Zygoptera (Odonata): biotic interactions and abiotic limitation. *Hydrobiologia* 427, 167–176.

Stoks, R. (1998) Effect of lamellae autotomy on survival and foraging success of the damselfly *Lestes sponsa* (Odonata: Lestidae). *Oecologia* 117, 443–448.

Stoks, R. (1999a) Autotomy shapes the trade-off between seeking cover and foraging in larval damselflies. *Behavioral Ecology and Sociobiology* 47, 70–75.

Stoks, R. (1999b) Natural and sexual selection in the damselfly *Lestes sponsa*. PhD thesis, University of Antwerp, Belgium.

Stoks, R. (2001) Food stress and predator-induced stress shape developmental performance in a damselfly. *Oecologia* 127, 222–229.

Stoks, R. and Johansson, F. (2000) Trading off mortality risk against foraging effort in damselflies that differ in life cycle length. *Oikos* 91, 559–567.

Stoks, R. and McPeek, M.A. (2003a) Predators and life histories shape *Lestes* damselfly assemblages along a freshwater habitat gradient. *Ecology* 84, 1576–1587.

Stoks, R. and McPeek, M.A. (2003b) Antipredator behavior and physiology determine *Lestes* species turnover along the pond-permanence gradient. *Ecology* 84, 3327–3338.

Stoks, R., De Block, M., Van Gossum, H., Valck, F., Lauwers, K., Verhagen, R., Matthysen, E. and De Bruyn, L. (1999a) Lethal and sublethal costs of autotomy and predator presence in damselfly larvae. *Oecologia* 120, 87–91.

Stoks, R., De Block, M., Van Gossum, H. and De Bruyn, L. (1999b) Phenotypic shifts caused by predation: selection or life-history shifts? *Evolutionary Ecology* 13, 115–129.

Stoks, R., McPeek, M.A. and Mitchell, J.L. (2003) Evolution of prey behavior in response to changes in predation regime: damselflies in fish and dragonfly lakes. *Evolution* 57, 574–585.

Suhling, F. and Lepkojus, S. (2001) Differences in growth and behaviour influence asymmetric predation among early-instar dragonfly larvae. *Canadian Journal of Zoology* 79, 854–860.

Sultan, S.E. and Spencer, H.G. (2002) Metapopulation structure favors plasticity over local adaptation. *American Naturalist* 160, 271–283.

Thornhill, R. and Alcock, J. (1983) *The Evolution of Insect Mating Systems*. Harvard University Press, Cambridge, Massachusetts.

Tollrian, R. and Harvell, C.D. (1999) *The Ecology and Evolution of Inducible Defenses*. Princeton University Press, Princeton, New Jersey.

Tonn, W.M. and Magnuson, J.J. (1982) Patterns in the species composition and richness of fish assemblages in northern Wisconsin lakes. *Ecology* 63, 1149–1166.

Van Buskirk, J. and Arioli, M. (2002) Dosage response of an induced defense: how sensitive are tadpoles to predation risk? *Ecology* 83, 1580–1585.

Van Buskirk, J. and Saxer, G. (2001) Delayed costs of an induced defense in tadpoles? Morphology, hopping, and development rate at metamorphosis. *Evolution* 55, 821–829.

Van Tienderen, P.H. (1991) Evolution of generalists and specialists in spatially heterogeneous environments. *Evolution* 45, 1317–1331.

Vannote, R.L. and Sweeney, B.W. (1980) Geographical analysis of thermal equlibria: a conceptual model for evaluating the effect of natural and modified thermal regimes on aquatic insect communities. *American Naturalist* 115, 677–695.

Weisser, W.W., Braendle, C. and Minoretti, N. (1999) Predator-induced morphological shift in the pea aphid. *Proceedings of the Royal Society of London, Series B* 266, 1175–1181.

Westman, A., Johansson, F. and Nilsson, A.N. (2000) The phylogeny of the genus *Leucorrhinia* and the evolution of larval spines (Anisoptera: Libellulidae). *Odonatologica* 29, 129–136.

Wiklund, C. and Fagerström, T. (1977) Why do males emerge before females? A hypothesis to explain incidence of protandry in butterflies. *Oecologia* 31, 153–158.

Wiseman, S.W., Cooper, S.D. and Dudley, T.L. (1993) The effects of trout on epibenthic odonate naiads in stream pools. *Freshwater Biology* 30, 133–145.

12 The Peppered Moth: Decline of a Darwinian Disciple

MICHAEL E.N. MAJERUS

Department of Genetics, Downing Street, Cambridge, UK

1. Introduction

1.1 The rise of the melanic moth

The rise of the black, *carbonaria* form of the peppered moth, *Biston betularia*, in response to changes in the environment caused by the industrial revolution in the UK, is one of the best-known examples of evolution in action. The reasons for the prominence of this example are three-fold.

1. The rise was spectacular, occurred in the recent and well-documented past, and was timely: the first record of an individual of *carbonaria* being published by Edleston in 1864, just 5 years after the *Origin of Species* (Darwin, 1859).
2. The difference in the forms was obvious. While many other examples of natural selection have been documented since the peppered moth story became widely known, including the evolution of beak size in Darwin's finches and numerous cases of antibiotic and pesticide resistance, these cases do not have the visual impact of the peppered moth.
3. The major mechanism through which *carbonaria* rose is easy to both relate and understand.

The story, in brief, is this. The non-melanic peppered moth is a white moth, liberally speckled with black scales (Fig. 12.1). In 1848, a black form, f. *carbonaria* (Fig. 12.2), was recorded in Manchester (Edleston, 1864). By 1864, the majority of male peppered moths that assembled to virgin females were *carbonaria* (Edleston, 1864), and by 1895, 98% of the Mancunian population were black. The *carbonaria* form spread to many other parts of the UK, reaching high frequencies in industrial centres and regions downwind (Kettlewell, 1958, 1973).

© Royal Entomological Society 2005. *Insect Evolutionary Ecology*
(eds M.D.E. Fellowes, G.J. Holloway and J. Rolff)

Fig. 12.1. The non-melanic form of the peppered moth. Note, the moth was found in the wild in this position.

Fig. 12.2. The *carbonaria* form of the peppered moth. Note, the moth was found in the wild in this position.

In 1896, the lepidopterist J.W. Tutt hypothesized that the increase in *carbonaria* was the result of differential bird predation in polluted regions. Bernard Kettlewell obtained evidence in support of this hypothesis in the 1950s, with his predation experiments in polluted and unpolluted woodlands (Kettlewell, 1955a, 1956). The results of this work showed that in polluted woodland, the pale nominate form, f. *betularia*, was more heavily predated than was *carbonaria*, the reverse being the case in the unpolluted woodland. It was the reciprocal nature of Kettlewell's data in the two woodlands, allied to mark–release–recapture work in the two woods, and extensive survey work showing a strong positive correlation between *carbonaria* frequency and industrial pollutants (Kettlewell, 1958), that made the case so persuasive. It became the classical example of Darwinian evolution in action (Wright, 1978; Majerus, 2002).

Over the next 40 years many other studies were carried out on the peppered moth in the UK, across Europe, in the USA and in Japan. These unearthed many other details of the peppered moth's biology, and, while some of these have been at odds with elements of the abridged peppered moth story related in numerous biology textbooks, none seriously undermined the veracity of Tutt's hypothesis, or Kettlewell's evidence. Perhaps the zenith of the peppered moth's popularity as an example of Darwinian evolution came in 1996, when, reporting work carried out in England and the USA showing that similar changes in melanic frequencies had occurred on both sides of the Atlantic (Grant *et al.*, 1996), *The New York Times* depicted the peppered moth on the front page of its science section (Yoon, 1996).

1.2 The peppered moth in decline

Yet, since Kettlewell's experiments, the black peppered moth has suffered two declines.

- First, following the enactment of anti-pollution legislation during the 1950s and subsequently, *carbonaria* frequency has declined dramatically in the UK (Clarke *et al.*, 1985; Brakefield and Lees, 1987; Cook *et al.*, 1990; Mani and Majerus, 1993; West, 1994; Grant *et al.*, 1995) and elsewhere (Brakefield, 1990; Grant *et al.*, 1996, 1998).
- Second, the reputation of the peppered moth as an example of Darwinian evolution in action, has suffered a severe decline. The cause of this decline can be sourced to the publication of my book on melanism (Majerus, 1998), or more accurately, a review of it in *Nature* by Professor Jerry Coyne (1998).

It is the second of these declines that this chapter addresses. I first relate how the decline in the peppered moth's reputation came about. I briefly discuss whether criticisms of the story are justified and consider the peppered moth's status as an example of evolution. I also consider the accusations of fraud and conspiracy theory aimed by some commentators at Kettlewell, Ford and evolutionary biologists in the UK. Thereafter, I give a personal view of why I feel reasonably qualified to discuss the behaviour, ecology and evolution of the peppered moth, and I briefly give my own view of the rise and fall of the black peppered moth. Finally, I suggest two major pieces of work that need to be done. If achieved, they should clarify some of the current uncertainties in the case, and may redeem the reputation of the peppered moth as an example of evolution in action, at least to those who are open-minded on the subject.

1.3 *Melanism: Evolution in Action*

Melanism: Evolution in Action (Majerus, 1998) was commissioned by Oxford University Press to be published 25 years after Kettlewell's book on melanism, *The Evolution of Melanism* (Kettlewell, 1973). The mandate that I was given was to critically appraise the phenomenon of melanism amongst animals in an evolutionary context and to 'update' Kettlewell. In the book, two chapters are devoted to the peppered moth. The first describes the basic peppered moth story as it is outlined in many biology textbooks, and outlines Kettlewell's work. The second dissects the story, looking at each of the seven component parts of the basic story (see Box 12.1), critically assessing the evidence for each, and discussing additional factors pertinent to the case, such as UV visual sensitivity by birds, and morph-specific resting site selection by peppered moths.

1.4 Coyne's review and *The Sunday Telegraph* article

Professor Coyne's review of *Melanism: Evolution in Action* was published on 5 November 1998, under the title 'Not black and white'. I read the review with mounting dismay. Generally the review was positive. Indeed, Coyne wrote:

> Occupying a quarter of the book, the *Biston* analysis is necessary reading for all evolutionists, as are the introductory chapters on the nature of melanism, its distribution among animals, and its proposed causes.

However, the message from the review was that the peppered moth case is fatally flawed as an example of Darwinian evolution. Coyne writes:

> ...for the time being we must discard *Biston* as a well-understood example of natural selection in action...

The passage that caused me most personal concern was:

> Majerus concludes, reasonably, that all we can deduce from this story is that it is a case of rapid evolution, probably involving pollution and bird predation. I would, however, replace 'probably' with 'perhaps'.

Box 12.1. Components of the 'text book' peppered moth story (from Majerus, 1998).

---

1. The peppered moth has two distinct forms; one, the typical form, being white with black speckling, the other, f. *carbonaria*, being almost completely black.
2. The two forms of the peppered moth are genetically controlled by a single gene, the *carbonaria* allele being completely dominant to the *typica* [*betularia*] allele.
3. Peppered moths fly at night and rest on tree trunks by day.
4. Birds find peppered moths on tree trunks and eat them.
5. The ease with which birds find peppered moths on tree trunks depends on how well the peppered moths are camouflaged.
6. Typical peppered moths are better camouflaged than melanics on lichen-covered tree trunks in unpolluted regions, while melanics are better camouflaged in industrial regions where tree trunks have been denuded of lichens and blackened by atmospheric pollution.
7. The frequencies of melanic and typical peppered moths in a particular area are a result of the relative levels of bird predation on the two forms in that area, and migration into the area of peppered moths from regions in which the form frequencies are different.

I checked my own book to see where I had concluded 'probably'. In this context, I could not find the word!

Coyne's (1998) review was followed up by an article in *The Sunday Telegraph* by Robert Matthews (1999); entitled 'Scientists pick holes in Darwin moth theory'. This article begins:

> Evolution experts are quietly admitting that one of their most cherished examples of Darwin's theory, the rise and fall of the peppered moth, is based on a series of scientific blunders. Experiments using the moth in the Fifties and long believed to prove the truth of natural selection are now thought to be worthless, having been designed to come up with the 'right' answer.

This opening was a surprise to me. I know most of those who have experimented with the peppered moth, and do not know any who would subscribe to this view. Moreover, if evidence were obtained that seriously undermined the qualitative accuracy of the case, it would be of such importance in academic circles that I cannot imagine any scientist speaking of it 'quietly'. The Matthews article is littered with numerous scientific inaccuracies and misquotations, but then many press reports, particularly of science, are. However, one would not expect misrepresentation in a book review in *Nature*. I leave judgement of whether Coyne's 1998 review was a misrepresentation of my book to Donald Frack, an American scientist, who has long wrestled in the USA with creationists and intelligent design advocates (see Box 12.2).

1.5 *Of Moths and Men*

Following Matthews' article, many papers and articles on the peppered moth appeared. On the anti-evolution side were titles such as 'Second thoughts about peppered moths' (Wells, 2001), 'Darwinism in a flutter: Did a moth show evolution in action?' (Smith, 2002), 'The moth that failed' (Raeburn, 2002), 'Staple of evolutionary teaching may not be a textbook case' (Wade, 2002) and 'Moth-eaten statistics' (Wells, 2002). From peppered moth workers and scientific philosophers have come a series of papers either adding data on the decline of *carbonaria* (Cook *et al.*, 1999; Grant and Clarke, 1999; Cook, 2000a; Grant and Wiseman, 2002), considering other minutiae of the case (Cook and Grant, 2000), or commenting on the criticisms of the case (Grant, 1999; Hagen, 1999; Rudge, 1999, 2003; Cook, 2000b; Allchin, 2001; Majerus, 2002).

Then came the publication of *Of Moths and Men: Intrigue, Tragedy and the Peppered Moth* (Hooper, 2002), which is, according to the front cover, 'A riotous story of ambition and deceit'.

This book, which purports to give 'the untold story of science and the peppered moth', is essentially an attack on the peppered moth case, those who have worked on the evolution of melanism in this species, lepidopterists in general and Kettlewell and Ford in particular. As Grant (2002) puts it in reviewing the book for *Science*:

> What it delivers is a quasi-scientific assessment of the evidence for natural selection in the peppered moth (*Biston betularia*), much of which is cast in doubt by the author's relentless suspicion of fraud.

Box 12.2. Extracts from an article posted by Donald Frack on the anticreation@talk origins.org website, 30 March 1999.

A while back creationists on the lists I subscribe to, and elsewhere in the 'real' world and cyberspace, began crowing over the death of the peppered moth as 'an example of evolution'. The references cited are a book review by Jerry Coyne in *Nature* of Michael Majerus's *Melanism: Evolution in Action*, and a later article from interviews with Majerus and Coyne in the on-line version of *The Telegraph*. These documents have been quite literally flaunted to show that evolutionists have been the willingly blind victims of everything from poor research to outright fraud, and that this famous example of natural selection has been abandoned by 'knowledgeable scientists'.

... I finally became fed up with this newest creationist claim, and the fact that no-one seemed to refer to the actual book upon which Coyne's review was based, but simply to the review itself. To evaluate the situation I located the seemingly notorious book by Majerus at UC Riverside.

... I opened Majerus' book anticipating a bashing for Kettlewell. ...From twenty years of reading anti-evolution literature, as well as advocacy of non-mainstream science views, I think I can pretty often see the attack coming in the form of qualifying with 'supposed evidence', etc. and confrontational discussions throughout the text. I expected this from Majerus.

... Throughout the chapter 'The Peppered Moth Story', Majerus gives not the slightest hint of the bomb I was waiting for. His discussion of Kettlewell's experiments, and those of others, are so fairly and complimentarily done that I was amazed at the thought that he was about to destroy it all. ...How was Majerus going to unhinge the discussion in 'The Peppered Moth Story Dissected'? And why did he lead his readers on so cruelly without a hint that they were being given trash data? I read to the end of the second chapter like it was a whodoneit [*sic*].

... If you're waiting for the punch line, here it is. There is essentially no resemblance between Majerus' book and Coyne's review of it. If you pick through the book, you might be able to argue for Coyne's accuracy – but only at the expense of completely ignoring the majority of the text and all of Majerus' intent. If I hadn't known differently, I would have thought the review was of some other book.

Coyne (2002) in a review in *Nature* goes further. He criticizes her 'flimsy conspiracy theory', her theme of 'ambitious scientists who will ignore the truth for the sake of fame and recognition', by which 'she unfairly smears a brilliant naturalist', and her lack of criticality when she champions Ted Sargent's 'phenotypic induction' theory, as 'she conveniently glosses over the simple and unassailable fact that the light and dark alleles of *Biston* segregate as Mendelian variants when tested under uniform experimental conditions'. Coyne concludes:

> This issue matters, at least in the United States, because creationists have promoted the problems with *Biston* as a refutation of evolution itself. Even my own brief critique of the story [his 1998 review] has become grist for the creationists' mill. By peddling innuendo and failing to distinguish clearly the undeniable *fact* of selection from the contested *agent* of selection, Hooper has done the scientific community a disservice.

Coyne's reference to the United States is interesting. The anti-Darwinian lobby has had considerable success in the USA in ensuring, through legislation and litigation, that creationism and intelligent design theories are given equal time to Darwinian evolution in biology teaching in schools. In the UK, we may have thought that we were immune to such action. However, in August 2002, an English Local Education Authority passed a similar mandate.

The peppered moth story is taught widely in UK schools. Indeed, it is an integral part of many biology courses. For example, the Oxford, Cambridge and RSA Examinations Board (OCR) syllabus for GCSE Biology 2003, contains under the subsection on evolution: 'Describe how the process of natural selection may result in changes within a species, as illustrated by the peppered moth'.

Coyne's 'at least in the United States' is too parochial. The peppered moth story matters – period!

Hooper's book, *Of Moths and Men*, is so strewn with errors, misrepresentations, and misinterpretations that it is impossible to enumerate them all here. The writings of Wells and some of the other critics of the peppered moth story are similarly plagued. However, they cannot be simply dismissed, as many of the readers of these critics are not armed with the knowledge of evolutionary biology, genetics and ecological entomology necessary to perceive the errors and manipulations within these works. Their writings are lively and readable, and their arguments can be persuasive to those with limited or no training in evolutionary genetics or entomology. Furthermore, few of their readers will have the time to refer to original and review papers on the peppered moth, written by those who have worked with the peppered moth, to judge the veracity of the words of people such as Hooper and Wells.

2. The Peppered Moth's Place in Evolution

Three important questions of the peppered moth case should thus be addressed:

1. Does it provide proof of biological evolution (changes in the frequencies of heritable genetic variants over time)?

2. Does it provide proof of Darwinian evolution (changes in the frequencies of heritable variants over time through the mechanism of selection)?

3. Is the main agent of selection differential bird predation?

2.1 Proof of biological evolution

Biological evolution may be defined as a change in the frequency of an allele through time. The *carbonaria* form of the peppered moth differs from f. *betularia* with respect to the alleles of a single gene (Bowater, 1914; Lees, 1981). The frequency of the *carbonaria* allele did increase during the 19th and first half of the 20th century, and is now declining. This is irrefutable proof of biological evolution.

2.2 Evidence of Darwinian evolution

We may take certain observations of the peppered moth as fact.

1. From numerous breeding experiments, both published and unpublished, it is incontrovertible that the forms of the peppered moth are inherited according to Mendel's laws of inheritance.

2. The frequencies of f. *carbonaria* and f. *swettaria* (the melanic form of the peppered moth in North America) have varied both temporally and spatially (e.g. Kettlewell, 1958; Bishop, 1972; Brakefield and Lees, 1987; Mani and Majerus, 1993; Grant *et al.*, 1996; Majerus, 1998).

3. There has been and is a correlation between *carbonaria* frequency and pollution levels, particularly sulphur dioxide levels (e.g. Kettlewell, 1973; Lees, 1981; Brakefield and Lees, 1987; Cook, 2000b).

4. The observed changes in the frequencies of forms of the peppered moth, both in the 19th century and currently, are too rapid to be accounted for by random genetic drift (Haldane, 1924; Ford, 1964; Mani and Majerus, 1993; Grant *et al.*, 1995).

These factual observations are sufficient to provide evidence that selection has had a major role in the rise and fall of *carbonaria*.

As Coyne (2002) points out, even Hooper (2002) cannot find an alternative to selection to cause the striking directional changes observed in the peppered moth. He highlights that:

> Hooper's grudging admission of this fact occupies but one sentence: 'It is reasonable to assume that natural selection operates in the evolution of the peppered moth'.
>
> (Hooper, 2002, p. 312).

2.3 Is the agent of selection differential bird predation?

This question is more difficult to address. Critics of Tutt's differential bird predation hypothesis rely heavily on design flaws in Kettlewell's experiments in the 1950s. Hooper (2002), for example, bases much of her case on design changes that Kettlewell described himself. Wells (2000, 2001) takes valid criticisms made by evolutionary biologists – e.g. Brakefield, Clarke, Cook, Creed, Grant, Howlett, Sheppard and myself – out of context, and reinterprets them without consideration of the known natural history of the moth, or, for that matter, of practical and logistical expediency. Despite their detailed, if somewhat selective, dissection of Kettlewell's work, neither Wells nor Hooper assesses the eight subsequent independent field predation studies (reviewed by Cook, 2000a). They fail to comment on variations in the methodologies used, many of which were designed to correct for precisely the deficiencies in Kettlewell's procedures that, they argue, undermine the peppered moth case.

These eight studies, and indeed Kettlewell's own experiments, consistently show that the fitness of a morph is correlated to the frequencies of the forms in a particular area and to concurrent changes in these frequencies (Cook, 2000b;

Grant, 2002). Were I giving my view of the procedures in each of these nine studies, including my own (Howlett and Majerus, 1987), I would certainly criticize each for artificiality in some respect. However, the cause of the artificiality varies between studies – using dead moths, moths at unnatural frequencies, moths at unnatural densities, not allowing moths to take up natural resting sites, and so on. Reviewing all these studies, it is difficult to believe that the artificiality in each case just happens, by chance, to provide results that support Tutt's bird predation hypothesis. Yet this is what some critics would have us believe.

Other mechanisms have been proposed to account for the rise in *carbonaria*. These include direct mutagenic effects of pollutants (Harrison, 1927; Sargent *et al.*, 1998), which may be dismissed on the basis of the wealth of data, spanning almost a hundred years, showing Mendelian segregations of the forms of the peppered moth, when reared under controlled conditions, and that *carbonaria* has an inherent physiological advantage (Ford, 1937, 1940; Hooper, 2002; but see Ford, 1964), which is difficult to reconcile with the recent decline in *carbonaria*. Neither has any empirical support from studies of the peppered moth. Indeed, at present, only the agent of differential bird predation has any experimental support.

In summary, the situation is this. The case of the peppered moth provides irrefutable proof of biological evolution through the process of selection. While there is considerable evidence that differential bird predation is the main agent of selection, this evidence is only circumstantial.

3. The Nature of Criticisms of the Peppered Moth Case

What then can we say of criticisms of the peppered moth story and of Kettlewell's experiments in particular? The criticisms seem to me to have differing tones and can thus be split into three categories.

1. Scientific criticisms of artificiality (e.g. 'bird-table effect', morphs not released at natural frequencies, translocated moths may have different behaviours, bred and wild-caught moths may act differently, moths released in the day so that they do not select resting sites at night).
2. Pseudo-scientific criticisms (e.g. predation by bats is probably higher than bird predation).
3. Data fudging and/or fraud.

3.1 Criticisms of artificiality in Kettlewell's experimental procedures

Many criticisms have been aimed at the experiments conducted by Kettlewell in the 1950s. Some of these were first proposed by Kettlewell himself, others by scientists who worked on peppered moths. Most may have some validity. The major criticisms have been:

1. The densities of moths in Kettlewell's predation and mark–release–recapture experiments were too great.

2. Kettlewell released moths on to tree trunks. The evidence that exists suggests that although some peppered moths naturally rest in exposed positions on tree trunks, this is not their preferred resting site.

3. In his mark–release–recapture experiments, Kettlewell released moths during the day. Peppered moths prompted to fly during the day will settle on the first substrate that they encounter, and generally remain still thereafter. Thus, moths released during daylight will not select the same sites as those that settle at the end of night flight. It is improbable that the degree of crypsis secured by Kettlewell's released moths would have been as high as that of moths in the wild.

4. Kettlewell used mixtures of wild-caught and laboratory-bred specimens, which may have behaved differently.

5. The moths that Kettlewell released in Birmingham and Dorset may not have originated in the same locations, and so may have had local behavioural adaptations.

In addition, many later workers glued dead moths on to trees in 'life-like' positions, selecting sites that maximized their crypsis. I have tried to do this by very carefully gluing moths on to birch tree trunks, and releasing a similar number of live moths on to the trunks soon after dawn. A class of students then assessed the degree of crypsis of the moths by walking towards the trunks and saying when they could see any moth. For all forms, the live moths were more cryptic than the glued moths.

With one exception, one or more of the studies subsequent to Kettlewell's avoids each of these sources of artificiality, and shows that they do not affect the differential bird predation hypothesis. The only criticism that can be aimed at all the predation studies conducted to date is that the moths available for predation did not take up their own resting positions during the pre-dawn flight that characterizes this species. This criticism should be addressed in future predation experiments.

3.2 Pseudo-scientific criticisms

It has been pointed out that most of the critics of the peppered moth case as an example of evolution in action (with the notable exception of Sargent) have never worked on the moth, nor are most of them experienced field biologists or have trained in evolutionary genetics (Frack, 1999; Majerus, 2002). Thus one of the problems with many of these critics is that they do not have a thorough understanding of how selection operates, or any understanding of the moth itself. A trivial example illustrates the point. Kettlewell (1955a) wrote of releasing moths for his experiments, that: 'It was important to see that the sun did not shine on the moths'. He does not say why this was important. However, Hooper (2002) comes up with a reason, writing: 'He scored each moth … after making sure that the sun did not shine directly on the insect, which would have fried it'. This is not the reason. Any experienced lepidopterist knows that night-active cryptic moths, which rest on bark by day, avoid the sun to reduce shadow effects and overheating. Most of them rest on the north-facing side of trees or in full shadow. However, a moth in a position that becomes directly sunlit as the day progresses does not 'fry'. It simply moves across the bark ahead of the sun.

Although Hooper's lack of evolutionary and ecological knowledge is revealed on numerous occasions in both *Of Moths and Men* and in an e-mail that she sent to me prior to the publication of her book (Hooper to Majerus, 16 November 2000), there is a persuasiveness to much of the pseudo-scientific nonsense that she peddles. One example, concerning the predation of moths by bats, will serve as illustration.

3.3 Bats versus birds

The questions that Hooper asked me about bat predation in her e-mail are given verbatim in Box 12.3.

My responses were given in a lengthy telephone conversation the following Sunday. I do not recall precisely what I said, but my preparatory notes on her e-mail are that Kettlewell's reasoning is correct, that the two forms are unlikely to differ in palatability or smell, but that scale types and pigments might affect sonar. I also explained Kettlewell's reasoning, and the flaw in Hooper's, in detail, by theoretical example. My example is cited by Hooper (2002).

Box 12.3. Extract from an e-mail from Judith Hooper to Mike Majerus, 16/11/00, asking questions about the peppered moth.

---

11. BAT PREDATION: In Kettlewell's time, various people who challenged him (including Heslop Harrison) said that bat predation was a more likely source of selection than bird predation. Kettlewell himself admitted that bats accounted for 90% of the mortality of the moths but said that this didn't matter because it wasn't selective—ergo, even if only 10% of the predation was by birds hunting by sight, that 10% is what makes the difference and drives evolution. It seems to me that there are several flaws in this reasoning.

(a) If you had only 10% of moth mortality effected by birds hunting selectively (and 90% by bats, totally random) wouldn't you see a different statistical outcome? How would you get the robust 2:1 and 3:1 advantages that Kettlewell got in three different experiments?

(b) Wouldn't it be wrong to ASSUME that bat selection was totally random? Would a good scientist need to do an experiment to rule out selective predation by bats, esp. if bats were responsible for 90% of adult mortality? Isn't it possible that one genotype might be more palatable, or smell different, or something? Of course, this would have been a problem for Kettlewell's model, because bats hunt at night, preying on moths that are flying, so crypsis against tree trunks would not be the issue. You'd be back to square one.

I'd love to hear your thoughts on this?

Question 12 was a similar enquiry relating to pre-adult mortality.

Say three hundred eggs are originally laid. Once you get to the adult stage, maybe you have ten left. Of these more than half are killed by things not hunting by sight, so say you have four moths left – two typical and two *carbonaria*. You must be prepared to say that none of the mortality prior to this is due to selection on colour pattern, no pleiotropic effects of alleles, no differences in palatability, no greater energetic costs in producing black pigment and so on. If so, then despite 296 moths being killed up to that point, if those two typicals are eaten by birds, you've increased *carbonaria* by a hundred per cent at one go.

Hooper (2002) then asks:

Can we really be sure that bat predation is *not* selective, that there is not some yet unidentified difference between melanics and typicals that makes one morph more vulnerable to bats? Certain night-flying moths can dodge or jam bat sonar, according to several studies, and it is not known whether this ability is equally distributed.

This passage stretches the bounds of probability. Following this line of reasoning, the assumptions that we would have to make are that not only could bats distinguish between the forms by sonar, smell, taste or behaviour, but that the form that was taken more would vary geographically, and that this geographic variation was, by chance, strongly correlated to pollution levels.

Although it seems unlikely to anyone who has observed bats feeding on moths around moth traps that the bats could be behaving differently towards different colour forms of a species, the test of Hooper's question about the 'need to do an experiment to rule out selective predation by bats' is not difficult to address. Thus, in June 2003, I conducted such an experiment.

3.4 Non-selective predation of peppered moths by bats

A sample of 400 laboratory-reared male peppered moths were released sequentially between 11 p.m. and 3 a.m. over five nights, 20 m from a mercury-vapour light, in the grounds of the Genetics Field Station, Cambridge, being attended by pipistrelle bats. Equal numbers of f. *betularia* and f. *carbonaria* were released. The moths, which had all eclosed earlier on the day of release, were kept individually in Perspex boxes. These were numbered randomly and moths were released in numerical order. The bats were flying above the trap, taking moths flying in the area. Up to seven bats were observed feeding at a time. At 10-min intervals, five boxes were laid on the ground and opened. Moths were watched as they took flight, and followed by eye, with the help of night glasses, until they were lost from view, or were seen to be caught by a bat.

The results are given in Table 12.1. There is no significant difference in the numbers of the two forms that were caught by the bats. Bats do catch and eat peppered moths flying at night, but they do so randomly with respect to the forms of the moth.

Table 12.1. Predation of the *carbonaria* and *betularia* forms of *Biston betularia* by pipistrelle bats. Equal numbers of the two forms were released in the vicinity of a mercury vapour light.

Form	Flew and lost from view	Did not take flight within 10 min	Caught by bats
carbonaria	114	35	51
betularia	107	39	54

3.5 Data fudging and/or fraud?

One of the most damaging criticisms of Kettlewell's work is the reported increase in recapture rates that occurred in Birmingham from 1 July 1955. Hooper (2002) notes that after recaptures running in low single digits for the first six days (in fact it was five days – see Kettlewell, 1955a), on the morning of 1 July, 23 marked moths were recaptured, and the increased recapture rates were maintained thereafter. She ties in the increase in recapture rates with a letter from E.B. Ford to Kettlewell dated 1 July, in which Ford wrote: 'It is disappointing that the recoveries are not better … However, I do not doubt that the results will be very worth while …'. Hooper gives her own translation of this passage as: 'Now I do hope you will get hold of yourself and deliver up some decent numbers'. Hooper makes large of 1 July. She writes: 'what happened between the last day of June and the first day of July 1953 to turn the tide', implying of course, that it was the arrival of Ford's letter.

Three points should be made. First, while I do not know when Kettlewell received Ford's letter, it is notable that the recapture rates had certainly risen on the night before Ford wrote the letter. Kettlewell was certainly disappointed by the low initial recapture rates, and increased the number of marked moths released on 30 June, not after receiving Ford's letter. Hooper's interpretation that the change in Kettlewell's data was a response to Ford's letter is factually wrong. One wonders whether the errors Hooper makes over the data she draws from Table 5 in Kettlewell's 1955a paper result from her poor ability to read a table of scientific data, or are a deliberate attempt to mislead.

Second, from my own experience of moth trapping over 40 years, I am aware that moth trap catch sizes vary greatly, both in respect of total catch and for individual species. The variations are not always predictable. Factors such as temperature, cloud cover, and wind speed and direction can have a very marked effect, as can something as innocuous as a slight repositioning of a trap, a point made by Coyne (2002).

Third, we should examine the increase in recapture rates that Hooper is concerned about. Kettlewell's own work shows that the number of recaptures declines very rapidly after the first night. If the recapture levels for the two nights following days when no moths had been released (nights of 26/27 June and 29/30 June) are excluded, the average proportion of released moths recaptured up to the night of 30 June/1 July is 0.117. (Excluding recaptures on these two nights makes the pre-1 July data more strictly comparable with the post-1 July data, and gives a more conservative, i.e. lower, recapture rate for the first period.) Thereafter the proportion is 0.267. This is certainly an increase.

Hooper writes:

> The average number released prior to 30 June was 30.8, while the average for 30 June–4 July was 92.5. Was this why he recaptured more? Or was there some other reason as well?

There may have been some other reason. The threefold increase in the number of moths released may have effectively flooded the area with moths, to an extent where the predators of the peppered moth in the area were at least partially satiated, leading to an increase in the survival of the released moths and so to increased recapture rates. Despite criticizing the high densities of Kettlewell's releases, Hooper does not seem to have considered the possible effects of this flaw in Kettlewell's procedure in sufficient depth to realize that it could answer her own question.

It is interesting, but perhaps not surprising, that Hooper does not try to answer her own question. Most anti-evolution critics of the peppered moth story seem keen to simply discredit the peppered moth case, and in particular Kettlewell. They do not seem prepared to seek alternative explanations or interpretations of data. Those that do offer alternatives tender such ill-conceived hypotheses, based on the most tenuous evidence, and frequently showing little understanding of genetics, evolutionary processes, or the behaviour of the subject material, that they can be rapidly dismissed (e.g. Coyne, 2002; Grant, 2002; Cook, 2003). However, in the wealth of data that has been accumulated on the peppered moth, there are inconsistencies. Critics of the peppered moth case are quick to stress these inconsistencies, and aim accusations of fudged data or fraud. They rarely seek a scientific explanation of inconsistencies. Kettlewell's work on morph-specific resting site selection provides an example.

3.6 Morph-specific resting site selection?

Kettlewell (1955b) reports the results of experiments carried out in 1954 on background recognition in the peppered moth. In this experiment, Kettlewell lined a large cider barrel with alternate black and white strips of cloth or rough paper, all of identical texture. A sheet of glass was placed on top of the cylinder and was then covered with white muslin. The barrel was situated outside, but out of direct sunlight. Each evening, up to six peppered moths of the same sex, with *carbonaria* and *betularia* in equality, were released into the barrel. At dawn, the resting positions of the moths were scored. Moths that rested on the floor, or across two backgrounds (80 of 198 moths released), were excluded from analysis. The results (Table 12.2) showed a significant difference in the behaviour of the forms, with almost two-thirds of *carbonaria* resting on black surfaces and two-thirds of *betularia* resting on white.

Kettlewell (1955b) proposed that a peppered moth, after landing on a surface, but before clamping down, will select a position where it is out of the

Table 12.2. The resting positions of the *betularia* and *carbonaria* forms of the peppered moth in Kettlewell's barrel experiments, in which the moths were presented with a choice of black and white surfaces of equal area (from Kettlewell, 1955b).

	f. *betularia*	f. *carbonaria*	Totals
Black background	20	38	58
White background	39	21	60
Totals	59	59	118

sun, where it can align its body with a groove in the bark, and where the contrast between the colour of the substrate and the moth's circumocular tufts is minimized.

Various authors have subsequently investigated resting site selection in the peppered moth, using a range of experimental approaches, including attempted approximate replications of Kettlewell's experiments, and various manipulations of the circumocular tufts of moths (Howlett and Majerus, 1987; Grant and Howlett, 1988; Howlett, 1989; Jones, 1993). Resting site preferences have been reported in some studies (e.g. Table 12.3), but not in others. However, none has found morph-specific resting site preferences within a population.

The failure to replicate Kettlewell's results has brought veiled accusations of fraud by Kettlewell (Wells, 2000; Hooper, 2002). However, neither Wells nor Hooper attempts to take the various datasets at face value and seek a biological explanation to reconcile variations in them. Yet such an explanation exists and has been published (Majerus, 1989, 1998).

Howlett (1989) modelled the rise of a mutant allele, unlinked to the colour pattern locus, that induced a preference for peppered moths to select dark homogeneous backgrounds, rather than pale heterogeneous backgrounds, to rest upon. He assumed that the fitness of *carbonaria* would be increased by the expression of such an allele, and that of *betularia* would be reduced by it. The model showed that the allele would only increase in frequency in populations in which *carbonaria* was already common. Majerus (1989) argued that this might account for the morph-specific resting site selection reported by Kettlewell (1955b). His hypothesis is based upon the premise that a heritable preference to prefer to rest on dark homogeneous substrates would have evolved in regions

Table 12.3. The resting preferences of the *betularia*, *insularia* and *carbonaria* forms of the peppered moth, when presented with a choice of black and white surfaces of equal area in cylinders (from Howlett and Majerus, 1987).

	f. *betularia*	f. *insularia*	f. *carbonaria*	Totals
Black side	58	30	70	158
White side	20	7	14	41
Floor	21	5	36	62
Totals	99	42	120	261

with high *carbonaria* frequency, but not where *carbonaria* is rare, as Howlett's model suggested. Where *carbonaria* is rare, moths would retain the ancestral resting site choice, which Majerus assumes is for pale heterogeneous surfaces. There is some support for these assumptions, both Grant and Howlett (1988) and Jones (1993) finding variation in the resting site preferences of moths from different populations, with preferences for dark backgrounds being found in all populations with high melanic frequency. The thesis is then that the moths that Kettlewell used in his barrel experiments were drawn from different populations, the f. *carbonaria* from an industrial population and the f. *betularia* from a rural one.

The source of the moths that Kettlewell used in his barrels is not known, despite exhaustive enquiries (Majerus, 1998), so it is not possible to verify this explanation. However, were this explanation correct, it would explain the disparity between Kettlewell's results and those of others, without having to resort to unverifiable accusations of fraud.

4. A Personal View From the Horse's Mouth!

It is dangerous to open one's mouth, to put pen to paper, or to start tapping away at the keyboard. If you have opinions and offer these to others, you run the risk of being misquoted, misrepresented when your words are taken out of context, or indeed having words falsely attributed to you.

Over the peppered moth, I have suffered all three since writing *Melanism: Evolution in Action*. Some of these misrepresentations are repeated over and over again in anti-Darwinian literature, possibly because citing a Cambridge University evolutionary geneticist appears to give authority to the subjective private agendas of pseudo-scientists or journalists.

Some of the interpretations and statements falsely attributed to me should, I feel, be exposed, for some are damaging to good science, or to the reputation of individual scientists, or to the public's image of how good science is generally conducted.

Good science is conducted largely through observation, hypothesis formation, prediction and experimentation. Much of the work on the peppered moth has been conducted in this way. An impressive array of evidence has been accumulated. Many of the experiments, particularly those conducted in the field, are not without their flaws. Moreover, predicted outcomes have not always been realized. However, if you wade through the 200+ papers written about industrial melanism in moths, it is difficult to come to any conclusion other than that natural selection through the agent of differential bird predation is largely responsible for the rise and fall of the *carbonaria* form of the peppered moth.

That said, my own conviction that bird predation is largely responsible is not based purely on empirical data from experiments published in the literature. I know that Tutt's differential bird predation hypothesis is correct because I know about peppered moths. For those who have never seen a peppered moth in the wild, which is almost everybody; for those with anti-Darwinian agendas; and for scientists, well-trained in rigour, stringency and experimental controls; for differing reasons this statement must seem insufficient, if not heretical. However, I stick by it.

The biography of the great geneticist Barbara McLintock was titled *A Feel for the Organism* (Keller, 1993). I think that I have a feel for some organisms. My credentials are these: I caught my first butterfly when I was 4 years old (Majerus, 1994). I learnt the basics of Mendelian genetics when I was 10 years old (Berry, 1990). For 45 years I have bred, collected, photographed and recorded moths, butterflies and ladybirds in the UK, across Europe and latterly around the world. I have run one or more moth traps almost nightly for 40 years. This experience has given me something of a feel for the organisms that I observe.

I bred my first broods of the peppered moth in 1964, following Ford's (1955) advice on careful separation of broods and writing notes on all procedures used. I found my first peppered moth at rest in the wild in the same year (Howlett and Majerus, 1987). As far as I am aware, I have found more peppered moths at rest in their natural resting position than any other person alive. I admit to being, in part, a moth man.

In the first chapter of *Of Moths and Men*, Hooper (2002) assassinates the character of 'moth men', who have 'stunted social skills of the more monomaniacal computer hackers, going about with misbuttoned shirts and uncombed hair, spouting taxonomic Latin'. She cites Ted Sargent, who considers moth collectors to be weirder than butterfly collectors. According to Hooper, Sargent is awed by moth enthusiasts who 'can go up to a streetlight and start naming these things … It's an extraordinary talent'.

But it isn't extraordinary. Hundreds of thousands of children across the world can recognize hundreds of different Pokemon characters, and provide details of their characteristics, their evolutionary potential and their powers in contest. How is this different from a 12-year-old who can recognize several hundred species of macro moth, know when they fly, and what their larvae feed upon? Calling out names to a group of people around a moth trap, the names I use are English, not Latin, for I learnt them, out of interest and fascination, when I was a child, and the English names were easier.

I know the peppered moth, and I know that J.W. Tutt was essentially correct in his explanation of the rise of *carbonaria*. However, for those who do not 'know' the peppered moth, whether they are scientists, teachers, or members of the public, this should not and **cannot** be enough. So, what is needed to prove whether changes in frequencies of the peppered moth are indeed the result of differential bird predation? And can the declining reputation of the peppered moth be reversed?

5. Two Evidences for Proof

In my view, two pieces of evidence are critical. The first is that birds eat a greater proportion of one form than the other to an extent consistent with monitored changes in the frequencies of the forms. The second is that a connection should be made between the genotype and phenotype.

5.1 The genotype–phenotype link

Taking the second point first, it is an unfortunate omission that the multiple allelic gene that controls melanism in the peppered moth in the UK has not been identified and sequenced. The critical step of connecting genotype with phenotype has thus not been accomplished in this classical case of Darwinian evolution in action. However, this step has recently been accomplished in another case of adaptive melanism involving crypsis (Nachman *et al.*, 2003). The rock pocket mouse, *Chaetodipus intermedius*, varies in coat colour. Strong correlations between coat colour and substrate have been shown for this and related species in south western USA (Dice, 1930; Blair, 1943). The most obvious correlations occur where black volcanic lava rocks abut white granitic rocks. The similarity of the dorsal pelage of mice and their substrate is adaptive, providing cryptic protection against birds of prey, particularly owls (Sumner, 1934; Dice, 1947). Nachman *et al.* (2003), using candidate genes from the many that affect coat colour in laboratory mice, identified, by association analysis, the mutation causing melanism in a population of mice on a lava bed in Arizona. The mutation is in the melanocortin-1-receptor gene (*MC1R*). Other mutations of this gene are known to be associated with melanic phenotypes in jaguars, *Panthera onca*, and jaguarundis, *Herpailurus yaguarondi* (Eizirik *et al.*, 2003). In these species showing melanic polymorphism in the wild, morph frequencies appear to be at equilibrium. Work on a system with an established genotype–phenotype characterization, and in which the frequencies of forms are changing directionally, i.e. under selection, would allow changes in adaptive mutations to be tracked. The peppered moth is the obvious candidate species. A similar association analysis, using candidate genes from *Drosophila* (Hollocher *et al.*, 2000; Wittkopp *et al.*, 2003), *Manduca sexta* (Hiruma and Riddiford, 1990, 1993) or *Papilio glaucus* (Koch *et al.*, 2000), should be rewarding. It could provide the genotype–phenotype link in the most celebrated example of Darwinian evolution in action. Furthermore, it would lay to rest the phenotypic induction hypotheses (Harrison, 1927; Sargent *et al.*, 1998; Hooper, 2002); allow determination of whether f. *carbonaria* in the UK and f. *swettaria* in the USA are caused by the same mutation, and permit more stringent examination of any pleiotropic effects of melanic mutations that may affect fitness.

The investigation of whether the mutations responsible for the *carbonaria* and *swettaria* forms – in the UK and the USA, respectively – are the same or different is critical to the evaluation of another criticism of the peppered moth case. The correlations between melanic frequency and pollution levels are not precisely the same in the UK and the USA (Sargent *et al.*, 1998; Wells, 2001; Hooper, 2002). Interpretation of the data from either side of the Atlantic leads to an extreme dichotomy of views. While some take the broad similarities in the patterns of change to be strong support of the industrial melanism thesis (Grant *et al.*, 1995, 1996; Yoon, 1996; Majerus, 1998), critics take the differences in the two instances to falsify the industrial melanism hypothesis (Sargent *et al.*, 1998; Hooper, 2002).

The real question is whether we should expect the situation to be precisely the same on both sides of the Atlantic. It is pertinent to note that peppered moths in the USA are not the same as those in the UK, being a different subspecies, *Biston betularia cognataria* (not a different species, *Biston cognataria*, as stated by Hooper, 2002), and while the melanics *carbonaria* and *swettaria* are very similar, the non-melanics, f. *betularia* in the UK and f. *cognataria* in the USA, are rather different. Presumably this difference is a result of differences in the detailed manner in which selection has acted on the colour patterns of these non-melanic forms in Europe and America in the distant past. Given this, it would indeed be surprising if the situation on both sides of the Atlantic was identical. Indeed, one can go further, for the situation in the Netherlands, while again qualitatively similar to that in the UK, with melanic frequencies being correlated to pollution levels, is again quantitatively different, with melanic frequencies never reaching the 90+% levels common in the UK when *carbonaria* was at its zenith (Brakefield, 1990).

The mutations responsible for the melanic forms in these geographically isolated populations have not been identified and need not be the same. The recent finding of different mutations controlling melanic forms of rock pocket mice on lava flows in Arizona and New Mexico (Hoekstra and Nachman, 2003), demonstrates that phenotypic similarity need not be based upon genotypic similarity. In the fine detail, it would be more surprising if the situations in these isolated populations of the peppered moth were the same. The next step in addressing these differences is obviously the identification of the mutation or mutations involved.

5.2 A new predation experiment

To determine whether changes in *carbonaria* frequency can be accounted for by differential bird predation requires a predation experiment that avoids the suggested flaws in those carried out by Kettlewell and others. Such an experiment was designed in 2001. The design took account of the criticisms aimed at previous experiments, plausibility of procedure and methods of statistical analysis. Initial testing of release procedures, the trees that peppered moths rest upon in the area, the visibility of subjects during experiments and levels of predation, and was undertaken in 2001, in Madingley Wood, to assess feasibility.

The experimental design is as follows:

Location: A 1-ha garden plot. The garden is surrounded by a mature hawthorn / cherry / dog-rose / sallow / ash / oak / elder / sloe / leylandii / privet boundary hedge. Approximately half of the plot comprises old orchard (apple, pear, plum) and mature deciduous trees (oak, birch, willow, goat sallow). Agricultural land, mature deciduous hedges and small areas of deciduous woodland surround the garden.

Release positions: 103 branches were identified and numbered. Branches vary in their angle to the horizontal from 79° to −37°, sourcing from the trunk. All release positions were on native trees. The height above ground of the positions varies from 1.8 to 26 m. Lower positions are reached by ladders. Climbing aids have been attached to trees (or adjacent trees) to allow access to higher positions.

The method of release uses 0.5–1-m-long black netting sleeves placed around branches. Wire rings are set around the branches to produce a netting cylinder with the netting proud of the branch by approximately 20 cm. The ends of the cylinders are tapered and elasticated so that they fit tightly around the branch. Each evening, 12 such cylinders are constructed. The positions for the cylinders are chosen by random numbers from the 103 release sites. One peppered moth, either *betularia* or *carbonaria*, is released into each cylinder within the hour prior to sunset. The cylinders are carefully removed the following morning, during the 40 minutes prior to sunrise. The positions of the moths are noted. Moths disturbed during this procedure, or not resting on the branches, are removed. This removal, on average, comprises about one-third of the moths, leaving a release density of about 8 moths/ha. Period of release is during the period that the peppered moth is on the wing in the area (determined as the average of the earliest and latest record in light traps over the previous 20 years), i.e. 11 May until 19 August.

The moths are all of Cambridge origin (within 5 km of the release site). Only *carbonaria* and *betularia* forms are used. Moths are used at the relative frequencies that they occurred in the area in the previous summer (determined by running a set of four light traps and four pheromone traps, at Madingley Wood, 1.9 km from the release site). The moths used are of four types, used in approximately equal numbers:

1. Moth-trap-caught males.
2. Pheromone-trap-caught males.
3. Bred males that had eclosed on the day of use.
4. Bred females that had eclosed on the day prior to use and had mated the previous night in cages hung outside at another location (Genetics Field Station). The main dispersal flight of females takes place on the second night after eclosion, having mated the first night (Liebert and Brakefield, 1987).

The genotypes of bred males and females used are known. The genotypes of the moths released are in the approximate proportions that the genotypes occur in the area.

Recording of predation: 58 of the release positions are visible from the house at the centre of the garden. As moths do not always rest on a part of the branch visible from the house, the number of released moths that can be kept under direct observation varied, but over the test period (2001) it was 29%, and over the first two years of the experiment (2002–2003) it has been 26% of those released. Observation is aided by binoculars or a telescope. Any moths taken by birds are recorded, as is the species of bird. Any moths seen to change position are also recorded.

Observations commence immediately after the last cylinder is removed, and continue for between 3 and 4 h. Approximately 4 h after sunrise, each release site is visited, and a record made of whether the moth is present or absent. All remaining moths are removed to prevent an accumulated increase in the density of moths in the garden.

Analysis of the results will consider the relative rates of predation of the two forms. The question of whether these rates can account for the declining frequencies of *carbonaria* in the Cambridge area will be addressed. In addition, data from the four types of moth used will be compared, as will be data from observed and unobserved disappearances.

The design of this experiment tries to circumvent as many of the criticisms aimed at Kettlewell's predation experiments as possible. Thus the design uses moths released at low density, at natural frequencies, on to natural resting sites in such a way that they choose their own resting position, albeit in a restricted arena. The design also only uses moths originating from the area of the release site, and should allow determination of differences between both male and female moths and light-trap-caught, pheromone-trap-caught and bred moths.

Due to the low frequency of *carbonaria*, and the current rate of decline of *carbonaria*, which gives a fitness disadvantage for *carbonaria* of approximately 0.15, the experiment will run for 5 years.

Airing the experimental design here has the aim of avoiding future accusations, of the type aimed at Kettlewell (Wells, 2000; Hooper, 2002), of altering the design as the work progresses.

5.3 New data on the natural resting sites of peppered moths

The work has already yielded one interesting piece of data. The slowest accumulating dataset that I have is of the resting positions of peppered moths I have found in the wild since 1964. This dataset, first published in 1987 (Howlett and Majerus, 1987), has continued to build. The set up to 2001 is given in Table 12.4a, and consists of just 59 moths, a rate of 1.55 moths located per year. While constructing or removing release sleeves in the trees, I have found, by eye, a considerable number of moths, 27 of which have been peppered moths (Table 12.4b). The rate of find has thus risen to 13.5 moths

Table 12.4. The natural resting positions of peppered moths: (a) moths found in the wild, 1964–2001 (all locations); (b) moths found while climbing trees or working in the canopy to construct or remove release sleeves, 2002–2003 (Springfield, Cambridge).

	f. *betularia*	f. *insularia*	f. *carbonaria*
(a)			
Exposed trunk	4	1	2
Unexposed trunk	2	1	4
Trunk/branch joint	11	5	7
Branches	9	5	8
(b)			
Exposed trunk	1	0	0
Unexposed trunk	1	0	0
Trunk/branch joint	2		1
Branches (> 5 cm diameter)	7	1	2
Branches/twigs (< 5 cm diameter)	6	2	0

The remaining four comprised two mating pairs (one *betularia* female × *carbonaria* male, the other both *betularia*), both on the side/underside of branches with a diameter > 5 cm.

per year. Furthermore, all of these moths were more than 2 m above ground, most were in the upper half of the trees that they were on and only five were on the main trunks of the trees.

Critics of the peppered moth have often pointed to a statement made by Clarke *et al.* (1985): '... In 25 years we have only found two *betularia* on the tree trunks or walls adjacent to our traps, and none elsewhere'. The reason now seems obvious. Few people spend their time looking for moths up in the trees. That is where peppered moths rest by day.

6. Endnote

The peppered moth provides irrefutable scientific proof of biological evolution through the Darwinian mechanism of selection. Although this case is one of the most celebrated and visible examples of Darwinian evolution in action, it is only one of many examples of evolutionary change resulting from selection. The numerous other examples, including pesticide resistance, antibiotic resistance and over a hundred other examples of industrial melanism in the Lepidoptera, should not be ignored. Even if the mechanism of selection in the peppered moth were not differential bird predation, this would not disprove Darwin's theory, which is supported by so many other examples. Criticisms of Kettlewell's work only criticized the agent of selection, they do not invalidate the role of selection in the evolution of melanism in this species.

If molecular analysis does provide the link between genotype and phenotype, and if the predation experiment does supply evidence fulfilling the predicted differences in bird predation of the forms to account for the current rate of decline in *carbonaria*, will the anti-evolution lobby be convinced, and redeem the reputation of the peppered moth as the exemplar *par excellence* of evolution in action? Sadly, I doubt it.

For my part, for the present, I stand by my view, given in the conclusion of Chapter 6 of *Melanism: Evolution in Action* (Majerus, 1998, p. 155):

> My view of the rise and fall of the melanic peppered moth is that differential bird predation in more or less polluted regions, together with migration, are primarily responsible, almost to the exclusion of other factors.

References

Allchin, D.S. (2001) Kettlewell's missing evidence, a study in black and white. *Journal of College Science Teaching* 31, 240–245.

Berry, R.J. (1990) Industrial melanism and peppered moths (*Biston betularia* (L.)). *Biological Journal of the Linnean Society* 39, 301–322.

Bishop, J.A. (1972) An experimental study of the cline of industrial melanism in *Biston betularia* (L.) (Lepidoptera) between urban Liverpool and rural North Wales. *Journal of Animal Ecology* 41, 209–243.

Blair, W.F. (1943) Ecological distribution of mammals of the Tularosa Basin, New Mexico. *Contributions from the Laboratory of Vertebrate Biology: University of Michigan* 20, 20–24.

Bowater, W. (1914) Heredity of melanism in Lepidoptera. *Journal of Genetics* 3, 299–315.

Brakefield, P.M. (1990) A decline of melanism in the peppered moth, *Biston betularia* in the Netherlands. *Biological Journal of the Linnean Society* 39, 327–334.

Brakefield, P.M. and Lees, D.R. (1987) Melanism in *Adalia* ladybirds and declining air pollution in Birmingham. *Heredity* 59, 273–277.

Clarke, C.A., Mani, G.S. and Wynne, G. (1985) Evolution in reverse: clean air and the peppered moth. *Biological Journal of the Linnean Society* 26, 189–199.

Cook, L.M. (2000a) A century and a half of peppered moths. *Entomologists Record and Journal of Variation* 112, 77–82.

Cook, L.M. (2000b) Changing views on melanic moths. *Biological Journal of the Linnean Society* 69, 431–441.

Cook, L.M. (2003) The rise and fall of the carbonaria form of the peppered moth. *Quarterly Review of Biology* 78, 1–19.

Cook, L.M. and Grant, B.S. (2000) Frequency of *insularia* during the decline in melanics in the peppered moth *Biston betularia* in Britain. *Heredity* 85, 580–585.

Cook, L.M., Rigby, K.D. and Seaward, M.R.D. (1990) Melanic moths and changes in epiphytic vegetation in north-west England and north Wales. *Biological Journal of the Linnean Society* 39, 343–354.

Cook, L.M., Dennis, H.L. and Mani, G.S. (1999) Melanic morph frequency in the peppered moth in the Manchester area. *Proceedings of the Royal Society of London, Series B* 266, 293–297.

Coyne, J.A. (1998) Not black and white. *Nature* 396, 35–36.

Coyne, J.A. (2002) Evolution under pressure. *Nature* 418, 19–20.

Darwin, C.R. (1859) *On the Origin of Species by Means of Natural Selection, or the Preservation of Favoured Races in the Struggle for Life.* John Murray, London.

Dice, L.R. (1930) Mammal distribution in the Alanogordo region, New Mexico. *Occasional Papers of the Museum of Zoology, University of Michigan* 213, 1–32.

Dice, L.R. (1947) Effectiveness of selection by owls on deermice (*Permycus maniculatus*) which contrast in colour with their background. *Contributions from the Laboratory of Vertebrate Biology: University of Michigan* 34, 1–20.

Edleston, R.S. (1864) No title (first *carbonaria* melanic of moth *Biston betularia*). *Entomologist* 2, 150.

Eizirik, E., Yuhki, N., Johnson, W.E., Menotti-Raymond, M., Hannah, S.S. and O'Brien, S.J. (2003) Molecular genetics and evolution of melanism in the cat family. *Current Biology* 13, 448–453.

Ford, E.B. (1937) Problems of heredity in the Lepidoptera. *Biological Reviews* 12, 461–503.

Ford, E.B. (1940) Genetic research in the Lepidoptera. *Annals of Eugenics* 10, 227–252.

Ford, E.B. (1945) *Butterflies: New Naturalist Series 1.* Collins, London.

Ford, E.B. (1955) *Moths: New Naturalist Series 30.* Collins, London.

Ford, E.B. (1964) *Ecological Genetics.* Methuen, London.

Frack, D. (1999) Peppered moths – in black and white. Posting to the Evolution List at evolution-owner@lists.calvin.edu, hosted by Calvin College: archive now available at http://www.asa3.org/archive/evolution/199903 (March, 1999) and http://www.asa3.org/archive/evolution/199903 (April, 1999).

Grant, B.S. (1999) Fine tuning the peppered moth paradigm. *Evolution* 53, 980–984.

Grant, B.S. (2002) Sour grapes of wrath. *Science* 297, 940–941.

Grant, B.S. and Clarke, C.A. (1999) An examination of intraseasonal variation in the incidence of melanism in peppered moths, *Biston betularia* (Geometridae). *Journal of the Lepidopterists Society* 53, 99–103.

Grant, B.S. and Howlett, R.J. (1988) Background selection by the peppered moth (*Biston betularia* Linn.): individual differences. *Biological Journal of the Linnean Society* 33, 217–232.

Grant, B.S. and Wiseman, L.L. (2002) Recent history of melanism in American peppered moths. *Journal of Heredity* 89, 465–471.

Grant, B.S., Owen, D.F. and Clarke, C.A. (1995) Decline of melanic moths. *Nature* 373, 565.

Grant, B.S., Owen, D.F. and Clarke, C.A. (1996) Parallel rise and fall of melanic peppered moths in America and Britain. *Journal of Heredity* 87, 351–357.

Grant, B.S., Cook, A.D., Clarke, C.A. and Owen, D.F. (1998) Geographic and temporal variation in the incidence of melanism in peppered moth populations in America and Britain. *Journal of Heredity* 89, 465–471.

Hagen, J.B. (1999) Retelling experiments: H.B.D. Kettlewell's studies of industrial melanism in peppered moths. *Biology and Philosophy* 14, 39–54.

Haldane, J.B.S. (1924) A mathematical theory of natural and artificial selection. *Transactions of the Cambridge Philosophical Society* 23, 19–41.

Harrison, J.W.H. (1927) The induction of melanism in the Lepidoptera and its evolutionary significance. *Nature* 119, 127–129.

Hiruma, K. and Riddiford, L.M. (1990) Regulation of dopa decarboxylase gene expression in the larval epidermis of the tobacco hornworm by 20-hydroxyecdysone and juvenile hormone. *Developmental Biology* 110, 509–513.

Hiruma, K. and Riddiford, L.M. (1993) Molecular mechanisms of cuticular melanisation in the tobacco hornworm, *Manduca sexta* (L) (Lepidoptera: Sphingidae). *International Journal of Insect Morphology and Embryology* 22, 103–117.

Hoekstra, H.E. and Nachman, M.W. (2003) Different genes underlie adaptive melanism in different populations of rock pocket mice. *Molecular Ecology* 12, 1185–1194.

Hollocher, H., Hatcher, J.L. and Dyreson, E.G. (2000) Genetic developmental analysis of abdominal pigmentation differences across species in the *Drosophila dunni* subgroup. *Evolution* 54, 2057–2071.

Hooper, J. (2002) *Of Moths and Men: Intrigue, Tragedy and the Peppered Moth*. Fourth Estate, London.

Howlett, R.J. (1989) The Genetics and Evolution of Rest Site Preference in the Lepidoptera. PhD thesis, Cambridge University, UK.

Howlett, R.J. and Majerus, M.E.N. (1987) The understanding of industrial melanism in the peppered moth (*Biston betularia*) (Lepidoptera: Geometridae). *Biological Journal of the Linnean Society* 30, 31–44.

Jones, C.W. (1993) Habitat Selection in Polymorphic Lepidoptera. PhD thesis, Cambridge University, UK.

Keller, E.F. (1993) *A Feel for the Organism: The Life and Work of Barbara McLintock*. W.H. Freeman, Basingstoke, UK.

Kettlewell, H.B.D. (1955a) Selection experiments on industrial melanism in the Lepidoptera. *Heredity* 9, 323–342.

Kettlewell, H.B.D. (1955b) Recognition of appropriate backgrounds by the pale and black phases of the Lepidoptera. *Nature* 175, 943–944.

Kettlewell, H.B.D. (1956) Further selection experiments on industrial melanism in the Lepidoptera. *Heredity* 10, 287–301.

Kettlewell, H.B.D. (1958) A survey of the frequencies of *Biston betularia* L. (Lep.) and its melanic forms in Britain. *Heredity* 12, 51–72.

Kettlewell, H.B.D. (1973) *The Evolution of Melanism*. Clarendon Press, Oxford.

Koch, P.B., Behnecke, B. and ffrench-Constant, R.H. (2000) The molecular basis of melanism and mimicry in a swallowtail butterfly. *Current Biology* 10, 591–594.

Lees, D.R. (1981) Industrial melanism: genetic adaptation of animals to air pollution. In: Bishop, J.A. and Cook, L.M. (eds) *Genetic Consequences of Man Made Change*. Academic Press, London, pp. 129–176.

Liebert, T.G. and Brakefield, P.M. (1987) Behavioural studies on the peppered moth *Biston betularia* and a discussion of the role of pollution and epiphytes in industrial melanism. *Biological Journal of the Linnean Society* 31, 129–150.

Majerus, M.E.N. (1989) Melanic polymorphism in the peppered moth *Biston betularia* and other Lepidoptera. *Journal of Biological Education* 23, 267–284.

Majerus, M.E.N. (1994) *Ladybirds: New Naturalist Series 81*. HarperCollins, London.

Majerus, M.E.N. (1998) *Melanism: Evolution in Action*. Oxford University Press, Oxford.

Majerus, M.E.N. (2002) *Moths*. HarperCollins, London.

Mani, G.S. and Majerus, M.E.N. (1993) Peppered moth revisited: analysis of recent decreases in melanic frequency and predictions for the future. *Biological Journal of the Linnean Society* 48, 157–165.

Matthews, R. (1999) Scientists pick holes in Darwin moth theory. *The Sunday Telegraph*, 14 March.

Nachman, M.W., Hoekstra, H.E. and D'Agostino, S.L. (2003) The genetic basis of adaptive melanism in pocket mice. *Proceedings of the National Academy of Sciences* 100, 5268–5273.

Raeburn, P. (2002) The moth that failed. *The New York Times*, 25 July, section 7, p. 3.

Rudge, D.W. (1999) Taking the peppered moth with a grain of salt. *Biology and Philosophy* 14, 9–37.

Rudge, D.W. (2003) The role of photographs and films in Kettlewell's popularisations of the phenomenon of industrial melanism. *Science and Education* 12, 261–287.

Sargent, T.D., Millar, C.D. and Lambert, D.M. (1998) The 'classical' explanation of industrial melanism: assessing the evidence. In: Hecht, M.K., MacIntyre, R.J. and Clegg, M.T. (eds) *Evolutionary Biology*, Vol. 30. Plenum Press, New York, pp. 299–322.

Smith, P.D. (2002) Darwinism in a flutter: Did a moth show evolution in action? *The Guardian*, 11 May, p. 11.

Sumner, F.B. (1934) Does 'protective coloration' protect? Results of some experiments with fishes and birds. *Proceedings of the National Academy of Sciences* 20, 559–564.

Tutt, J.W. (1896) *British Moths*. George Routledge.

Wade, N. (2002) Staple of evolutionary history may not be a textbook case. *New York Times*, 18 June, section F, p. 3.

Wells, J. (2000) *Icons of Evolution: Science or Myth? Why Much of What We Teach About Evolution is Wrong*. Regnery Press, Washington, DC.

Wells, J. (2001) Second thoughts about peppered moths: this classical story of evolution by natural selection needs revising. *The True Origin Archive*, http://trueorigin. org/pepmoth1.htm

Wells, J. (2002) Moth-eaten statistics: A reply to Kenneth R. Miller. *Discovery Institute: Centre for Renewal of Science and Culture – Article Database*, http://www. discovery.org/viewDB/index.php3?command=view&id=1147&program=CRSC

West, B.K. (1994) The continued decline of melanism in *Biston betularia* L. (Lep.: Geometridae) in N.W. Kent. *Entomologists Record and Journal of Variation* 106, 229–232.

Wittkopp, P.J., Williams, B.L., Selegue, J.E. and Carroll, S.B. (2003) *Drosophila* pigmentation evolution: divergent genotypes underlying convergent evolution. *Proceedings of the National Academy of Sciences USA* 100, 1808–1813.

Wright, S. (1978) *Evolution and the Genetics of Populations*, Vol. 4: *Variability Within and Among Natural Populations*. University of Chicago Press, Chicago, Illinois.

Yoon, C.K. (1996) Parallel plots in classic of evolution. *New York Times: Science Times*, 12 November, pp. C1, C7.

13 Insecticide Resistance in the Mosquito *Culex pipiens*: Towards an Understanding of the Evolution of *ace* Genes

M. WEILL,[1] P. LABBE,[1] O. DURON,[1] N. PASTEUR,[1] P. FORT[2] AND M. RAYMOND[1]

[1] *Institut des Sciences de l'Évolution (UMR 5554), Laboratoire Génétique et Environnement, Université Montpellier II, Montpellier, France;* [2] *Centre de Recherche en Biochimie des Macromolécules (FRE2593), CNRS, Montpellier, France*

1. Introduction

Acetylcholinesterase (AChE, EC 3.1.1.7) terminates synaptic transmission at cholinergic synapses in the central nervous system (CNS) of insects, by rapid hydrolysis of the neurotransmitter acetylcholine (Toutant, 1989). Numerous studies have focused on insect AChE because it is the target of organophosphates (OPs) and carbamates, two major classes of pesticides used for pest management in agriculture and public health. Target (AChE) insensitivity has been described in many insect species (see review in Fournier and Mutéro, 1994). To identify the mutation(s) reducing target sensitivity and thus conferring insecticide resistance, genes encoding AChE (i.e. *ace* genes) have been cloned and sequenced. The first invertebrate *ace* gene was cloned in *Drosophila melanogaster*, and several mutations involved in AChE insensitivity were identified (Fournier *et al.*, 1993; Fournier and Mutéro, 1994; Mutéro *et al.*, 1994). Independent studies indicated that only one *ace* gene was coding for AChE in cholinergic synapses in this species, e.g. segmental aneuploidy technique, mutagenesis, and rescue of lethal mutations by germline minigene transformation (Hall and Kankel, 1976; Greenspan *et al.*, 1980; Hoffmann *et al.*, 1992; Fournier and Mutéro, 1994). From this work on *Drosophila*, it was assumed that only one *ace* gene was present in insects.

The gene coding for the AChE insecticide target in mosquitoes was not easy to find because of erroneous assumptions introduced by the *Drosophila melanogaster* model. In the case of *Culex pipiens*, the gene cloned by homology with the *Drosophila* insecticide target (now named *ace-2*) was sex-linked and not involved in resistance, suggesting that another *ace* gene was coding for the main synaptic AChE in mosquitoes (Bourguet *et al.*, 1996b; Malcolm *et al.*, 1998). Indeed, the recent release of the full *Anopheles gambiae* genome disclosed the presence of two distinct and divergent *ace* genes (47% divergence at the amino acid level), one of which, *ace-1*, having no homologue in *D. melanogaster*, was responsible for resistance in the mosquito *C. pipiens* (Weill *et al.*, 2002, 2003).

The cloning of *ace-1* in the mosquito *C. pipiens* was an opportunity to study in detail adaptive evolution at this locus. A gene duplication of the *ace-1* locus was detected, increasing in frequency in natural populations. This interesting situation in the mosquito *C. pipiens* shed some light on the puzzling situation found in Diptera, which could not be considered as resulting from a neutral process.

2. Insecticide Resistance and *ace-1* in the Mosquito *Culex pipiens*

Insensitive AChE is one of the possible resistance mechanisms (for a general review, see McKenzie, 1996). In the mosquito, *C. pipiens*, it was first detected in southern France in a 1978 sample (Raymond *et al.*, 1986). This trait was monofactorial and autosomal. Homozygous mosquitoes for this trait are highly resistant to OPs and carbamates (100-fold to chlorpyrifos, >9000-fold to propoxur), due to the inability of the insecticide to inhibit the mutated AChE. As a result, the resistance allele (called *ace-1^R*) coding this insensitive AChE has spread rapidly in natural populations. However, its frequency remained low in adjacent untreated areas connected by migration, indicating that there is a fitness cost associated with *ace-1^R*. The accumulation of data from field and laboratory studies (including the recent cloning of *ace-1* in mosquitoes and the identification of mutations responsible for high insecticide resistance) has allowed us to develop a better understanding of this fitness cost.

2.1 The fitness cost of *ace-1^R*

The cloning of *ace-1* in mosquitoes (Weill *et al.*, 2002) was a first step to identifying the mutations responsible for the AChE insensitivity. A single non-synonymous point mutation was found (GGC to AGC at position 119, using the *Torpedo* nomenclature), giving a glycine to serine substitution (G119S). Moreover, within the *C. pipiens* species, this mutation arose at least twice independently, once in the temperate form (*C. p. pipiens*) and once in the tropical form (*C. p. quinquefasciatus*) (Weill *et al.*, 2003). This mutation, located in the oxyanion hole, alters the enzyme active site, and the serine at position 119 reduces the access to the catalytic triads (Weill *et al.*, 2004). The access reduction is substantial for the insecticide (leading to the insensitivity), but also affects the binding of acetylcholine, the natural substrate. Thus insensitivity towards the insecticide is associated with a reduction in the normal enzyme function. In *C. pipiens*, this reduction is about 60% (Bourguet *et al.*, 1997a).

Apparently, AChE has been naturally selected for its particularly high enzymatic activity, as it is one of the fastest enzymes known so far: up to 10^4 substrate molecules are hydrolysed per second per enzyme molecule (Quinn, 1987). This suggests that a slight reduction in activity could be somehow translated to a significant fitness cost, through suboptimal values for one or several life-history traits. Thus the observed 60% reduction of AChE activity should convert to substantial phenotypic differences, probably amenable to empirical measurements. Several life-history traits were investigated in absence of insecticide, either in the field or in the laboratory, and most display a negative change when the gene *ace-1^R* is present (Table 13.1). An overall fitness cost cannot be calculated from individual estimates of the various life-history traits, as the ecological importance of each trait in the species life cycle is insufficiently known. Alternatively, a global measure of the fitness cost is feasible by studying the cline of *ace-1^R* frequency at the boundary between a treated and a non-

Table 13.1. Effect of *ace-1^{RR}* on several life-history traits, relative to *ace-1^{SS}*, measured in *Culex pipiens* from the field or in the laboratory.

Life history trait	Effect	Reference
Field		
Female overwintering survival:		
End of winter	Mortality 7% per day	(Gazave *et al.*, 2001)
All winter	Mortality 51–69%[a]	(Chevillon *et al.*, 1997; Lenormand *et al.*, 1999; Lenormand and Raymond, 2000)
Developmental time	+8.2%	(Bourguet *et al.*, 2004)
Larval survival	Generally decreased[b]	(Bourguet *et al.*, 2004)
Wing size	−2.2%	(Bourguet *et al.*, 2004)
Fluctuating asymmetry	None	(Bourguet *et al.*, 2004)
Laboratory		
Developmental time	*ace-1^{RR}* slower than *ace-1^{RS}*	(Raymond *et al.*, 1985)
Mating competition	3.2× reduction	(Berticat *et al.*, 2002a)
Larval predation by:		
Sigara lateralis	+1.3×	(Berticat *et al.*, 2004)
Guignotus pusillus	+2.2×	(Berticat *et al.*, 2004)
Hydrometra stagnorum	+2.1×	(Berticat *et al.*, 2004)
Plea minutissima	+1.9–2.7×	(Berticat *et al.*, 2004)
Adult predation by:		
Holocnemus pluchei	None	(Berticat *et al.*, 2004)
Wolbachia density:		
Larvae	Higher density (4.6×)	(Berticat *et al.*, 2002b)
Males	Higher density (9.8×)	(Berticat *et al.*, 2002b)
Females	Higher density (2.1×)	(Berticat *et al.*, 2002b)

[a] Effect of *ace-1^R* relative to *ace-1^S*.
[b] Variable with larval density.

treated area. This provides a cost of approx. 11% per generation during the breeding season, and 50–60% for survival during the overwintering season (Lenormand et al., 1998, 1999; Lenormand and Raymond, 2000).

The cost associated with a new adaptive gene is expected to evolve, as any genetic change reducing it will be selected (Fisher, 1958; Orr, 1998). Despite theoretical work, due to insufficient empirical data, this process is not well understood. In *Lucilia cuprina*, cost reduction operates through the selection of a modifier (not yet identified molecularly), independent from the resistance gene itself (McKenzie and Purvis, 1984; McKenzie and Game, 1987). For *Culex pipiens*, cost reduction for the *Ester* locus (providing resistance through esterase overproduction) operates through allele replacement, i.e. less costly resistance alleles progressively replace more costly ones (Guillemaud et al., 1998; Raymond et al., 1998). For *ace-1^R*, our actual understanding of the cholinergic synapse suggests that the primary factor generating the cost is an insufficient activity of the insensitive AChE (Bourguet et al., 1997a,b). Thus cost reduction could operate by increasing AChE activity. This could result from the occurrence of another mutation at *ace-1*, compensating for the detrimental effect of G119S on AChE activity without substantially affecting insensitivity, or from increased *ace-1^R* expression (Charpentier and Fournier, 2001), through, for example, increased transcription, translation, or mRNA stability. In the case of *C. pipiens*, however, increased AChE activity is achieved through an unusual mechanism: a duplication of *ace-1* generating a haplotype with an *ace-1^S* and an *ace-1^R* copy.

2.2 Duplication as a cost modifier

A haplotype corresponding to a duplication (*ace-1$^{R\cdot S}$*) has been described in *C. pipiens* from Martinique and southern France, corresponding to two independent occurrences (Bourguet et al., 1996a; Lenormand et al., 1998; Weill et al., 2003). In both cases, a copy encoding the sensitive and the insensitive AChE are both present in the same haplotype. This haplotype occurred around 1993 in southern France, and has spread tremendously quickly; from being undetectable in 1993 it reached a frequency of ~30% in 1995, partially replacing *ace-1^R* (Lenormand et al., 1998). Its advantage over *ace-1^R* is not due to higher insensitivity, as the '*R* copy' of *ace-1$^{R\cdot S}$* is also characterized with the same G119S mutation. The overall AChE activity of *ace-1$^{R\cdot S}$* is higher than that displayed by *ace-1^R*, suggesting a lower cost and explaining this replacement.

Much more remains to be discovered about this haplotype, such as its exact genomic organization and the presence of possible dosage compensation. Nevertheless, it represents a rare example of an incipient occurrence of a new function through gene duplication, driven by selection. Mosquitoes possessing the *ace-1$^{R\cdot S}$* haplotype express two types of AChE in the cholinergic synapses, with distinct enzymatic parameters, each one coded by a distinct locus. This situation is prone to evolution in various directions, potentially shedding some light on older duplications of the *ace* gene in animals, and on the puzzling situation found for *ace* genes in Diptera.

3. The Situation of *ace* Genes in Other Mosquitoes

The gene *ace-1* was found in all mosquito species investigated (including the genera *Culex*, *Aedes* and *Anopheles*); the gene *ace-2*, homologous to the *Drosophila ace* gene, was also found in mosquito species from the above genera (Weill *et al.*, 2002). In the complete genome of *Anopheles gambiae*, only two *ace* genes were found (*ace-1* and *ace-2*). Two mosquito species of the *Anopheles* genus have developed a highly insensitive AChE: *A. gambiae ss* (the main malaria vector in Africa) and *A. albimanus* (a malaria vector in Central America). In both cases, the mutation G119S was found and was the same as in *C. pipiens* (Weill *et al.*, 2003, 2004). Thus the same mutation (G119S in *ace-1*) explaining a high AChE insensitivity has occurred independently at least four times in mosquitoes (twice in *C. pipiens*, and twice in *Anopheles*). This situation suggests that there are few possible sites, perhaps only one, to greatly enhance AChE1 insensitivity towards OP or carbamate insecticides without drastically reducing its activity. Moderate levels of AChE insensitivity, described in *Culex pipiens* from Cyprus and in a Middle-Eastern malaria vector, *Anopheles sacharovi* (Hemingway *et al.*, 1985; Bourguet *et al.*, 1997a), might be explained by alternative mutations, which remain to be identified.

Due to the identical G119S mutation found in *Anopheles*, a high fitness cost similar to that observed in *C. pipiens* is to be expected. Preliminary results obtained with field samples from the Ivory Coast indicate that this is probably the most likely case for *A. gambiae* (F. Chandre, France, 2003, personal communication). Incidentally, the dramatic decrease in *A. albimanus* with an insensitive AChE1 following the reduction of pesticide treatment in Central America (Hemingway *et al.*, 1997), is consistent with the hypothesis of high fitness cost of the G119S mutation in this species.

In conclusion, two *ace* genes exist in the Culicidae family; *ace-1* providing the main synaptic AChE and displaying mutations explaining AChE insensitivity, and *ace-2* with an unknown function. All the available data are consistent with a severe fitness cost associated with an *ace-1* mutation generating a high AChE insensitivity.

4. The Situation of *ace* Genes in Other Diptera

The number of *ace* genes is easily identified in species in which the genome is completely sequenced. There is only one *ace* gene (now called *ace-2*) in *Drosophila melanogaster* (Weill *et al.*, 2002). This gene is expressed in cholinergic synapses, and displays mutations responsible for AChE insensitivity and insecticide resistance (Fournier *et al.*, 1993; Mutéro *et al.*, 1994). Two other dipteran species use *ace-2* for their cholinergic synapses, as indicated by *ace-2* mutations providing OP or carbamate resistance: the housefly *Musca domestica* and the olive fruit fly *Bactrocera oleae* (Kozaki *et al.*, 2001; Walsh *et al.*, 2001; Vontas *et al.*, 2002). No information is currently available for the eventual presence or absence of *ace-1* in these species. They belong to distinct families (Drosophilidae, Muscidae and Tephritidae), all included in the suborder Brachycera of the Diptera (and more precisely in the Muscomorpha infraorder). The other suborder, the Nematocera, contains the Culicidae family, where ace-1 is the locus coding for insensitive AChE.

Thus within the Diptera, there are species where cholinergic AChE insensitivity is provided by a mutation either on ace-1 or on ace-2 (Weill et al., 2002). These two ace genes result from an ancient duplication probably older than the emergence of arthropods (Fig. 13.1). In other insects (at least one Lepidoptera and five Hemiptera species), the gene homologous to ace-2 is not involved in resistance (Tomita et al., 2000; Ren et al., 2002; Javed et al., 2003). This is also probably the case for a coleopteran (*Leptinotarsa decemlineata*, see Weill et al., 2002). Thus the ancestral character in Diptera was probably the simultaneous presence of both ace-1 and ace-2, with ace-1 providing the main synaptic AChE, as it is now the case in the Culicidae. However, in a group within the Diptera which remains to be clearly circumscribed, ace-1 was lost and ace-2 acquired the function of providing the main synaptic AChE. How could such a dramatic change have taken place?

5. Changing the Function of an *ace* Gene

5.1 Necessary conditions

The change from ace-1 to ace-2 in Diptera first implies that both proteins were expressed in cholinergic synapses before the loss of ace-1, and have (or have had) some degree of compensatory function for acetylcholine (ACh) hydrolysis.

The compensatory function of two ace genes (ace-1 and ace-2) in cholinergic synapses is known in the nematode *Caenorhabditis elegans*: although ace-1 ace-2 double null mutants display abnormal, uncoordinated locomotion (the Ace-Unc phenotype), strains carrying a single mutation in either ace-1 or ace-2 have little apparent defect in behaviour, growth, or reproduction (Culotti et al., 1981; Johnson et al., 1981). The lack of phenotypic defect in the single mutants might be explained if the specific functions of each gene are non-essential in laboratory conditions (for details, see Johnson et al., 1988; Grauso et al., 1998). The expression pattern of ace-1 and ace-2 in C. elegans is similar in the related species C. briggsae (Combes et al., 2000, 2001). These two species diverged 40 million years ago (Kennedy et al., 1993), suggesting that the actual expression pattern of ace-1 and ace-2 is relatively stable. It is thus possible that the presence of two cholinergic AChE with distinct enzymatic properties, and with different expression patterns (Combes et al., 2001), is adaptive in nematodes. Another example of compensatory function has been described in mice, where AChE nullizygotes are non-lethal, due to ACh hydrolysis by BChE (or butyrylcholinesterase, which belongs to the AChE family) in glial cells surrounding the synapse (Melusam, 2002). Are ace-1 and ace-2 in insects co-expressed, even in unequal quantities, in cholinergic synapses? Preliminary data are promising (ace-2 is expressed in the brain of *Apis mellifera*; Shapira et al., 2001), although this question cannot be answered yet.

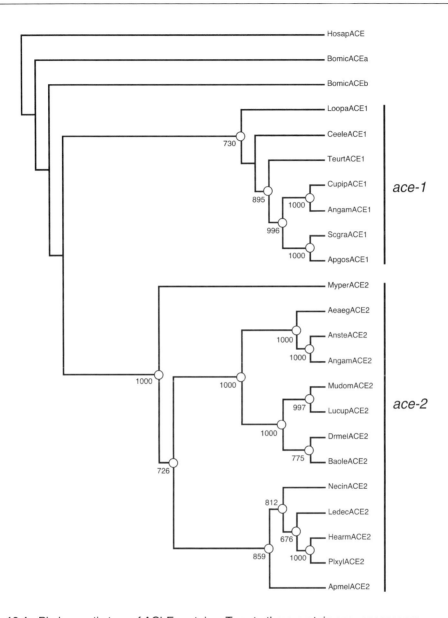

Fig. 13.1. Phylogenetic tree of AChE proteins. Twenty-three protein sequences were retrieved from the ESTHER database (http://www.ensam.inra.fr/cgi-bin/ace/index). Sequences were aligned and a bootstrapped unrooted tree was constructed. Only nodes supported by > 50% bootstraps (i.e. scores above 500) are indicated. Scale bar represents 10% divergence. Aeaeg: *Aedes aegypti*; Angam: *Anopheles gambiae*; Anste: *Anopheles stephensi*; Apgos: *Aphis gossypi*; Apmel: *Apis mellifera*; Baole: *Bactroceras oleae*; Bomic: *Boophilus microplus*; Ceele: *Caenorhabditis elegans*; Cupip: *Culex pipiens*; Drmel: *Drosophila melanogaster*; Hearm: *Helicoverpa armigera*; Hosap: *Homo sapiens*; Ledec: *Leptinotarsa decemlineata*; Loopa: *Loligo opalescens*; Lucup: *Lucilia cuprina*; Mudom: *Musca domestica*; Myper: *Myzus persicae*; Necin: *Nephotettix cincticeps*; Plxyl: *Plutella xylostella*; Scgra: *Schizaphis graminum*; Teurt: *Tetranychus urticae*.

The change from *ace-1* to *ace-2* in Diptera implies also that a cost was transiently generated, at least while the changed expression of one *ace* gene was not totally compensated for by a corresponding change in the other. As both a reduction or an increase of AChE activity generates a cost, the first event could have been either a reduced expression of *ace-1*, or an increased expression of *ace-2* in synapses. As the *C. pipiens* situation shows, the advantage must be greater than the cost for this change to take place (despite its cost, *ace-1^R* was spreading in treated areas).

5.2 A possible scenario

The *C. pipiens* example indicates that natural selection was responsible for the spread of the initial *ace-1* gene duplication. This is because in an environment where challenge with AChE inhibitors (here insecticides) is frequent, the co-expression in the synapse of both gene products of the *ace-1^{R-S}* haplotype is adaptive. Once this duplication is established (which is not yet the case for *C. pipiens*), each gene starts to diverge in sequence, and both might acquire distinct non-cholinergic functions. A new environmental change could occur, removing the need for the synaptic co-expression of two distinct AChE. In such a case, there is the possibility that only one gene will contribute to the main synaptic AChE. This is because deleterious mutations at one genetic locus could be compensated for by overexpression of the other gene. This process would lead, with sufficient evolutionary time, to one gene being mainly expressed in synapses. The other gene could be maintained in the genome if it had non-cholinergic functions, otherwise it will be lost. AChE have functions other than neurotransmitter hydrolysis in cholinergic synapses (Massoulié *et al.*, 1993); for example, there are striking cases of non-neuronal AChE in parasitic nematodes (Lee, 1996; Hussein *et al.*, 1999). Globally, our knowledge on the non-cholinergic function of AChE is very limited, including for *ace-2* in insects.

5.3 *ace* duplication

The process of *ace* duplication has apparently occurred many times in evolution, in the various animal lineages. For example, in Chordates at least two independent duplications occurred, one giving the AChE and BChE in vertebrates, and the other giving two *ace* genes in Cephalochordates (Weill *et al.*, 2002). Another example is found in nematodes: a duplication of the ancestor of *ace-2* has occurred, the new copy being subject to a later duplication (before the divergence of *C. elegans* and *C. briggsae*), giving *ace-3* and *ace-4* (Grauso *et al.*, 1998). Once a duplication has occurred, there is the possibility of a diversification of non-cholinergic functions, as explained above. For example, AChE and BChE in vertebrates have different names because their functions are very distinct: BChE is no longer expressed in cholinergic synapses (although it can contribute to ACh hydrolysis, see above), and its exact function, perhaps blood detoxification, is not settled (for a discussion, see

Massoulié *et al.*, 1993). The selective advantage of the initial *ace* duplication in vertebrates is not known. However, the possible detoxification function of BChE suggests that the presence of cholinergic inhibitors in the environment (a frequent challenge for herbivores) could represent an interesting candidate.

6. Conclusion

We are witnessing the occurrence of a new duplication (*ace-1^{R-S}*) in *C. pipiens*. This appeared approximately 10 years ago. The duplication spreads due to its advantage over other alleles (*ace-1^R* or *ace-1^S*), and it co-expresses two different AChE in synapses. It is probably the only published example of a duplication which has been observed occurring in a natural population, and it indicates that non-neutral processes are at play. It is tempting to speculate that the initial duplication giving *ace-1* and *ace-2*, which took place at least 5×10^8 years ago, was also driven by natural selection, and that the co-expression of AChE1 and AChE2 in synapses was an adaptation. From such a situation, it was possible to preferentially express only one of them in synapses, the other one being expressed preferentially in other locations or even sometimes lost. Some Diptera kept *ace-1* for the synaptic AChE, others *ace-2*. Data for many other insects and other arthropods will probably soon be available, extending our understanding of the evolution of *ace* genes on a broader scale. In addition, the *ace-1^{R-S}* haplotype in *C. pipiens* represents a unique model to enhance our understanding of how adaptation constructs new functions at the beginning of the process. Monitoring its evolution in natural populations represents a fascinating project.

Acknowledgements

We are grateful to C. Bernard, A. Berthomieu and C. Berticat for technical help, and V. Durand for help in the literature search. This investigation received financial support from the Ministère de la Recherche (PAL+ No 2002–45) and CNRS 'Post-séquençage anophèle'. This is publication ISEM 2004.006.

References

Berticat, C., Boquien, G., Raymond, M. and Chevillon, C. (2002a) Insecticide resistance genes induce a mating competition cost in *Culex pipiens* mosquitoes. *Genetical Research* 79, 41–47.

Berticat, C., Rousset, F., Raymond, M., Berthomieu, A. and Weill, M. (2002b) High *Wolbachia* density in insecticide resistant mosquitoes. *Proceedings of the Royal Society of London, Series B* 269, 1413–1416.

Berticat, C., Duron, O., Heyse, D. and Raymond, M. (2004) Insecticide resistance genes confer a predation cost on mosquitoes, *Culex pipiens*. *Genetical Research* 83, 189–196.

Bourguet, D., Raymond, M., Bisset, J., Pasteur, N. and Arpagaus, M. (1996a) Duplication of the *Ace.1* locus in *Culex pipiens* from the Caribbean. *Biochemical Genetics* 34, 351–362.

Bourguet, D., Raymond, M., Fournier, D., Malcolm, C.A., Toutant, J.P. and Arpagaus, M. (1996b) Existence of two acetylcholinesterases in the mosquito *Culex pipiens* (Diptera: Culicidae). *Journal of Neurochemistry* 67, 2115–2123.

Bourguet, D., Lenormand, T., Guillemaud, T., Marcel, V. and Raymond, M. (1997a) Variation of dominance of newly arisen adaptive genes. *Genetics* 147, 1225–1234.

Bourguet, D., Raymond, M., Berrada, S. and Fournier, D. (1997b) Interaction between acetylcholinesterase and choline acetyltransferase: a hypothesis to explain unusual toxicological responses. *Pesticide Science* 51, 276–282.

Bourguet, D., Guillemaud, T., Chevillon, C. and Raymond, M. (2004) Fitness costs of insecticide resistance in natural breeding sites of the mosquito *Culex pipiens*. *Evolution* 58, 128–135.

Charpentier, A. and Fournier, D. (2001) Levels of total acetylcholinesterase in *Drosophila melanogaster* in relation to insecticide resistance. *Pesticide Biochemistry and Physiology* 70, 100–107.

Chevillon, C., Bourguet, D., Rousset, F., Pasteur, N. and Raymond, M. (1997) Pleiotropy of adaptive changes in populations: comparisons among insecticide resistance genes in *Culex pipiens*. *Genetical Research* 68, 195–203.

Combes, D., Fedon, Y., Grauso, M., Toutant, J.P. and Arpagaus, M. (2000) Four genes encode acetylcholinesterases in the nematodes *Caenorhabditis elegans* and *Caenorhabditis brigsae*: cDNA sequences, genomic structures, mutations and *in vivo* expression. *Journal of Molecular Biology* 300, 727–742.

Combes, D., Fedon, Y., Toutant, J.-P. and Arpagaus, M. (2001) Acetylcholinesterase genes in the nematode *Caenorhabditis elegans*. *International Review of Cytology* 209, 207–239.

Culotti, J.G., Von Ehrenstein, G., Culotti, M.R. and Russell, R.L. (1981) A second class of acetylcholinesterase-deficient mutants of the nematode *Caenorhabditis elegans*. *Genetics* 97, 281–305.

Fisher, R.A. (1958) *The Genetical Theory of Natural Selection*. Dover, New York.

Fournier, D. and Mutéro, A. (1994) Modification of acetylcholinesterase as a mechanism of resistance to insecticides. *Comparative Biochemistry and Physiology* 108C, 19–31.

Fournier, D., Mutéro, A., Pralavorio, M. and Bride, J.M. (1993) *Drosophila* acetylcholinesterase: mechanisms of resistance to organophosphates. *Chemico-Biological Interactions* 87, 233–238.

Gazave, E., Chevillon, C., Lenormand, T., Marquine, M. and Raymond, M. (2001) Dissecting the cost of insecticide resistance genes during the overwintering period of the mosquito *Culex pipiens*. *Heredity* 87, 441–448.

Grauso, M., Culetto, E., Combes, D., Fedon, Y., Toutant, J.-P. and Arpagaus, M. (1998) Existence of four acetylcholinesterase genes in the nematodes *Caenorhabditis elegans* and *Caenorhabditis briggsae*. *FEBS Letters* 424, 279–284.

Greenspan, R.J., Finn, J.A. and Hall, J.C. (1980) Acetylcholinesterase mutants in *Drosophila* and their effects on the structure and function of the central nervous system. *Journal of Comparative Neurology* 189, 741–774.

Guillemaud, T., Lenormand, T., Bourguet, D., Chevillon, C., Pasteur, N. and Raymond, M. (1998) Evolution of resistance in *Culex pipiens*: allele replacement and changing environment. *Evolution* 52, 430–440.

Hall, J.C. and Kankel, D.R. (1976) Genetics of acetylcholinesterase in *Drosophila melanogaster*. *Genetics* 83, 517–535.

Hemingway, J., Malcolm, C.A., Kissoon, K.E., Boddington, R.G., Curtis, C.F. and Hill, N. (1985) The biochemistry of insecticide resistance in *Anopheles sacharovi*:

comparative studies with a range of insecticide susceptible and resistant *Anopheles* and *Culex* species. *Pesticide Biochemistry and Physiology* 24, 68–76.

Hemingway, J., Penulla, R.P., Rodriguez, A.D., James, B., Edge, W., Rogers, H. and Rodriguez, M.H. (1997) Resistance management strategies in malaria vector mosquito control: a large-scale field trial in southern Mexico. *Pesticide Science* 51, 375–382.

Hoffmann, F., Fournier, D. and Spierer, P. (1992) Minigenes rescues acetylcholinesterase lethal mutations in *Drosophila melanogaster*. *Journal of Molecular Biology* 223, 17–22.

Hussein, A.S., Chacón, M.R., Smith, A.M., Acevedo, R.T. and Selkirk, M.E. (1999) Cloning, expression, and properties of nonneuronal secreted acetycholinesterase from the parasitic nematode *Nippostrongylus brasiliensis*. *Journal of Biological Chemistry* 274, 9312–9319.

Javed, N., Viner, R., Williamson, M.S., Field, L.M., Devonshire, A.L. and Moores, G.D. (2003) Characterization of acetylcholinesterases, and their genes, from the hemipteran species *Myzus persicae* (Sulzer), *Aphis gossypii* (Glover), *Bemisia tabaci* (Gennadius) and *Trialeurodes vaporariorum* (Westwood). *Insect Molecular Biology* 12, 613–620.

Johnson, C.D., Duckett, J.G., Culotti, J.G., Herman, R.K., Meneely, P.M. and Russell, R.L. (1981) An acetylcholinesterase-deficient mutant of the nematode *Caenorhabditis elegans*. *Genetics* 97, 261–279.

Johnson, C.D., Rand, J.B., Herma, R.K., Stern, B.D. and Russell, R.L. (1988) The acetylcholinesterase genes of *C. elegans*: identification of a third gene (*ace-3*) and mosaic mapping of a synthetic lethal phenotype. *Neuron* 1, 165–173.

Kennedy, B.P., Aamodt, E.J., Allen, F.L., Chung, M.A., Heschl, M.F.P. and McGhee, J.D. (1993) The gut esterase gene (*ges-1*) from the nematodes *Caenorhabditis elegans* and *Caenorhabditis briggsae*. *Journal of Molecular Biology* 229, 890–908.

Kozaki, T., Shono, T., Tomita, T. and Kono, Y. (2001) Fenitroxon insensitive acetylcholinesterases of the housefly, *Musca domestica* associated with point mutations. *Insect Biochemistry and Molecular Biology* 31, 991–997.

Lee, D.L. (1996) Why do some nematode parasites of the alimentary tract secrete acetylcholinesterase? *International Journal for Parasitology* 26, 499–508.

Lenormand, T. and Raymond, M. (2000) Clines with variable selection and variable migration: model and field studies. *American Naturalist* 155, 70–82.

Lenormand, T., Guillemaud, T., Bourguet, D. and Raymond, M. (1998) Appearance and sweep of a gene duplication: adaptive response and potential for a new function in the mosquito *Culex pipiens*. *Evolution* 52, 1705–1712.

Lenormand, T., Bourguet, D., Guillemaud, T. and Raymond, M. (1999) Tracking the evolution of insecticide resistance in the mosquito *Culex pipiens*. *Nature* 400, 861–864.

Malcolm, C.A., Bourguet, D., Ascolillo, A., Rooker, S.J., Garvey, C.F., Hall, L.M.C., Pasteur, N. and Raymond, M. (1998) A sex-linked *ace* gene, not linked to insensitive acetylcholinesterase-mediated insecticide resistance in *Culex pipiens*. *Insect Molecular Biology* 7, 107–120.

Massoulié, J., Pezzementi, L., Bon, S., Krejci, E. and Vallette, F.M. (1993) Molecular and cellular biology of cholinesterases. *Progress in Neurobiology* 41, 31–91.

McKenzie, J.A. (1996) *Ecological and Evolutionary Aspects of Insecticide Resistance*. R.G. Landes Company and Academic Press, Georgetown, Texas.

McKenzie, J.A. and Game, A.Y. (1987) Diazinon resistance in *Lucilia cuprina*: mapping of a fitness modifier. *Heredity* 59, 371–381.

McKenzie, J.A. and Purvis, A. (1984) Chromosomal localisation of fitness modifiers of diazinon resistance genotypes of *Lucilia cuprina*. *Heredity* 53, 625–634.

Melusam, M. (2002) Cholinesterases in health and disease. In: *Seventh International Meeting on Cholinesterases*. Pucon, Chile, p. 13.

Mutéro, A., Pralavorio, M., Bride, J.M. and Fournier, D. (1994) Resistance-associated point mutations in insecticide-insensitive acetylcholinesterase. *Proceedings of the National Academy of Sciences USA* 91, 5922–5926.

Orr, H.A. (1998) The population genetics of adaptation: the distribution of factors fixed during adaptive evolution. *Evolution* 52, 935–949.

Quinn, D.M. (1987) Acetylcholinesterase: enzyme structure, reaction dynamics, and virtual transition states. *Chemical Reviews* 87, 955–979.

Raymond, M., Pasteur, N., Fournier, D., Cuany, A., Bergé, J. and Magnin, M. (1985) Le gène d'une acétylcholinestérase insensible au propoxur détermine la résistance de *Culex pipiens* à cet insecticide. *Comptes Rendus de l'Académie de Sciences de Paris, Série III* 300, 509–512.

Raymond, M., Fournier, D., Bride, J.-M., Cuany, A., Bergé, J., Magnin, M. and Pasteur, N. (1986) Identification of resistance mechanisms in *Culex pipiens* (Diptera: Culicidae) from southern France: insensitive acetylcholinesterase and detoxifying oxidases. *Journal of Economic Entomology* 79, 1452–1458.

Raymond, M., Chevillon, C., Guillemaud, T., Lenormand, T. and Pasteur, N. (1998) An overview of the evolution of overproduced esterases in the mosquito *Culex pipiens*. *Philosophical Transactions of the Royal Society of London, Series B* 353, 1–5.

Ren, X., Han, Z. and Wang, Y. (2002) Mechanisms of monocrotophos resistance in cotton bollworm, *Helicoverpa armigera* (Hübner). *Archives of Insect Biochemistry and Physiology* 51, 103–110.

Shapira, M., Thompson, C.K., Soreq, H. and Robinson, G.E. (2001) Changes in neuronal acetylcholinesterase gene expression and division of labor in honey bee colonies. *Journal of Molecular Neuroscience* 17, 1–12.

Tomita, T., Hidoh, O. and Kono, Y. (2000) Absence of protein polymorphism attributable to insecticide-insensitivity of acetylcholinesterase in the green rice leafhopper, *Nephotettix cincticeps*. *Insect Biochemistry and Molecular Biology* 30, 325–333.

Toutant, J.P. (1989) Insect acetylcholinesterase: catalytic properties, tissue distribution and molecular forms. *Progress in Neurobiology* 32, 423–446.

Vontas, J.G., Hejazi, M.J., Hawkes, N.J., Cosmidis, N., Loukas, M. and Hemingway, J. (2002) Resistance-associated point mutations of organophosphate acetylcholinesterase, in the olive fruit fly *Bactrocera oleae*. *Insect Molecular Biology* 11, 329–336.

Walsh, S.B., Dolden, T.A., Moores, G.D., Kristensen, M., Lewis, T., Devonshire, A.L. and Williamson, M.S. (2001) Identification and characterization of mutations in housefly *(Musca domestica)* acetylcholinesterase involved in insecticide resistance. *Biochemical Journal* 359, 175–181.

Weill, M., Fort, P., Berthomieu, A., Dubois, M.-P., Pasteur, N. and Raymond, M. (2002) A novel acetylcholinesterase gene in mosquitoes' codes for the insecticide target and is non-homologous to the *ace* gene in *Drosophila*. *Proceedings of the Royal Society of London, Series B* 269, 2007–2016.

Weill, M., Lutfalla, G., Mogensen, K., Chandre, F., Berthomieu, A., Berticat, C., Pasteur, N., Philips, A., Fort, P. and Raymond, M. (2003) Insecticide resistance in mosquito vectors. *Nature* 423, 136–137.

Weill, M., Malcolm, M., Chandre, F., Mogensen, K., Berthomieu, A., Marquine, M. and Raymond, M. (2004) The unique mutation in *ace-1* giving high insecticide resistance is easily detectable in mosquito vectors. *Insect Molecular Biology* 13, 1–7.

14 Molecular and Ecological Differentiation of Species and Species Interactions Across Large Geographic Regions: California and the Pacific Northwest

JOHN N. THOMPSON[1] AND RYAN CALSBEEK[2]

[1]*Department of Ecology and Evolutionary Biology, University of California, Santa Cruz, California, USA;* [2]*Center for Tropical Research, University of California, Los Angeles, California, USA*

1. Introduction

Through a combination of molecular and ecological studies in recent decades, we now know that many insect species are collections of genetically differentiated populations. Few insect species are panmictic in their molecular markers throughout their geographical ranges. Moreover, insect populations often differ in their phenotypic traits, including their feeding preferences and performance on different hosts, their ways of searching for hosts, and their defences against enemies (e.g. Singer and Thomas, 1996; Nielsen, 1997; Thompson, 1998; Kraaijeveld and Godfray, 1999; Craig *et al.*, 2000; de Jong *et al.*, 2000; Fellowes and Godfray, 2000; Althoff and Thompson, 2001).

One of the current major challenges in evolutionary ecology is therefore to link the geographic scales of molecular and ecological differentiation. Much of the pattern of molecular differentiation may result from historical patterns of population subdivision, accompanied by random genetic drift on molecular markers or limited dispersal (Irwin, 2002). In contrast, much of the pattern of geographic differentiation in ecological traits may result from complex patterns of divergent and convergent selection on phenotypic traits. Different molecular and phenotypic markers can sometimes suggest different geographic patterns of population subdivision (Althoff and Thompson, 2001; Ballard *et al.*, 2003). As molecular tools become a standard part of analyses in insect evolutionary ecology, it is becoming increasingly important that we understand how the scales of molecular and ecological differentiation compare.

Such comparisons have also become important in our understanding of interspecific interactions. Interacting species often exhibit patterns of genetic differentiation across landscapes, and we need to be able to interpret the combined information provided by molecular and ecological studies. The geographic mosaic theory of coevolution argues that:

1. The outcomes of interactions differ among environments.
2. Strong reciprocal selection acts shape interactions only in some environments (coevolutionary hotspots).
3. Gene flow, random genetic drift, and metapopulation dynamics continually remix coevolving traits across landscapes (Thompson, 1994, 1999; Gomulkiewicz et al., 2000; Nuismer et al., 2000).

Within this framework, molecular studies have the potential to provide a comparative geographical template for the analysis of coevolving traits, but we need to know how the scale of this molecular template compares with the scale of phenotypic differentiation driven by the coevolutionary process.

Geographically widespread insect species are potentially important tools for understanding the spatial scales of molecular and ecological differentiation. Insect taxa as different as swallowtail butterflies and the parasitoids of *Drosophila* include species that cross multiple biogeographic regions. Each species and its pairwise interactions with other species can be studied individually to get some sense of the geographical patterns of diversification, but interpretation of these studies will be stronger if we can combine phylogeographic and ecological studies of many species within and across biogeographic regions. These multispecific templates will allow the disentangling of general processes that affect all taxa within and across regions from those specific to particular species or particular interspecific interactions.

In this chapter, we first summarize what is known about the geographical pattern of molecular and ecological differentiation in the interaction between *Greya politella* moths and their host plants in western North America. We then place those results in the context of patterns of molecular differentiation in other taxa across the same geographical region.

2. *Greya* and its Host Plants

The interaction between *Greya politella* and its saxifragaceous host plants is one of the most widely distributed plant/insect interactions in western North America. The moths range from southern British Columbia to southern California and east to the Rocky Mountains. *Greya politella* is restricted to a small number of host plants in the genera *Lithophragma* and *Heuchera*, and local populations of the moths feed on one, or occasionally two, hosts. The moths are close relatives of yucca moths and, like some yucca moths, pollinate their host plants while ovipositing into the flowers. Unlike yucca moths, however, pollination is passive rather than active. Pollen adhering to a female's abdomen rubs off on to the stigma when she inserts her ovipositor into a flower to lay eggs. The moths are mutualistic with their host plants in some populations but can be antagonistic in other populations where effective co-pollinators, especially bombyliid flies, occur (Thompson and Cunningham, 2002).

Molecular studies have shown that *G. politella* has differentiated into two major groups of populations along the western coast of North America. Populations in the Pacific Northwest (W-haplotypes) are genetically differentiated from populations within California (C-haplotypes) (Brown *et al.*, 1997). The California populations differ regionally in haplotypes, suggesting that these populations have been differentiating for a long time. Haplotypes differ between the northern and southern Coastal Ranges and between these ranges and the Sierra Nevada. In contrast, the Northwest populations show little molecular differentiation, with only a few base-pair differences among populations sampled so far. Populations in the Siskiyou Mountains of southern Oregon differ slightly from populations in the Interior Northwest, but no other molecular differentiation is apparent. These northwestern populations are mostly post-Pleistocene in origin, and many of the northwestern environments now inhabited by the moth were under or near glaciers 18,000 years ago. There are, then, two very different patterns of molecular differentiation in this species within western North America.

Most of these far-western populations of the moths feed on *Lithophragma*. Some of the Rocky Mountain populations use *Heuchera* (Thompson *et al.*, 1997; Nuismer and Thompson, 2001; Janz and Thompson, 2002), but the full pattern of molecular differentiation in the moths in the Rockies has not yet been evaluated. *Lithophragma* is a small but widespread genus in western North America that shows much regional variation in morphology within the genus. Although up to ten species have been described over the years, we now know that most populations fit within two monophyletic groups based upon analysis of chloroplast and nuclear DNA (Soltis *et al.*, 1992; Kuzoff *et al.*, 1999). Northwestern populations of the moths that feed on *Lithophragma* use only *L. parviflorum*. In California, however, the pattern is more complicated and is still only partially understood. *Lithophragma parviflorum* varies in morphology among the Interior Northwest, coastal Oregon, and California populations (Fig. 14.1), and some of the Oregon and California populations were previously split into separate species (Taylor, 1965). Although subsequent work has shown all these populations to be a single monophyletic clade (Soltis *et al.*, 1992; Kuzoff *et al.*, 1999), there are clear regional differences in the floral morphology of the plants that match what we currently know about the regional differences in mitochondrial DNA of the moths (Fig. 14.1).

At the southern end of the distribution of *L. parviflorum*, populations of this plant species are rare and the moths have shifted on to *Lithophragma cymbalaria*, a local endemic with yet another different floral morphology that fits within the other *Lithophragma* clade (Fig. 14.1). Some other populations in California feed on still other morphologically distinct populations of *Lithophragma*, but we know less about these populations.

There are, then, major differences in both the molecular and ecological structure of this plant/insect interaction between the Pacific Northwest and California. Some of these differences undoubtedly result from differences in the length of time available for genetic differentiation in the two regions. In the absence of comparative data, however, it has been difficult to evaluate the

Lithophragma floral morphology   Greya haplotypes

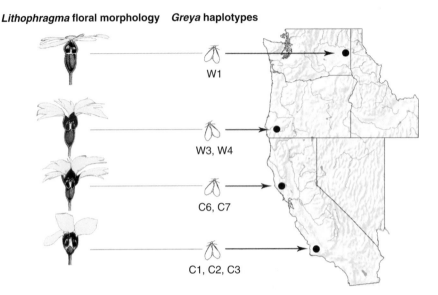

W1

W3, W4

C6, C7

C1, C2, C3

Fig. 14.1. Regional differentiation in floral morphology in the plant *Lithophragma parviflorum* (top three flowers), the yet different floral structure of *Lithophragma cymbalaria* (bottom flower), and coincident diversification in mitochondrial DNA haplotypes in the moth *Greya politella* in far-western North America. *Lithophragma parviflorum* and *G. politella* are widespread in California and the Pacific Northwest, whereas *L. cymbalaria* is endemic to a small region within southern California. The four highlighted regions are those showing high levels of differentiation in *G. politella* and its *Lithophragma* hosts within far-western North America. Other differences also occur in the subregions of California and the Pacific Northwest.

patterns. We therefore undertook an evaluation of the published phylogeographic studies for all plant and animal taxa from these two regions to provide a broader context to the scale and timing of differentiation in these interactions. Our goal was to provide both a background for evaluating regional differentiation in the interaction between *Greya* and *Lithophragma* (i.e. post-Pleistocene versus older populations) and a context for similar future studies of other interactions that cross regions differing in geological history.

3. Expectations from Contrasting Regions of the Western USA

As the number of phylogeographic studies accumulates for a region, so does the opportunity to better understand both the congruence and discord among patterns of population structure among different taxa (Avise *et al.*, 1998; Moritz and Faith, 1998; Walker and Avise, 1998.; Brunsfeld *et al.*, 2002). Traditionally, comparative phylogeographic studies have attempted to evaluate how regional biota may have been shaped by common historical events. There have been few attempts however, to evaluate how the geographical structure of populations has been shaped by different processes in different regions.

The degree of population differentiation within and among regions can serve as an indicator of the timescale on which phylogeographic patterns have been established. Major climatic events such as cycles of glaciation (Hays *et al.*, 1976) have created regions that differ in the lengths of time during which populations have undergone differentiation (Hewitt, 2000; Mila *et al.*, 2000). Pleistocene-scale processes are relatively recent events from a geological perspective, and levels of molecular diversification within Pleistocene-structured populations should be low (Taberlet and Bouvet, 1994; Demesure *et al.*, 1996). By contrast, speciation events associated with more ancient processes such as mountain building (Trewick and Wallis, 2001), palaeoclimate shifts (Schneider *et al.*, 1998), or closure of the isthmus of Panama (Lessios, 1998; Beu, 2001; Richardson *et al.*, 2001) may be associated with deeper levels of molecular diversification within and/or between lineages.

Several recent studies have underscored the importance of comparative phylogeography to our understanding of the processes that shape regional biodiversity. Hewitt's (1999) summary of population structure in European taxa demonstrated that post-glacial population structure across western Europe could be summarized using three general patterns of post-refugial recolonization. Comparative studies have also been useful for prioritizing conservation efforts in regions thought to be of particular significance in generating or maintaining biological diversity (Moritz *et al.*, 1997; Smith *et al.*, 2000). Comparative studies may also provide predictive power for understanding the population structure of local taxa whose genetic diversity has not yet been analysed (Arbogast and Kenagy, 2001; Calsbeek *et al.*, 2003).

In a previous analysis (Calsbeek *et al.*, 2003), we presented evidence that California's phylogeographic history was largely shaped by climate change and contemporaneous uplift of the Sierra Nevada, Transverse, and Coast Range mountains, which may have occurred roughly 2–5 million years ago (Huber, 1981). Those results reinforced earlier studies of the biogeographic of California and suggested that the geographical distribution of animal populations in California is related to the distribution of mountains across the state. The phylogeographic patterns imply a strong link between topography, changes in palaeoclimate, and population genetic structure of animal taxa. Although molecular diversification of plant taxa occurred on the same time scale as that of animals, the structure of plant biological diversity across the California landscape did not coincide with the geographic structure of mountain ranges.

Those results suggested two important points about the structure of species and species interactions in that region. First, subregions within California have been intact for long enough periods of time to allow for relatively stable patterns of differentiation among populations and, by inference, species interactions. Second, if in fact animal and plant taxa have responded differently to these processes, as suggested by the molecular results, then complex coevolutionary mosaics in species interactions will be almost inevitable. This would hold true to the extent that the molecular results provide some indication of differences in patterns of gene flow, population differentiation and speciation in animals and plants throughout the region.

We now extend these comparative analyses to the US Pacific Northwest (Washington, Oregon and Idaho) to compare the structure of phylogenetic histories within the two regions. We hypothesized that the contrasting climatological histories of California and the Pacific Northwest should have resulted in differences in phylogeographic structure between the two regions. In particular, we hypothesized that the average within-species sequence divergence is reduced in the Pacific Northwest relative to California, as a result of the relatively recent effects of Pleistocene glaciation. In contrast, we expected relatively greater levels of sequence divergence, and hence more population structure, among taxa in California relative to the Pacific Northwest. We tested this hypothesis by comparing the levels of sequence divergence within and between regions and by comparing the overall variance in the amounts of sequence divergence within and between regions. Comparison of total sequence divergence allows us to estimate differences in the age of phylogeographic structure in California and the Pacific Northwest, and estimation of the variance in sequence divergence among taxa may provide some insight into the continuity of phylogeographic structure. That is, if regional taxa have all been shaped to the same degree by processes such as the Pleistocene glaciation, then we would expect little variance in the amounts of sequence divergence among taxa. By contrast, more variable amounts of divergence would suggest the presence of both ancient and more recent lineages. Alternatively, glacial refugia may have generated greater variance in sequence divergence among taxa in the Pacific Northwest.

Our expectations came from knowledge of the geological history of these two regions. Both regions are topologically diverse and share a common history of volcanism that produced the Klamath and Coast Range mountains (Rhodes, 1987). However, the interior of the Pacific Northwest is dominated by the Rocky Mountains, an ancient chain dating back perhaps as far as the Eocene (Bird, 1988). California's central mountain range, the Sierra Nevada, has a much more recent orogeny, attaining the majority of its relief as recently as 2–5 million years ago (Huber, 1981).

Due in part to differences in climate, California and the Pacific Northwest represent vastly different sets of biological communities. The Pacific Northwest is characterized by a maritime coastal climate and, because of the rain shadow generated by the Cascades, a dry interior (summarized in Brunsfeld *et al.*, 2002). California has one of the most diverse climatological regimes of North America (Raven and Axelrod, 1978) encompassing 12 of North America's 48 different ecoregions (Bailey, 1994), and ranging from desert in the south to high alpine in the mountains. California is consequently home to extraordinary biological diversity and is generally considered to be one of the Earth's biodiversity hotspots (Mittermeier *et al.*, 1999; Myers *et al.*, 2000).

California and the Pacific Northwest have also experienced very different climatological pasts, particularly with reference to the past 2 million years. During the Pleistocene, glacial cycles (Milanovitch cycles) repeatedly drove the advance and retreat of ice sheets over the landscape of the Pacific Northwest (Hays *et al.*, 1976), reducing the availability of suitable habitat (Delcourt and Delcourt, 1993). Consequently, many local species are thought to have been driven into glacial refugia (Soltis *et al.*, 1997; Willis and Whittaker, 2000), pockets of habitat with suitable climate surrounded by otherwise uninhabitable glacial ice. Refugial populations tracked the glacial front, expanding out of refuges during interglacials and retreating back to the refuges when temperatures fell again (Willis and Whittaker, 2000; Knowles, 2001). Population expansion following isolation in a refuge should have genetic consequences analogous to founder events or population bottlenecks (Falconer and MacKay, 1996), and it is thought that low genetic diversity in populations derived from refugial populations is a signal of these past evolutionary events (Hewitt, 1996). Although nuclei of ice and snow descended from California's higher elevation mountains, glacial ice sheets did not move deep into California, and biological communities were not subject to the same degree of habitat loss during glacial cycles.

Thus, the geological and climatic processes that have structured California and the Pacific Northwest should have resulted in differences in population genetic structure as well. Despite the large differences in community structure and history of the two regions, some insect species have geographical ranges that span them, with populations genetically differentiated from each other. Examples include widespread species such as western anise swallowtails (*Papilio zelicaon*) (Wehling and Thompson, 1997; Thompson, 1998) and some *Greya* moths (Davis *et al.*, 1992; Thompson, 1997).

4. Analytical Approaches to Comparing Regions

Our overall approach was nested phylogenetically and geographically. We compared levels of molecular diversification between the two regions, but also separately within clades and subclades of species phylogenies from the two regions. Our interpretations were based on the application of molecular clocks to 20 different DNA sequence datasets. The use of molecular clocks has been criticized (Martin and Palumbi, 1993), owing to the fact that most lineages experience significant amounts of rate heterogeneity (i.e. different rates of molecular evolution across lineages). Nevertheless, we find that, despite the potential problems associated with evolutionary rate heterogeneity, the molecular-clock approach can still be a useful means of differentiating between processes that occur over timescales of tens to hundreds of thousands of years versus those that occurred over millions of years.

We attempted a comprehensive review of published molecular phylo-geographic studies in California and the Pacific Northwest based on DNA sequence data. New studies are appearing at an increasingly rapid rate. At the time we undertook these analyses we were able to include 12 studies from California and eight from the Pacific Northwest, including arthropods, mammals, reptiles and amphibians. These studies included only those species that had been sampled across a large part of California and/or the Pacific Northwest. The studies were based on a variety of gene sequence data including ND1/4, NADH dehydrogenase, mitochondrial cytochrome *b*, and cytochrome oxidase (Table 14.1). Unfortunately, there are still few DNA sequence data available for multiple populations of plant species in the interior

Table 14.1. Summary table of phylogeographic data based on DNA sequences reviewed in this study. 'Rate' refers to the rate of molecular evolution used to date nodes of interest (based on estimates reported in the literature for each group. For a complete description of methods see Calsbeek *et al.*, 2003). 'Region' refers to the biogeographic range across which taxa were sampled: California (CA), Pacific Northwest (PNW) or both regions (BOTH).

Taxon	Gene(s)	Rate	Region	Citation
Pituophis	NDH4/ND4	0.845	CA	(Rodriguez-Robles and De Jesus-Escobar, 2000)
Anniella	cyt b	0.845	CA	(Pearse and Pogson, 2000)
Crotalus	cyt b	0.845	CA	(Pook *et al.*, 2000)
Lampropeltis	NDH4/ND4	0.845	CA	(Rodriguez-Robles *et al.*, 1999)
Charina	NDH4/ND4	0.845	CA	(Rodriguez-Robles *et al.*, 2001)
Rana	COI/tRNAs	1.3	CA	(Macey *et al.*, 2001)
Greya	CO1	*0.95*	BOTH	(Brown *et al.*, 1997)
Tegeticula	CO1	0.95	CA	(Segraves and Pellmyr, 2001)
Sauromalus	cyt b	0.845	CA	(Petren and Case, 1997)
Taricha	cyt b	0.845	CA	(Tan and Wake, 1995)
Sorex	cyt b	2.185	PNW	(Demboski and Cook, 2001)
Tigriopus	CO 1	2.4	BOTH	(Edmands, 2001)
Microtus	cyt b	2.185	PNW	(Conroy and Cook, 2000)
Phyrnosoma	cyt b/ND4	0.845	PNW	(Zamudio *et al.*, 1997)
Ursus	cyt b	2.185	PNW	(Shields *et al.*, 2000)
Glaucomys	cyt b	2.185	PNW	(Arbogast, 1999)
Ascaphus	cyt b/NADH	1.3	PNW	(Nielson *et al.*, 2001)

Pacific Northwest (e.g. Segraves *et al.*, 1999), and we therefore excluded plant taxa from our comparative analyses. We obtained DNA sequences from GenBank using published accession numbers. All sequences were aligned by eye. We analysed all of the sequence data using the parsimony algorithm of the software package PAUP, version 4.0d64 for Macintosh (Swofford, 1998). We conducted tree searches under the equal and unordered weights criterion (Fitch parsimony) with 1000 random sequence additions and TBR branch swapping.

We applied individual rates of molecular evolution to each taxonomic group to maintain conservative estimates of divergence times. In some cases, rates of molecular evolution were estimated directly from sequence divergence of taxa on islands compared with the mainland (e.g. *Pituophis*; Rodriguez-Robles and De Jesus-Escobar, 2000). Other estimates come from rates published in the literature (Table 14.1). We dated nodes by calculating the average number of substitutions between the node of interest and all terminal taxa from that node (see Calsbeek *et al.*, 2003, for complete molecular methods).

We estimated total molecular diversification by calculating the average percentage sequence divergence within a species, between subclades and within subclades. Use of the term 'subclade' here refers to the largest phylogenetic group nested within a species tree that shares a common node. We also measured the variance in total sequence divergence within and between subclades for each study. We compared the total amounts of molecular diversification between

California and the Pacific Northwest using ANOVA. We used F_{max} tests (Sokal and Rohlf, 1995) to compare the variance in sequence divergence. Thus our analyses are nested at three different levels of molecular diversification: total average amounts of sequence divergence among taxa in California versus the Pacific Northwest, divergence between subclades of each region, and divergence within subclades of each region.

Initial analyses indicated that phylogenetic trees for most taxa from the two regions had characteristic topologies that are generalized in Fig. 14.2. Overall, California taxa were characterized by clades in the northern and southern portions of the state that were delineated by mountain ranges (Calsbeek et al., 2003). Differentiation was also evident between some coastal clades and Sierran clades. In contrast, taxa from the Pacific Northwest typically formed coastal and inland subclades with very low levels of genetic differentiation between them. Hence, the overall patterns for all other taxa were similar to that observed for *Greya* moths.

Total molecular diversification (% sequence divergence) in California was significantly greater than in the Pacific Northwest (ANOVA $F_{1, 18} = 5.03$, $P < 0.03$), as was the average onset of species diversification (mean age in CA = 3.59 million years versus PNW = 0.8 million years; $F_{1, 18} = 12.97$, $P < 0.002$). Again, these overall patterns were similar to that observed in *Greya*. At other scales, total sequence divergence did not differ between regions. That is, the average amounts of within-subclade and between-subclade sequence divergence did not differ between California and the Pacific Northwest (ANOVA $F_{1, 37} = 0.02$, $P = 0.89$ and $F_{1, 18} = 0.26$, $P = 0.61$ for within- and between-subclade differences, respectively). Degrees of freedom in these analyses differ because the number of clade-based comparisons differs among studies.

Comparing the variance in total sequence divergence, California taxa had more than 3.5 times as much variance in within-subclade divergence as Pacific Northwest taxa, but this difference was only marginally significant ($F_{max} = 3.53$, d.f. = 11,7, critical value$_{0.05} = 4.71$, $P < 0.1$). The between-subclade differences were highly significant, however. California taxa had significantly greater variance in sequence divergence between subclades than did taxa from the Pacific Northwest ($F_{max} = 12.72$, d.f. = 11,7, critical value$_{0.005} = 10.4$, $P < 0.005$).

5. Implications for Interpretation of *Greya/Lithophragma* Interactions

This is one of the first comparative studies of phylogeography across multiple taxa and regions, and it suggests considerable incongruence between adjacent regions. The relatively high molecular and ecological complexity found in California and low complexity found in the Pacific Northwest in the interaction between *Greya* and *Lithophragma* fits well within the broader pattern of diversification found in the comparative phylogeographic analyses. Hence, the complex geographic mosaic found in the *Greya/Lithophragma* interactions results in part from more general patterns of geographical differentiation that have affected multiple taxa in these two regions.

(A)

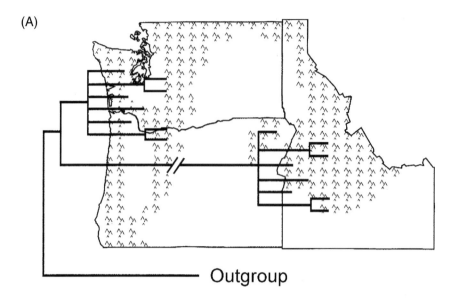

(B)

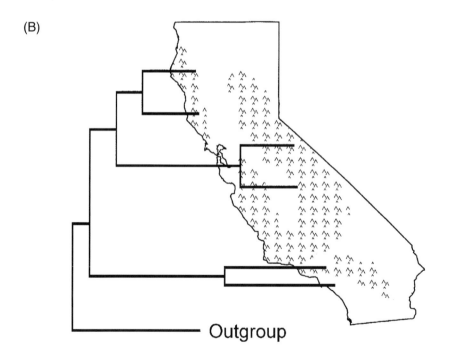

Fig. 14.2. Generalized pattern of phylogeographic differentiation in California (B) and the Pacific Northwest (A), based upon the combined differences observed across multiple taxa. Individual species differ from these generalized patterns. The Pacific Northwest shows much less regional differentiation than California. The lengths of the branches are intended only to convey a sense of overall patterns found across taxa within the two regions and are not quantitative measures of branch lengths.

The processes that have shaped California's biota are likely to have occurred over the past 2–5 million years. In contrast, very low levels of sequence divergence within the Pacific Northwest are more consistent with a strong response to events that occurred during the Pleistocene (i.e. < 2 million years ago). The significant difference in average molecular diversification between California and the Pacific Northwest suggests that the two regions have experienced very different phylogeographic histories. Moreover, these data suggest that, for species matching these overall patterns, Pacific Northwest populations may either have been derived from Californian populations or experienced long-term bottlenecks in genetic diversity. That is the pattern evident in *G. politella* and *Lithophragma*. Molecular-clock analyses of differentiation within *Greya* suggest that *G. politella* originated at least several million years ago (Pellmyr *et al.*, 1998; Pellmyr, 2003), and it currently shows much more regional differentiation in California than in the Pacific Northwest. Similarly, the *L. parviflorum* species complex shows much more molecular differentiation in California than in the Pacific Northwest (Kuzoff *et al.*, 1999).

Both for *Greya* and for other animal taxa, the degree of molecular diversification between subclades is also significantly more variable in California than in the Pacific Northwest. In other words, species distributions in California represent a combination of both recent and more ancient divergences between lineages. In the Pacific Northwest, however, low variance in between-subclade sequence divergence suggests that all divergences between lineages in the region are of recent origin. This result is again consistent with the interpretation that taxa in the Pacific Northwest experienced bottleneck effects owing to reduced population sizes in glacial refugia. California, however, did not experience the same regional effects of glaciation and many taxa maintain more ancient genetic separations.

In a previous study we reported that although animal taxa in California demonstrate striking patterns of congruent population structure, plant populations do not (Calsbeek *et al.*, 2003). We did not undertake a systematic analysis of plant studies in our analysis of the Pacific Northwest because there was only one study available at the time that incorporated sequence data in the Interior Northwest (Segraves *et al.*, 1999). Although this makes it impossible for us to speculate on the congruence between plant and animal diversification at the level of DNA sequence, molecular fragment-based analyses of plant populations from the Pacific Northwest do raise some interesting questions. For example, across five angiosperms and one fern species, there is a strong north/south differentiation of haplotypes from Alaska to northern California (Soltis *et al.*, 1997). Similar patterns show up in *Tellima grandiflora* (Soltis *et al.*, 1991), *Tolmiea menziesii* (Soltis *et al.*, 1989) and numerous other plant species of the Pacific Northwest (reviewed in Brunsfeld *et al.*, 2002). These populations therefore show evidence of post-Pleistocene northward movement of a small number of haplotypes, reinforcing the pattern found in *Greya* and other animal taxa.

The overall results provide a historical context for evaluating the geographic mosaic of coevolution between *Greya* and its host plants. Within California, the moths and plants are evolving in subclades that have been separated for long enough periods of time to be evident in distinct regions of molecular differentiation. Under such conditions, regional differences in the interaction could result from a complex set of evolutionary and coevolutionary forces. In the Pacific Northwest, the moth and plant populations have been differentiating over a relatively short period of time. Nonetheless, ecological research has shown differences among populations in the outcomes of the interaction (Thompson and Cunningham, 2002), the behaviour of the moths (Janz and Thompson, 2002) and the life history and morphological traits of the plants (J. Thompson, B. Cunningham and C. Fernandez, unpublished). This rapid differentiation in traits important to the interaction over the past 10,000–12,000 years suggests a strong role for natural selection in diversification of these populations.

The two regions, California and the Pacific Northwest, may each represent fragile ecosystems. However, the fragility of regional biodiversity appears to have different explanations in these two regions. As one of the earth's 25 biodiversity hotspots (Mittermeier *et al.*, 1999; Myers *et al.*, 2000), California represents a rich but endangered centre for biological diversity. Nearly 50% of its taxa are restricted to California (Myers *et al.*, 2000). Moreover, the richness of California's phylogeographic structure indicates that cryptic genetic variation abounds within the state (Riddle *et al.*, 2000), and this variation should be a conservation priority. In contrast, the Pacific Northwest harbours lower levels of molecular genetic diversity, at least among animal taxa.

The geographic mosaic of the interaction between *Greya* and *Lithophragma* therefore has a regional structure that results from a combination of major historical differences in different regions and ongoing ecological differences in the structure and the dynamics of the interaction. The intersection of all these forces across the ever-changing geographical ranges of these species points to the inevitability of geographic mosaics in widespread interspecific interactions.

The comparative approach to phylogeography is increasingly becoming an important tool for evaluating the structure and dynamics of biotas. These studies are crucial to our understanding of the general processes that generate and maintain population structure (Arbogast and Kenagy, 2001) and should also be useful as a means of prioritizing conservation efforts for taxa whose genetical structure is still unknown (Moritz *et al.*, 1997; Moritz and Faith, 1998). For example, Hewitt's (1999) study showed that the recolonization patterns of 12 taxa following European glaciation could be summarized into three patterns of re-invasion from ice-age refugial areas in the south. Comparative studies in Australia suggest that known patterns of population structure and sequence divergence reflect a common history shared by many Australian taxa (Schneider *et al.*, 1998; Stuart-Fox *et al.*, 2001). Similarly, the comparative approach has been used to demonstrate that geological processes have influenced the spatial distribution of river drainages and therefore population structure of fishes and turtles of the American Southeast (Walker and Avise, 1998). Such studies will inform future evaluations of the geographic mosaic of coevolution among taxa worldwide by providing templates on which to evaluate how historical processes and coevolution interact to shape adaptation and diversification in interspecific interactions.

Acknowledgements

We thank Catherine Fernandez, Katherine Horjus, Grant Pogson and James Richardson for helpful discussions that improved the manuscript. The research was supported by grants from the National Science Foundation and the Packard Foundation.

References

Althoff, D.M. and Thompson, J.N. (2001) Geographic structure in the searching behaviour of a specialist parasitoid: combining molecular and behavioural approaches. *Journal of Evolutionary Biology* 14, 406–417.

Arbogast, B.S. (1999) Mitochondrial DNA phylogeography of the new world flying squirrels (*Glaucomys*): implications for Pleistocene biogeography. *Journal of Mammalogy* 80, 142–155.

Arbogast, B.S. and Kenagy, G.J. (2001) Comparative phylogeography as an integrative approach to historical biogeography. *Journal of Biogeography* 28, 819–825.

Avise, J.C., Walker, D. and Johns, G.C. (1998) Speciation durations and Pleistocene effects on vertebrate phylogeography. *Proceedings of the Royal Society of London, Series B* 265, 1707–1712.

Bailey, R.G. (1994) *Description of Ecoregions of the United States.* US Department of Agriculture Miscellaneous Publication. USDA Forest Service, Washington, DC.

Ballard, J.W.O., Chernoff, B. and James, A.C. (2003) Divergence of mitochondrial DNA is not corroborated by nuclear DNA, morphology, or behavior in *Drosophila simulans*. *Evolution* 56, 527–545.

Beu, A.G. (2001) Gradual Miocene to Pleistocene uplift of the Central American isthmus: evidence from tropical American tonnoidean gastropods. *Journal of Paleontology* 75, 706–720.

Bird, P. (1988) Formation of the Rocky Mountains, western United States: a continuum computer model. *Science* 239, 1501–1507.

Brown, J.M., Leebens-Mack, J.H., Thompson, J.N., Pellmyr, O. and Harrison, R.G. (1997) Phylogeography and host association in a pollinating seed parasite *Greya politella* (Lepidoptera: Prodoxidae). *Molecular Ecology* 6, 215–224.

Brunsfeld, S.J., Sullivan, J., Soltis, D.E. and Soltis, P.S. (2002) Comparative phylogeography of north-western North America: a synthesis. *Symposia of the British Ecological Society* 14, 319–339.

Calsbeek, R., Thompson, J.N. and Richardson, J.E. (2003) Patterns of molecular evolution and diversification in a biodiversity hotspot: the California Floristic Province. *Molecular Ecology* 12, 1021–1029.

Conroy, C.J. and Cook, J.A. (2000) Phylogeography of a post-glacial colonizer: *Microtus longicaudus* (Rodentia: Muridae). *Molecular Ecology* 9, 165–175.

Craig, T.P., Itami, J.K., Shantz, C., Abrahamson, W.G., Horner, J.D. and Craig, J.V. (2000) The influence of host plant variation and intraspecific competition on oviposition preference and offspring performance in the host races of *Eurosta solidaginis*. *Ecological Entomology* 25, 7–18.

Davis, D.R., Pellmyr, O. and Thompson, J.N. (1992) Biology and systematics of *Greya* Busck and *Tetragma*, new genus (Lepidoptera: Prodoxidae). *Smithsonian Contributions to Zoology* 524, 1–88.

de Jong, P.W., Frandsen, H.O., Rasmussen, L. and Nielsen, J.K. (2000) Genetics of resistance against defences of the host plant *Barbarea vulgaris* in a Danish flea beetle population. *Proceedings of the Royal Society of London, Series B* 267, 1663–1670.

Delcourt, P.A. and Delcourt, H.R. (1993) Paleoclimates, paleovegetation, and paleofloras of North America during the late Quaternary. In: *Flora of North America*. Oxford University Press, New York, pp. 71–94.

Demboski, J.R. and Cook, J.A. (2001) Phylogeography of the dusky shrew, *Sorex monticolus* (Insectivora, Soricidae): insight into deep and shallow history in northwestern North America. *Molecular Ecology* 10, 1227–1240.

Demesure, B., Comps, B. and Petit, R.J. (1996) Chloroplast DNA phylogeography of the common beech (*Fagus sylvatica* L.) in Europe. *Evolution* 50, 2515–2520.

Edmands, S. (2001) Phylogeography of the intertidal copepod *Tigriopus californicus* reveals substantially reduced population differentiation at northern latitudes. *Ecology* 10, 1743–1750.

Falconer, D.S. and MacKay, T.F.C. (1996) *Introduction to Quantitative Genetics*. Longman, Essex, UK.

Fellowes, M.D.E. and Godfray, H.C.J. (2000) The evolutionary ecology of resistance to parasitoids by *Drosophila*. *Heredity* 84, 1–8.

Gomulkiewicz, R., Thompson, J.N., Holt, R.D., Nuismer, S.L. and Hochberg, M.E. (2000) Hot spots, cold spots, and the geographic mosaic theory of coevolution. *American Naturalist* 156, 156–174.

Hays, J.D., Imbrie, J. and Shanckleton, N.J. (1976) Variations in the Earth's orbit: pacemaker of the ice ages. *Science* 194, 1121–1132.

Hewitt, G.M. (1996) Some genetic consequences of ice ages, and their role in divergence and speciation. *Biological Journal of the Linnean Society* 58, 247–286.

Hewitt, G.M. (1999) Post-glacial re-colonization of European biota. *Biological Journal of the Linnean Society* 68, 87–112.

Hewitt, G.M. (2000) The genetic legacy of the Quaternary ice ages. *Nature* 405, 907–913.

Huber, N.K. (1981) *Amount and Timing of Late Cenozoic Uplift and Tilt of the Central Sierra Nevada, California: Evidence from the Upper San Joaquin River Basin*. US Geological Survey Professional Paper 1197.

Irwin, D.E. (2002) Phylogeographic breaks without geographic barriers to gene flow. *Evolution* 56, 2383–2394.

Janz, N. and Thompson, J.N. (2002) Plant polyploidy and host expansion in an insect herbivore. *Oecologia* 130, 570–575.

Knowles, L.L. (2001) Did the Pleistocene glaciations promote divergence? Tests of explicit refugial models in montane grasshoppers. *Molecular Ecology* 10, 691–701.

Kraaijeveld, A.R. and Godfray, H.C.J. (1999) Geographic patterns in the evolution of resistance and virulence in *Drosophila* and its parasitoids. *American Naturalist* 153, S61–S74.

Kuzoff, R.K., Soltis, D.E., Hufford, L. and Soltis, P.S. (1999) Phylogenetic relationships with *Lithophragma* (Saxifragaceae): hybridization, allopolyploidy and ovary diversification. *Systematic Botany* 24, 598–615.

Lessios, H.A. (1998) The first stage of speciation as seen in organisms separated by the Isthmus of Panama.

Macey, J.R., Strasburg, J.L., Brisson, J.A., Vredenburg, V.T., Jennings, M. and Larson, A. (2001) Molecular phylogenetics of western North American frogs of the *Rana boylii* species group. *Molecular Phylogenetics and Evolution* 19, 131–143.

Martin, A.P. and Palumbi, S.R. (1993) Body size, metabolic rate, generation time, and the molecular clock. *Proceedings of the National Academy of Sciences of the USA* 90, 4087–4091.

Mila, B., Girman, D.J., Kimura, M. and Smith, T.B. (2000) Genetic evidence for the effect of a postglacial population expansion on the phylogeography of a North American songbird. *Proceedings of the Royal Society of London, Series B* 267, 1033–1040.

Mittermeier, R.A., Myers, N., Robles Gil, P. and Mittermeier, C.G. (1999) *Hotspots: Earth's Biologically Richest and Most Endangered Terrestrial Ecoregions*. CEMEX and Conservation International, Mexico City.

Moritz, C. and Faith, D.P. (1998) Comparative phylogeography and the identification of genetically divergent areas for conservation. *Molecular Ecology* 7, 419–429.

Moritz, C., Joseph, L., Cunningham, M. and Schneider, C. (1997) *Molecular Perspectives on Historical Fragmentation of Australian Tropical and Subtropical Rainforests: Implications for Conservation*. University of Chicago Press, Chicago, Illinois.

Myers, N., Mittermeier, R.A., Mittermeier, C.G., da Fonseca, G.A.B. and Kent, J. (2000) Biodiversity hotspots for conservation priorities. *Nature* 403, 853–858.

Nielsen, J.K. (1997) Variation in defences of the plant *Barbarea vulgaris* and in counteradaptations by the flea beetle *Phyllotreta nemorum*. *Entomologia Experimentalis et Applicata* 82, 25–35.

Nielson, M., Lohman, K. and Sullivan, J. (2001) Phylogeography of the tailed frog (*Ascaphus truei*): implications for the biogeography of the Pacific Northwest. *Evolution* 55, 147–160.

Nuismer, S.L. and Thompson, J.N. (2001) Plant polyploidy and non-uniform effects on insect herbivores. *Proceedings of the Royal Society of London, Series B* 268, 1937–1940.

Nuismer, S.L., Thompson, J.N. and Gomulkiewicz, R. (2000) Coevolutionary clines across selection mosaics. *Evolution* 54, 1102–1115.

Pearse, D.E. and Pogson, G.H. (2000) Parallel evolution of the melanic form of the California legless lizard, *Anniella pulchra*, inferred from mitochondrial DNA sequence variation. *Evolution* 54, 1041–1046.

Pellmyr, O. (2003) Yuccas, yucca moths, and coevolution: a review. *Annals of the Missouri Botanical Garden* 90, 35–55.

Pellmyr, O., Leebens-Mack, J.H. and Thompson, J.N. (1998) Herbivores and molecular clocks as tools in plant biogeography. *Biological Journal of the Linnean Society* 63, 367–378.

Petren, K. and Case, T.J. (1997) A phylogenetic analysis of body size evolution and biogeography in chuckwallas (*Sauromalus*) and other iguanines. *Evolution* 51, 206–219.

Pook, C.E., Wuster, W. and Thorpe, R.S. (2000) Historical biogeography of the western rattlesnake (Serpentes: Viperidae: *Crotalus viridis*), inferred from mitochondrial DNA sequence information. *Molecular Phylogenetics and Evolution* 15, 269–282.

Raven, P.H. and Axelrod, D.I. (1978) *Origin and Relationships of the California Flora*. University of California Press, Berkeley, California.

Rhodes, P.T. (1987) Historic glacier fluctuations at Mount Shasta, Siskiyou County. *California Geology* 40, 205–209.

Richardson, J.E., Pennington, R.T., Pennington, T.D. and Hollingsworth, P.M. (2001) Rapid diversification of a species-rich genus of neotropical rain forest trees. *Science* 293, 2242–2245.

Riddle, B.R., Hafner, D.J., Alexander, L.F. and Jaeger, J.R. (2000) Cryptic vicariance in the historical assembly of a Baja California Peninsular Desert biota. *Proceedings of the National Academy of Sciences USA* 97, 14438–14443.

Rodriguez-Robles, J.A. and De Jesus-Escobar, J.M. (2000) Molecular systematics of New World gopher, bull, and pinesnakes (*Pituophis*: Colubridae), a transcontinental species complex. *Molecular Phylogenetics and Evolution* 14, 35–50.

Rodriguez-Robles, J., DeNardo, D.F. and Staub, R. (1999) Phylogeography of the California mountain kingsnake, *Lampropelits zonata* (Colubridae). *Molecular Ecology* 8, 1923–1934.

Rodriguez-Robles, J.A., Stewart, G.R. and Papenfuss, T.J. (2001) Mitochondrial DNA-based phylogeography of North American rubber boas, *Charina bottae* (Serpentes: Boidae). *Molecular Phylogenetics and Evolution* 18, 227–237.

Schneider, C.J., Cunningham, M. and Moritz, C. (1998) Comparative phylogeography and the history of endemic vertebrates in the West Tropics rainforests of Australia. *Molecular Ecology* 7, 787–498.

Segraves, K.A. and Pellmyr, O. (2001) Phylogeography of the yucca moth *Tegeticula maculata*: the role of historical biogeography in reconciling high genetic structure with limited speciation. *Molecular Ecology* 10, 1247–1253.

Segraves, K.A., Thompson, J.N., Soltis, P.S. and Soltis, D.E. (1999) Multiple origins of polyploidy and the geographic structure of *Heuchera grossulariifolia*. *Molecular Ecology* 8, 253–262.

Shields, G.F., Adams, D., Garner, G., Labelle, M., Pietsch, J., Ramsay, M., Schwartz, C., Titus, K. and Williamson, S. (2000) Phylogeography of mitochondrial DNA variation in brown bears and polar bears. *Molecular Phylogenetics and Evolution* 15, 319–326.

Singer, M.C. and Thomas, C.D. (1996) Evolutionary responses of a butterfly metapopulation to human- and climate-caused environmental variation. *American Naturalist* 148, S9–S39.

Smith, T.B., Holder, K., Girman, D., O'Keefe, K., Larison, B. and Chan, Y. (2000) Comparative avian phylogeography of Cameroon and Equatorial Guinea mountains: implications for conservation. *Molecular Ecology* 9, 1505–1516.

Sokal, R.R. and Rohlf, J.F. (1995) *Biometry: The Principles and Practice of Statistics in Biological Research*. Freeman, New York.

Soltis, D.E., Soltis, P.S., Ranker, T.A. and Ness, B.D. (1989) Chloroplast DNA variation in a wild plant, *Tolmiea menziesii*. *Genetics* 121, 819–826.

Soltis, D.E., Mayer, M.S., Soltis, P.S. and Edgerton, M. (1991) Chloroplast-DNA variation in *Tellima grandiflora* (Saxifragaceae). *American Journal of Botany* 78, 1379–1390.

Soltis, D.E., Soltis, P.S., Thompson, J.N. and Pellmyr, O. (1992) Chloroplast DNA variation in *Lithophragma* (Saxifragaceae). *Systematic Botany* 17, 607–619.

Soltis, D.E., Gitzendanner, M.A., Strenge, D.D. and Soltis, P.S. (1997) Chloroplast DNA intraspecific phylogeography of plants from the Pacific Northwest of North America. *Plant Systematics and Evolution* 206, 353–373.

Stuart-Fox, D.M., Schneider, C.J., Moritz, C. and Couper, P.J. (2001) Comparative phylogeography of three rainforest-restricted lizards from mid-east Queensland. *Australian Journal of Zoology* 49, 119–127.

Swofford, D.L. (1998) *Phylogenetic Analysis Using Parsimony (PAUP) 4.0b2*. Sinauer, Sunderland, Massachusetts.

Taberlet, P. and Bouvet, J. (1994) Mitochondrial DNA polymorphism, phylogeography, and conservation genetics of the brown bear *Ursus arctos* in Europe. *Proceedings of the Royal Society of London, Series B* 255, 195–200.

Tan, A.M. and Wake, D.B. (1995) MtDNA phylogeography of the California Newt, *Taricha torosa* (Caudata, Salamandridae). *Molecular Phylogenetics and Evolution* 4, 383–394.

Taylor, R.L. (1965) *The Genus Lithophragma (Saxifragaceae)*. University of California Publications in Botany 37, Berkeley, California.

Thompson, J.N. (1994) *The Coevolutionary Process*. University of Chicago Press, Chicago, Illinois.

Thompson, J.N. (1997) Evaluating the dynamics of coevolution among geographically structured populations. *Ecology* 78, 1619–1623.

Thompson, J.N. (1998) The evolution of diet breadth: monophagy and polyphagy in swallowtail butterflies. *Journal of Evolutionary Biology* 11, 563–578.

Thompson, J.N. (1999) Specific hypotheses on the geographic mosaic of coevolution. *American Naturalist* 153, S1–S14.

Thompson, J.N. and Cunningham, B.M. (2002) Geographic structure and dynamics of coevolutionary selection. *Nature* 417, 735–738.

Thompson, J.N., Cunningham, B.M., Segraves, K.A., Althoff, D.M. and Wagner, D. (1997) Plant polyploidy and insect/plant interactions. *American Naturalist* 150, 730–743.

Trewick, S.A. and Wallis, G.P. (2001) Bridging the 'beech-gap': New Zealand invertebrate phylogeography implicates Pleistocene glaciation and Pliocene isolation. *Evolution* 55, 2170–2180.

Walker, D. and Avise, J.C. (1998) Principles of phylogeography as illustrated by freshwater and terrestrial turtles in the southeastern United States. *Annual Review of Ecology and Systematics* 29, 23–58.

Wehling, W.F. and Thompson, J.N. (1997) Evolutionary conservatism of oviposition preference in a widespread polyphagous insect herbivore, *Papilio zelicaon*. *Oecologia* 111, 209–215.

Willis, K.J. and Whittaker, R.J. (2000) The refugial debate. *Science* 287, 1406–1407.

Zamudio, K.R., Jones, K.B. and Ward, R.H. (1997) Molecular systematics of short-horned lizards: biogeography and taxonomy of a widespread species complex. *Systematic Biology* 46, 284–305.

15 The Genetic Basis of Speciation in a Grasshopper Hybrid Zone

DAVID M. SHUKER,[1] TANIA M. KING,[2] JOSÉ L. BELLA[3] AND ROGER K. BUTLIN[4]

[1]School of Biological Sciences, University of Edinburgh, Ashworth Laboratories, Edinburgh, UK; [2]Department of Zoology, University of Otago, Dunedin, New Zealand; [3]Departamento de Biología, Facultad de Ciencias, Universidad Autónoma de Madrid, Madrid, Spain; [4]Department of Animal and Plant Sciences, University of Sheffield, Sheffield, UK

1. Introduction

Speciation has always been an important area of research in evolutionary biology, highlighted by its place in the Modern Synthesis (Dobzhansky, 1937; Mayr, 1942). Throughout, insects have played a major role in the development and testing of speciation theory, in particular *Drosophila*. Nevertheless, many questions remain and research on mechanisms of speciation by both evolutionary ecologists and geneticists continues to gather momentum (e.g. Otte and Endler, 1989; Howard and Berlocher, 1998; Schluter, 2000; Barton, 2001a, and papers therein). Currently, this impetus is being maintained by advances in our understanding of the genetic basis of speciation. The availability of novel molecular genetic and genomic approaches that attempt to identify and characterize 'speciation genes' has meant that non-model organisms can begin to contribute to our understanding of speciation in ways that until recently were only possible in model organisms like *Drosophila*. Importantly, these techniques allow speciation research to come out of the laboratory, to provide an interface between the genetics and ecology of speciation in both insects and other taxa. A good example of an organism ideally placed to take advantage of these latest techniques is the meadow grasshopper, *Chorthippus parallelus*, and here we consider what *C. parallelus* can tell us about speciation genetics in an ecological context.

Speciation genetics has two components:

1. The genetics of reproductive isolation, meaning the genetic basis of the actual traits that produce pre- or postzygotic reproductive isolation between divergent populations.
2. The genetic basis of other species differences, for example quantitative traits that are not implicated in maintaining isolation.

© Royal Entomological Society 2005. *Insect Evolutionary Ecology*
(eds M.D.E. Fellowes, G.J. Holloway and J. Rolff)

Due to the difficulty of studying speciation as it occurs, it is often problematic to distinguish whether traits that differ between two species were involved in the original speciation process, or whether they have evolved after isolation of the two gene pools (see e.g. Orr, 2001). Clearly, behavioural or morphological differences that prevent interbreeding may have evolved after the formation of the two reproductively isolated species, and therefore independently of the role they now have in maintaining the isolation and integrity of the species (Etges, 2002). Analysis of different *Drosophila* species complexes has yielded much insight into the genetic basis of both pre- and postzygotic isolation (e.g. Coyne and Orr, 1989a, 1997, 1999; Ting *et al.*, 2001) yet many of these species groups are evolutionarily old enough to make it difficult to disentangle the order or causality of the genetic changes. Other more indirect methods have to be used to estimate which forms of isolation evolved first, and what genes were involved (Coyne and Orr, 1999; Ting *et al.*, 2000; Noor *et al.*, 2001).

Systems in which the genetic basis of reproductive isolation (actually 'making' a species) and the genetics of species differences can be disentangled are particularly valuable. Species that form hybrid zones can offer just these opportunities, and can be used to analyse the genetic basis of reproductive isolation between partially diverged populations. Temperate hybrid zones tend to be evolutionarily young, as they are formed by the secondary contact of populations separated geographically during the last glaciation (Hewitt, 1999, 2001). Hybrid zones therefore offer real-life genetic introgression experiments under natural conditions. The effect of these introgressed genes in hybrids can either be positive or negative, or of course selectively neutral (Barton, 2001b; Burke and Arnold, 2001). The challenge for hybrid-zone geneticists is to characterize the crucial parts of the genome, or the individual genes themselves, that are involved in hybrid fitness and reproductive isolation in the field, in the same way that *Drosophila* geneticists do in the laboratory. This will involve moving from phenotypic assays of natural hybrids in their current ecological context to detailed molecular genetics.

Here we review developments in the theory concerning the genetics of reproductive isolation and recent empirical work, particularly in *Drosophila*. We then outline the progress being made to meet the challenge of achieving comparable resolution in our 'non-model' organism, the European meadow grasshopper, *Chorthippus parallelus*. Two subspecies of *C. parallelus* now meet and form hybrid zones in the Pyrenees, following a period of geographical separation during recent glaciation events. Hybrids in the wild are fully fertile, yet crosses between the subspecies yield F_1 male hybrids suffering hybrid sterility, through testes dysfunction and disruption of meiosis (Hewitt *et al.*, 1987; Bella *et al.*, 1990). Female F_1 hybrids appear to have full fertility, and neither sex suffers reduced viability. Since males are the heterogametic sex (they are XO), this hybrid F_1 male sterility is an example of Haldane's rule. In addition to this postzygotic reproductive isolation, there are also behavioural and morphological differences that contribute to prezygotic reproductive isolation (Butlin, 1998; Tregenza *et al.*, 2000a).

First, we briefly review the empirical work on *C. parallelus*, highlighting the biogeographical history of the species group, the patterns of divergence across

Europe, and more specifically the nature of the divergence across the hybrid zones. This will provide the context for the rest of the chapter. Second, we review the field of speciation genetics, concentrating primarily on the recent theoretical advances made in understanding hybrid incompatibilities. Third, we review the genetics of hybrid zones, linking with the previous section through the work of Gavrilets (1997a). Last, we outline the approaches that we are currently applying to *C. parallelus*, spanning theoretical, quantitative and molecular genetic methods, to characterize the genetic basis of hybrid male sterility.

2. *Chorthippus parallelus*: Biogeography and the Patterns of Divergence and Hybridization

Chorthippus parallelus is a common flightless gomphocerine grasshopper of wet meadows, distributed widely across Europe. It is found at altitudes from sea level up to approximately 2000 m. There are a number of geographic races across Europe (Hewitt, 1999), of which the Spanish race is morphologically, chromosomally and behaviourally distinct enough to be recognized as a subspecies (*C. parallelus erythropus*). Historically, the distribution of this species, as for many others across Europe, has been strongly influenced by cycles of glaciation, with the spread of the ice sheets restricting the grasshopper's range ever more to the south (Hewitt, 1993, 1996, 1999). Current Spanish populations of *C. p. erythropus* are derived from populations that persisted through the most recent glaciation in a refuge in the south of Spain, whilst the other *C. p. parallelus* populations across Europe descend from a number of refugia in Italy, the Balkans and Turkey (and perhaps further east: Hewitt, 1999). Evidence from both nuclear and mitochondrial genetic markers shows that grasshoppers from the Balkan refuge colonized much of northern and western Europe, somewhere between 6000 and 8000 years ago (Cooper *et al.*, 1995; Lunt *et al.*, 1998). These grasshoppers reached southern France, Switzerland and Austria before grasshoppers from either the Spanish or Italian refugia could cross the Pyrenees or the Alps, respectively.

The *C. p. parallelus* and *C. p. erythropus* subspecies now form hybrid zones in the Pyrenees (reviewed by Butlin, 1998), and recently similar hybrid zones have been discovered in the Alps between populations derived from the Balkan and Italian refugia (Flanagan *et al.*, 1999). Generally the hybrid zones are quite narrow given the broad range of the species, with the area containing phenotypically intermediate individuals typically being less than 50 km wide (Butlin, 1998). However, this is wide relative to the estimated dispersal distance of about 30 m per generation (Virdee and Hewitt, 1990).

There have been numerous cycles of glaciation influencing the biogeographic patterns across Europe, so we can ask when the divergence between *C. p. parallelus* and *C. p. erythropus* commenced. A number of molecular studies suggest that the two subspecies have been diverging for approximately 500,000 years (= 500,000 generations), although such estimates are always problematic (Cooper and Hewitt, 1993; Cooper *et al.*, 1995; Lunt *et al.*, 1998). This means that the two subspecies have been diverging for the time between the last 4–6 ice ages (Hewitt, 1999). Hewitt (1996) puts forward the hypothesis that a particularly severe glaciation event around half a million years ago removed all *C. parallelus* from southern Europe, leaving individuals only in refugia further east, such as Turkey, to recolonize as it became warmer. Since then the Spanish, Italian and Balkan refugial populations have only been in sporadic contact, exchanging genes in a limited way between glaciations. With every new glaciation event, hybrid individuals located in the contact zones over the high mountain cols would be the first to die out. The evolutionary processes in the hybrid zones are therefore temporary. This suggests that what we see in a *C. parallelus* hybrid zone is not a process of parapatric speciation, but rather what happens when populations that diverge in allopatry, to a point of partial reproductive isolation, come back into contact (Butlin, 1998). Therefore we have the opportunity to observe allopatric speciation in action and see how genes from divergent populations interact. Genes that contribute to reproductive isolation in these circumstances are clearly contributing to the speciation process rather than being accumulated after speciation has been completed.

Most research has concentrated on the differences between *C. p. parallelus* and *C. p. erythropus*, and the examination of the hybrid zones between them in the Pyrenees. There are a number of differences between the subspecies (reviewed by Hewitt, 1993; Butlin, 1998). In terms of morphology, *C. p. erythropus* has distinct red hind tibiae, a greater number of stridulatory pegs in males (situated on the inner surface of the hind femur and used in the production of acoustic mating signals) and slight differences in several other traits; discriminant function analysis has shown that peg number most reliably separates the subspecies (Butlin and Hewitt, 1985a; Butlin *et al.*, 1991). In terms of behaviour, there are differences between the two subspecies in male mate-finding strategies, with *C. p. parallelus* males tending to remain stationary for longer periods, producing extended bouts of calling song, whilst *C. p. erythropus* males tend to move around more, singing less frequently and for shorter periods (Butlin and Hewitt, 1985b; Neems and Butlin, 1993). *C. p. erythropus* also has slightly slower calling songs and a more structured courtship song (Butlin and Hewitt, 1985b).

There are also important chromosomal and molecular differences, including a difference in the number of nucleolar organizing regions (NORs: Gosálvez *et al.*, 1988). *C. p. parallelus* has three NORs, with one on the X chromosome and one on each of chromosomes 2 and 3, whilst *C. p. erythropus* has only the two autosomal NORs. The NOR cline is displaced to the Spanish side of the Pyrenean hybrid zones, meaning that the X-linked NOR has penetrated further than many other characters into the *C. p. erythropus* genome (Ferris *et al.*, 1993). There are also differences in the heterochromatin patterning of the X chromosome associated with the differential presence of the NOR region, and hybrid individuals can be found that have a unique interstitial heterochromatin band halfway along the X chromosome (Bella *et al.*, 1993; Serrano *et al.*, 1996; Gosálvez *et al.* 1997). Cytogenetic analysis of Spanish populations of *C. p. erythropus* away from the Pyrenees has revealed the existence of chromosomal variation amongst Iberian populations: central and southern populations within this area of endemism are clearly different from the northern and Pyrenean populations (J.L. Bella *et al.*, in preparation).

Interestingly, there are analogous NOR patterns in the hybrid zones in the Alps, where Balkan-derived *C. p. parallelus* populations meet Italian-derived *C. p. parallelus* (Flanagan *et al.*, 1999). Whilst there are not the obvious morphological differences observed between *C. p. parallelus* and *C. p. erythropus* across the Pyrenees, the similarity in the pattern of cytogenetics is striking. In addition, whilst hybrids in the contact zones are fertile (including males), similar patterns of male hybrid sterility are observed if populations at either end of the hybrid zones are crossed (Flanagan *et al.*, 1999). This suggests that the Italian and Spanish races may be more closely related to each other than to the Balkan-derived northern European races, and that the chromosomal and testes function differences pre-date the morphological differences now observed between Spanish and Italian populations.

Two Pyrenean hybrid zones have been examined in particular detail: at the Col du Portalet in the west, and at the Col de la Quillane in the east (e.g. Butlin *et al.*, 1991; see also Buño *et al.*, 1994, for a comparison with a hybrid zone at the western end of the Pyrenees). Generally the centres of the clines for a large number of traits are in the same region, albeit not precisely coincident. The widths of the clines vary enormously, from less than 1 km to over 20 km (Butlin, 1998). The cline for NOR number is an exception, as discussed above. Another interesting exception was thought to be the cline for differences in cuticular hydrocarbon blend (Neems and Butlin, 1994). However, recent work has questioned the existence of this cline (Buckley *et al.*, 2003). There are clear differences between the subspecies in cuticular hydrocarbon composition but the environment influences these differences. Unlike the mating songs, cuticular hydrocarbon composition also varies substantially among populations within subspecies (Tregenza *et al.*, 2000a,b). Displacement of clines for the genes underlying hybrid sterility will be discussed below.

Wide clines for individual characters suggest weak selection on those characters (Barton and Hewitt, 1985). Differences in the positions and widths of clines for different traits further suggest that overall selection against hybrids is weak. This is because strong selection against hybrids maintains linkage disequilibrium among alleles derived from the parental populations and this, in turn, results in clines that are coincident and concordant (Barton and Gale, 1993). We also see little or no elevation in the variance of quantitative traits, or the covariances between traits, in the centre of the hybrid zone (Shuker *et al.*, submitted). This is in contrast to hybrid zones with strong selection or strong assortative mating (e.g. *Bombina*, Nürnberger *et al.*, 1995; *Chorthippus brunneus/jacobsi*, Bridle and Butlin, 2002).

Within the *C. parallelus* hybrid zones, all individuals are intermediates (they are unimodal hybrid zones: Jiggins and Mallett, 2000), rather than a mix of parentals and hybrid F_1s, F_2s, and early generation backcross individuals. Importantly, all individuals within the hybrid zone appear to be fully fertile and fully viable (Ritchie *et al.*, 1992). There is also no evidence for differential adaptation of the two subspecies that might lead to 'extrinsic' postzygotic isolation (i.e. isolation due to loss of adaptation in hybrids rather than intrinsic genetic incompatibility; Virdee and Hewitt, 1990). This fits with the inference of weak selection but not with the observation that F_1 hybrid males are sterile. We will return to this issue below.

In addition to the intrinsic postzygotic reproductive isolation, there is also substantial prezygotic isolation between *C. p. parallelus* and *C. p. erythropus*, mediated through mating behaviour and mate choice. For example, mate choice tests using grasshoppers from populations on either side of the hybrid zone showed positive assortative mating (Ritchie *et al.*, 1989). Females use song to distinguish between the two subspecies (Butlin and Hewitt, 1985b) but also mate assortatively if song is prevented, with the use of other cues seemingly more important for *C. p. erythropus* (Ritchie, 1990). The differences between the subspecies in cuticular hydrocarbon blend have been suggested as the most likely candidate (Ritchie, 1990; Neems and Butlin, 1995; Butlin, 1998; Tregenza *et al.*, 2000b). Importantly, there is no evidence for reinforcement of mating signals or mating preferences in the hybrid zone (reviewed in Butlin, 1998). This no longer seems surprising, given the weak selection operating within the hybrid zone.

3. Genetics of Speciation

Reproductive isolation is a rather special sort of trait, as it is a product of two species and the interaction between their two divergent genomes (Coyne, 1994). In principle, isolation could result from within-locus interactions, such as underdominance, or from between-locus epistatic interactions. The problem with underdominance is the difficulty of arriving at populations fixed for distinct alleles (DD and D'D') when the heterozygote (DD') has reduced fitness (see, for example, Barton and Charlesworth, 1984). The evolution of reproductive isolation involving negative epistatic interactions between alleles at different loci can avoid this hurdle. The classic formulation is the Dobzhansky–Muller (D-M) model (Orr, 1996). The most straightforward representation is a two-locus model. Imagine an ancestral species that has a two-locus genotype *aabb* and that this species becomes separated for some reason into two populations. In one population a mutant allele *A* becomes fixed (by drift or selection) at the first locus. Assume that the *Aa* genotype is at least as fit as the *aa* genotype, and also assume that the *A* allele and the *b* allele are genetically compatible, since natural selection would otherwise have acted against the *A* allele. In the second population a mutant allele *B* becomes fixed at the second locus; in a similar way we assume that *a* and *B* alleles are genetically compatible and *Bb* is as fit as *bb*. If the two populations come back into contact the F_1 hybrids will have the genotype *AaBb*, and the derived alleles at the two loci that have never interacted (*A* and *B*) will be placed together in the same individual. If the *A* and *B* alleles interact deleteriously, postzygotic isolation through inviability or sterility can result.

In this scenario, incompatibilities arise because new allelic combinations that have not been screened by natural selection are created in hybrids. Within a population, derived alleles with negative epistatic interactions with alleles at other loci would be removed by natural selection, but since the derived alleles spread to fixation in independent populations this barrier does not have to be overcome. There is a series of (intrinsically) fit genotypes that connect the two diverged populations via the ancestral genotype (*AAbb, Aabb, aabb, aaBb, aaBB*) – equivalent to a ridge of high fitness connecting two populations in an adaptive landscape, with hybrids falling into an adaptive valley (Barton, 1988; Gavrilets, 1997a,b).

The evolution of hybrid incompatibilities via the D-M model has been considered in detail by Orr (1995; Orr and Orr, 1996). Several important points emerged from his analysis. First, hybrid incompatibilities involving a pair of loci are initially asymmetric (i.e. only derived alleles at the two loci can be incompatible). Also, derived alleles are involved in more potential incompatibilities than ancestral alleles, and the effect of a single locus will increase over time as it interacts with more loci. Finally, later substitutions can cause more incompatibilities than earlier ones because there are more opportunities for negative interactions between more differentiated genotypes. This last point suggests that the number of incompatibilities, and thus the reproductive isolation, may evolve at a rate faster than linearly with time. Orr (1995) showed that this is the case, with the probability of speciation increasing at least as fast as the square of the time since divergence commenced, a process he termed 'snowballing'. This, of course, means that with time it becomes increasingly difficult to work out which incompatibilities initiated the reproductive isolation, and increasingly likely that the number of genes involved will be overestimated. In addition, Orr extended the analysis to consider interactions between three or more loci. Perhaps unsurprisingly, it becomes easier to evolve isolation if incompatibilities between multiple loci can occur.

The forces fixing derived alleles in their diverging populations are not important for the basic D-M model, but when population subdivision varies they can become important. Orr and Orr (1996) showed that if the new alleles are fixed by drift, then the population size of the diverging populations does not influence the time required for a particular threshold number of incompatibilities to arise ('time to speciation'). However, if the substitutions in the diverging populations are the result of selection, then the time to speciation is influenced by the pattern of population subdivision, with speciation occurring more rapidly if there are two large populations, as opposed to many smaller populations. This result runs counter to some intuitive arguments about the role of population subdivision in speciation (Orr and Orr, 1996).

The D-M model is most widely discussed in the context of incompatibilities between loci within individuals, resulting in intrinsic postzygotic isolation through sterility or inviability. However, the same underlying principles apply to the evolution of prezygotic isolation and extrinsic postzygotic isolation. Two isolated populations may diverge in mating signals or in ecologically important traits. Natural and sexual selection ensure that the derived characters function well together within populations but do not guarantee that they are compatible between populations. On secondary contact, signals and responses may be incompatible, resulting in prezygotic isolation, or new combinations of ecological traits may have poor competitive ability or low protection against predation, resulting in extrinsic postzygotic isolation. Therefore the same principles that underlie sterility and inviability might be expected to apply to other aspects of speciation as well.

Below we consider the current ideas and data on the genetics of both prezygotic and postzygotic isolation, initially in experimental systems and then in hybrid zones. Some patterns of reproductive isolation in *C. parallelus* will be discussed, but the analysis of sterility genes in the hybrid zone will be left to a separate section.

3.1 Prezygotic reproductive isolation

Prezygotic isolation occurs when populations become behaviourally or ecologically distinct in some way, such that mating between individuals from the different populations becomes less likely. It may also result from interactions after mating, such as differential sperm survival or gamete recognition (e.g. Howard and Gregory, 1993), an effect that has been reported in *C. parallelus* (Bella *et al.*, 1992). The underlying differences can arise as a result of natural selection on traits such as habitat preference or feeding ecology ('ecological speciation', recently reviewed by Schluter, 1998, 2000, 2001). Alternatively, the difference may be the result of the evolution of differences in mating behaviour, for instance divergence of mating signals and the mate preferences for those signals, driven by sexual or natural selection (Panhuis *et al.*, 2001). Sexual selection has a rich theoretical background (Andersson, 1994) and the potentially arbitrary nature of the evolution of mating signals and preferences is often highlighted in terms of a possible role in speciation (e.g. Lande, 1981; Pomiankowski and Iwasa, 1998). The importance of models of sexual selection for speciation has recently been reviewed by Turelli *et al.* (2001). In truth, natural and sexual selection are likely to interact (and perhaps oppose each other) as populations diverge ecologically and behaviourally. In addition, traits that diverge and lead to prezygotic isolation can also give rise to postzygotic isolating mechanisms, if the alleles at the relevant loci selected in diverging populations produce genetic incompatibilities in a hybrid genetic background, leading to either extrinsic or intrinsic reproductive isolation (Wu and Davis, 1993; Civetta and Singh, 1998).

The role of prezygotic isolation in speciation is often considered to be a secondary one. This is partly because of the attention given to the debate about reinforcement. Reinforcement is the evolution of prezygotic isolation (typically through assortative mating) in response to the reduced fitness of hybrids. By definition, reinforcement theory therefore assumes that postzygotic isolation comes first. If hybrids are unfit, due to deleterious genetic incompatibilities, selection favouring mechanisms that prevent the formation of hybrids in the first place seems logical. However, there are problems with reinforcement and the field remains contentious even though recent models suggest that the process is more plausible than it once seemed (Liou and Price, 1994; Kelly and Noor, 1996; Noor, 1999; Kirkpatrick, 2000; Servedio, 2000; Sadedin and Littlejohn, 2003). The pattern expected following reinforcement is greater divergence in mating signals and the responses to them, or stronger assortative mating, in sympatry relative to allopatry; in other words, reproductive character displacement. This pattern is common (Howard, 1993; Coyne and Orr, 1997; Noor, 1999; Turelli *et al.*, 2001), but it can be generated by mechanisms other than reinforcement (Butlin and Ritchie, 1994). Few examples are really convincing, probably the best being in a hybrid zone between pied and collared flycatchers (Saetre *et al.*, 1997). In *Chorthippus parallelus*, there is no evidence for reinforcement despite a variety of different tests for its effects (reviewed by Butlin, 1998). We now know that this is because the hybrid zone was modified by selection to remove the hybrid incompatibilities causing the production of unfit hybrids (see below). This removes the selection pressure that is needed to drive reinforcement and it is a possible outcome that has rarely been considered in either theoretical or empirical studies.

In fact, there are good reasons to think that prezygotic isolation may often initiate speciation (Butlin and Ritchie, 1994). Sexual selection can drive divergence between isolated populations more rapidly than natural selection and even initiate speciation in sympatry (Panhuis *et al.*, 2001). Divergence in mating signals may cause the reduced hybrid fitness that drives reinforcement (Coyne and Orr, 1989a) or ecological divergence may cause prezygotic isolation as a pleiotropic effect (Via, 2001). In either case, prezygotic isolation has a role right from the start of the process, and comparative studies suggest that prezygotic isolation evolves at least as fast as postzygotic isolation (Coyne and Orr, 1989a, 1997). In *C. parallelus*, prezygotic isolation of approximately 0.5 (measured as the isolation index from assortative mating experiments; Ritchie *et al.*, 1989) has evolved over about 0.5 million years. This is broadly in line with the expectation for allopatric divergence based on Coyne and Orr's *Drosophila* data. However, Tregenza *et al.* (2000c) showed that almost as strong assortative mating has evolved between some populations derived from the same glacial refuge that presumably shared a common ancestor only about 10,000 years ago. Therefore, prezygotic isolation can evolve very rapidly in some circumstances. In *C. parallelus*, Tregenza *et al.* suggest that some aspect of the process of rapid northward colonization promoted evolutionary changes in mating behaviour.

The genetics of prezygotic isolation have recently been very thoroughly reviewed by Ritchie and Phillips (1998). Progress in this area has been overshadowed by studies of postzygotic isolation, which is easier to study because the traits involved, especially sterility, are more clear-cut than the behavioural or morphological traits involved in prezygotic isolation. However, we do need to know about the genetic basis of sexually selected traits, not least in order to make reasonable genetic assumptions in models of reinforcement or speciation by sexual selection. We still have little idea of how many genes are likely to be involved in traits of interest for prezygotic isolation, either in terms of the sexually selected traits involved in mating behaviour, or in terms of naturally selected traits involved in habitat preferences or foraging. Does divergence between populations arise from selection on standing variation or from novel major gene mutations? This again is important if we want to understand why some taxa speciate rapidly and some do not. In addition, what is the role of sex-linked genes in the evolution of prezygotic isolation? Without this information it is difficult to estimate how long divergence takes or whether particular kinds of genes tend to be involved in controlling/initiating the critical changes.

Ritchie and Phillips (1998) summarized the empirical data but found few patterns. They concluded, for example, that sex-linked genes are commonly involved in sexual isolation in the Lepidoptera (see also Pashley, 1998) but are not predominant in other taxa. New data obtained since this review still fail to reveal much common ground (see also Coyne and Orr, 1999). The pheromonal mating signal system of *Drosophila melanogaster* has revealed a gene of large effect (Takahashi *et al.*, 2001). A short deletion in the regulatory region of the desaturase gene *desat2* appears to control a cuticular hydrocarbon difference that is a key component of the partial prezygotic isolation between cosmopolitan, M-type, populations and 'Zimbabwe', Z-type, populations. The deletion is associated with mating discrimination between M and Z (Fang *et al.*, 2002) but overall isolation is known to involve other courtship elements and the total number of loci involved is at least 15 (Ting *et al.*, 2001). Analysis of recombinant inbred lines derived from a cross between strains of *D. melanogaster* identified three quantitative trait loci of relatively large effect accounting for more than half of the total genetic variance in a male courtship song character (Gleason *et al.*, 2002). However, the remaining 46% of the variation could be highly polygenic.

Other comparisons, between rather than within *Drosophila* species, also suggest that individual loci can have large effects but that multiple loci are also involved. For example, Doi *et al.* (2001) found that replacing a single section of chromosome II made *Drosophila pallidosa* females accept *D. ananassae* males as readily as their own females do. In the *D. virilis* group, a gene (or genes) that makes a large contribution to courtship song differences between the species is located on the X chromosome but interacts with autosomal loci (Päällysaho *et al.*, 2003). The X chromosome region involved shows complex rearrangements between species, echoing a pattern of association between speciation genes and chromosomal modifications that restrict recombination which has been observed elsewhere (*Drosophila*, Noor *et al.*, 2001; *Helianthus* sunflowers, Rieseberg *et al.*, 1999; see Ortiz-Barrientos *et al.*, 2002). Beyond insects, the rapid speciation of the cichlid fish of the African Great Lakes, apparently driven by sexual selection, has been linked to switches between a limited set of colour patterns (Seehausen *et al.*, 1999; Lande *et al.*, 2001). These colour patterns have a superficially simple genetic basis, but they are linked to the sex chromosomes and so to areas of suppressed recombination and also to sex-reversal genes. It may be that it is these associations that allow the spread of novel colour morphs, which would normally be opposed by female mating preferences, in parallel with a transition from male heterogamety to female heterogamety (Lande *et al.*, 2001).

The emerging pattern is actually quite similar to those for the genetics of postzygotic isolation (see below) and for species differences in other traits (Orr, 2001). It is common for many loci to be involved but, amongst these loci, there may be alleles of large effect. Preferential associations with the sex chromosomes do exist but are not universal, whilst the occurrence of speciation genes in areas of suppressed recombination is a common observation. Clearly, many more studies are needed before such generalizations are really secure. In *C. parallelus*, we know only that the song and morphological differences have a broadly polygenic basis (Butlin and Hewitt, 1988), although the clear chromosomal differences between the subspecies are of interest, concentrated as they are on the X chromosome.

3.2 Postzygotic reproductive incompatibilities

There is now a good deal of empirical evidence supporting the role of D-M epistasis in postzygotic reproductive isolation (Coyne and Orr, 1999; Johnson, 2000; Wade, 2000; Turelli *et al.*, 2001). Much of this is based upon a series of introgression experiments using *Drosophila* species complexes (see Coyne and Orr, 1999), but significant support also comes from hybrid-zone studies (see below). There is also a renewed interest in epistasis in evolution more generally (Wolf *et al.*, 2000).

A pervasive pattern in postzygotic reproductive isolation is that when only one sex suffers hybrid inviability or sterility, it is generally the heterogametic sex; this pattern is known as Haldane's rule (reviewed by Wu *et al.*, 1996; Laurie, 1997; Orr, 1997). The pattern holds true for 99% of 223 examples of heterogametic sex-only hybrid sterility, and for 90% of 115 examples of heterogametic sex-only hybrid inviability (Turelli, 1998). Many empirical studies of postzygotic isolation have focused on Haldane's rule, partly because it is such a strong pattern but partly because the unaffected homogametic sex permits backcrossing. Genetic analyses that would be impossible if both sexes were inviable or sterile are therefore possible. After much controversy, two major theories have emerged as likely explanations of Haldane's rule: the dominance hypothesis and the faster male hypothesis. The dominance hypothesis is a derivative of the Dobzhansky–Muller model. It posits that the heterogametic sex is more likely to show hybrid dysfunction because some of the underlying deleterious epistatic genetic incompatibilities are effectively X-linked partial recessives. Since the heterogametic sex has only one X chromosome, any X-linked loci that have alleles incompatible with other parts of the hybrid genetic background will lead to a loss of fitness. In the homogametic sex, with two copies of each X-linked locus, compatible alleles may be present and, if dominant, will prevent any reduction in hybrid fitness. This partial recessivity is assumed to be due to loss-of-function (Orr, 1993). The faster male hypothesis, on the other hand, posits that male genes evolve faster than female genes, perhaps due to the effects of sexual selection acting more strongly on males (Wu and Davis, 1993). Obviously this hypothesis can only explain Haldane's rule when males are the heterogametic sex (i.e. not butterflies and birds).

There is good empirical evidence from insects that both processes can play a part: the pattern of hybrid sterility and inviability between pairs of *Drosophila* species in terms of X chromosome size and genetic distance are consistent with the dominance hypothesis (Turelli and Begun, 1997). Evidence for the faster male hypothesis comes from the observation of Haldane's rule in *Aedes* mosquitoes (Presgraves and Orr, 1998). They have single locus sex determination and functionally equivalent X and Y chromosomes (so the X-linked loci are not hemizygous in males), and so the dominance hypothesis cannot act. However, the sex bias is not as strong as in *Anopheles* mosquitoes, which have heterochromatic Y chromosomes. The difference between the genera is greater for sterility than for inviability. Presgraves and Orr (1998) conclude that the two processes act together in male heterogametic species and in opposition in female heterogametic species, and that the faster male hypothesis is associated more with sterility than with inviability.

The dominance hypothesis has been analysed theoretically by Turelli and Orr (1995, 2000), building on the framework of accumulating genetic incompatibilities developed by Orr (1993, 1995) discussed above. Their analyses have yielded a number of important insights. For example, the dominance hypothesis provides a simple explanation for the 'large X effect', commonly seen in examples of Haldane's rule. With incompatibilities involving recessive alleles, the hemizygous X placed in a hybrid background will cause more problems than a heterozygous autosome. The model results are also consistent with the temporal lag between the evolution of heterogametic and homogametic postzygotic isolation, and Turelli and Orr (1995) suggest that to explain the data from *Drosophila*, alleles affecting hybrid fitness must be recessive. They also clarify the role of the dominance hypothesis for both hybrid sterility and inviability. Genes influencing sterility are likely to differ from those influencing viability (Wu and Davis, 1993). Within species there is good evidence that alleles influencing sterility tend to act on one sex only (e.g. 'testes' genes, or 'ovary' genes), whilst lethal alleles tend to influence both sexes. This means that the hybrid sterility observed in males that results from a hemizygous X chromosome does not necessarily have to appear in females that are experimentally given two copies of the 'sterile' X because these genes do not necessarily influence fertility in females. Observations such as this do not therefore represent a challenge to the dominance theory (cf. Coyne and Orr, 1989b). Unfortunately though, the 1995 paper suggests it is not straightforward to calculate the number of incompatible genes from empirical data.

In their 2000 paper, Turelli and Orr extend the model further, concentrating on the influence of epistatic interactions between different types of loci. Of key concern for Haldane's rule is the input of incompatibilities due to interactions between heterozygous loci and interactions between heterozygous and homozygous (or hemizygous) loci. Haldane's rule will arise if the latter are more than twice as severe as the former (reiterating the results from before that for Haldane's rule to arise via dominance, the incompatibilities act as partial recessives). They also consider the input of interactions between homozygous loci and conclude that they too need to be considered when evaluating the influence of X chromosomes on postzygotic isolation. Comparisons of the effects of dominance and faster male evolution show that both can act together in male-heterogametic species, but that dominance must overcome faster male evolution in female-heterogametic species, requiring extreme recessivity of incompatibilities. That dominance can overcome faster male evolution is clear from female heterogametic insects such as the Lepidoptera, where female hybrids are generally sterile, despite faster male evolution (Laurie, 1997; Pashley, 1998).

As noted above, extrinsic postzygotic isolation can also evolve as a result of negative epistatic interactions, in this case dependent on the environment for their expression. If the fitness outcomes of the genetic interactions in hybrids are influenced by the environment (adding a $g \times e$ component to fitness), these gene-by-environment interactions can also influence the evolution and nature of the hybrid incompatibilities, for instance increasing the number of possible incompatibilities (Bordenstein and Drapeau, 2001). Thus, 'intrinsic' and 'extrinsic' causes of postzygotic isolation may not be easily separable. In addition, some models of chromosomal speciation also fit the D-M model. An example from outside the insects is the phenomenon of 'monobrachial homology' between products of independent Robertsonian fusion events in shrews and other mammals (Searle, 1993). The individual fusion events cause little, if any, fitness reduction in hybrids, but the interaction between different fusions in hybrids between divergent races can cause a major fitness cost.

Introgression experiments in *Drosophila* suggest that many parts of the genome contain loci contributing to hybrid inviability and sterility (Wu and Palopoli, 1994; Wu, 2001). However, making a single locus or small group of loci homozygous on the 'wrong' genetic background can cause complete inviability or sterility, and single locus mutations are known that can 'rescue' inviable or sterile hybrids. This means that the question of the number of genes required for speciation remains open because it suggests that many incompatibilities may have accumulated since complete isolation was achieved. Two other critical questions need to be answered:

1. *What is the normal function of the genes involved?* Remember that these genes are identified on the basis of a phenotype (hybrid inviability or sterility) that cannot have been important in their evolution.
2. *Was the divergence at these loci due to selection or genetic drift?* If it was due to selection, was this selection for local adaptation or a result of sexual selection?

Clearly, these questions are related and can be tackled by identifying and characterizing individual 'speciation genes'. At present, there is little guidance from theory about what to expect, although one recent suggestion is that gene duplication might underlie D-M incompatibilities. If the redundancy created by gene duplication is resolved by the loss of function of one copy in one population and of the other copy in a different population, hybrids could end up with no functional copy (Lynch and Conery, 2000; Lynch and Force, 2000).

Progress in reaching this level of genetic resolution has accelerated rapidly over the last few years but, nevertheless, molecular identification and characterization of only three 'speciation genes' has been achieved, two of which are in insects (Noor, 2003; Wu and Ting, 2004). In *Drosophila*, the *OdsH* gene occurs at the locus where a sterility gene, *Odysseus*, has been mapped

(Ting *et al.*, 1998), and *Nup96* has been shown to cause inviability when the *D. simulans* allele interacts with a *D. melanogaster* X chromosome (Presgraves *et al.*, 2003). In platyfish, the *Xmrk-2* gene is overexpressed in *Xiphophorus maculatus* × *X. helleri* hybrids, causing potentially lethal tumours (Walter and Kazianis, 2001). Both of the *Drosophila* loci show strong evidence, from sequence analysis, for divergence under selection (see also Wu and Ting, 2004). The *OdsH* gene contains a homeobox sequence, suggesting a role in development, and is specifically expressed in testes, unlike homologous sequences in other species. However, its function and the likely form of selection causing divergence are still speculative. *Nup96* is a nucleoporin, a component of a ubiquitous and highly conserved complex of proteins involved in movement of RNA and proteins between the nucleus and the cytoplasm. It is very hard to imagine what might have caused it to diverge rapidly under selection.

The patterns of postzygotic isolation and their evolutionary significance in *C. parallelus* have been examined by Tregenza *et al.* (2002), following their work on the evolution of prezygotic isolation. The evolution of the hybrid male sterility in 0.5 million years of divergence is fast but not exceptional in comparison with *Drosophila* (Coyne and Orr, 1997). The analysis of populations from across Europe, including the *C. p. erythropus* subspecies in Spain, showed that long-term isolation in the various refugia was the source of most of the incompatibility. Postzygotic isolation (measured, as in the hybrid zone studies, as testes follicle length: Hewitt *et al.*, 1987) was not promoted by the colonization effects associated with the movement out of the glacial refugia, unlike some aspects of prezygotic isolation (see above). Instead, postzygotic isolation was only weakly correlated with genetic distance, with crosses to the Spanish population tending to yield dysfunctional male testes, whilst there was also some reduction in testes follicle length amongst crosses between populations derived from the Balkan refugia, primarily involving the British population. Interestingly, separate work on isolated Spanish populations of *C. p. erythropus* has also revealed incompatibilities among these populations (J.L. Bella, unpublished data). The work by Tregenza and co-workers therefore suggests that pre- and postzygotic isolation in *C. parallelus* have evolved independently. Within-species comparisons of this sort are needed from many more taxa (Tregenza, 2002), not only to test different hypotheses about the origins of reproductive isolation, but also to allow us to compare how different forms of reproductive isolation evolve. Are the patterns from *C. parallelus* unique or common?

4. Hybrid Zones and the Evolution of Reproductive Isolation

The study of hybrid zones has proved very successful in exploring how speciation works, even if processes at work within hybrid zones themselves are often not driving (parapatric) speciation, but reflecting aspects of reproductive isolation following secondary contact (Harrison, 1993; Butlin, 1998). Instead of crossing species in the laboratory, we have natural hybrids to work with. Complicated recombinant genotypes are created and tested by natural selection in ways that may vary spatially across the zone, over many generations, with the input of wild-type genotypes by migration. This allows us to consider ecology and genetics simultaneously. In this section, we first briefly introduce the genetic study of hybrid zones in terms of the population genetic treatment of clines. We then follow Gavrilets (1997a) and consider what happens when the evolution of hybrid incompatibilities following the Dobzhansky–Muller model are allowed to play out in a hybrid zone.

Hybrid zones can be simple contact zones in which individuals of the two hybridizing populations meet each other in a relatively straightforward way through homogeneous habitat, producing smooth clines for traits that differ between the two populations. The hybrid zones of *C. p. parallelus* and *erythropus* are typical of this kind. Alternatively, the contact zones may be complex, for instance if the interacting populations are adapted to different habitats that are patchily distributed within the wider landscape matrix, giving rise to mosaic hybrid zones (Harrison and Rand, 1989). A good example of an insect mosaic hybrid zone is found in northern Spain where two other *Chorthippus* species, *C. brunneus* and *C. jacobsi*, meet and hybridize (Bridle *et al.*, 2001), although in this example the interaction between habitat structure and grasshopper distribution is not straightforward, suggesting a historical component to the current pattern (Bridle *et al.*, 2002; Bailey *et al.*, 2004). Hybridizing species can exhibit both kinds of contact zone across their geographic range as a result of differences in habitat structure (e.g. *Bombina* toad hybrid zones, MacCallum *et al.*, 1998).

The *C. parallelus* hybrid zones are unimodal tension zones, where the zone is maintained by the balance between selection against hybrids and the immigration of wild-type genotypes into the zone (Barton and Hewitt, 1985, 1989). In unimodal tension zones we see smooth sigmoid clines in traits as they change from one population, through intermediates, to the other. The clines in morphological and behavioural traits also represent the underlying gene frequencies across the hybrid zone. The population genetics of clines are well understood (e.g. Felsenstein, 1976; Endler, 1977; Barton, 1983, 1986, 1999; Barton and Bengtsson, 1986; Gavrilets and Cruzan, 1998). One aspect of cline theory that is often considered is the barrier to gene flow, calculated as the distance of free habitat (with no selection of any sort against hybrid genotypes) that would cause the same delay to the passage of a neutral allele.

The genetic analysis of tension zones has been reviewed by Barton and Gale (1993). Regardless of the form of selection on either single-locus or quantitative traits, cline width is proportional to the ratio between dispersal and the square root of selection ($\sigma/\sqrt{s}$). In other words, the width of a tension zone is determined by the balance between dispersal in the zone and selection against hybrids (see also Barton and Hewitt, 1985, 1989). This relationship can be used to examine selection pressures, but it has led to perhaps rather unrealistically weak selection estimates (Barton and Hewitt, 1985), probably due to poorly estimated dispersal rates (Barton and Gale, 1993). The shape of a cline is only weakly dependent on the form of selection, although epistasis between reasonably large numbers of genes can distort cline shape away from a sigmoid curve (which should be linear on a logit scale). The effects of epistasis are of interest, however, for traits related to reproductive isolation, for the reasons outlined in Section 3 (see also Barton and Shpak, 2000).

Linkage disequilibrium is important in hybrid zones, with theory predicting an increase in linkage disequilibrium in the centre of the hybrid zone due to the continued influx of parental genotypes. Barton and Gale (1993) showed how to estimate linkage disequilibria from marker and quantitative traits, and also showed that patterns of linkage disequilibrium can be used to estimate dispersal and selection from natural hybrid zones, using zones in *Podisma* grasshoppers and *Bombina* toads as examples. Generally, both segregation and linkage disequilibria will elevate the genetic variance of traits in the centre of a hybrid zone (see also Barton, 1999), and since the equation they derive for the variance at the centre of the zone contains a term for the effective number of loci (n_e, from their equation 5b), theoretically it is possible to estimate the effective number of loci underlying a trait difference between parental populations. This is based on the Castle–Wright–Lande method of estimating the number of genes, and as such has low statistical power, and is sensitive to non-additive genetic effects, including epistasis. However, it does show that patterns of trait variance across a hybrid zone can provide estimates of the underlying genetic basis of a trait.

Durrett *et al.* (2000) added spatial structure to a model of a hybrid zone in order to explore possible differences between hybrid zones formed by secondary contact of already divergent populations (as in *C. parallelus*), or those formed by primary intergradation (as in models of parapatric speciation), where selection for divergence occurs within the geographic range. Hybrid zone theory has suggested that it is very difficult, if not impossible, to tell these different scenarios apart (Barton and Hewitt, 1985), but they suggest that under certain circumstances signals of primary intergradation versus secondary contact can persist for thousands of generations.

Barton and Hewitt (1981) pioneered the analysis of hybrid zones to make inferences about the genetic basis of hybrid incompatibilities. In the alpine grasshopper, *Podisma pedestris*, they observed that the fitness of grasshoppers from the centre of the hybrid zone was lower than the fitness of F_1 hybrids when both sets of offspring were generated in the laboratory. F_1 hybrids are heterozygous at all loci differentiating the races but have a complete set of genes for each race (at least in the homogametic females). Natural hybrids from the zone centre have, on average, 50% of their genes from each race but are typically heterozygous only for some differentiated loci and do not have a complete set of genes from either race. Thus, the low fitness of natural hybrids relative to F_1s suggests a major contribution of negative epistasis as opposed to underdominance. Measuring the dispersal of the grasshoppers and the width of the zone of reduced fitness further enabled Barton and Hewitt to estimate the average selection pressure per locus (~3%) and the number of loci involved (~150) in lowering hybrid fitness.

In some hybrid zones, strong selection against hybrids leads to narrow clines. This, in turn, creates strong linkage disequilibrium in the centre of the hybrid zone. The disequilibrium prevents selection from operating independently on individual loci and also causes a greater barrier to the diffusion of neutral alleles across the cline (Barton and Gale, 1993). The contact between *Bombina bombina* and *B. variegata* toads in Poland is of this type (Szymura and Barton, 1991). Here, too, analysis of cline widths and linkage disequilibria suggest that selection against hybrids is due to weak selection ($\sim$2%) at each of many loci ($\sim$55).

Other than in *Chorthippus parallelus*, hybrid sterility has not been analysed in detail in other insect hybrid zones. In terms of examples from plants, in the sunflowers *Helianthus annuus* and *H. petiolaris*, male sterility of hybrids is a major barrier to gene exchange. Rieseberg *et al.* (1999) compared the introgression of alleles at 88 marker loci in three replicate hybrid zones and identified 16 chromosomal regions significantly associated with male sterility with less introgression than expected under neutrality. These regions were strongly associated with chromosomal rearrangements (inversions and translocations). This makes it difficult to proceed further with the identification of individual loci or, indeed, to distinguish between genic and chromosomal causes of sterility (Rieseberg, 2001). Rieseberg (2001) argues that the major effect of chromosomal rearrangements is to spread the barrier to gene exchange caused by a single selected locus to a larger number of surrounding genes by reducing recombination. The same argument has been made for chromosomal inversions in *Drosophila* (Noor *et al.*, 2001).

Overall, the hybrid-zone data are compatible with the results of introgression experiments: epistasis seems to be more important than underdominance and many loci are involved. However, individual loci of large effect have not been detected in hybrid-zone studies to date.

Of particular interest to the present discussion and the hybrid zone of *C. parallelus* is the model of Gavrilets (1997a). He considered the behaviour of Dobzhansky–Muller incompatibilities in a hybrid zone using a stepping-stone model. As mentioned earlier, in the simple two-locus model of D-M incompatibilities, there is a series of fit recombinant genotypes that connects the now divergent populations, even though F_1 hybrids are unfit due to genetic incompatibilities between derived alleles fixed in the two populations. Gavrilets (1997a) showed that just such a series of genotypes can be generated in a hybrid zone, with fit recombinant genotypes coming to predominate in the centre of the zone. Importantly, this process leads to the displacement of the clines for the derived alleles at the two loci causing incompatibility. These clines are asymmetric because introgression in one direction is opposed by strong incompatibility selection, whereas introgression in the other direction is only opposed by the weak advantage assumed to have caused the initial replacement of the ancestral by the derived allele. As a result of the displacement, few incompatible genotypes appear in the hybrid zone and all populations have high mean fitness (Fig. 15.1). In addition, clines of other traits, unlinked to loci influencing hybrid fitness, are expected to be symmetric and broadly concordant. In the model, Gavrilets followed a neutral marker locus and showed that introgression is only weakly impeded unless it is very closely linked to the selected loci. This model establishes important principles for the outcome of D-M epistasis following secondary contact but it leaves open many specific questions. For example, what would be the fate of sex-linked incompatibility loci? How would the behaviour of the clines be influenced by increasing the number of loci involved or altering the pattern of interaction (to

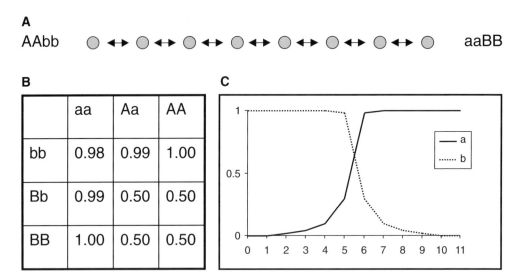

Fig. 15.1. The Gavrilets' (1997a) two-locus model of Dobzhansky–Muller epistasis in a hybrid zone. **A** Two divergent populations with derived alleles (A and B) fixed at alternative loci are separated by a linear array of habitat patches. **B** The fitness matrix for all possible two-locus genotypes. Genotypes where the derived alleles are placed together have reduced fitness. A ridge of high fitness genotypes exists linking the two divergent populations, via the ancestral genotype (*aabb*). **C** Following migration into the hybrid zone and selection on hybrids, fit recombinant genotypes exist in the centre of the zone, such that A and B alleles rarely co-occur. The clines in allele frequencies are displaced and asymmetrical.

make them recessive, for example)? We are currently investigating these issues using an individual-based simulation model (Shuker, Butlin and Bullock, in preparation).

5. Hybrid Male Sterility in *C. parallelus*

As we have seen, the most striking pattern of subspecies divergence in *C. parallelus* is the strong hybrid male dysfunction that occurs when subspecies are crossed (both in the Pyrenees and Alps; Hewitt *et al.*, 1987; Butlin, 1998; Flanagan *et al.*, 1999; Tregenza *et al.*, 2002). However, despite the behavioural and morphological intermediate phenotypes seen in the hybrid zones, no male hybrid sterility has been found, intermediate or otherwise, in the hybrid zones studied to date (Ritchie *et al.*, 1992; Shuker, unpublished data). We can ask two related questions about these patterns of postzygotic reproductive isolation between *C. p. parallelus* and *C. p. erythropus*:

1. What is the genetic basis of the isolation?
2. How has the hybrid male dysfunction been ameliorated amongst naturally occurring hybrids in the wild?

One can determine the positions of clines for genes underlying hybrid male sterility by crossing naturally occurring hybrids to parental individuals from either end of the hybrid zone. Crosses between more distant locations are expected to produce less-fertile offspring and the transition in offspring fertility marks the cline in alleles from *C. p. erythropus* that cause sterility against a *C. p. parallelus* background, when the latter subspecies is used as the reference parent (Virdee and Hewitt, 1994). Measured in this way, clines for sterility genes are broadly coincident with the clines for the majority of other traits across the hybrid zones, but the cline produced by crossing to *C. p. parallelus* is displaced from the cline produced by crossing to *C. p. erythropus* by about one cline-width (about 5 km; Underwood, 1994; Butlin, 1998; Fig. 15.1). This is precisely the pattern predicted by the Gavrilets' (1997a) model and it fully explains the absence of hybrid sterility in the field because populations at the zone centre have universally compatible, recombinant genotypes. Gavrilets' model predicts that clines will be asymmetrical but the data available to date are insufficient to distinguish this shape difference. The width of the cline suggests that the selection favouring derived over ancestral alleles at the loci causing sterility is very weak (a cline width of 5 km and a dispersal distance of 30 m per generation implies selection $<<1\%$). One proviso is that the crossing procedure always places a recombined genome from the zone, including the X chromosome, in a male with a complete genome of the reference parent subspecies. This means that only X-linked or dominant incompatibilities will be revealed.

How can further progress be made in understanding the genetic basis of postzygotic isolation in *Chorthippus parallelus*? We are currently using two complementary approaches. First, there is potential for further quantitative analysis to improve the estimates of the cline widths, and thus address the important question of the selection favouring the spread of derived alleles, and to provide estimates of the numbers of loci involved. As noted above, segregation in the centre of a hybrid zone increases the variance in quantitative traits in inverse proportion to the number of loci accounting for differentiation between the parental populations. In the crosses used to map sterility clines, this principle can be applied to variation in offspring sterility (measured as testis follicle length). The interactions with the reference genome present in each offspring individual can be viewed as combining additively. Variation in the positions and widths of clines in morphological and behavioural traits (Butlin *et al.*, 1991), and the little or no elevation in variance or covariance for these traits in the zone centre (Shuker *et al.*, submitted), suggest that there is negligible linkage disequilibrium. Therefore, the elevation in variance of follicle length at the cline centre can be used to infer the number of sterility genes. This can be done independently for the two directions of cross because there is no reason to suppose that the same number of derived alleles will have been fixed in both subspecies.

Quantitative genetic approaches will not lead to the identification of sterility loci at the molecular level, however. Fortunately, there are now molecular tools that provide alternative routes to gene discovery in a non-model insect such as *Chorthippus parallelus*. Our second approach is to exploit these tools. Research on *Drosophila* has shown that pairs of closely related species with low levels of sequence divergence can have different patterns of gene expression at a high proportion of loci (approximately half of 4776 loci analysed for *D. melanogaster* and *D. simulans* that differ at about 3.8% of nucleotides; Ranz et al., 2003). Even more strikingly, 83% of the loci that show changes in expression have altered patterns of sex-bias in expression levels. Many genes with male-biased expression are expressed in the testes (Parisi et al., 2003) and a large set of testis-specific co-expressed loci was identified in a general survey of expression patterns in *Drosophila* (Arbeitman et al., 2002). Michalak and Noor (2003) have recently identified differences in the expression of numerous genes in pure and hybrid males of *D. simulans* and *D. mauritiana*, including genes associated with spermatogenesis. Observations of this sort suggest that one might begin a search for sterility genes by isolating sequences expressed in testes that show altered levels of expression between subspecies. Applying this approach in *C. parallelus*, it is possible to sort differentially expressed transcripts on the basis of the position and width of the cline in expression: this should match the phenotypically determined sterility cline.

In practice, we have followed this outline by extracting mRNA from testes of the two subspecies, making cDNA and then isolating differentially expressed transcripts using the technique of suppression subtractive hybridization (SSH; Diatchenko et al., 1996). This PCR-based approach is a very efficient method for isolating differentially expressed sequences (Bole-Feysot et al., 2000). Cloned candidate sequences have then been checked for differential expression by screening dot blots with cDNA from either subspecies. Sequences with strong expression differences can then be tested for a clinal pattern using 'virtual Northern' hybridizations (Guiguen et al., 1999). cDNA prepared from several individuals per population is run on a gel, blotted on to a nylon membrane and probed with the candidate sequence. This reveals variation within and among populations. This strategy is providing some very promising sterility gene candidates (King, unpublished data). When sufficient sequence data are available, these candidates can be identified by homology or by structural comparisons, their patterns of evolution among related species can be tested for evidence of divergence under selection, and other aspects of expression variation can be measured, such as tissue and developmental stage specificity. Importantly, given the clear chromosomal differences between the two subspecies, the chromosomal location of these candidate sequences can be ascertained by fluorescent *in situ* hybridization (FISH), and then analysed for their crossing-over patterns and meiotic behaviour. The correlation between expression and sterility can also be determined in experimental crosses. Although the methods for critically testing the role of a candidate gene in sterility available in *Drosophila*, such as complementation (Presgraves et al., 2003), cannot be applied to *Chorthippus*, it is still possible to build a strong case.

We believe that a full appreciation of the genetic basis of reproductive isolation can only be achieved when the conclusions reached in experimental studies of model organisms have been tested in a wider range of species. Insect hybrid zones allow us to address questions that are not easily tackled in the laboratory, particularly concerning the role of selection in generating the divergence that underlies isolation. Until recently, indirect approaches have been necessary, but now they can be complemented with more direct molecular methods, and this combination is likely to lead to rapid progress in broadening our understanding of the genetics and ecology of speciation.

Acknowledgements

We are very grateful to Bill Jordan and Tom Tregenza for help and advice, and to Godfrey Hewitt and Mike Ritchie for many years of fruitful interaction. Funding was provided by NERC (R.K.B.) and the Spanish BOS-2002-00232 (J.L.B.).

References

Andersson, M. (1994) *Sexual Selection*. Princeton University Press, Princeton, New Jersey.

Arbeitman, M.N., Furlong, E.E.M., Imam, F., Johnson, E., Null, B.H., Baker, B.S., Krasnow, M.A., Scott, M.P., Davis, R.W. and White, K.P. (2002) Gene expression during the life cycle of *Drosophila melanogaster*. *Science* 297, 2270–2275.

Bailey, R.I., Thomas, C.D. and Butlin, R.K. (2004) Premating barriers to gene exchange and their implications for the structure of a mosaic hybrid zone between *Chorthippus brunneus* and *C. jacobsi* (Orthoptera: Acrididae). *Journal of Evolutionary Biology* 17, 108–119.

Barton, N.H. (1983) Multilocus clines. *Evolution* 37, 454–471.

Barton, N.H. (1986) The effects of linkage and density-dependent regulation on gene flow. *Heredity* 57, 415–426.

Barton, N.H. (1988) Speciation. In: Myers, A.A. and Giller, P.S. (eds) *Analytical Biogeography*. Chapman and Hall, London, pp. 185–218.

Barton, N.H. (1999) Clines in polygenic traits. *Genetical Research, Cambridge* 74, 223–236.

Barton, N.H. (2001a) Speciation. *Trends in Ecology and Evolution* 16, 325.

Barton, N.H. (2001b) The role of hybridization in evolution. *Molecular Ecology* 10, 551–568.

Barton, N. and Bengtsson, B.O. (1986) The barrier to genetic exchange between hybridising populations. *Heredity* 56, 357–376.

Barton, N.H. and Charlesworth, B. (1984) Genetic revolutions, founder effects, and speciation. *Annual Review of Ecology and Systematics* 15, 133–164.

Barton, N.H. and Gale, K.S. (1993) Genetic analysis of hybrid zones. In: Harrison, R.G. (ed.) *Hybrid Zones and the Evolutionary Process*. Oxford University Press, New York, pp. 13–45.

Barton, N.H. and Hewitt, G.M. (1981) The genetic basis of hybrid inviability between two chromosomal races of the grasshopper *Podisma pedestris*. *Heredity* 47, 367–383.

Barton, N.H. and Hewitt, G.M. (1985) Analysis of hybrid zones. *Annual Review of Ecology and Systematics* 16, 113–148.

Barton, N.H. and Hewitt, G.M. (1989) Adaptation, speciation and hybrid zones. *Nature* 341, 497–503.

Barton, N.H. and Shpak, M. (2000) The effect of epistasis on the structure of hybrid zones. *Genetical Research, Cambridge* 75, 179–198.

Bella, J.L., Gosálvez, J. and Hewitt, G.M. (1990) Meiotic imbalance in laboratory produced hybrid males of *Chorthippus parallelus parallelus* and *Chorthippus parallelus erythropus. Genetical Research, Cambridge* 56, 43–48.

Bella, J.L., Butlin, R.K., Ferris, C. and Hewitt, G.M. (1992) Asymmetrical homogamy and unequal sex ratio from reciprocal mating-order crosses between *Chorthippus parallelus* subspecies. *Heredity* 68, 345–352.

Bella, J.L., Serrano, L., Hewitt, G.M. and Gosálvez, J. (1993) Heterochromatin heterogeneity and rapid divergence of the sex chromosomes in *Chorthippus parallelus parallelus* and *Chorthippus parallelus erythropus* (Orthoptera). *Genome* 36, 542–547.

Bole-Feysot, C., Perret, E., Roustan, P., Bouchard, B. and Kelly, P.A. (2000) Analysis of prolactin-modulated gene expression profiles during the Nb2 cell cycle using differential screening techniques. *Genome Biology* 1, research 0008.1–0008.15.

Bordenstein, S.R. and Drapeau, M.D. (2001) Genotype-by-environment interaction and the Dobzhansky–Muller model of postzygotic isolation. *Journal of Evolutionary Biology* 14, 490–501.

Bridle, J.R. and Butlin, R.K. (2002) Mating signal variation and bimodality in a mosaic hybrid zone between *Chorthippus* grasshopper species. *Evolution* 56, 1184–1198.

Bridle, J.R., Baird, S.J.E. and Butlin, R.K. (2001) Spatial structure and habitat variation in a grasshopper hybrid zone. *Evolution* 55, 1832–1843.

Bridle, J.R., Vass-De-Zomba, J. and Butlin, R.K. (2002) Fine-scale ecological and genetic variation in a *Chorthippus* grasshopper hybrid zone. *Ecological Entomology* 27, 499–504.

Buckley, S.H., Tregenza, T. and Butlin, R.K. (2003) Transitions in cuticular composition across a hybrid zone: historical accident or environmental adaptation? *Biological Journal of the Linnean Society* 78, 193–201.

Buño, I., Torroja, E., López-Fernández, C., Butlin, R.K., Hewitt, G.M. and Gosálvez, J. (1994) A hybrid zone between two subspecies of the grasshopper *Chorthippus parallelus* along the Pyrenees: the west end. *Heredity* 73, 625–634.

Burke, J.M. and Arnold, M.L. (2001) Genetics and the fitness of hybrids. *Annual Review of Genetics* 35, 31–52.

Butlin, R.K. (1998) What do hybrid zones in general, and the *Chorthippus parallelus* zone in particular, tell us about speciation? In: Howard, D.J. and Berlocher, S.H. (eds) *Endless Forms: Species and Speciation.* Oxford University Press, Oxford, UK, pp. 367–378.

Butlin, R.K. and Hewitt, G.M. (1985a) A hybrid zone between *Chorthippus parallelus parallelus* and *Chorthippus parallelus erythropus* (Orthoptera: Acrididae): morphological and electrophoretic characters. *Biological Journal of the Linnean Society* 26, 269–285.

Butlin, R.K. and Hewitt, G.M. (1985b) A hybrid zone between *Chorthippus parallelus parallelus* and *Chorthippus parallelus erythropus* (Orthoptera: Acrididae): behavioural characters. *Biological Journal of the Linnean Society* 26, 287–299.

Butlin, R.K. and Hewitt, G.M. (1988) Genetics of behavioural and morphological differences between parapatric subspecies of *Chorthippus parallelus* (Orthoptera: Acrididae). *Biological Journal of the Linnean Society* 33, 233–248.

Butlin, R.K. and Ritchie, M.G. (1994) Mating behaviour and speciation. In: Slater, P.J.B. and Halliday, T.R. (eds) *Behaviour and Evolution.* Cambridge University Press, Cambridge, UK, pp. 43–79.

Butlin, R.K., Ritchie, M.G. and Hewitt, G.M. (1991) Comparisons among morphological characters and between localities in the *Chorthippus parallelus* hybrid zone

(Orthoptera: Acrididae). *Philosophical Transactions of the Royal Society of London, Series B* 334, 297–308.

Civetta, A. and Singh, R.S. (1998) Sex related genes, directional sexual selection, and speciation. *Molecular Biology and Evolution* 15, 901–909.

Cooper, S.J.B. and Hewitt, G.M. (1993) Nuclear DNA sequence divergence between parapatric subspecies of the grasshopper *Chorthippus parallelus*. *Insect Molecular Biology* 2, 1–10.

Cooper, S.J.B., Ibrahim, K.M. and Hewitt, G.M. (1995) Post-glacial expansion and genome subdivision in the European grasshopper *Chorthippus parallelus*. *Molecular Ecology* 4, 49–60.

Coyne, J.A. (1994) Ernst Mayr and the origin of species. *Evolution* 48, 19–30.

Coyne, J.A. and Orr, H.A. (1989a) Patterns of speciation in *Drosophila*. *Evolution* 43, 362–381.

Coyne, J.A. and Orr, H.A. (1989b) Two rules of speciation. In: Otte, D. and Endler, J.A. (eds) *Speciation and its Consequences*. Sinauer Associates, Sunderland, Massachusetts, pp. 180–207.

Coyne, J.A. and Orr, H.A. (1997) Patterns of speciation in *Drosophila* revisited. *Evolution* 51, 295–303.

Coyne, J.A. and Orr, H.A. (1999) The evolutionary genetics of speciation. In: Magurran, A.E. and May, R.M. (eds) *Evolution of Biological Diversity*. Oxford University Press, Oxford, UK, pp. 1–36.

Diatchenko, L., Lau, Y.F.C., Campbell, A.P., Chenchik, A., Moqadam, F., Huang, B., Lukyanov, S., Lukyanov, K., Gurskaya, N., Sverdlov, E.D. and Siebert, P.D. (1996) Suppression subtractive hybridization: a method for generating differentially regulated or tissue-specific cDNA probes and libraries. *Proceedings of the National Academy of Sciences USA* 93, 6025–6030.

Dobzhansky, T. (1937) *Genetics and the Origin of Species*. Columbia University Press, New York.

Doi, M., Matsuda, M., Tomaru, M., Matsubayashi, H. and Oguma, Y. (2001) A locus for female discrimination behavior causing sexual isolation in *Drosophila*. *Proceedings of the National Academy of Sciences USA* 98, 6714–6719.

Durrett, R., Buttel, L. and Harrison, R. (2000) Spatial models for hybrid zones. *Heredity* 84, 9–19.

Endler, J.A. (1977) *Geographic Variation, Speciation, and Clines*. Princeton University Press, Princeton, New Jersey.

Etges, W.J. (2002) Divergence in mate choice systems: does evolution play by rules? *Genetica* 116, 151–166.

Fang, S., Takahashi, A. and Wu, C.-I. (2002) A mutation in the promoter of desaturase 2 is correlated with sexual isolation between *Drosophila* behavioral races. *Genetics* 162, 781–784.

Felsenstein, J. (1976) The theoretical population genetics of variable selection and migration. *Annual Review of Genetics* 10, 253–280.

Ferris, C., Rubio, J.M., Serrano, L., Gosálvez, J. and Hewitt, G.M. (1993) One way introgression of a subspecific sex chromosome marker in a hybrid zone. *Heredity* 71, 119–129.

Flanagan, N.S., Mason, P.L., Gosálvez, J. and Hewitt, G.M. (1999) Chromosomal differentiation through an Alpine hybrid zone in the grasshopper *Chorthippus parallelus*. *Journal of Evolutionary Biology* 12, 577–585.

Gavrilets, S. (1997a) Hybrid zones with Dobzhansky-type epistatic selection. *Evolution* 51, 1027–1035.

Gavrilets, S. (1997b) Evolution and speciation on holey adaptive landscapes. *Trends in Ecology and Evolution* 12, 307–312.

Gavrilets, S. and Cruzan, M.B. (1998) Neutral gene flow across single locus clines. *Evolution* 52, 1277–1284.

Gleason, J.M., Nuzhdin, S.V. and Ritchie, M.G. (2002) Quantitative trait loci affecting a courtship signal in *Drosophila melanogaster*. *Heredity* 89, 1–6.

Gosálvez, J., López-Fernandez, C., Bella, J.L. and Hewitt, G.M. (1988) A hybrid zone between *Chorthippus parallelus parallelus* and *Chorthippus parallelus erythropus* (Orthoptera: Acrididae): chromosomal differentiation. *Genome* 30, 656–663.

Gosálvez, J., Mason, P.L. and López-Fernández, C. (1997) Differentiation of individuals, populations and species of Orthoptera: the past, present and future of chromosome markers. In: Gangwere, S.K., Muralirangan, M.C. and Muralirangan, M. (eds) *The Bionomics of Grasshoppers, Katydids and their Kin.* CAB International, Wallingford, UK, pp. 355–383.

Guiguen, Y., Baroiller, J.F., Ricordel, M.J., Iseki, K., McMeel, O.M., Martin, S.A.M. and Fostier, A. (1999) Involvement of estrogens in the process of sex differentiation in two fish species: the rainbow trout (*Oncorhynchus mykiss*) and a tilapia (*Oreochromis niloticus*). *Molecular Reproduction and Development* 54, 154–162.

Harrison, R.G. (ed.) (1993) *Hybrid Zones and the Evolutionary Process.* Oxford University Press, Oxford, UK.

Harrison, R.G. and Rand, D.M. (1989) Mosaic hybrid zones and the nature of species boundaries. In: Otte, D. and Endler, J.A. (eds) *Speciation and its Consequences.* Sinauer Associates, Sunderland, Massachusetts, pp. 111–134.

Hewitt, G.M. (1993) After the ice: *Parallelus* meets *Erythropus* in the Pyrenees. In: Harrison, R.G. (ed.) *Hybrid Zones and the Evolutionary Process.* Oxford University Press, Oxford, UK, pp. 140–164.

Hewitt, G.M. (1996) Some genetic consequences of ice ages, and their role in divergence and speciation. *Biological Journal of the Linnean Society* 58, 247–276.

Hewitt, G.M. (1999) Post-glacial re-colonization of European biota. *Biological Journal of the Linnean Society* 68, 87–112.

Hewitt, G.M. (2001) Speciation, hybrid zones and phylogeography – or seeing genes in space and time. *Molecular Ecology* 10, 537–549.

Hewitt, G.M., Butlin, R.K. and East, T.M. (1987) Testicular dysfunction in hybrids between parapatric subspecies of the grasshopper *Chorthippus parallelus*. *Biological Journal of the Linnean Society* 31, 25–34.

Howard, D.J. (1993) Reinforcement: the origin, dynamics, and fate of an evolutionary hypothesis. In: Harrison, R.G. (ed.) *Hybrid Zones and the Evolutionary Process.* Oxford University Press, New York, pp. 46–69.

Howard, D.J. and Berlocher, S.H. (eds) (1998) *Endless Forms: Species and Speciation.* Oxford University Press, Oxford.

Howard, D.J. and Gregory, P.G. (1993) Post-insemination signalling systems and reinforcement. *Philosophical Transactions of the Royal Society of London, Series B* 340, 231–236.

Jiggins, C.D. and Mallet, J. (2000) Bimodal hybrid zones and speciation. *Trends in Ecology and Evolution* 15, 250–255.

Johnson, N.A. (2000) Gene interactions and the origin of species. In: Wolf, J.B., Brodie, E.D., III and Wade, M.J. (eds) *Epistasis and the Evolutionary Process.* Oxford University Press, Oxford, UK, pp. 197–212.

Kelly, J.K. and Noor, M.A.F. (1996) Speciation by reinforcement: a model derived from studies of *Drosophila*. *Genetics* 143, 1485–1497.

Kirkpatrick, M. (2000) Reinforcement and divergence under assortative mating. *Proceedings of the Royal Society of London, Series B* 267, 1649–1655.

Lande, R. (1981) Models of speciation by sexual selection on polygenic traits. *Proceedings of the National Academy of Sciences USA* 78, 3721–3725.

Lande, R., Seehausen, O. and van Alphen, J.J.M. (2001) Mechanisms of rapid sympatric speciation by sex reversal and sexual selection in cichlid fish. *Genetica* 112, 435–443.

Laurie, C.C. (1997) The weaker sex is heterogametic: 75 years of Haldane's Rule. *Genetics* 147, 937–951.

Liou, L.W. and Price, T.D. (1994) Speciation by reinforcement of premating isolation. *Evolution* 48, 1451–1459.

Lunt, D.H., Ibrahim, K.M. and Hewitt, G.M. (1998) MtDNA phylogeography and post-glacial patterns of subdivision in the meadow grasshopper, *Chorthippus parallelus*. *Heredity* 80, 633–641.

Lynch, M. and Conery, J.S. (2000) The evolutionary fate and consequences of duplicate genes. *Science* 290, 1151–1155.

Lynch, M. and Force, A.G. (2000) The origin of interspecific genomic incompatibility via gene duplication. *American Naturalist* 156, 590–605.

MacCallum, C.J., Nürnberger, B., Barton, N.H. and Szymura, J.M. (1998) Habitat preferences in the *Bombina* hybrid zone in Croatia. *Evolution* 52, 227–239.

Mayr, E. (1942) *Systematics and the Origin of Species.* Columbia University Press, New York.

Michalak, P. and Noor, M.A.F. (2003) Genome-wide patterns of expression in *Drosophila* pure species and hybrid males. *Molecular Biology and Evolution* 20, 1070–1076.

Neems, R.A. and Butlin, R.K. (1993) Divergence in mate finding behaviour between two subspecies of the meadow grasshopper, *Chorthippus parallelus*. *Journal of Insect Behavior* 6, 421–430.

Neems, R.A. and Butlin, R.K. (1994) Variation in cuticular hydrocarbons across a hybrid zone in the grasshopper *Chorthippus parallelus*. *Proceedings of the Royal Society of London, Series B* 257, 135–140.

Neems, R.A. and Butlin, R.K. (1995) Divergence in cuticular hydrocarbons between parapatric subspecies of the meadow grasshopper, *Chorthippus parallelus* (Orthoptera: Acrididae). *Biological Journal of the Linnean Society* 54, 139–149.

Noor, M.A.F. (1999) Reinforcement and other consequences of sympatry. *Heredity* 83, 503–508.

Noor, M.A.F. (2003) Genes to make new species. *Nature* 423, 699–700.

Noor, M.A.F., Grams, K.L., Bertucci, L.A. and Reiland, J. (2001) Chromosomal inversions and the reproductive isolation of species. *Proceedings of the National Academy of Sciences, USA* 98, 12084–12088.

Nürnberger, B., Barton, N., MacCallum, C., Gilchrist, J. and Appleby, M. (1995) Natural selection on quantitative traits in the *Bombina* hybrid zone. *Evolution* 49, 1224–1238.

Orr, H.A. (1993) A mathematical model of Haldane's Rule. *Evolution* 47, 1606–1611.

Orr, H.A. (1995) The population genetics of speciation: the evolution of hybrid incompatibilities. *Genetics* 139, 1805–1813.

Orr, H.A. (1996) Dobzhansky, Bateson, and the genetics of speciation. *Genetics* 144, 1331–1335.

Orr, H.A. (1997) Haldane's Rule. *Annual Review of Ecology and Systematics* 28, 195–218.

Orr, H.A. (2001) The genetics of species differences. *Trends in Ecology and Evolution* 16, 343–350.

Orr, H.A. and Orr, L.H. (1996) Waiting for speciation: the effect of population subdivision on the time to speciation. *Evolution* 50, 1742–1749.

Ortiz-Barrientos, D., Reiland, J., Hey, J. and Noor, M.A.F. (2002) Recombination and the divergence of hybridizing species. *Genetica* 116, 167–178.

Otte, D. and Endler, J.A. (eds) (1989) *Speciation and its Consequences.* Sinauer Associates, Sunderland, Massachusetts.

Päällysaho, S., Aspi, J., Liimatainen, J.O. and Hoikkala, A. (2003) Role of X chromosomal song genes in the evolution of species-specific courtship songs in *Drosophila virilis* group species. *Behavior Genetics* 33, 25–32.

Panhuis, T.M., Butlin, R.K., Zuk, M. and Tregenza, T. (2001) Sexual selection and speciation. *Trends in Ecology and Evolution* 16, 364–371.

Parisi, M., Nuttall, R., Naiman, D., Bouffard, G., Malley, J., Andrews, J., Eastman, S. and Oliver, B. (2003) Paucity of genes on the *Drosophila* X chromosome showing male-biased expression. *Science* 299, 697–700.

Pashley, D.P. (1998) Sex linkage and speciation in Lepidoptera. In: Howard, D.J. and Berlocher, S.H. (eds) *Endless Forms: Species and Speciation.* Oxford University Press, Oxford, UK, pp. 309–319.

Pomiankowski, A. and Iwasa, Y. (1998) Runaway ornament diversity caused by Fisherian sexual selection. *Proceedings of the National Academy of Sciences USA* 96, 5106–5111.

Presgraves, D.C. and Orr, H.A. (1998) Haldane's Rule in taxa lacking a hemizygous X. *Science* 282, 952–954.

Presgraves, D.C., Balagopalan, L., Abmayr, S.M. and Orr, H.A. (2003) Adaptive evolution drives divergence of a hybrid inviability gene between two species of *Drosophila. Nature* 423, 715–719.

Ranz, J.M., Castillo-Davis, C.I., Meiklejohn, C.D. and Hartl, D.L. (2003) Sex-dependent gene expression and evolution of the *Drosophila* transcriptome. *Science* 300, 1742–1745.

Rieseberg, L.H. (2001) Chromosomal rearrangements and speciation. *Trends in Ecology and Evolution* 16, 351–358.

Rieseberg, L.H., Whitton, J. and Gardner, K. (1999) Hybrid zones and the genetic architecture of a barrier to gene flow between two sunflower species. *Genetics* 152, 713–727.

Ritchie, M.G. (1990) Are differences in song responsible for assortative mating between subspecies of *Chorthippus parallelus* (Orthoptera: Acrididae)? *Animal Behaviour* 39, 685–691.

Ritchie, M.G. and Phillips, S.D.F. (1998) The genetics of sexual isolation. In: Howard, D.J. and Berlocher, S.H. (eds) *Endless Forms: Species and Speciation.* Oxford University Press, Oxford, UK, pp. 291–308.

Ritchie, M.G., Butlin, R.K. and Hewitt, G.M. (1989) Assortative mating across a hybrid zone in *Chorthippus parallelus* (Orthoptera: Acrididae*). Journal of Evolutionary Biology* 2, 339–352.

Ritchie, M.G., Butlin, R.K. and Hewitt, G.M. (1992) Fitness consequences of potential assortative mating within and outside a hybrid zone in *Chorthippus parallelus* (Orthoptera: Acrididae): implications for reinforcement and sexual selection theory. *Biological Journal of the Linnean Society* 45, 219–234.

Sadedin, S. and Littlejohn, M.L. (2003) A spatially explicit individual-based model of reinforcement in hybrid zones. *Evolution* 57, 962–970.

Saetre, G.P., Moum, T., Bures, S., Kral, M., Adamjan, M. and Moreno, J. (1997) A sexually selected character displacement in flycatchers reinforces premating isolation. *Nature* 387, 589–592.

Schluter, D. (1998) Ecological causes of speciation. In: Howard, D.J. and Berlocher, S.H. (eds) *Endless Forms: Species and Speciation.* Oxford University Press, Oxford, UK, pp. 114–129.

Schluter, D. (2000) *The Ecology of Adaptive Radiation.* Oxford University Press, Oxford, UK.

Schluter, D. (2001) Ecology and the origin of species. *Trends in Ecology and Evolution* 16, 372–380.

Searle, J.B. (1993) Chromosomal hybrid zones in eutherian mammals. In: Harrison, R.G. (ed.) *Hybrid Zones and the Evolutionary Process.* Oxford University Press, New York, pp. 309–351.

Seehausen, O., Mayhew, P.J. and van Alphen, J.J.M. (1999) Evolution of colour patterns in East African cichlid fish. *Journal of Evolutionary Biology* 12, 514–534.

Serrano, L., García de la Vega, C., Bella, J.L., López-Fernández, C., Hewitt, G.M. and Gosálvez, J. (1996) A hybrid zone between two subspecies of *Chorthippus parallelus*: X-chromosome variation through a contact zone. *Journal of Evolutionary Biology* 9, 173–184.

Servedio, M.R. (2000) Reinforcement and the genetics of non-random mating. *Evolution* 54, 21–29.

Szymura, J.M. and Barton, N.H. (1991) Genetic analysis of a hybrid zone between the fire-bellied toads, *Bombina bombina* and *B. variegata*, near Cracow in southern Poland. *Evolution* 40, 1141–1159.

Takahashi, A., Tsaur, S.C., Coyne, J.A. and Wu, C.-I. (2001) The nucleotide changes governing cuticular hydrocarbon variation and their evolution in *Drosophila melanogaster*. *Proceedings of the National Academy of Sciences USA* 98, 3920–3925.

Ting, C.-T., Tsaur, S.C., Wu, M.L. and Wu, C.-I. (1998) A rapidly evolving homeobox at the site of a hybrid sterility gene. *Science* 282, 1501–1504.

Ting, C.-T., Tsaur, S.C. and Wu, C.-I. (2000) The phylogeny of closely related species as revealed by the genealogy of a speciation gene, *Odysseus*. *Proceedings of the National Academy of Sciences USA* 97, 5313–5316.

Ting, C.-T., Takahashi, A. and Wu, C.-I. (2001) Incipient speciation by sexual isolation in *Drosophila*: concurrent evolution at multiple loci. *Proceedings of the National Academy of Sciences USA* 98, 6709–6713.

Tregenza, T. (2002) Divergence and reproductive isolation in the early stages of speciation. *Genetica* 116, 291–300.

Tregenza, T., Pritchard, V.L. and Butlin, R.K. (2000a) Patterns of trait divergence between populations of the meadow grasshopper, *Chorthippus parallelus*. *Evolution* 54, 574–585.

Tregenza, T., Buckley, S.H., Pritchard, V.L. and Butlin, R.K. (2000b) Inter- and intra-population effects of sex and age on epicuticular composition of meadow grasshopper, *Chorthippus parallelus*. *Journal of Chemical Ecology* 26, 257–278.

Tregenza, T., Pritchard, V.L. and Butlin, R.K. (2000c) The origins of premating reproductive isolation: testing hypotheses in the grasshopper *Chorthippus parallelus*. *Evolution* 54, 1687–1698.

Tregenza, T., Pritchard, V.L. and Butlin, R.K. (2002) The origins of postmating reproductive isolation: testing hypotheses in the grasshopper *Chorthippus parallelus*. *Population Ecology* 44, 137–144.

Turelli, M. (1998) The causes of Haldane's Rule. *Science* 282, 889–891.

Turelli, M. and Begun, D.J. (1997) Haldane's Rule and X-chromosome size in *Drosophila*. *Genetics* 147, 1799–1815.

Turelli, M. and Orr, H.A. (1995) The dominance theory of Haldane's Rule. *Genetics* 140, 389–402.

Turelli, M. and Orr, H.A. (2000) Dominance, epistasis and the genetics of postzygotic isolation. *Genetics* 154, 1663–1679.

Turelli, M., Barton, N.H. and Coyne, J.A. (2001) Theory and speciation. *Trends in Ecology and Evolution* 16, 330–343.

Underwood, K.L. (1994) Mating pattern in a grasshopper hybrid zone. PhD thesis, The University of Leeds, Leeds, UK.

Via, S. (2001) Sympatric speciation in animals: the ugly duckling grows up. *Trends in Ecology and Evolution* 16, 381–390.

Virdee, S.R. and Hewitt, G.M. (1990) Ecological components of a hybrid zone in the grasshopper *Chorthippus parallelus* (Zetterstedt) (Orthoptera: Acrididae). *Boletin de Sanidad Vegetal (Fuera de Serie)* 20, 299–309.

Virdee, S.R. and Hewitt, G.M. (1994) Clines for hybrid dysfunction in a grasshopper hybrid zone. *Evolution* 48, 392–407.

Wade, M.J. (2000) A gene's eye view of epistasis, selection and speciation. *Journal of Evolutionary Biology* 15, 337–346.

Walter, R.B. and Kazianis, S. (2001) *Xiphophorus* interspecies hybrids as genetic models of induced neoplasia. *Institute for Laboratory Animal Research Journal* 42, 299–321.

Wolf, J.B., Brodie, E.D. III and Wade, M.J. (eds) (2000) *Epistasis and the Evolutionary Process*. Oxford University Press, Oxford, UK.

Wu, C.-I. (2001) The genic view of the process of speciation. *Journal of Evolutionary Biology* 14, 851–865.

Wu, C.-I. and Davis, A.W. (1993) Evolution of postmating reproductive isolation: the composite nature of Haldane's Rule and its genetic basis. *American Naturalist* 142, 187–212.

Wu, C.-I. and Palopoli, M.F. (1994) Genetics of postmating reproductive isolation in animals. *Annual Review of Genetics* 27, 283–308.

Wu, C.-I. and Ting, C.-T. (2004) Genes and speciation. *Nature Reviews Genetics* 5, 114–122.

Wu, C.-I., Johnson, N.A. and Palopoli, M.F. (1996) Haldane's Rule and its legacy: why are there so many sterile males? *Trends in Ecology and Evolution* 11, 281–284.

16

Assortative Mating and Speciation as Pleiotropic Effects of Ecological Adaptation: Examples in Moths and Butterflies

CHRIS D. JIGGINS,[1] IGOR EMELIANOV[2] AND JAMES MALLET[3]

[1]Institute of Evolutionary Biology, School of Biological Sciences, University of Edinburgh, Edinburgh, UK; [2]Plant and Invertebrate Ecology Division, Rothamsted Research, Harpenden, UK; [3]The Galton Laboratory, Department of Biology, University College London, London, UK

1. Introduction

Where divergent ecological adaptation also leads to assortative mating between populations, speciation is facilitated. We have extensively studied two examples from among the Lepidoptera, in which ecologically selected traits have pleiotropic effects on the mating system. Based on our understanding of mating behaviour and pleiotropy derived from these empirical examples, we here review pleiotropic effects in speciation and propose a classification depending on effects on the mating system. This classification helps us to interpret the way in which pleiotropy has led to assortative mating in nature, but will also, we hope, be of interest to theoreticians. Certain models of mating behaviour rarely considered in models of sympatric speciation seem to be those that are most empirically justifiable and also likely to be involved in speciation in the real world. We conclude that pleiotropy might be a rather more general route to speciation than previously supposed.

1.1 Definition

Pleiotropy is the phenomenon whereby a single gene affects several different aspects of the phenotype. In evolution, pleiotropy is commonly used to refer to phenotypic effects of a gene other than those originally favoured by natural selection. These pleiotropic effects are also, and perhaps more correctly, 'side-effects' of evolution at genes under selection. In the context of speciation we are particularly interested in how genes favoured by ecological selection might cause reproductive isolation as a side-effect.

Here we concentrate on the effects on pre-mating isolation (mate choice), which can be readily interpreted in the light of pleiotropy. Instead, most recent commentary has been focused on pleiotropic effects on post-mating isolation (hybrid inviability and sterility), particularly on the genetic bases of Dobzhansky–Müller incompatibilities (Turelli and Orr, 1995, 2000; Orr and Turelli, 2001). A great deal is known about the genetics of post-mating isolation, but we have little idea of the original roles of genes causing hybrid breakdown (but see Barbash *et al.*, 2003; Presgraves *et al.*, 2003). Sterility and inviability must be secondary pleiotropic effects, because low fitness cannot be favoured within populations. In contrast, traits affecting mating behaviour are more visible, and their original role in ecological adaptation might be more easily deduced. In this chapter we explore the pleiotropic effects acting on mate choice, and show how it is likely to be common for ecologically selected genes to have incidental effects on pre-mating reproductive isolation.

1.2 Pleiotropic effects in sympatric and allopatric speciation

Speciation is one example of the process by which one population splits into two genetically distinguishable daughter populations. Under almost any species definition, an important characteristic of a species is the ability to coexist with others without fusion into a single gene pool. Stable coexistence is possible only if closely related sexual species are somewhat reproductively isolated and also ecologically distinct. Perhaps the major challenge in understanding speciation is therefore to explain how genes for ecological distinctness become associated with genes that prevent gene flow.

In allopatric speciation, associations between genes are caused in part by geography, such that ecological traits, mating traits and mating preferences will often diverge in concert between two populations. The speed of speciation in allopatry will be enhanced if genes involved in ecological divergence also cause reproductive isolation. As recognized by Dobzhansky (1937), the common occurrence of 'ecological and seasonal isolation' implies that genes for habitat choice often result in reproductive isolation. Hence, a pleiotropic role of ecological adaptation in causing reproductive isolation is implied, even if not explicitly stated in the 'modern synthesis' view of allopatric speciation.

In sympatric speciation, divergent selection on its own must lead to reproductive isolation. It has long been realized that recombination rapidly breaks down associations between genes under selection and genes for mate choice, making sympatric speciation difficult (Felsenstein, 1981). Unsurprisingly therefore, pleiotropy, which ensures an inherent association between assortative mating genes and ecological traits, could play an important role in sympatric and parapatric speciation (Bush, 1975; Moore, 1981; Doebeli, 1996). In summary, therefore, both sympatric and allopatric speciation may involve pleiotropic effects of genes under natural selection as a cause of reproductive isolation. Their importance is supported by laboratory experiments showing that the evolution of reproductive isolation as a by-product of adaptation to distinct environments is likely (Rice and Hostert, 1993). Although in neither case are such effects necessary, speciation will be easier and more rapid when ecological selection contributes to reproductive isolation.

1.3 A classification of pleiotropic effects

Although pleiotropy is sometimes invoked in theoretical models of speciation, there has been little attempt to distinguish the ways in which pleiotropic effects might act. Based on our contemplation of empirical examples, we here propose a classification of pleiotropic effects according to their effects on the mating system. The aim of this classification is twofold. First, we believe that it will help in understanding empirical examples. Second, we hope to stimulate a greater realism in theoretical models of speciation. As we discuss below, some of the most common forms of pleiotropy in nature are rarely considered in theoretical models of sympatric speciation.

This is intended to be complementary to a recent classification of speciation models (Kirkpatrick and Ravigné, 2002) that did not directly consider 'pleiotropy' in our sense. The distinct classes listed below are an extension of the subcategories of prezygotic isolating mechanisms, or 'Element II' of Kirkpatrick and Ravigné. Our classification distinguishes between mating cues and mating preferences. Perhaps most commonly these will be male traits and female preferences, although it need not be that way around – in the *Heliconius* example below, male preference is more important than female preference.

1. *Pleiotropic assortment traits.* A single trait under divergent ecological selection also causes assortative mating, essentially by affecting both male and female components of behaviour. For example, habitat choice or the time of adult emergence can affect temporal or spatial location of both sexes. This is typified by host races, where adaptation to a new host leads directly to assortative mating.

2. *Pleiotropic mating cues.* Traits under divergent selection are also used as cues in mating. Here, genes under divergent ecological selection are not directly involved with assortative mating, but instead form part of a suite of traits chosen during mate selection. Speciation therefore requires evolution of mate preference subsequent to the initial divergence of mating cues caused by disruptive natural selection. An example would be divergence in butterfly wing patterns between two populations. If such patterns are also used in mate recognition, then mate preferences may also diverge between populations, and so generate assortative mating.

3. *Pleiotropic mating preferences.* Genes under divergent selection also affect mate preference. Female preferences may diverge under natural selection, perhaps due to adaptation of the sensory system to a novel habitat. Subsequently, male traits coevolve to exploit the novel female sensory system. For example, the spectral sensitivity of a visual predator may be locally adapted to divergent light environments, leading to females in different populations becoming responsive to different signals. Subsequent sexual selection of male signals in response to changes in female sensitivity might lead to assortative mating between divergent populations.

Any combination of pleiotropy types 1–3 is of course possible. Alternatively, we may have:

4. *No pleiotropy.* Neither preference nor the traits chosen during mating are under direct ecological selection.

We now discuss each category in more detail.

1.3.1 Pleiotropic assortment traits – where traits affect both male and female behaviour

In some circumstances niche choice may equate to mate choice. The classic case in which pleiotropic effects cause reproductive isolation in this manner is during host shifts among phytophagous insects (Bush, 1969). Clearly, given the great diversity of host-plant relationships found among the insects, and the phylogenies of the insects and of their host plants, host-plant use is a trait that is evolutionarily labile (Mitter and Brooks, 1983). Furthermore, some insects use their host plants as a cue in mating behaviour, so that individuals choosing to exploit a novel host will only encounter mates who have made a similar host choice. The classic example is *Rhagoletis pomonella*, in which males patrol host fruits and wait for and court ovipositing females. Release experiments have shown that apple and hawthorn-associated host races show a strong preference for alighting on their native host. Since mating occurs on hosts, this leads to a reduction in cross-mating of over 90% between host races (Feder *et al.*, 1994). However, *Rhagoletis* is perhaps an extreme example, where mate-finding is based almost exclusively on habitat cues. In other cases mates are guided by a mixture of habitat cues and habitat-independent signals, such as sex pheromones, a good example being the host races of *Zeiraphera* moths described below.

Even where mating does not occur directly on the host plant, host phenology may induce assortment between ecotypes. Host plants may differ in nutritional suitability through the year, leading to differences in emergence time between races found on different hosts. If mating receptivity is sufficiently short-lived then this can lead to assortative mating, as in *Enchenopa* treehoppers (Wood and Keese, 1990). Phenology switching may even drive speciation in the absence of host specialization or host choice (Butlin, 1990). Similarly, adaptation to latitudinal differences in climate can lead to different numbers of generations per year, which can also lead to genetic isolation due to a lack of seasonal overlap in parapatric populations, as in *Pieris* and *Aricia* butterflies (Held and Speith, 1999; I.R. Wynne, R.J. Wilson, A.S. Burke, F. Simpson, A.S. Pullin, J. Mallet and C.D. Thomas, unpublished). Hence genes initially selected for adaptation to host phenology produce assortative mating between ecotypes.

In fact, any system where there is divergent habitat choice and where mating occurs in the habitat chosen provides a pleiotropic assortment trait. Mosaic hybrid zones provide a good example, where habitat choice leads to assortative mating on a regional scale (Howard and Waring, 1991; MacCallum *et al.*, 1998). Nonetheless, few habitats are as clearly defined in space as the host plants of herbivorous insects, which perhaps explains why studies of sympatric speciation have concentrated on these organisms (Drès and Mallet, 2002). In conclusion, there are many excellent examples of pleiotropic assortative mating traits, whereby adaptation to a novel niche leads directly to appreciable pre-mating isolation. Indeed, it seems likely to be much more common than hitherto believed.

Despite many good examples of pleiotropic assortment traits in nature, they are rarely considered in theoretical models. In one of the earliest sympatric speciation models, Maynard Smith specifically dealt with the case 'habitat selection' in which mating occurred within a chosen habitat (Maynard Smith, 1966: 643). In contrast, subsequent arguments about sympatric speciation were based on models that ignored this scenario (e.g. Felsenstein, 1981). These models led to a general belief that the major difficulty for sympatric speciation is that associations between mating and ecological traits are broken down by recombination. However, in mating systems that involve pleiotropic assortment traits, mating and ecological adaptation are affected by the same genes, so this argument does not apply. The applicability of sympatric speciation theory therefore depends heavily on the specifics of the mating system.

One reason why pleiotropic assortment traits have been ignored in sympatric models of speciation is that adaptation to particular habitats can be regarded as a form of partial allopatry or parapatry. To reverse the argument of Kirkpatrick and Ravigné (2002), genes for pleiotropic assortment traits are identical in their effects on gene flow to an allopatric distribution. Models of sympatric speciation have typically focused on 'pure sympatry', where individuals are as likely to meet and can potentially mate with the other incipient species as they are with their own type, whatever their genetic constitution (Kondrashov and Mina, 1986). In nature, habitat choice may cause a restriction of gene flow and therefore, arguably, parapatric speciation is more generally applicable (Gavrilets, 2003). In our view, treating pleiotropic assortment traits as a form of allopatry would, however, confuse cases of reduced gene flow due to genetic constitution of the individuals involved (pleiotropic assortment) with cases where reduced gene flow is due entirely to environmental factors beyond genetic control (true allopatry), and it is this distinction that is at the crux of the argument about allopatric versus sympatric speciation.

1.3.2 Pleiotropic mating cues – where traits chosen during mating are under ecological selection

Pleiotropic mating cues may be more difficult to detect, as it is necessary to study both traits under divergent selection and mating preferences. Perhaps for this reason there are fewer well-studied examples. Butterfly wing patterns are used for hiding from or signalling to predators, as well as signalling to conspecifics (Vane-Wright, 1979; Nijhout, 1991; Vane-Wright and Boppré, 1993; Brunton and Majerus, 1995). Hence, divergence in wing pattern initiated by selection for better defensive capability may lead to divergent selection on mate preferences (Vane-Wright and Boppré, 1993; Jiggins et al., 2001b). Below, we review our work in *Heliconius* butterflies demonstrating the coevolution of mate preferences with mimetic colour patterns and habitat preferences, leading to ecological and reproductive isolation between populations with distinct mimetic patterns.

In cactophilic *Drosophila*, larvae raised on different hosts emerge as adults with distinct cuticular hydrocarbon compositions. These hydrocarbons influence pre-mating isolation between populations (Etges, 1992; Brazner and Etges, 1993; Etges and Ahrens, 2001). It seems plausible, therefore, that speciation in this system might occur through shifts in host-plant use leading to divergent mating signals, followed by the evolution of preferences involving the novel hydrocarbon signals. Nonetheless, a significant proportion of the hydrocarbon differences between populations are unrelated to host use (Etges and Ahrens, 2001) and there is no clear evidence as yet for adaptation of mate preferences to locally occurring hydrocarbons. Hence the evidence in this case remains equivocal. Another example is 'Darwin's finches', in which beak size is adapted to exploiting particular foods, but divergence in beak size leads to a correlated change in song produced by males (Podos, 2001). Presumably there may be subsequent adaptation of female preferences to these novel songs, although learnt mate choice by female nestlings based on their father's song can preclude the need for genetic evolution of mating preferences and lead more directly to assortment between individuals with divergent beak sizes (Grant, 1986).

When pleiotropy is considered in theoretical treatments of speciation, it is often as a pleiotropic mating cue. A particular trait is imagined to diverge under ecological selection. This trait is involved in the mating system such that female preferences subsequently diverge, leading to assortative mating. For example, Maynard Smith considered a scenario that involved '*modifier genes*' in which a novel allele B at a modifier locus causes assortative mating depending on the ecologically selected locus A (Maynard Smith, 1966). Therefore b has no effect on mating, but once B is fixed, A always mates with A, and a with a. The fixation of B can be favoured in a process analogous to reinforcement because of selection against Aa heterozygotes, which tend to be found in an inappropriate environment due to hybridization. This is an example of what Felsenstein called a 'one-allele' model, in which a single allele increases assortativeness of mating based on the trait locus (Felsenstein, 1981; Kirkpatrick and Ravigné, 2002). However, divergence of mate preferences could also be modelled in a two-allele scenario in which allele b might represent preference for a and allele B preference for A. In general, two-allele models seem far more likely. In butterflies, for example, it is easier to imagine one phenotype being a preference for red males and the alternative phenotype being a preference for blue males. The corresponding, and rather improbable, one-allele scenario would be a single 'choosiness' phenotype that caused red to mate with red and blue with blue. However, two-allele models make speciation much harder, as alternative alleles must go to fixation in each of the divergent ecotypes, and any hybridization may cause recombination to act against speciation (Felsenstein, 1981). Hence, there are various ways in which speciation might occur subsequent to divergence in a pleiotropic mating cue.

1.3.3 Pleiotropic mating preferences – where mate preferences are under ecological selection

Pleiotropic mating preferences can arise where the environment chosen by an organism constrains or otherwise influences intraspecific communication. This is implied in the 'sensory drive' hypothesis (Endler, 1992; Endler and Basolo,

1998; Boughman, 2002; *see also* West-Eberhard, 1983). Boughman described three related processes that affect signal evolution:

1. *Signal transmission*. Signals must pass through the local environment before being received and are thus subject to alteration by properties of that environment.

2. *Perceptual tuning*. Individuals receiving signals will have their receptor mechanisms adapted to the local environment for foraging, predator detection etc., as well as sexual selection, and thus are likely to be more receptive to certain kinds of signals. This adaptation of the sensory system to the local environment is likely to lead to changes in the kinds of signals most readily detected by females. Hence genes for adaptation of the sensory system to a local habitat are likely to have pleiotropic effects on mate recognition.

3. *Signal matching*. The signals themselves will become adapted to maximize 'visibility' to the particular receptor system of the local population. This third stage refers to selection on the mating cue itself, driven by prior changes in female preferences, and therefore represents a form of sexual selection that occurs subsequent to pleiotropic divergence of female preferences.

One of the best-studied examples involves stickleback ecotypes that are adapted to habitats with differing degrees of water clarity (Boughman, 2001). Males in clear limnetic habitat have more red coloration, those living in murkier waters are blacker, these colours being the most contrasting signals in the respective habitats. Females have corresponding differences in their detection thresholds for red light. One possible sequence of events is that female visual perceptivity changed to maximize foraging ability in the distinct habitats, and was followed by divergence in the male trait to enhance conspicuousness to the perceptual systems of locally occurring females (Boughman, 2001). As far as we are aware there are no well-studied examples of sensory drive in insects, but this may be because their sensory systems are poorly understood.

2. Empirical Examples

Here, we discuss examples studied extensively by us which provide evidence that pleiotropic assortment traits and pleiotropic mating cues are generally important in natural systems.

2.1 Speciation in the larch budmoth

Phytophagous insects that use long-distance pheromones to attract mates are seemingly the last kinds of insects one might expect to show pleiotropy between host use and mating behaviour. In many night-flying Lepidoptera, females 'call' for males by producing pheromones. Calling females can attract males from distances far greater than from within a single host individual. Hence, unlike the situation in *Rhagoletis* (see above), host choice does *not* equate to mate choice. For example, larch- and pine-adapted host races of the larch budmoth *Zeiraphera diniana* occur sympatrically and are genetically differentiated, implying significant reproductive isolation (Emelianov *et al.*, 1995). Assortative mating is primarily due to major blend differences in the pheromones between races, which are unrelated to host use. Hence we did not expect to find pleiotropy between host use and mate choice. However, our recent work has shown that even here there is strong evidence that host choice is acting as a pleiotropic assortment trait: host choice has a very strong influence on cross-mating between the host races, in spite of the predominant effect of pheromone composition on mate choice.

Z. diniana is distributed widely across the Palaearctic, where it feeds on a variety of conifers in the family *Pinaceae*, particularly pine (*Pinus*), larch (*Larix*) and spruce (*Picea*). We have studied this species in the Swiss and French Alps, where two forms are often found sympatrically, feeding on European larch (*Larix decidua*) and cembran pine (*Pinus cembra*) where they grow in mixed forest stands. The two forms of the moth are strongly differentiated in colour pattern of larvae, pheromone blend, and at a number of genetic marker loci (Baltensweiler *et al.*, 1977; Emelianov *et al.*, 1995). The larch form is well known for its 9–10-year outbreak cycles near the tree-line in the Alps, where populations can cycle through 10^5-fold changes in density, a phenomenon that has been studied for many years (Baltensweiler *et al.*, 1977).

Although the two forms are strongly differentiated, none of the genetic differences is completely fixed, with colour polymorphisms, pheromone blends and allozyme loci all showing some evidence of transfer between the two forms via hybridization. For this reason, the two forms are considered to be 'host races' rather than separate species. When we studied the probability of hybridization in the field and laboratory, we found that assortative mating mediated by pheromones was not perfect. We baited traps with live females to attract wild males, using an allozyme-based identification procedure, and we found that the probability of cross-attraction between the two forms varied between 3% and 38% (Emelianov *et al.*, 2001). In the laboratory at close range, cross-mating was readily achievable (20–30% of matings), occurring only slightly less often than at random (50%) (Drès, 2000). In fact, males will attempt to copulate with any moth-like object and homosexuality is quite common (Benz, 1973). Thus there is ample possibility for cross-mating at close range, and pheromone attraction is the most important selective agent in assortative mating. Overall, we expect the species to hybridize at the rate of a few per cent per generation in sympatric populations (Emelianov *et al.*, 2003), and we now have good genetic data to show the existence of hybrids in the wild, and for gene flow between the two host races in some regions of the genome (Emelianov *et al.*, 2004). The implication is that selection must counterbalance the homogenizing effect of gene flow and maintaining between-race divergence in certain genomic regions (Emelianov *et al.*, 2004).

We were interested in whether there could be an influence of the host plant on reproductive isolation between the two races. To test this, traps baited with live females were put out on different hosts (Fig. 16.1). Larch-race

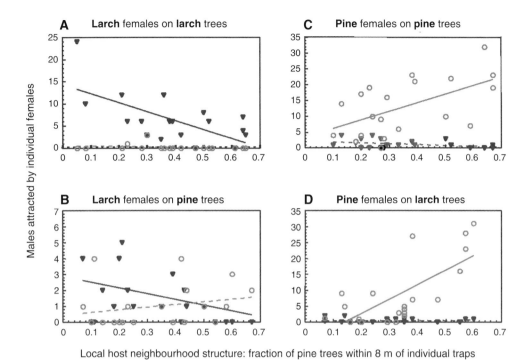

Local host neighbourhood structure: fraction of pine trees within 8 m of individual traps

Fig. 16.1. Effect of host neighbourhood on assortative attraction. Panels **A–D** represent the four types of traps (larch females on larch trees, larch females on pine trees, pine females on pine trees and pine females on larch trees). In each panel, the numbers of males of each race attracted by individual females are plotted against the fraction of pine trees in local host neighbourhoods (i.e. 8 m radius circular patches of larch–pine forest surrounding individual traps). Larch male data are shown as solid triangles, and pine male data are shown as hollow circles. Solid regression lines represent males of the same race as the calling female (that is, larch males in panels **A** and **B**, and pine males in panels **C** and **D**). Dotted lines represent catches of alien males.

females tended to attract significantly more pine-race males when the traps were placed on pine than when placed on their own host. However, the reciprocal effect was not evident; pine females attracted slightly fewer larch males when placed on larch than when placed on pine, although the effect was not significant.

A more striking result was the influence of the immediate surroundings of the trap. As is true for most plant species, the distributions of larch and pine trees in mixed forests are significantly clumped, with fractions of larch and pine trees varying between individual patches within the forest. Females cross-attracted many more males of the other host race when placed in an environment surrounded by the other host than if surrounded by their own host species (Emelianov *et al.*, 2001). Thus, in spite of the very strong assortative attraction induced by the pheromone differences, assortativeness was strongly enhanced when females called from their own host tree, or from areas with a high proportion of their natural host; cross-attraction was much more likely if they called from the other host or in neighbourhoods dominated by the other host.

This effect of local host abundance on mate attraction might be caused solely by female host preferences, such that females will typically be calling from within clumps of their own host. Hence, each host clump would produce a clump of pheromone signals emitted by a 'chorus' of simultaneously calling same-race females. Such a clumped distribution of pheromone signals could amplify the power of the race-specific pheromone signal by combining 'voices' from individual members of the 'chorus'. This clumping might reduce the level of noise relative to signal, thus limiting the probability of male cross-attraction even in the absence of male host preferences. This would imply a role for pleiotropy in enhancing and maintaining *existing* racial differences, but it seems unlikely to have had a role before pheromone divergence.

Alternatively, specific host preference of calling females might be complemented by host-specific cues used by males searching for females. Host choice in insect parasites is usually thought of as a female trait, and there are no previously published data on host preference in male Lepidoptera. However, it would not be surprising if host choice is found in males as well as females: although males do not lay eggs, they do need to find females, and a good way to do this would be to use the inherited ability that females use to choose hosts for oviposition. To test for divergence of male and female host preference in *Zeiraphera*, we studied the distribution of adults in mixed forests. Emelianov *et al.* (2003) collected live adult *Zeiraphera* from branches in mixed forests early in the morning, when moths do not fly. Both sexes had a strongly biased tendency to rest on their own hosts in the field (86% and 82%, respectively, versus random settling of 50%). The distribution of males in the wild will then be affected both by their host and by locally calling females; for this reason, we also tested male and female host preferences in the laboratory and found a significant and heritable tendency for each sex and host race to alight on their own host (Emelianov *et al.*, 2003).

In *Zeiraphera*, therefore, a pleiotropic effect of host adaptation even today adds to mate choice chiefly regulated via pheromonal divergence, and there is no evidence that pheromone production or preferences are in any way dependent directly on host-plant use (Emelianov *et al.*, 2003). As host preference is expressed in both sexes, it is easy to imagine that host race formation began with a host choice shift, which led to some initial reproductive isolation via host choice/mate choice pleiotropy, enhanced by the clumped host distribution. Then, if even slightly divergent pheromonal signals and preferences become associated with the two host races, the degree of assortative mating, and thus host adaptation, would be enhanced in areas of sympatry, allowing further improvements in association of mate choice with host choice (see Johnson *et al.*, 1996, for a model of a similar scenario). This example therefore demonstrates how, in nature, assortative mating may often result from a complicated series of factors involving both pleiotropic effects of ecological divergence and non-pleiotropic mate choice.

2.2 Speciation in *Heliconius*

Brightly coloured and slow-flying heliconiine butterflies are among the most striking insects in neotropical rainforests. Similar colour patterns shared by heliconiine, ithomiine and pierid butterflies were initially explained by Bates (1862) as a deceptive strategy whereby rare palatable or unpalatable species gain from looking similar to common unpalatable 'model' species. Later, Müller (1879) showed how both members of a pair of unpalatable species might benefit from mimicry, because the cost of teaching predators to associate a particular pattern with unpalatability is thereby shared. *Heliconius* feed on *Passiflora* and related host plants from which they sequester cyanogenic glycosides (Brown *et al.*, 1991). In addition, the adults can synthesize the same chemicals *de novo* (Nahrstedt and Davis, 1983; Engler *et al.*, 2000). All species in the genus *Heliconius* that have been tested are distasteful and are avoided by bird predators such as jacamars (Brower *et al.*, 1963; Chai, 1986, 1988). Hence, pattern convergence between *Heliconius* species is most likely to be mutualistic Müllerian mimicry (Mallet, 1999).

The wing patterns of *Heliconius* are under strong mimetic selection (Benson, 1972; Mallet and Barton, 1989; Kapan, 2001) that leads to local convergence of pattern between 'mimicry rings' consisting of two or more species. At first sight, mimicry, a theory of convergence, is an unlikely candidate for a causative agent in speciation. However, in nature there is a great diversity of patterns at both a local and a regional scale. There are sympatric mimicry rings with distinct patterns in most areas of the neotropics, whose distinctness is maintained by segregation in both vertical and horizontal dimensions (Mallet and Gilbert, 1995; Beccaloni, 1997; Joron *et al.*, 1999). On a broader spatial scale, there are striking switches in the dominant mimetic patterns between regional faunas, both within and between species (Brown, 1979).

The phylogenetic distribution of mimicry shifts among species seems to follow the classic pattern of 'adaptive radiation', as though closely related species can avoid competition by switching to a new mimicry ring (Turner, 1976). This suggests that mimetic adaptation and diversification might play a role in speciation. Here we review our work on diversification in *Heliconius*, from the formation of colour pattern races through to good species.

2.2.1 How do novel patterns arise?

The diversity of mimicry patterns both within and between regions gives rise to ample opportunity for populations to interact with novel potential model species. Wherever a population finds itself sympatric with a mimicry ring that is numerically either more abundant or more toxic than its own, there will be selection favouring a switch to a new pattern, albeit across an adaptive valley. There are many examples of this process having occurred. For example, *Heliconius hecale* mimics the ithomiine species *Melinaea idae idae* in Panama, but switches to a different ithomiine, *Tithorea tarricina*, in Costa Rica, where *M. idae* does not occur (DeVries, 1986).

A second possibility that must have occurred frequently in *Heliconius* is that switches occur to entirely novel patterns. The frequency-dependent nature of mimetic selection means that any pattern, provided it is suitably effective in being seen and remembered by predators, could potentially go to fixation once it has become sufficiently common in a population. Presumably, the initial rise of a novel pattern has to occur through drift and/or temporary local relaxation of selection, but once established will be maintained by frequency-dependent selection. For example, the Ecuadorean Pacific coast races, *Heliconius erato cyrbia* and *H. melpomene cythera* have a highly idiosyncratic pattern not seen in any other species (Brown, 1979). This most probably arose completely *de novo* in one of the species which was then mimicked by the other (Mallet, 1999). Newly evolved races of *Heliconius* have been considered as some of the best examples of Wright's shifting balance (Mallet, 1993). Distinct patterns clearly lie on alternative adaptive peaks separated by a fitness valley, caused by selection against hybrid patterns. Furthermore, recent evidence suggests that hybrid zones between races are mobile, providing evidence for the key and contested Phase III of the shifting balance (Barton, 1992; Blum, 2002).

In some cases particular colours or patterns may be favoured in particular environments. For example, the iridescent blue colour of *H. cydno chioneus* in Panama produces a polarized signal that may be more conspicuous in dark understorey habitats (Sweeney *et al.*, 2003). Similarly, Amazonian forest habitats have a large mimicry ring of yellow and orange rayed patterns, while surrounding savanna regions to the north and south, in Venezuela, French Guiana and Brazil, more commonly have bold red and yellow patterns (Benson, 1982). Nonetheless, there are many exceptions to such generalizations and hybrid zones are rarely associated with obvious habitat features (Mallet, 1993), suggesting that the different patterns are acting as arbitrary alternative signals, each with similar overall memorability for predators.

Some of the genetic variability for this diversification might derive from hybridization (Gilbert, 2003). As expected under Müllerian mimicry, patterns are generally monomorphic within *Heliconius* populations. However, where divergent races meet in hybrid zones, recombination produces novel gene combinations giving rise to a high diversity of phenotypic patterns. Much rarer hybrids also occur between species and might similarly contribute variation on an evolutionary timescale (Mallet *et al.*, 1998; Gilbert, 2003). Furthermore, there are some obvious candidates for 'hybrid species', where naturally occurring taxa appear to share pattern elements from other species. For example, *H. hermathena* has a red forewing band similar to that of *H. erato* and narrow yellow fore- and hindwing bands very similar to *H. charithonia* (Brown and Benson, 1977). Similarly, the red and yellow forewing band of *H. heurippa* can be recreated in crosses between *H. melpomene* and *H. cydno* (Linares, 1989; Naisbit *et al.*, 2003). Hence it seems likely that hybridization plays a role in generating pattern diversity. Obviously, novel mutations must also be important, and separating the relative contributions of mutation and hybridization will only become possible with molecular characterization of switch gene loci.

2.2.2 Coevolution of pattern with mate preferences

Mimetic switches therefore usually generate parapatric populations with divergent colour patterns, maintained distinct by frequency-dependent mimetic selection (Turner, 1971). Most of these populations interbreed freely and so speciation is not considered to have occurred. Since colour pattern genes are generally in Hardy–Weinberg equilibrium where races of *Heliconius* meet, it has been assumed that there is no reproductive isolation between such races, aside from that generated by selection against colour pattern hybrids (Mallet, 1993; Mallet *et al.*, 1998).

None the less, it has long been known that colour pattern is an important cue in courtship behaviour in *Heliconius* (Crane, 1955), so mimetic patterns seem likely to be used as cues in mate detection, and as such are candidates as ecologically selected traits with pleiotropic effects on mate choice. To test this, we collected several races of *H. melpomene* with very distinct patterns and investigated their colour pattern preferences in the laboratory (Jiggins *et al.*, 2004). Butterfly courtship generally involves an initial phase in which males detect females using visual cues, followed by close-range interactions between the sexes that involve visual, pheromonal and tactile communication (Vane-Wright and Boppré, 1993). In *Heliconius*, females mate when teneral and have little ability to reject males. As a result, mate choice is mainly exerted by males. We were primarily interested in the role of colour patterns in courtship and so concentrated on the use of visual cues by males. In order to test the preferences of males for different colour patterns, moving colour pattern models were presented to population cages of males in paired trials, consisting of a control model of the same pattern as the males being tested, and an experimental model with the pattern of a different population. Models were presented sequentially, with the order of presentation randomized. In total, five distinct races were tested in 20 pairwise comparisons. In virtually all the trials, males preferred to court their own patterns, with only a few comparisons showing a significant preference for another race (Fig. 16.2).

The experiments were initially carried out with wings dissected from female butterflies. To confirm that preferences were based on colour pattern and not some other aspect of the wing such as cuticular pheromones, we repeated comparisons showing significant preferences with paper models printed using a standard inkjet printer. Paper models produced results that were in general strikingly concordant with those from the real wings (Fig. 16.2; Jiggins *et al.*, 2004). The level of responsiveness to models was generally reduced compared with that observed with dissected wings, and we rarely observed any courtship. Nonetheless, the fact that attraction to a model can be replicated using crude paper models provides strong support for the idea that initial preference is indeed based largely on visual cues.

In *H. melpomene* there was an exception to the general rule that populations prefer their own patterns: the broad red forewing band of the 'postman' pattern in *H. m. melpomene* and *H. m. rosina* was attractive to males of all races, leading, in a few cases, to significant preference for a pattern other than their own (Fig. 16.2). This might represent a constraint on the

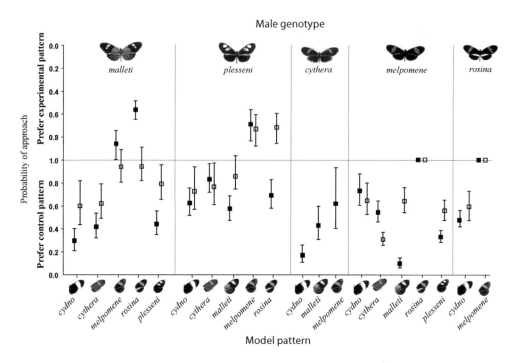

Fig. 16.2. Mate preferences among races of *H. melpomene*. The data show the probability that males are attracted to experimental wing pattern model (model pattern), relative to a similar model with the same pattern as the males themselves. Filled points show experiments using real dissected butterfly wings, hollow points show experiments using paper models. In all comparisons except those involving the red forewing band 'postman' patterns, males show strong and significant preferences for their own pattern over that of other races. Probabilities were calculated using likelihood and bars show support limits, equivalent to 95% confidence intervals. Non-significant values from Jiggins *et al.* (2001a,b) are shown as values of 1 without support limits.

evolution of preferences, perhaps due to phylogenetically conserved neural physiology (Autumn *et al.*, 2002), or perhaps due to an ecological requirement for attraction to red and orange flowers, such as *Cephaelis tomentosa* and *Psiguria* spp., that provide pollen to adult *Heliconius melpomene*. Such constraints might therefore restrict or bias the direction of the evolution of mimicry.

2.2.3 Learning and speciation

In birds it has been suggested that imprinting might play an important role in speciation, by allowing species-specific recognition cues to be learnt through contact with parents or other conspecifics (Irwin and Price, 1999). *Heliconius* are known to have well-developed learning abilities and, for example, rapidly learn the position of floral and host-plant resources within their home ranges (Gilbert, 1975; Mallet, 1986). If colour-pattern-based mate preferences were similarly learnt, this would profoundly change the dynamics of preference evolution, perhaps making speciation more rapid. Experiments were therefore carried out to test for learnt pattern preferences. Males of *H. m. malleti* were raised in the laboratory and separated into two groups after eclosion. One was exposed to females of its own race and the other only to females and males of *H. m. plesseni* for a week before experiments were carried out. If males learnt the patterns of the females with whom they had contact, then a difference between the two groups in pattern preference was predicted. In fact both groups retained a strong preference for the *H. m. malleti* pattern, suggesting that preferences were genetic. The one possible caveat is that if males could somehow determine their own colour pattern, they may be able to learn and prefer it, irrespective of experience of females. A small number of *H. erato* individuals were tested by Crane (1955) to investigate this possibility, and provided no support for the idea; but clearly more experiments are warranted. Such self-learning seems unlikely, but if it were important it would make speciation rather easy, as novel colour patterns would automatically be preferred in mating; a similar possibility has already been suggested for Darwin's finches mentioned above.

Heliconius adults are highly social insects that interact strongly at times other than during courtship. Individuals of many species roost gregariously, and spend long periods of time hovering near one another prior to roosting (Mallet, 1986). Gregarious roosting is probably a means of clustering to avoid predators, and it seems likely that preferences measured in our experiments result from selection for social as well as sexual interactions. In fact, there is a significant tendency for individuals to roost at similar heights, in similar habitats, and even within the same roosts as their co-mimics, as well as of their own species (Mallet and Gilbert, 1995). Hence, colour preferences might in part be selected to increase gregariousness between mimetic species. What is important from the point of view of speciation, however, is that such preferences do lead straightforwardly to assortative mating (Jiggins *et al.*, 2001b, 2004).

2.2.4 From races to species

In summary, colour pattern races of *H. melpomene* differ in mimetic colour patterns, and the same patterns are also used as cues in mate finding. Mate choice by males coevolves strongly with these colour patterns. Hence wing pattern differences generate a degree of both pre- as well as post-mating isolation (because of selection against non-mimetic hybrids). However, these parapatric and conspecific populations hybridize freely, and therefore have not yet speciated. The pre-mating isolation measured in our experiments is clearly not sufficient to maintain stable sympatric coexistence. In order to understand how coexisting 'good' species arise, we have also studied two species pairs that still hybridize occasionally in nature, one in each of the two major clades within the genus *Heliconius* (Jiggins *et al.*, 1996, 2001b; Mallet *et al.*, 1998). In both cases species differ in colour pattern, such that rare intermediate hybrid colour patterns are selected against due to mimetic selection mediated by predators. Furthermore, both species pairs also show strong assortative mating that is responsible for the rarity of hybrids. In the case of *H. melpomene* and *H. cydno*, mimetic patterns play a similar role in mate recognition as between races (Jiggins *et al.*, 2001b, 2004). Hence the pleiotropic effects of mimetic pattern on mate choice are similar between species and between races, and contribute to reproductive isolation between these species in the wild.

Apart from Haldane's-rule sterility in male hybrids between *H. melpomene* and *H. cydno* (Jiggins *et al.*, 2001a), and likely close-range pheromone differences in both pairs, a key factor that makes them different from races seems to be habitat and host-plant use. For example, *H. himera* and *H. erato* are parapatric species which hybridize relatively infrequently in narrow hybrid zones where they meet (Jiggins *et al.*, 1996; Jiggins and Mallet, 2000). Hybrid zones are associated with a transition from the wet forest habitat of *H. erato* to the much dryer regions where *H. himera* is found. This ecological shift appears to be the main difference between *H. himera* and the other races of *H. erato*, suggesting that ecological selection may play a major role in maintaining the species differences in the hybrid zone. However, these species have almost identical host-plant use, which probably explains their inability to coexist in broad sympatry (Jiggins *et al.*, 1997). In the *H. melpomene* group, the closest sister taxon to *H. melpomene* is the *H. cydno* species complex, with mtDNA suggesting a divergence time of 1.5 million years, similar to that between *H. himera* and *H. erato* (Beltrán *et al.*, 2002). However, unlike the latter pair, *H. melpomene* and *H. cydno* are broadly sympatric. *Heliconius melpomene* and *H. cydno* differ chiefly in microhabitat preference and host-plant use, rather than in adaptations to non-overlapping habitats, such as wet and dry forest as in *H. erato* and *H. himera*. Hence their ecologies allow sympatric ecological segregation and broad overlap. This ecological segregation also maximizes the advantages of mimicry: *H. melpomene* occurs with its co-mimic, *H. erato*, in open areas, while *H. cydno* and co-mimic *H. sapho* are both found in closed canopy forest (Gilbert, 1991; Estrada and Jiggins, 2002). Thus, genes for habitat preference are co-adapted with those for colour pattern.

If mimetic patterns are also used as mating cues, this raises the question of how co-mimics with identical patterns recognize their own species (Brower *et al.*, 1963; Vane-Wright and Boppré, 1993). When male *H. melpomene* were presented with their own patterns and those of their sympatric co-mimic, *H. erato*, they could not distinguish between them (C. Estrada and C.D. Jiggins, unpublished). Similarly, in the wild it is common to see mimetic butterflies of different species, such as *H. cydno* and *H. sapho*, approach one another and interact briefly before flying on. It seems likely, therefore, that confusion does occur between co-mimics in nature, leading to time wasted in pursuit of unsuitable partners. Time-wasting may be minimized because mimicry is most common between rather distantly related species in the genus *Heliconius* (Turner, 1976). Distantly related species may have diverged sufficiently in close-range pheromonal or other mating signals so that conspecifics can be readily distinguished from co-mimics.

2.2.5 What drives divergence in mate preference?

Having reviewed the empirical evidence, we now attempt to place the *Heliconius* data in the context of the speciation theory described at the beginning of the chapter. This requires a certain amount of speculation regarding the genetic basis of mate preference, but the assumptions are testable. Assuming that preference is genetic and due to alleles for a behavioural rule such as 'prefer individuals with orange rays', they will follow a 'two-allele' model whereby alleles evolve to different frequencies in each population. So how do preference genes become associated with their respective colour pattern? Given the strong selection on pattern, and monomorphism of patterns in populations outside narrow hybrid zones, it seems most likely that when a novel pattern arises it goes to fixation very rapidly in the local population.

Thus, a novel pattern phenotype, *A*, goes to fixation locally and is followed by an increase in frequency of *B*, a choice phenotype causing individuals to prefer pattern *A* over the ancestral pattern, *a*. Is *B* favoured by direct or indirect selection? Since hybrid zones are very narrow relative to the range of each race, and there is no evidence for assortative mating in hybrid-zone populations, reinforcement seems unlikely, at least initially. This implies that preferences are favoured not by selection to reduce the production of unfit hybrids, but rather by direct selection on preference alleles. This most probably occurs through improved mate-finding efficiency (Dawkins and Guilford, 1996). A population that has recently acquired a novel pattern *A* still has the ancestral choice phenotype *b* that prefers the *a* pattern as a mating cue. Individuals therefore waste time courting the wrong species, or waiting for individuals with *a* patterns. Thus, *B* is favoured, since less time is wasted finding potential mates. This process will be driven by sexual selection, since *B* individuals are likely to achieve more matings than *b*. It could also be driven purely by natural selection if mating success is similar for *B* and *b*, but the former has more time to feed or avoid predators. Some combination of the two is likely. When considered in this way, the fixation of alternative preference alleles between populations with distinct patterns seems almost inevitable.

We have some data about the form of selection on mate preference from experiments on hybrid matings (Naisbit *et al.*, 2001). Hybrids between *H. melpomene* and *H. cydno* suffer a mild mating disadvantage; matings are more probable within parental types than between hybrids and either parental. In addition, F_1 hybrids prefer other F_1 hybrids, suggesting that the genetic basis of mate preference is largely additive. These experiments also demonstrate that, once strong mating preferences have evolved, disruptive sexual selection against rare hybrids acts as an additional pleiotropic effect that further contributes to reproductive isolation (Naisbit *et al.*, 2001). In contrast, between *H. erato cyrbia* and *H. himera*, we could not detect disruptive sexual selection against hybrids (McMillan *et al.*, 1997). The differences between these two systems may be due to reinforcement, which appears to have narrowed the mating preferences of *H. melpomene* where it overlaps with *H. cydno* (Jiggins *et al.*, 2001a,b).

Fixation of alternative habitat preference alleles should similarly be favoured between parapatric populations in order to maximize overlap with co-mimics. Speciation of *H. melpomene* and *H. cydno* is therefore likely to have been parapatric, with abutting populations diverging in mimicry, colour-based mate preference, and habitat preference. Genes affecting all of these traits would be strongly epistatic. Hence, mate-preference alleles for white and iridescent blue colour patterns and habitat-preference alleles for living in forest understorey would have been favoured in the incipient *H. cydno* population that was mimetic with forest-dwelling *H. sapho*. Meanwhile, in the adjacent incipient *H. melpomene* population the red-band mate preference alleles and open habitat ecological preference alleles would have been favoured to maximize overlap with its co-mimic, *H. erato*.

The order in which these changes occurred is perhaps impossible to determine, but the study of *H. melpomene* races described above suggests that perhaps the first step in divergence is a change in colour pattern and associated mate preferences. The races of *H. melpomene* have not yet speciated, perhaps because they all mimic *H. erato*, with correspondingly similar preferences for open habitat.

In summary, habitat preference, mate preference and mimetic pattern form co-adapted gene complexes in *Heliconius* populations, in a manner similar to the co-adaptation of communication systems to the environment described in the 'sensory drive' hypothesis (Endler, 1992; Endler and Basolo, 1998). Our proposed model for speciation in *Heliconius* therefore requires an initial leap across an adaptive valley, represented by the evolution of a novel colour pattern form. Subsequent adaptation of habitat preference and mate preference alleles can occur through the gradual accumulation of additive genetic changes, in a manner similar to that proposed in the Dobzhansky–Müller model for the evolution of post-mating incompatibilities.

3. Conclusions

Our results from these two lepidopteran systems suggest that genes selected by particular ecological conditions may commonly have some pleiotropic effect on mating. We have highlighted the distinction between pleiotropic assortment traits, where a single ecologically selected trait leads automatically to assortative mating between divergent populations; pleiotropic mating cues, where the ecologically selected loci also control a trait chosen during mating; and pleiotropic mating preferences, where mate preferences are influenced by ecological selection. In the latter two, the ecologically selected traits influence the mating system but do not lead directly to assortative mating; subsequent coevolution of mate choice and mating cues performs this task. There are many examples of pleiotropic assortment traits, as any form of habitat choice where mating occurs in the habitat chosen (including temporal habitat, or phenology) will lead to a degree of assortative mating. In the pheromone-signalling host races of *Zeiraphera* a pleiotropic assortment trait is combined with habitat-independent mating cues. In contrast, good examples of pleiotropic mating cues and pleiotropic mate preferences are fewer, presumably because they require a more detailed understanding of the genetic basis of ecological traits, mating signals and mating preferences. *Heliconius* butterflies provide a good example of pleiotropic mating cues, where mimetic patterns are used as cues in mate-finding, leading to coevolution of mate choice and assortative mating. However, divergence in *Heliconius* mimicry does not automatically lead to completion of speciation; for that, further ecological divergence seems necessary. In both cases, more information is needed on the genetics of mate preferences to fully understand their origin. Nonetheless, we hope that our review and the current focus on the importance of ecology in speciation will stimulate further studies to illuminate the role of pleiotropic effects in species formation.

Acknowledgements

We thank Michele Drès, Werner Baltensweiler and Catalina Estrada for help with field and laboratory work, Mark Kirkpatrick, Angeles de Cara and Jitka Polechová for comments on the manuscript, and NERC, BBSRC, the National Geographical Society, the British Ecological Society, the Smithsonian Tropical Research Institute and the Royal Society for funding.

References

Autumn, K., Ryan, M.J. and Wake, D.B. (2002) Integrating historical and mechanistic biology enhances the study of adaptation. *Quarterly Review of Biology* 77, 383–408.

Baltensweiler, W., Benz, G., Bovey, P. and Delucchi, V. (1977) Dynamics of larch bud moth populations. *Annual Review of Entomology* 22, 79–100.

Barbash, D.A., Siino, D.F., Tarone, A.M. and Roote, J.A. (2003) A rapidly evolving Myb-related protein causes species isolation in *Drosophila*. *Proceedings of the National Academy of Sciences of the USA* 100, 5302–5307.

Barton, N.H. (1992) On the spread of new gene combinations in the third phase of Wright's shifting balance. *Evolution* 46, 551–557.

Bates, H.W. (1862) Contributions to an insect fauna of the Amazon valley, Lepidoptera: Heliconidae. *Transactions of the Linnean Society of London* 23, 495–566.

Beccaloni, G. (1997) Vertical stratification of ithomiine butterfly (Nymphalidae: Ithomiinae) mimicry complexes: the relationship between adult flight height and larval host-plant height. *Biological Journal of the Linnean Society* 62, 313–341.

Beltrán, M., Jiggins, C.D., Bull, V., McMillan, W.O., Bermingham, E. and Mallet, J. (2002) Phylogenetic discordance at the species boundary: gene genealogies in *Heliconius* butterflies. *Molecular Biology and Evolution* 19, 2176–2190.

Benson, W.W. (1972) Natural selection for Müllerian mimicry in *Heliconius erato* in Costa Rica. *Science* 176, 936–939.

Benson, W.W. (1982) Alternative models for infrageneric diversification in the humid tropics: tests with passion vine butterflies. In: Prance G.T. (ed.) *Biological Diversification in the Tropics*. Columbia University Press, New York, pp. 608–640.

Benz, G. (1973) Role of sex pheromone and its significance for heterosexual and homosexual behaviour of the larch bud moth. *Experientia* 29, 553.

Blum, M.J. (2002) Rapid movement of a *Heliconius* hybrid zone: evidence for phase III of Wright's shifting balance theory? *Evolution* 56, 1992–1998.

Boughman, J.W. (2001) Divergent sexual selection enhances reproductive isolation in sticklebacks. *Nature* 411, 944–948.

Boughman, J.W. (2002) How sensory drive can promote speciation. *Trends in Ecology and Evolution* 17, 571–577.

Brazner, J.C. and Etges, W.J. (1993) Premating isolation is determined by larval rearing substrates in cactophilic *Drosophila mojavensis*. II. Effects of larval substrates on time to copulation, mate choice and mating propensity. *Evolutionary Ecology* 7, 605–624.

Brower, L.P., Brower, J.V.Z. and Collins, C.T. (1963) Experimental studies of mimicry. 7. Relative palatability and Müllerian mimicry among Neotropical butterflies of the subfamily Heliconiinae. *Zoologica* 48, 65–84.

Brown, K.S. (1979) *Ecologia Geográfica e Evoluçâo nas Florestas Neotropicais*. Universidade Estadual de Campinas, Campinas, Brazil.

Brown, K.S. and Benson, W.W. (1977) Evolution in modern Amazonian non-forest islands: *Heliconius hermathena*. *Biotropica* 9, 95–117.

Brown, K.S., Trigo, J.R., Francini, R.B., Barros de Morais, A.B. and Motta, P.C. (1991) Aposematic insects on toxic host plants: coevolution, colonization, and chemical emancipation. In: Price, P.W., Lewinsohn, T.M., Fernandes, G.W. and Benson, W.W. (eds) *Plant–Animal Interactions: Evolutionary Ecology in Tropical and Temperate Regions*. John Wiley, New York, pp. 375–402.

Brunton, C.F.A. and Majerus, M.E.N. (1995) Ultraviolet colors in butterflies: intraspecific or inter-specific communication? *Proceedings of the Royal Society of London, Series B* 260, 199–204.

Bush, G.L. (1969) Mating behaviour, host specificity, and the ecological significance of sibling species in frugivorous flies of the genus *Rhagoletis* (Diptera: Tephritidae). *American Naturalist* 103, 669–672.

Bush, G.L. (1975) Modes of animal speciation. *Annual Review of Ecology and Systematics* 6, 339–364.

Butlin, R.K. (1990) Divergence in emergence time of host races due to the differential gene flow. *Heredity* 65, 47–50.

Chai, P. (1986) Field observations and feeding experiments on the responses of rufous-tailed jacamars, *Galbula ruficauda*, to free-flying butterflies in a tropical rainforest. *Biological Journal of the Linnean Society* 29, 166–189.

Chai, P. (1988) Wing coloration of free-flying Neotropical butterflies as a signal learned by a specialized avian predator. *Biotropica* 20, 20–30.

Crane, J. (1955) Imaginal behaviour of a Trinidad butterfly, *Heliconius erato hydara* Hewitson, with special reference to the social use of color. *Zoologica* 40, 167–196.

Dawkins, M.S. and Guilford, T. (1996) Sensory bias and the adaptiveness of female choice. *American Naturalist* 148, 937–942.

DeVries, P. (1986) *The Butterflies of Costa Rica*. Princeton University Press, Princeton, New Jersey.

Dobzhansky, T. (1937) *Genetics and the Origin of Species*. Columbia University Press, New York.

Doebeli, M. (1996) A quantitative genetic competition model for sympatric speciation. *Journal of Evolutionary Biology* 9, 893–909.

Drès, M. (2000) Gene flow between host races of the larch budmoth *Zeiraphera diniana* (Lepidoptera: Tortricidae). PhD thesis, University of London, London.

Drès, M. and Mallet, J. (2002) Host races in plant feeding insects and their importance in sympatric speciation. *Philosophical Transactions of the Royal Society of London, Series B* 357, 471–492.

Emelianov, I., Mallet, J. and Baltensweiler, W. (1995) Genetic differentiation in *Zeiraphera diniana* (Lepidoptera: Tortricidae, the larch budmoth): polymorphism, host races or sibling species? *Heredity* 75, 416–424.

Emelianov, I., Dres, M., Baltensweiler, W. and Mallet, J. (2001) Host-induced assortative mating in host races of the larch budmoth. *Evolution* 55, 2002–2010.

Emelianov, I., Simpson, F., Narang, P. and Mallet, J. (2003) Host choice promotes reproductive isolation between host races of the larch budmoth. *Journal of Evolutionary Biology* 16, 208–218.

Emelianov, I., Marec, F. and Mallet, J. (2004) Genomic evidence for divergence with gene flow in host races of the larch budmoth. *Proceedings of the Royal Society of London, Series B* 271, 97–105.

Endler, J.A. (1992) Signals, signal conditions, and the direction of evolution. *American Naturalist* 139, S125–S153.

Endler, J.A. and Basolo, A.L. (1998) Sensory ecology, receiver biases and sexual selection. *Trends in Ecology and Evolution* 13, 415–420.

Engler, H.S., Spencer, K.C. and Gilbert, L.E. (2000) Preventing cyanide release from leaves. *Nature* 406, 144–145.

Estrada, C. and Jiggins, C.D. (2002) Patterns of pollen feeding and habitat preference among *Heliconius* species. *Ecological Entomology* 27, 448–456.

Etges, W.J. (1992) Premating isolation is determined by larval substrates in cactophilic *Drosophila mojavensis*. *Evolution* 46, 1945–1950.

Etges, W.J. and Ahrens, M.A. (2001) Premating isolation is determined by larval-rearing substrates in cactophilic *Drosophila mojavensis*. V. Deep geographic variation in epicuticular hydrocarbons among isolated populations. *American Naturalist* 158, 585–598.

Feder, J.L., Opp, S.B., Wlazlo, B., Reynolds, K., Go, W. and Spisak, S. (1994) Host fidelity is an effective premating barrier between sympatric races of the apple maggot fly. *Proceedings of the National Academy of Sciences USA* 91, 7990–7994.

Felsenstein, J. (1981) Skepticism towards Santa Rosalia, or why are there so few kinds of animals? *Evolution* 35, 124–138.

Gavrilets, S. (2003) Models of speciation: what have we learned in 40 years? *Evolution* 57, 2197–2215.

Gilbert, L.E. (1975) Ecological consequences of a coevolved mutualism between butterflies and plants. In: Gilbert, L.E. and Raven, P.R. (eds) *Coevolution of Animals and Plants*. University of Texas Press, Austin, Texas, pp. 210–240.

Gilbert, L.E. (1991) Biodiversity of a Central American *Heliconius* community: pattern, process, and problems. In: Price, P.W., Lewinsohn, T.M., Fernandes, T.W. and Benson, W.W. (eds) *Plant–Animal Interactions: Evolutionary Ecology in Tropical and Temperate Regions*. John Wiley & Sons, New York, pp. 403–427.

Gilbert, L.E. (2003) Adaptive novelty through introgression in *Heliconius* wing patterns: evidence for shared genetic 'tool box' from synthetic hybrid zones and a theory of diversification. In: Boggs, C.L., Watt, W.B. and Ehrlich, P.R. (eds) *Ecology and Evolution Taking Flight: Butterflies as Model Systems*. University of Chicago Press, Chicago, Illinois.

Grant, P.R. (1986) *Ecology and Evolution of Darwin's Finches*. Princeton University Press, Princeton, New Jersey.

Held, C. and Speith, H.T. (1999) First experimental evidence of pupal summer diapause in *Pieris brassicae* L.: the evolution of local adaptedness. *Journal of Insect Physiology* 45, 587–598.

Howard, D.J. and Waring, G.L. (1991) Topographic diversity, zone width, and the strength of reproductive isolation in a zone of overlap and hybridization. *Evolution* 45, 1120–1135.

Irwin, D.E. and Price, T. (1999) Sexual imprinting, learning and speciation. *Heredity* 82, 347–354.

Jiggins, C.D. and Mallet, J. (2000) Bimodal hybrid zones and speciation. *Trends in Ecology and Evolution* 15, 250–255.

Jiggins, C.D., McMillan, W.O., Neukirchen, W. and Mallet, J. (1996) What can hybrid zones tell us about speciation? The case of *Heliconius erato* and *H. himera* (Lepidoptera: Nymphalidae). *Biological Journal of the Linnean Society* 59, 221–242.

Jiggins, C.D., McMillan, W.O. and Mallet, J. (1997) Host plant adaptation has not played a role in the recent speciation of *Heliconius erato* and *Heliconius himera* (Lepidoptera; Nymphalidae). *Ecological Entomology* 22, 361–365.

Jiggins, C.D., Linares, M., Nasbit, R.E., Salazar, C., Yang, Z.H. and Mallet, J. (2001a) Sex-linked hybrid sterility in a butterfly. *Evolution* 55, 1631–1638.

Jiggins, C.D., Naisbit, R.E., Coe, R.L. and Mallet, J. (2001b) Reproductive isolation caused by colour pattern mimicry. *Nature* 411, 302–305.

Jiggins, C.D., Estrada, C. and Rodrigues, A. (2004) Mimicry and the evolution of premating isolation in *Heliconius melpomene*. *Journal of Evolutionary Biology* 17, 680–691.

Johnson, P.A., Hoppensteadt, F.C., Smith, J.J. and Bush, G. (1996) Conditions for sympatric speciation: a diploid model incorporating habitat fidelity and non-habitat assortative mating. *Evolutionary Ecology* 10, 187–205.

Joron, M., Wynne, I.R., Lamas, G. and Mallet, J. (1999) Variable selection and the coexistence of multiple mimetic forms of the butterfly *Heliconius numata*. *Evolutionary Ecology* 13, 721–754.

Kapan, D.D. (2001) Three-butterfly system provides a field test of Müllerian mimicry. *Nature* 409, 338–340.

Kirkpatrick, M. and Ravigné, V. (2002) Speciation by natural and sexual selection. *American Naturalist* 159, S22–S35.

Kondrashov, A.S. and Mina, M. (1986) Sympatric speciation: when is it possible? *Biological Journal of the Linnean Society* 27, 201–223.

Linares, M. (1989) Adaptive microevolution through hybridization and biotic destruction in the neotropics. PhD thesis, University of Texas, Austin, Texas.

MacCallum, C.J., Nürnberger, B., Barton, N.H. and Szymura, J.M. (1998) Habitat preference in the *Bombina* hybrid zone in Croatia. *Evolution* 52, 227–239.

Mallet, J. (1986) Gregarious roosting and home range in *Heliconius* butterflies. *National Geographic Research* 2, 198–215.

Mallet, J. (1993) Speciation, raciation, and color pattern evolution in *Heliconius* butterflies: evidence from hybrid zones. In: Harrison, R.G. (ed.) *Hybrid Zones and the Evolutionary Process*. Oxford University Press, New York, pp. 226–260.

Mallet, J. (1999) Causes and consequences of a lack of coevolution in Müllerian mimicry. *Evolutionary Ecology* 13, 777–806.

Mallet, J. and Barton, N.H. (1989) Strong natural selection in a warning color hybrid zone. *Evolution* 43, 421–431.

Mallet, J. and Gilbert, L.E. (1995) Why are there so many mimicry rings? Correlations between habitat, behaviour and mimicry in *Heliconius* butterflies. *Biological Journal of the Linnean Society* 55, 159–180.

Mallet, J., McMillan, W.O. and Jiggins, C.D. (1998) Mimicry and warning colour at the boundary between races and species. In: Howard, D.J. and Berlocher, S.H. (eds) *Endless Forms: Species and Speciation*. Oxford University Press, New York, pp. 390–403.

Maynard Smith, J. (1966) Sympatric speciation. *American Naturalist* 100, 637–650.

McMillan, W.O., Jiggins, C.D. and Mallet, J. (1997) What initiates speciation in passion vine butterflies? *Proceedings of the National Academy of Sciences USA* 94, 8628–8633.

Mitter, C. and Brooks, D.R. (1983) Phylogenetic aspects of coevolution. In: Futuyma, D.J. and Slatkin, M. (eds) *Coevolution*. Sinauer, Sunderland, Massachusetts.

Moore, W.S. (1981) Assortative mating genes selected along a gradient. *Heredity* 46, 191–195.

Müller, F. (1879) *Ituna* and *Thyridia*; a remarkable case of mimicry in butterflies. *Transactions of the Entomological Society of London* 1879, xx–xxix.

Nahrstedt, A. and Davis, R.H. (1983) Occurrence, variation and biosynthesis of the cyanogenic glucosides linamarin and lotaustralin in species of the Heliconiini (Insecta: Lepidoptera). *Comparative Biochemistry and Physiology* 75B, 65–73.

Naisbit, R.E., Jiggins, C.D. and Mallet, J. (2001) Disruptive sexual selection against hybrids contributes to speciation between *Heliconius cydno* and *Heliconius melpomene*. *Proceedings of the Royal Society of London, Series B* 268, 1849–1854.

Naisbit, R.E., Jiggins, C.D. and Mallet, J. (2003) Mimicry: developmental genes that contribute to speciation. *Evolution and Development* 5, 269–280.

Nijhout, H.F. (1991) *The Development and Evolution of Butterfly Wing Patterns*. Smithsonian Institution Press, Washington, DC.

Orr, H.A. and Turelli, M. (2001) The evolution of postzygotic isolation: accumulating Dobzhansky–Muller incompatibilities. *Evolution* 55, 1085–1094.

Podos, J. (2001) Correlated evolution of morphology and vocal signal structure in Darwin's finches. *Nature* 409, 185–188.

Presgraves, D.C., Balagopalan, L., Abmayr, S.M. and Orr, H.A. (2003) Adaptive evolution drives divergence of a hybrid inviability gene between two species of *Drosophila*. *Nature* 423, 715–719.

Rice, W.R. and Hostert, E.E. (1993) Laboratory experiments on speciation: what have we learned in 40 years? *Evolution* 47, 1637–1653.

Sweeney, A., Jiggins, C.D. and Johnson, S. (2003) Polarised light as a butterfly mating signal. *Nature* 423, 31–32.

Turelli, M. and Orr, H.A. (1995) The dominance theory of Haldanes Rule. *Genetics* 140, 389–402.

Turelli, M. and Orr, H.A. (2000) Dominance, epistasis and the genetics of postzygotic isolation. *Genetics* 154, 1663–1679.

Turner, J.R.G. (1971) Two thousand generations of hybridization in a *Heliconius* butterfly. *Evolution* 25, 471–482.

Turner, J.R.G. (1976) Adaptive radiation and convergence in subdivisions of the butterfly genus *Heliconius* (Lepidoptera: Nymphalidae). *Zoological Journal of the Linnean Society* 58, 297–308.

Vane-Wright, R.I. (1979) Towards a theory of the evolution of butterfly colour patterns under directional and disruptive selection. *Biological Journal of the Linnean Society* 11, 141–152.

Vane-Wright, R.I. and Boppré, M. (1993) Visual and chemical signalling in butterflies: functional and phylogenetic perspectives. *Philosophical Transactions of the Royal Society of London, Series B* 340, 197–205.

West-Eberhard, M.J. (1983) Sexual selection, social competition and speciation. *Quarterly Review of Biology* 58, 155–183.

Wood, T.K. and Keese, M.C. (1990) Host plant induced assortative mating in *Enchenopa* treehoppers. *Evolution* 44, 619–628.

17 Specializations and Host Associations of Social Parasites of Ants

JEREMY A. THOMAS, KARSTEN SCHÖNROGGE AND
GRAHAM W. ELMES

*Centre for Ecology and Hydrology (NERC), CEH Dorset,
Winfrith Technology Centre, Winfrith Newburgh, Dorchester, UK*

1. Introduction

Ant societies dominate most terrestrial ecosystems at the scales at which they function (Wilson, 1990), and numerous similar-sized organisms have evolved adaptations to coexist with them (Wasmann, 1894; Wheeler, 1910; Donisthorpe, 1927; Hölldobler and Wilson, 1990; Huxley and Cutler, 1991). The occurrence of myrmecophily has probably been much underestimated in the field, but drawing on Donisthorpe (1927) and other studies, Elmes (1996) calculated that as many as 100,000 species of myrmecophile exist among the insects. Since this figure is more than double the number of vertebrate species known on Earth, myrmecophily can be regarded as a common phenomenon in ecology as well as one that has evolved independently, and convergently, across many orders of insect (Hölldobler and Wilson, 1990).

Most known insect myrmecophiles are free-living commensals or mutualists, which live above ground within the foraging areas or territories of ants (Hölldobler and Wilson, 1990; Thomas, 1992a,b). They benefit from inhabiting these relatively enemy-free spaces (including freedom from hostile ants) and, in many cases, gain direct protection from their enemies. The functional and evolutionary ecology, and sheer diversity of the myrmecophilous adaptations of commensals and mutualists from several orders of insect are reviewed by Wasmann (1894), Donisthorpe (1927), Hinton (1951), Malicky (1969), Hölldobler and Wilson (1990), De Vries (1991a,b), Fiedler (1991, 1994, 1996, 1998), Pierce *et al.* (2002) and others. Although some adaptations are multifunctional, we loosely group them into three overlapping categories:

- **Evasive devices**, ranging from pacifying or non-threatening behaviours to morphological, behavioural or acquired-chemical disguise and camouflage.
- **Protection to withstand attack**, including some extreme morphological adaptations (e.g. highly sclerotized cuticles, onciform bodies with the vulnerable organs hidden underneath, setae, loose scales on wings, and many others).

- **Deliberate communication with ants** through sound, touch and, especially, through a diversity of chemical secretions that ranges from rewarding ants with food (sugars, amino acids etc.) to chemical mimicry (*sensu* Dettner and Liepert, 1994) or to the secretion of generalist allomones that either appease ants or alarm and agitate them. Chemical communication is the main means by which ant societies communicate with, and discriminate between, each other, and it is a major tool used by myrmecophiles from several insect orders to manipulate ant social behaviour; fine reviews are found in Hölldobler and Wilson (1990), Dettner and Liepert (1994) and Lenoir *et al.* (2001).

The interactions that have evolved between free-living commensal or mutualistic insect myrmecophiles and ants range from loose facultative associations to an obligate dependency for one or both partners. These interactions often appear complex and specialized (for butterflies alone, see Hinton, 1951; Malicky, 1969, 1970; Pierce and Elgar, 1985; Fiedler and Maschwitz, 1987, 1988, 1989; Pierce *et al.*, 1987, 2002; Pierce, 1989; De Vries and Baker, 1989; De Vries, 1990, 1991b; Fiedler and Hölldobler, 1992; Thomas, 1992a,b; Fiedler, 1996), yet compared with myrmecophiles that inhabit ant nests or the other fiercely protected niches of ant societies, most are relatively simple and most free-living mutualists or commensals are comparatively generalist, interacting with ant species from several subfamilies, although a few are restricted to a single ant genus or species (Hölldobler and Wilson, 1990; Thomas, 1992a,b; Pierce *et al.*, 2002).

In contrast, mutualistic species that inhabit ant nests are typically much more specialized and host-specific, whether they be the fungal gardens of *Atta* (leaf-cutting) ants (see Boomsma *et al.*, Chapter 6, this volume) or the subterranean domestic herds of aphids and coccids on whose secretions certain species of *Acanthomyops* and *Acropyga* (and probably other genera) appear wholly to depend for food (Way, 1953, 1963; Hölldobler and Wilson, 1990). Some (we suspect many) of these associations are host-specific and it seems increasingly probable that subspecific adaptations and dependencies have (co-)evolved. The recent demonstration of colony-level associations between different genotypes of fungi and leafcutter ants – in which winged *Atta* queens carry their (colony-)specific strain of fungus with them on flights, transmitting this extreme mutualism vertically when new ant colonies are founded (Boomsma *et al.*, Chapter 6, this volume) – has interesting parallels in *Cladomyrma* and *Acropyga* ant species, whose winged queens carry coccids taken from their natal nests in their mandibles during the nuptial flight (Hölldobler and Wilson, 1990).

2. Social Parasites of Ants

2.1 Definitions

We take a broad definition of the term social parasite to encompass all intruding arthropods that penetrate and inhabit, sometimes briefly, an ant society to exploit any resource that is valued and protected by the ant host, rather than the narrower use of the term when applied only to parasitic ant species (*see* Hölldobler and Wilson, 1990). In contrast to the distinction between mutualist and commensal, that between mutualist and social parasite is clear-cut. Hölldobler and Wilson (1990) found no evidence that these relationships were anything other than one-sided, with the parasite winning all measured benefits and the ants none. Nor have we, in our studies on the butterfly *Maculinea rebeli*, whose caterpillars live within the brood chambers of *Myrmica schencki* nests where they displace the ant larvae and are fed directly by nurse ants. Although we predicted that the copious secretions of *M. rebeli*, which are continuously drunk by the nurses, might transfer increased fitness to workers at the expense of their own brood, it transpired that workers that tended *M. rebeli* lost weight and were shorter-lived than those tending the ants' own brood, such were the demands imposed on the colony by the insatiable butterfly caterpillars (Wardlaw *et al.*, 2000). The only apparent exception to this generalization is the special case of *Myrmica ruginodis* microgyna queens (*see* Brian and Brian, 1951, 1955), which confer distinct ecological advantages on colonies supporting them but which may be the first step in the evolution of a social parasitic ant (Elmes, 1978, and below).

We estimate that around 10,000–20,000 morphospecies of insect have evolved as social parasites of ants, thus accounting for a significant proportion of the world's biodiversity (by comparison, about 10,000 species of bird and 4500 mammal species exist). But despite the many species, most ant social parasites are exceedingly rare or localized in comparison to the abundance and distributions not only of their ant hosts but also of the other symbionts that interact with ants (Thomas, 1980, 1995; Hölldobler and Wilson, 1990; Thomas and Morris, 1994). Before exploring this, we briefly examine which types of ant society are most susceptible to social parasite attack.

2.2 The type of ant society most vulnerable to attack

It has long been accepted that one common form of social organization is especially vulnerable to intrusion, namely those ants whose colonies contain variable, usually multiple, numbers of queens (polygyny) and inhabit multiple nest sites (polydomy) (e.g. Buschinger, 1970; Alloway *et al.*, 1982). Depending on the species, the number of queens in polygynous-polydomous colonies may range from zero to hundreds, and the local density of queens can vary intraspecifically between populations and over time. There is often a clear hierarchy in the social status of the queens: in the case of *Myrmica*, dominant alpha (α) queens occupy the central brood chamber where they receive maximum worker attention and lay most of the colony's eggs, while beta (β) queens are relegated to the outer chambers awaiting promotion to α status (Brian, 1968, 1986; Evesham, 1984). Somewhat similar social structures have been found for other polygynous species, such as *Leptothorax nylanderi* (Plateaux, 1970) and *Solenopsis invicta* (Ross, 1989). In these and other species, daughter queens may be re-absorbed by their natal colony after the nuptial flight (e.g. Elmes, 1973), or enter a closely related one nearby. However, colonies often recruit unrelated queens (e.g. Pearson, 1982; Elmes, 1987; Goodisman and Ross, 1998), which can be interpreted as unrelated queens penetrating the society and acting as temporary social parasites (Elmes, 1974). Colony spread, in this system, is usually by fission, with daughter colonies budding off from core nests to exploit vacant habitat; initially these offshoots tend to be queenless, but may later recruit daughter queens from the mother colony or nearby nests (Elmes, 1991; Pedersen and Boomsma, 1999). There can also be a strong seasonal element to polydomy, with offshoots spreading in spring–early summer and retracting to merge with the donor nest in autumn (e.g. Herbers, 1985).

Compared to the societies of monogynous ants, polydomous-polygynous systems are loose, flexible and dynamic, leading sometimes to the development of massive supercolonies occupying > 1 ha. This lifestyle imposes large reproductive costs to individuals, due to reduced kinship, but these are believed to be offset by shared benefits accrued at the colony level. Queens joining existing colonies obviously stand to benefit if their chances of individual colony foundation were small. Why the resident queens should tolerate joiners (even unrelated ones) is less clear, although some selective advantages have been suggested (Bourke and Franks, 1991, 1995). The ecological benefits at the colony level include an ability to withstand severe disturbances, resulting in colony fragmentation, and to spread rapidly, dominating new resources and pre-empting competitors, either seasonally or when fresh habitat becomes available following a perturbation. The greatest shared cost appears to be a periodic reduction in social cohesion, leading to increased vulnerability of the group to invasion by other organisms, including ant and non-ant social parasites that mimic the colony's recognition signals.

Hölldobler and Wilson (1990) list three further attributes that may predispose ant colonies to social parasitism: they live in dense populations; inhabit cool or arid climates; and learn the species' odour early in adult life, a theme developed by Lenoir *et al.* (2001). Our experiences suggest that the last two attributes seldom apply to the non-ant social parasites of ants (see also Pierce *et al.*, 2002).

2.3 Properties of social parasites

Our main generalizations are made at the conclusion of this chapter. Here we note that, in addition to being rare in comparison to hosts, most lines of social parasite are embedded among clades of mutualistic species, and are presumed to have evolved from similar ancestors which were pre-adapted to invading ant societies through the defensive, pacifying and communicative devices that had evolved to promote symbiosis.

It is also increasingly clear that social parasitism has evolved independently many times among myrmecophiles, often within closely related taxa. For example, the butterfly family Lycaenidae contains about 6000 species, of which most are mutualists and perhaps 300 (5%) species are social parasites; yet the latter are so polyphylitic that social parasitism is thought to have arisen independently among them at least 13 times (Fiedler, 1998; Elmes *et al.*, 2001; Pierce *et al.*, 2002).

A final generalization is that many remarkable examples of convergent evolution exist between different lines of social parasite from several orders of insect (Coleoptera, Diptera, Lepidoptera, Hymenoptera, and to a lesser extent Orthoptera and others). Although almost every niche within an ant colony has been exploited by one type of myrmecophile or another, certain distinct forms are apparent, most of which are described and illustrated by Wasman (1894, 1912), Donisthorpe (1927) and Hölldobler and Wilson (1990). Later, we extend Hölldobler and Wilson's (1990) argument that the comparatively few morphological and behavioural forms found among the estimated 10,000–20,000 species of social parasite relate to distinct functional types that exploit ant colonies in one of a relatively small number of ways.

3. Four Evolutionary Routes to Inquilism and Other Specialized Forms of Social Parasitism

Hölldobler and Wilson (1990) identify four evolutionary pathways leading to social parasitism, and in some cases ultimately to inquilism, in which the parasite is so embedded in its host society that it lives the entire life cycle there. Three involve the most pre-adapted organisms of all – other ants. The majority confirm Emery's rule (1909), which states that a parasitic species of ant is generally a close relative of its host ant species (see Savolainen and Vepsäläinen, 2003). One evolutionary route towards inquilism stems from ants that originally lived as commensals alongside or in the artefacts of other ant species, the eventual host typically being the larger and more dominant of the two. Emery's rule frequently applies because this relationship often involves close mimicry of the host by the parasite, and the morphology, behaviour, physiology and communication systems of closely related ants provided the most similar models in the first place.

A second distinctive route involves slave-making ants, which raid the nests of close relatives, carrying off brood into their own nests to rear to adulthood as slave labourers. The spectrum of behaviours between various slave-making types of ant and their hosts, synthesized into a convincing evolutionary pathway that culminates in social parasitism, are so fully reviewed by Buschinger (1986) and Hölldobler and Wilson (1990, pp. 452–464) that it is unnecessary to repeat them here. Worth noting, however, is the fact that slave-making ants employ a very different strategy for penetrating host colonies to any other known social parasite of ants. Whereas the vast majority of the latter employ some combination of evasion, stealth or mimicry to enter host nests, slave-makers are well armoured and solicit aggression by attacking their victim's workers, whilst releasing powerful alarm pheromones that cause the defending ants to panic and attack one another. This locks up much of the colony in combat, diminishing the protection otherwise available to the potential slave brood. Unlike typical alarm pheromones, which fleetingly trigger a wave of alarm across a colony, those of slave-making species are longer-chained, less volatile, and prolong the confusion for tens of minutes. Until recently, this strategy was thought to be unique to slave-making species. However, another striking example of convergent evolution was recently described in the parasitoid wasp *Ichneumon eumerus*, which oviposits in the final instar of the socially parasitic butterfly *Maculinea rebeli* inside *Myrmica* colonies. Whereas *M. rebeli* employs close mimicry to inhabit the brood chambers of its host, *Myrmica schencki*, the heavily sclerotized *I. eumerus* adults induce fierce fighting in *M. schencki*, immobilizing up to 80% of a colony and leaving the *M. rebeli* larvae temporarily unprotected (Thomas and Elmes, 1993). Interestingly, *I. eumerus* releases a cocktail of similar chemicals to those of slave-makers – some compounds attract the ants to the wasp to amplify contact, some generate extreme aggression, and others repel the ants so that they mainly attack each other – but they are even longer-chained and more persistent, allowing the single parasitoid hours or days to seek and attack its host (Thomas *et al.*, 2002).

The third route to social parasitism of ants by ants is very different. We have long argued that surplus β queens in polygynous ant colonies could, if not closely related to the α queens, result in conflicts of interest in which low-caste queens might benefit from cheating by producing sexual eggs. In this way surplus queens might be considered as social parasites of their own worker force (Buschinger, 1970, 1986, 1990; Elmes, 1974, 1976, 1978). With local reproductive isolation, this might eventually lead to the sympatric evolution of permanent parasitic queens within their own nests, before spreading to infest other colonies of the same species. Elmes (1978) illustrated a hypothetical evolutionary pathway using social parasites of *Myrmica* ants (reviewed by Radchenko and Elmes, 2003). Starting with highly polygynous populations producing miniature but fully functional microgynes that confer ecological advantages on their colonies (represented by the microgyna form of *M. ruginodis*; see Brian and Brian, 1951), non-functional microgynes, such as those found in some *M. rubra* colonies, might be produced, especially when there is no longer an ecological advantage for microgyny. Elmes (1978) envisaged that microgynes of *M. rubra*, which are morphologically similar to normal queens but which produce only sexuals and pseudogynes, were the first step in the evolution of a true social parasite; indeed, these were described later by Seifert (1993) as a separate species, *M. microrubra*. This might lead to specialized species, such as *Myrmica karavajevi* that has several host species, via intermediate host-specific forms such as *Myrmica hirsuta*, which is found occasionally (and only) in *M. sabuleti* colonies. Although the idea of rapid sympatric speciation contradicted the belief in the universality of slow allopatric speciation (e.g. Mayr, 1963), the concept is finding increasing favour (e.g. West-Eberhard, 1981; Hölldobler and Wilson, 1990). Recently, the hypothesis was given empirical support by Savolainen and Vepsäläinen (2003), who found that *M. microrubra* are most closely related to, and probably derived from, their parent populations, while *M. hirsuta* are close to their host but more similar to each other, possibly indicating a single speciation event.

The fourth route to social parasitism is the focus of our recent work (Thomas *et al.*, 1998a; Elmes *et al.*, 1999). It involves arthropods other than ants which, by definition, cannot fulfil Emery's rule. On the contrary, the same ant genus or species may be host to a diversity of convergent lines from very different insects. For example, we have found the Palaearctic ant *Myrmica scabrinodis* parasitized by *Maculinea teleius* and *M. alcon* (Lepidoptera), by *Microdon myrmicae* (Diptera) and by *Atemeles emarginatus* (Coleoptera) as well as by *Myrmica karavejevi* and *Myrmica vandeli*; indeed, in certain hotspots we have found *Maculinea teleius* and *Microdon myrmicae*, or *Microdon* and *Atemeles*, or *Atemeles* and *M. hirsuta* parasitizing the same individual nest.

Most of the pioneering studies of social parasites were concerned with behavioural interactions between host and parasite, including *Atemeles* and other staphylinid beetles (Wasmann, 1912; Donisthorpe, 1927; Hölldobler, 1967), *Maculinea* butterflies (Chapman, 1920; Frohawk, 1924), and *Leptothorax* ants. Ecological studies of social parasitism are much rarer, with the genus *Maculinea* providing the only examples of studies on the population dynamic interactions between a social parasite and its host, and of the ecological consequences of these (e.g. Thomas, 1980; Thomas *et al.*, 1998a; Elmes *et al.*, 1999). These studies – which were initiated to answer the nature-conservation question 'Why are most social parasites rare?' – showed that host specificity was a key factor for the survival of the social parasite (e.g. Thomas *et al.*, 1989; Elmes *et al.*, 1999; Schönrogge *et al.*, 2002).

4. Host Specificity in Social Parasites

4.1 Why are social parasites rare?

Despite many descriptions of the fine-tuning of specializations and the reconstruction of plausible evolutionary pathways leading to social parasitism (Hölldobler and Wilson, 1990), a key ecological question remains: why – when equipped with such successful mechanisms to trick and exploit ants, and when listed hosts are often among the commonest species in their zoogeographical regions – are most social parasites exceedingly rare in comparison to their hosts? Why, for example, does the UK *Red Data Book* (RDB) for insects list a disproportionately high proportion of social parasites from all orders among its most threatened species (Thomas and Morris, 1994)? Why did Heath's (1981) RDB for butterflies include all five socially parasitic species among only 15 species listed in the rarest category for Europe? Why do the 13 independently evolved lines of socially-parasitic lycaenid butterflies figure so prominently in the IUCN world list (Pierce *et al.*, 2002)? Why have we and our co-workers (A. Radchenko, personal communication), having excavated several thousand western Palaearctic *Myrmica* colonies in careers spanning 25–40 years, encountered ant social parasites, including microgynes, so infrequently? For example, *Myrmica hirsuta* appears to live only on sites with very high densities of *M. sabuleti*, where it infests < 1% of host nests (Elmes, 1994); and we have found *M. karavejevi* in only three localities, despite its host, *M. scabrinodis,* being arguably the commonest *Myrmica* species in western Europe. Other field myrmecophilists describe similar experiences (see e.g. Hölldobler and Wilson, 1990), and we give further examples later (see *Microdon mutabilis, M. myrmicae*).

Few infestation patterns have been quantified in the field; nevertheless, we tentatively recognize three population structures among social parasites:

1. High infestation, low transmission. The five European species of *Maculinea* butterfly and the hoverflies *Microdon mutabilis* and *M. myrmicae* typically live in discrete, closed, high-density populations, where they infest – and may destroy – up to 30–40% of host ant colonies in each generation (Thomas and Wardlaw, 1992; Thomas, 1995; Schönrogge *et al.*, 2002; our unpublished data). However, their populations are typically restricted to small (< 10–< 0.1 ha) patches that are occupied for many generations, whilst elsewhere in the landscape their hosts may be widespread (sometimes ubiquitous), resulting in fewer than 1 in 10,000 host nests being infested at this scale (our calculation).

2. Low infestation, high(ish) transmission. *Myrmica microrubra* (= *M. rubra* microgynes *sensu* Elmes, 1976), although more frequent in certain landscapes than others, are found widely throughout their host's distributions, but within few nests on any one site, to which little cost is apparent. Despite their sympatric origins (Savolainen and Vepsäläinen, 2003), we envisage that *M. rubra* microgynes spread from one host (super-)colony to infest neighbouring colonies in each region. If confirmed, they would be superior dispersers but poorer invaders of new host colonies than *Maculinea* and *Microdon*. *Anergates atratulus* (Hölldobler and Wilson, 1990) and, in our experience *Atelemes* spp., have a similar population structure.

3. Low infestation, very low transmission. In theory, this represents an evolutionarily unstable strategy, at least at the very low infection rates described. Nevertheless, in our experience, several European social parasites, including *Myrmica hirsuta* and *M. karavejevi*, exhibit this type of infestation, as may many other advanced species of social parasite. For example, Hölldobler and Wilson (1990), in a different context, comment thus on the incidence of ant–ant social parasites: 'The great majority of workerless [= highly integrated] parasites have been found at only one or two localities, and are extremely difficult to locate even when a deliberate search is made for them in the exact spots where they were first discovered. Usually they give the impression, quite possibly false, that they have no more than a toehold on their host populations and that they exist close to extinction.'

We addressed this conundrum first in a category 1 (high infestation–low transmission) species, the butterfly *Maculinea arion*, and attributed its rarity to a combination of factors: considerably greater host-ant specificity than had hitherto been realized, plus the need for two sequential hosts to coexist, the first (initial food-plant) sufficiently widely distributed to infest host ant colonies, the second (*Myrmica sabuleti*) in sufficient local abundance (> 55% of biotope foraged) to support a closed population of this sedentary species (Thomas, 1980, 1984a,b, 1991, 1995; Thomas *et al.*, 1989, 1998a; Elmes and Thomas, 1992; Hochberg *et al.*, 1992, 1994; Thomas and Wardlaw, 1992; Elmes *et al.*, 1994; Clarke *et al.*, 1997). More generally, Pierce *et al.* (2002) considered social parasitism to be an evolutionary dead-end within the Lycaenidae, resulting in such extreme specialization that species were unable to track rapid ecological perturbations. While this is essentially the argument advanced by Thomas (1980, 1984a, 1991) for the decline of *M. arion*, it does not explain why most social parasites were so rare in the first place, including in the ecologically stable tropics (e.g. Hölldobler and Wilson, 1990). But nor did our or co-workers' arguments concerning *Maculinea* explain the equal or greater rarity of the majority of other species that parasitize one common ant species without the constraint of a second host.

We will argue that even category 3 (low infestation–very low transmission) species may be successful exploiters of their hosts, but in order to succeed they have (co-)evolved such close adaptations to a local host genotype that, in extreme cases, each population may be restricted to exploiting it at the level of a single supercolony of ants.

4.2 Hypotheses to explain variation in the strength of host speciality in social parasites

We suggest two hypotheses that may help to explain variation in host specificity. Both are drawn partly from theory and partly from detailed knowledge of the functional biology of six social parasite species (described in detail in Section 6) and of a wider range of mutualistic myrmecophiles that has been studied in recent years.

Hypothesis 1

The deeper a social parasite penetrates towards the most protected and resource-rich niches within an ant society, the closer will it become adapted to or integrated with that society (Thomas and Elmes, 1998).

For this hypothesis we identify five zones (Fig. 17.1), four within the nest of a polygynous ant society, which are protected with increasing ferocity: foraging range/territory (score 1); outer nest/midden areas (score 2); outer nest chambers inhabited by β queens (score 3); main brood chambers (score 4); and inner chambers inhabited by α queens (score 5). Others exist – notably foraging columns which, although outside the nest, are well defended, and niches or chambers within or without nests where trophobionts (domestic aphids, coccids) or fungi are nurtured or seeds are stored – but five will suffice. Except under the most benign ecological conditions, when the cost of fighting outweighs the gain (e.g high worker:brood ratio, low queen:worker ratio, neither food nor space limiting), polygynous ant societies attack intruding adults (non-kin workers, queens) of their species if they enter nest zones 3–5, often fighting to the death (Winterbottom, 1981; Brian, 1988). Greater tolerance is shown to the larvae of non-kin colonies or other congeneric species, which may be nursed and fed if introduced to brood chambers. However, they are the first to be butchered and fed to the conspecific brood when

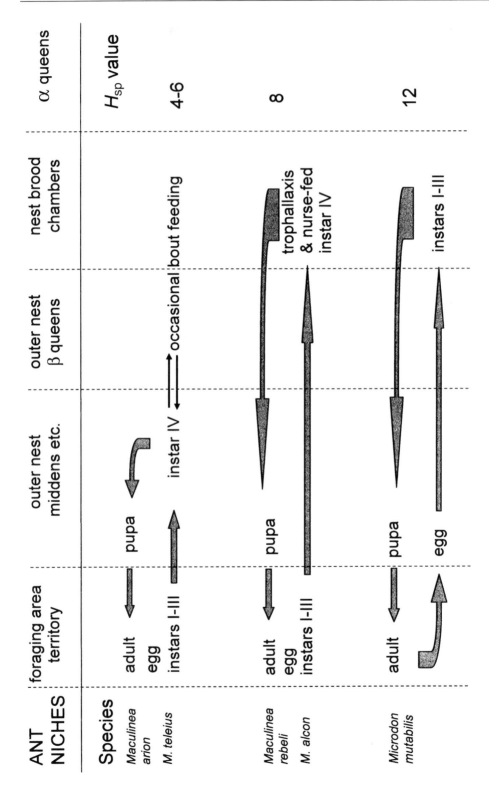

Fig. 17.1. Ant niches inhabited at different stages of the life cycle by three types of social parasite.

the colony experiences food shortages; further starvation results in the sacrifice of kin eggs, followed by kin larvae from the smallest upwards, to feed the largest of their sibling larvae and ultimately just the workers (Gerrish, 1994; Elmes *et al.*, 2004).

We suggest that any social parasite that depends solely on mimicking ant brood must achieve a closer match even than non-kin ants to its host's recognition codes, or must mimic some attribute of a high-ranked member of the society, if it is to gain sufficient status to induce direct feeding by workers ahead of their own kin larvae. Unless strong reinforcing cues are involved (e.g. mimicry of adult, especially queen, signals), we find no evidence that the mere acquisition of a host colony's 'gestalt' recognition odour (chemical camouflage *sensu* Dettner and Liepert, 1994; Vander Meer and Morel, 1998; Lenoir *et al.*, 2001) can deliver competitive supremacy over hungry kin brood; instead, recent studies suggest that the biosynthesis of copious mimetic allomones is involved (Elmes *et al.*, 2004; Schönrogge *et al.*, 2004). We restrict this deduction to the many social parasites, exemplified by *M. rebeli* (see Section 5.2), that divert substantial amounts of food from ant larvae in order to grow from juveniles to adulthood. Although mouth-to-mouth exchange of regurgitated food or secretions (trophallaxis) is an important part of this feeding, the exchange of small beads of liquid – which is mimicked by many adult social parasites — between the mouths of worker ants to reinforce social acceptance is another matter: this deprives the host colony of negligible food and we have no evidence that it leads to conflict.

An alternative (or additional) strategy evolved by many other social parasites is to combine exceptionally well-armoured cuticles with evasion and stealth. By definition, the latter two methods of infiltration are incompatible with direct feeding by worker ants. The capability of social parasites to deploy armour during different life stages led us to formulate Hypothesis 2.

Hypothesis 2

> The younger the stage of social parasite that exploits a host ant society, the more likely it will be closely adapted – and hence host-specific – to one host.

Here we divide social parasites into three periods of their lives, listed in the decreasing facility with which they can evolve both armour against attack and secretory organs to communicate with ants: adults of all orders plus the nymphs of Hemimetabola (score 1); pupae and late instars of Holometabola (score 2); young instars (and eggs) of Holometabola (score 3).

To elaborate, the adults of many social parasites are heavily sclerotized and able to withstand modest ant attacks, enabling them to penetrate colonies of > 1 ant species, especially if they move stealthily and acquire a weak camouflage of their host's gestalt odour. (Free-living adults that emerge from pupae in ant nests may use other devices during their brief escape from the colony, such as rewarding workers with secretions or employing loose sticky wing scales to baffle them; Pierce *et al.*, 2002.) The evolution of sclerotized cuticles appears also to be an adaptable trait in the nymphs of hemimetabolous insects; yet despite their preponderance among the domesticized mutualists of ants, remarkably few true social parasites are from the hemimetabolous orders of insect, *Myrmecophila* cricket species (Wasmann, 1901; Hölldobler, 1947; Akino *et al.*, 1996) being an obvious exception. This defence is not available to the larvae of holometabolous social parasites, whose cuticles remain flexible enough for continuous growth. We divided holometabolous myrmecophiles into two categories of vulnerability: the later instars that are often large enough to develop thick 'rubbery' cuticles that withstand nips by ants, though not concerted attacks (Malicky, 1969), and the first instars which, from the ants' perception, resemble thin-skinned packages of food conveniently located within their nest. Insect eggs, of course, can be heavily armoured, but hatching larvae take 2–3 days to bore out of the thickest-shelled examples, during which they are presumably vulnerable to ant attack (Thomas *et al.*, 1991). We know of no social parasite that oviposits into ant nests which has invested in thick-shelled eggs, although the unusually clumped population structure inherent in cuckoo *Maculinea* species has led to the evolution of the thickest chorions described for any (including overwintering) species of Lepidoptera, presumably to negate *Trichogamma* parasitoid above ground on their initial food-plants (Thomas *et al.*, 1991).

4.3 Ecological consequences of Hypotheses 1 and 2

Selection for greater host specificity (H_{sp}) due to increased penetration (P) or exploitation at a younger life stage (L_{stage}) seem unremarkable concepts in isolation. However, we suggest that each functions independently and that the net level of host specificity found in a myrmecophile will be described by a function of both variables:

$$H_{sp} = f(P).f(L_{stage}) \tag{1}$$

Although the functional dependencies are unknown, to illustrate the use of such an index we assume H_{sp} to be the simple product of P and L_{stage} from here on. Using the arbitrary scores given above to each category in Hypotheses 1 and 2, we obtain a highest value for host specificity of 15 (a social parasitoid that exploits or outcompetes resident α queens during its earliest instars) and a lowest value of 1 (an adult myrmecophile inhabiting the foraging range of an ant colony). In fact, we know of no myrmecophile that can outcompete α queens; slave-making ants that kill them are a different matter (Foitzik and Herbers, 2001). However, we predict that a holometabolous species of social parasite living as a young larva in ant brood chambers (H_{sp} value of 12) will evolve such close adaptation to (especially mimicry of) its host colony that it may be restricted to exploiting its host at a subspecific scale, rendering the parasite effective but rare. On the other hand, robust adult social parasites and perhaps the shallow-penetrating nymphs of *Myrmecophilla* crickets (H_{sp} values 3–4), should be able to exploit ants by employing more generalist, less restrictive adaptations, such as acoustic or tactile signalling (Hölldobler, 1947) and acquired 'camouflage odour' gained through exposure to the host colony, that enable them to eat less-valued items in nests but ill-equipped to solicit large quantities of food at the expense of the host's brood.

We do not expect these hypotheses to cover all variations in host specificity, but hope they may supplement the existing rules explaining patterns in social parasites. The population ecology and host specificity of many commensals and mutualists is known well enough from the field to test formula (1) on low-scoring (H_{sp} = 1–2) myrmecophiles (see Section 6) (Thomas, 1983, 1992a,b; Thomas and Lewington, 1991; Pierce *et al.*, 2002), but only six (out of perhaps 20,000) ant social parasite species have been studied in similar detail, of which five belong to one genus, *Maculinea* butterflies. Moreover, these six systems are not independent of our hypotheses, which were framed with some hindsight of their biology. Nevertheless, the six species encompass three very different types of system, with H_{sp} scores ranging from 4 to 12 (Fig. 17.1). Since many other social parasites exhibit strong convergent evolution in form towards one or other of these types, we tentatively suggest there may be a similar convergence in function.

In Section 5 we describe case histories of how one or two examples function, inasmuch as is known, in each of the three model systems.

5. Case Histories of Host Specificity and the Degree of Social Integration in Three Social Parasite Systems

5.1 *Maculinea arion* and *Maculinea teleius* (Lepidoptera, Lycaenidae)

High-virulence, low-transmission predators of ant brood, which enter host colonies in their final larval instar and mainly inhabit the outer nest chambers

The biology of these butterflies is similar (see Fig. 17.1) apart from their use of different food-plants and *Myrmica* ant hosts. Both are low-scoring (H_{sp} 4–6), comparatively unspecialized social parasites with two sequential larval hosts: a specific initial food-plant (*M. arion*: *Thymus* spp. or *Origanum vulgare*; *M. teleius*: *Sanguisorba officinalis*) followed by parasitism of *Myrmica* ants in the final instar. Like all *Maculinea* species, the adults live above ground in mid–late summer and show no interaction with ants after eclosion from pupae in the outer cells of *Myrmica* nests, from which they emerge while the ants are quiescent in the early morning. Van Dyck *et al.* (2000) and Wynhoff (2001) suggest otherwise, but we find no evidence for ant-mediated oviposition in any *Maculinea* species (Thomas, 1984b, 2002; Thomas and Elmes, 2001), although *M. teleius* and *M. nausithous*, which share *S. officinalis* as a food-plant and often coexist on the same site, oviposit preferentially on different growth forms of *Sanguisorba* that coincide with the optimum ecological niches of their respective ant host species (Thomas, 1984b; Thomas *et al.*, 1989, 2002).

M. arion and *M. teleius* eggs are laid between flower buds (Thomas *et al.*, 1991) on which the first three larval instars feed. Each develops quickly for 3 weeks but gains little weight, following a shallow growth trajectory (Elmes *et al.*, 2001). The fourth and final instar inhabits a *Myrmica* ant nest, where it lives for at least 10 months and acquires 98–99% (*M. arion*) or 92–93% (*M. teleius*) of its ultimate biomass without further moults (Thomas and Wardlaw, 1992). Other known *Maculinea* species show the same striking deviation from Dyar's law of constant insect growth between instars, and consequent loss of allometry between head and body size, a deviation that has evolved independently in *Lepidochrysops*, another genus of lycaenid social parasites that feeds initially on plants (Elmes *et al.*, 1991a,b, 2001). This is perhaps explained by selection for small size at the time of entry into ant nests, to facilitate initial mimicry of ant larvae and transport by workers, balanced by the development in the (much larger) final instar of mutualist lycaenids – from which these social parasites clearly evolved (Fiedler, 1998; Pierce *et al.*, 2002) – of high densities of myrmecophilous glands and, perhaps crucially, of cuticles about 20 times thicker than in typical non-myrmecophilous lepidopteran larvae (Malicky 1969; Elmes *et al.*, 2001).

After the third and final moult, the 1–2 mg (*M. arion*) caterpillar remains on its food-plant until about 18.00 h, when it drops 2–3 cm to the ground and awaits discovery by an ant. This places it in the optimum foraging niche of all *Myrmica* species at their time of peak foraging, enhancing its chances of discovery by a *Myrmica* worker rather than by those of other genera by > 200-fold (Thomas, 2002). Early reports that *M. arion* (Frohawk, 1924) and *M. teleius* (Schroth and Maschwitz, 1984) caterpillars actively seek host ant nests before adoption derive from experimental artefacts (Thomas, 1984a,b, 2002; Fiedler, 1990); instead, the larvae await discovery by a *Myrmica* worker within 1–2 cm of their food-plant. The classic (Frohawk, 1924; Purefoy, 1953) hour-long adoption interaction between *M. arion* and *Myrmica* then ensues, when the caterpillar's dorsal nectary organ (DNO) and secretory pores are constantly 'milked' by the worker and the caterpillar perhaps acquires some of the ant's gestalt odour (Thomas, 2002). Eventually, the caterpillar rears up, probably mimicking a *Myrmica* larva (Cottrell, 1984), which causes the *Myrmica* worker to carry it into the ant nest and place it among the brood. Adoption of *M. teleius* appears identical (Thomas, 1984b; Fiedler, 1990).

M. arion caterpillars quickly leave the brood to spin a silk pad on the roof of an outer chamber, where they rest for most of the next 10 months. In form, they possess the typical flattened body of the final instar of a mutualistic lycaenid, which conceals the mouth and other vulnerable parts (Fig. 17.2a). Most other adaptations of mutualists are recognizable, apart from tentacle organs that agitate ants, but modified for social parasitism. Thus the cuticle of *M. arion* is even thicker than that of a typical lycaenid; the churring sounds produced by all *Maculinea* species differ greatly from those of typical lycaenids and mimic the adult stridulations of the genus *Myrmica*, but not any particular *Myrmica* species (De Vries *et al.*, 1993); and the secretions, which are abundant, coat the body in a cocktail of hydrocarbons that mimic the recognition signals of *Myrmica* (K. Schönrogge, E. Napper, G.W. Elmes, J.A. Pickett and J. A.Thomas, unpublished).

M. arion and *M. teleius* are obligate predators of ant brood that rely much on armour, evasion and stealth to exploit *Myrmica* colonies. *M. arion* feeds

a.   PREDACIOUS SPECIES (e.g. *M. arion*)   CUCKOO SPECIES (e.g. *M. rebeli*)

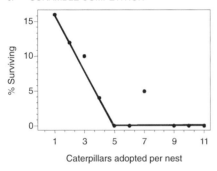

b.   INEFFICIENT
     1 butterfly per 250 ants (laboratory)
     1.2 butterflies per ant nest (field)

EFFICIENT
1 butterfly per 50 ants (laboratory)
5.6 butterflies per ant nest (field)

c.   SCRAMBLE COMPETITION

CONTEST COMPETITION

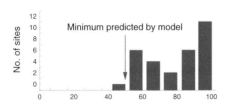

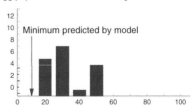

d.   Erratic 1-year growth

1 and 2 year developers

e.   Destroys ant colony

Weakens ant colony

f.   3–6.5 higher survival with host *Myrmica*

24-∞ higher survival with host *Myrmica*

g.   Minimum coexistence with host ants for
     population to persist

Model: high (55%)

low (7%)

Field: % egg population in host ant's range

% egg population in host ant's range

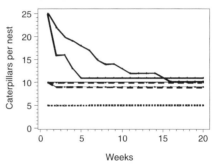

h.   Small/erratic populations

large/stable populations

i.   Host-specificity

can use host ant populations widely
across Europe

adapted to local host ant population in Europe:
restricted range
local speciation?

Fig. 17.2. A comparison of the traits, costs and benefits in two systems of social parasitism –
predatory and cuckoo feeding – within the genus *Maculinea*.

erratically, gliding slowly to the brood chamber, then briefly 'binge feeding' on large larvae before returning to its pad where it may rest for up to 10 days digesting the meal. *Myrmica* colonies are small compared to those of many ants, but caterpillars use this resource in the most efficient way possible for a predator of larvae (Thomas and Wardlaw, 1992). By eating the largest available prey, they initially kill only those larvae that will soon pupate and be lost as food. At the same time, the fixed number of larvae in the second (overwintering) cohort of *Myrmica* brood is left to grow larger before it is killed. When this occurs, large individuals are again selected, leaving small ones to grow on. With the range of colony sizes available on a typical site, Thomas and Wardlaw (1992) calculated that this increased the carrying capacity of nests by 124%, although for 2 weeks after adoption it incurs a cost of increased mortality in nests that contain queen ants, due to the probability of contamination by gyne pheromones from the largest larvae, while the caterpillars grow from their mean adoption weight of 1.3 mg to 8 mg (Thomas and Wardlaw, 1990). It is perhaps to reduce this vulnerable period that *M. teleius*, for which no queen effect is reported (E. Figurny, 2000, personal communication), enters ant nests at three times the size (mean weight 4.3 mg) of any other *Maculinea* species.

M. arion also gains an estimated 31% ergonomic advantage by refraining to feed for several weeks after its host breaks hibernation in spring, during which the caterpillars lose about 6% in weight while their food supply (ant larvae) grows by 27% (Thomas and Wardlaw, 1992). An ability to starve or fast during normal growth periods appears to be a third ergonomic adaptation. Large caterpillars, which exhaust their food supply in spring, exert such low attraction over the broodless workers that the colony frequently deserts, leaving the social parasite behind. In due course, an offshoot of this polydomous ant genus colonizes the vacant nest, importing a fresh supply of ant larvae. Thomas and Wardlaw (1992) estimated that this occurs in 80% of the nests that successfully produce a *M. arion* adult. Nevertheless, overcrowding is a major cause of mortality in *M. arion*, which experiences steep scramble competition, generally resulting in 100% mortality when > 4 larvae are adopted into a single *Myrmica* nest (see Fig. 17.2c, taken from Thomas and Wardlaw, 1992).

Before our studies it was believed that *M. arion* was equally successful as a parasite of any *Myrmica* species, and some suggested that other ant genera, including *Lasius*, were secondary hosts (e.g. Ford, 1945; Hinton, 1951). We found that although oviposition was indiscriminate and caterpillars were adopted with equal facility into any of up to five *Myrmica* species that foraged under the initial food-plant, the mean survival of *M. arion* in the field was 6.4 times higher in colonies of *M. sabuleti* than with any other *Myrmica* species (Fig. 17.3; Thomas, 1977, 1980, 2002; Thomas *et al.*, 1989, 1998a; Thomas and Elmes, 1998), perhaps because they secrete species-specific mimetic allomones while invading the brood chambers to feed (see *M. rebeli*). Similarly, *M. teleius* survival was 2.9 times higher with *M. scabrinodis* (which quickly kills *M. arion*) than with any other *Myrmica* species that adopted it in the field (Fig. 17.3; Thomas *et al.*, 1989; Elmes *et al.*, 1998; Thomas and Elmes, 1998). Models based on 6-year life tables and natality measurements predict that a

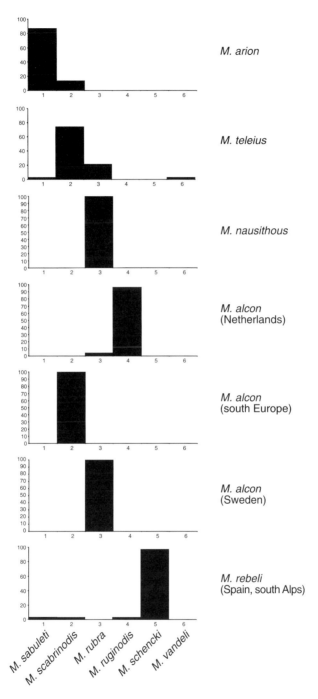

Fig. 17.3. Host specificity in five species of *Maculinea* butterfly, with data for the cuckoo species *M. alcon* presented for populations from three regions of Europe. Data are shown as the proportion of individuals surviving (total = 100) in each of six ant species that commonly adopt *Maculinea* larvae (from Thomas *et al.*, 1989; Thomas and Elmes, 1998; unpublished results).

population of *M. arion* species will persist (λ > 1) only if > 55% of the larval
population is adopted by *M. sabuleti*, and declines rapidly to extinction if other
(secondary) host species predominate (Fig. 17.2g; Thomas, 1991; Thomas *et
al.*, 1998a): field data from across Europe support this prediction (Fig. 17.2g;
Thomas, 1995, 1999; Thomas and Elmes, 1998; Thomas *et al.*, 1998a). In
other words, although individual *M. arion* occasionally survive with 'non-host'
Myrmica species, it is species-specific at the population level.

There is no evidence, however, of subspecific adaptation by predatory
Maculinea species. On the contrary, following extinction in the UK and the
Netherlands, *M. arion* and *M. teleius* have respectively been re-established using
donor populations from Sweden and Poland (Thomas, 1995; Wynhoff, 2001).
Moreover, observed population dynamic interactions between Swedish *M. arion*
and UK *M. sabuleti* colonies on six sites have closely followed model predictions
of annual fluctuations and equilibrium levels for up to 22 generations, despite
parameterization from interactions with old UK populations of *M. arion*
(Thomas, 1995, 1999, unpublished results). This suggests that neither local
adaptation by the social parasite, nor local resistance by the host, nor a
population level co-evolutionary arms-race between both partners have evolved
to any significant extent in these predacious *Maculinea* systems.

5.2 *Maculinea rebeli* and *Maculinea alcon* (Lepidoptera, Lycaenidae)

*High-virulence, low-transmission cuckoo-feeders in ant brood chambers,
which enter host colonies in their last larval instar*

Like the other *Maculinea* (see Fig. 17.1), these close relatives oviposit on
specific food-plants (*M. rebeli*: xerophytic *Gentiana* species, mainly *G. cruciata*;
M. alcon: wet grass or heathland *Gentiana*, mainly *G. pneumonanthe*), but the
lifestyle diverges radically after abandoning their food-plant. Caterpillars are
again retrieved by any *Myrmica* species to encounter them beneath gentians,
but are picked up and carried into the ant's nest within 1–2 s of being found,
without the elaborate interactions of *M. arion* or *M. teleius* (Elmes *et al.*,
1991a). Deep inside the nest they inhabit brood chambers and become
intimately integrated with the host society, receiving frequent grooming and, on
begging (Fig. 17.2a), being fed by the nurse ants with trophic eggs, solid food
and by trophallaxis (cuckoo feeding). For a month caterpillars may also eat
some ant brood, but have little direct impact on numbers: thereafter, owing to
Myrmica nurses' segregation of all brood by size, the > 10 mg caterpillars are
kept apart from the brood and fed entirely by adult ants (Elmes *et al.*,
1991a,b). As they grow, their morphological resemblance to a predatory or
phytophagous lycaenid diminishes, for they develop thinner-skinned cylindrical
bodies (Fig. 17.2a) with very high densities of secretory pores (Malicky, 1969).

Two atypical populations apart (Als *et al.*, 2001, 2002), both cuckoo
species exhibit much higher levels of host specificity at the individual and
population level than their predacious congeners (Figs 17.2f, 17.3; Thomas
and Elmes, 1998). Working mainly with *M. rebeli* from the Pyrenees and

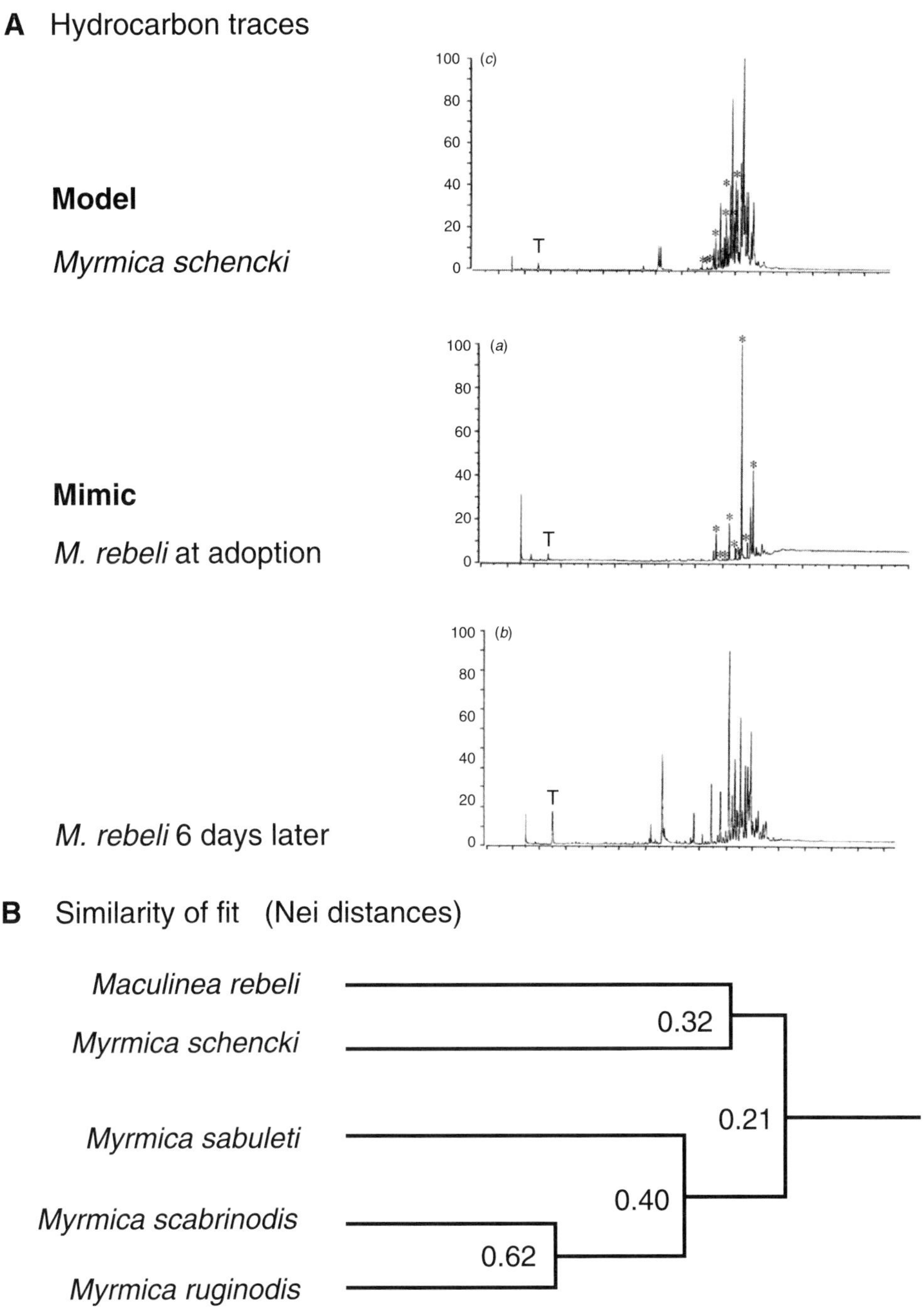

A Hydrocarbon traces

Model

Myrmica schencki

Mimic

M. rebeli at adoption

M. rebeli 6 days later

B Similarity of fit (Nei distances)

Fig. 17.4. Chemical mimicry of *Myrmica schencki* by *Maculinea rebeli* larvae. **A** Hydrocarbon profiles of *M. rebeli* after leaving food-plant before exposure to ant (a); after 6 days with *M. rebeli* (b); and of the model, *M. schencki* larvae (c). **B** Similarity between hydrocarbon profiles of *M. rebeli* and five *Myrmica* species that commonly adopt larvae, including its natural host, *M. schencki* (from Akino *et al.*, 1999).

southern Alps, where *Myrmica schencki* is host (Thomas *et al.*, 1989), we have come to understand some of the processes whereby this system functions.

On leaving its gentian, *M. rebeli* secretes a simple mixture of surface hydrocarbons that weakly mimic those of *Myrmica* (Fig. 17.4A; Akino *et al.*, 1999; Elmes *et al.*, 2002). Its profile is closest to *M. schencki,* and vice versa (Fig. 17.4B), but is sufficiently similar to all *Myrmica* species for the caterpillar to be quickly retrieved by the first worker to touch it (Akino *et al.*, 1999). At this stage we doubt whether it is possible for a *Maculinea* larva to evolve a species-specific retrieval cue, since a *Myrmica* forager will gently retrieve the larva of any congener that is artificially placed in its territory (Gerrish, 1994).

On reaching the domain of the nurse ants, greater discrimination occurs. About 90% of caterpillars survive the first 48 h in the brood chambers of their closest model, *Myrmica schencki,* but only 45–60% survive on average with other 'non-host' *Myrmica* species (Fig. 17.5; Schönrogge *et al.*, 2004). Thereafter, individuals that pass this initial period of integration usually survive well with any *Myrmica* species, so long as the colony remains well-fed and benign. But if, as occurs often in the field (but seldom in the laboratory), the colony experiences food shortage or other stress, we find that *M. rebeli* continues to survive well with *M. schencki,* at the expense of the ant brood and workers, whereas its caterpillars are cut up and fed to the ant larvae in colonies of 'non-host' *Myrmica* species (Elmes *et al.*, 2004; Fig. 17.5).

How is such discrimination achieved? We knew already that after 6 days with *M. schencki*, *M.rebeli* possesses a more complex profile of surface hydrocarbons that closely matches its host (Fig. 17.4; Akino *et al.*, 1999). We originally

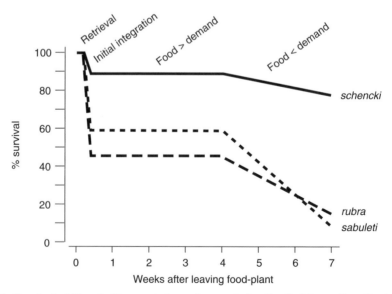

Fig. 17.5. Survival of *M. rebeli* larvae in four phases during the first 3 months after leaving its initial food-plant, when coexisting with three species of *Myrmica*: *M. schencki* (natural host, solid line); *M. sabuleti* (dotted line); *M. rubra* (dashed line) (from Elmes *et al.*, 1991a, 2004; Schönrogge *et al.*, 2004).

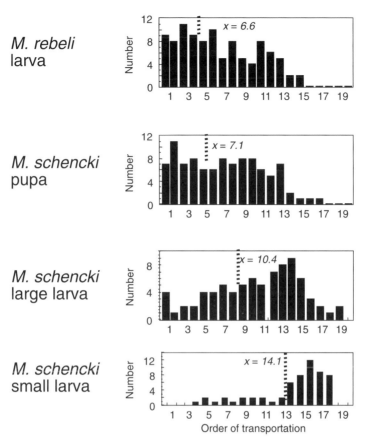

Fig. 17.6. The order within the hierarchy in *Myrmica schencki* societies in which *Maculinea rebeli* larvae are retrieved after perturbation, 6 days after entry to ant nest (from Thomas *et al.*, 1998b).

assumed that this resulted from the absorption by the caterpillar of background odours from the nest (chemical camouflage). This now seems naïve, for if *M. rebeli* can absorb the odour of one *Myrmica* colony, it should be equally successful at parasitizing any species after the initial period of integration.

It is now clear that *M. rebeli* caterpillars synthesize additional hydrocarbons after adoption that improve and amplify their mimicry of *M. schencki* and are thought to encompass cues from higher castes, possibly pupae or queens (Schönrogge *et al.*, 2004). Whatever the exact model, the effect is to elevate the status of the parasite in *M. schencki*'s hierarchy, so much so that, whereas *M. rebeli* is the last item to be retrieved by agitated nurses if a nest is disturbed within a few hours of the caterpillar's retrieval, the same caterpillars – although no larger than medium-sized ant brood – are among the first items to be rescued, alongside pupae and well ahead of large and small *M. schencki* larvae, if the disturbance is repeated 6 days later (Fig. 17.6; Thomas *et al.*, 1998b). *M. rebeli* behaves differently in the nests of non-host *Myrmica* species:

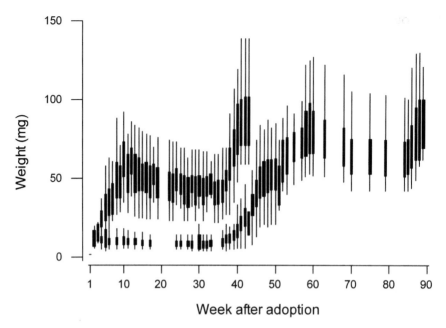

Fig. 17.7. Polymorphic growth (1- and 2-year development) by larvae of *Maculinea rebeli* in *Myrmica schencki* ant societies (from Thomas *et al.*, 1998b).

caterpillars switch off their ill-matching secretions and rely solely on absorbing the ants' background odour (Schönrogge *et al.*, 2004). This presumably results in a weak signal because their social status remains low, so much so that when the disturbance bioassay was repeated using apparently well-integrated 6-day Spanish *M. rebeli* in *M. sabuleti* nests, its caterpillars were the last items to be rescued (if at all) after perturbation (J.A. Thomas *et al.*, unpublished). It is not surprising, therefore, that they are first to be sacrificed when a non-host species is short of food (Fig. 17.5; Elmes *et al.*, 2004). Indeed, the hungry parasite may experience a double jeopardy because – presumably to win attention – it switches on its *schencki*-like secretions when starving, at exactly the time when the non-host *Myrmica* workers are most xenophobic (Elmes *et al.*, 2004; Schönrogge *et al.*, 2004).

We can now compare some costs and benefits of the predacious and cuckoo systems in *Maculinea* social parasites (Fig. 17.2). At first sight, the cuckoo species enjoy most benefits. By inducing direct feeding by worker ants (Fig. 17.2a), they jump the food chain compared with brood predators, increasing the capacity of a standard-sized *Myrmica* colony to support these 120–150 mg butterflies by four to fivefold (Fig. 17.2b; Elmes *et al.*, 1991a,b; Thomas and Wardlaw, 1992). Moreover, when caterpillars are overcrowded inside nests (a common event in the field; Thomas *et al.*, 1998a), the predacious species experience severe scramble competition, often resulting in 100% mortality despite their ability to draw a neighbouring colony into vacant nests (Fig. 17.2c; Thomas and Wardlaw, 1992). Cuckoo species, on the other hand, merely experience contest competition at high densities, because the nurse ants select the maximum number of individuals they can feed, which survive at the expense of the others (Fig. 17.2C; Thomas *et al.*, 1993). Moreover, both *M. rebeli* (Thomas *et al.*, 1998b) and *M. alcon* (Schönrogge *et al.*, 2000; Als *et al.*, 2002) have evolved a remarkable fixed polymorphism in larval growth that further smoothes density-dependent starvation: one-third of caterpillars grow quickly and pupate after 10 months in the ant nest, while two-thirds of individuals feed slowly, taking 22 months before pupation (Figs 17.2d, 17.7).

Moving from individuals to populations, the predatory species have an apparent benefit of higher (but not very high) survival in non-host *Myrmica* species (Fig. 17.2f), which theoretically provide refuges if local conditions become catastrophic for the host species (Thomas, 1999). Against this, in both theory and practice, cuckoo species can inhabit more sites within a region because, due to their efficient exploitation of and lower damage to hosts (Fig. 17.2f), they can persist in suboptimal patches where host densities are very low (Fig. 17.2g; Hochberg *et al.*, 1992, 1994; Elmes *et al.*, 1996; Clarke *et al.*, 1998; Thomas and Elmes, 1998; Thomas *et al.*, 1997, 1998b). Furthermore, the over-compensating effect of density-dependent 'scramble' mortalities means that populations of predatory species are not only smaller but also more erratic than those of cuckoo species, and hence more prone to extinction (Fig. 17.2h; Thomas *et al.*, 1998b).

The main cost of the cuckoo lifestyle is that increased specialization restricts each social parasite to a narrower, regional part of its host's range. We have seen in *M. arion* and *M. teleius* that any population of their respective host species seems suitable, at least across western Europe. However, local adaptations are apparent in *M. rebeli* and *M. alcon*. For example, *M. rebeli* from the Pyrenees is more (100%) host-specific to *M. schencki* than those from the Hautes-Alpes (J.A. Thomas *et al.*, unpublished), and host shifts occur over greater distances. Thus *M. rebeli* exploits *Myrmica sabuleti* from the north Alps eastwards (Meyer-Hozak, 2000; Steiner *et al.*, 2003), an ant with which it experiences 100% mortality in the Pyrenees, and in the central High Alps, *M. rebeli* apparently exploits *Myrmica sulcinodis* (Steiner *et al.*, 2003). Similar host shifts are found in *M. alcon* (Elmes *et al.*, 1994): in south and east Europe it exclusively parasitizes *Myrmica scabrinodis*, but from the Pays Bas northwards it exploits *M. ruginodis* (Belgium, Netherlands, Finland) or *M. rubra* (northern Finland, Sweden). Some of these represent substantial evolutionary shifts because – as with pathogens, true parasites and parasitoids – they involve not only a change of food but also of the social parasite's main enemy and environment. In our examples, *Myrmica schencki* has the most different hydrocarbon profile from *M. sabuleti* of any *Myrmica* species studied (Elmes *et al.*, 2002), and kills *Maculinea rebeli* from northeast Europe as quickly as *M. sabuleti* kills *M. rebeli* from southwest Europe (J.A. Thomas *et al.*, unpublished). Similarly, *M. scabrinodis* is chemically and behaviourally very different from *M. ruginodis* and *M. rubra* (which are rather similar) (Elmes *et al.*, 2002), and those *M. alcon* populations that exploit *M. scabrinodis* seem to be wholly incompatible with the latter two species, and vice versa for northern *M. alcon* (Elmes *et al.*, 1994; Thomas and Elmes, 1998; Als *et al.*, 2001, 2002).

These major functional types, each restricted to a different region of Europe, may be well down the path towards cryptic speciation.

Host-switching across *M. alcon*'s range has been studied in Denmark, straddling the boundaries of the *ruginodis* and *rubra* populations (Gadeberg and Boomsma, 1997; Als *et al.*, 2001, 2002). Here, Als *et al.* (2001, 2002) found isolated populations that were specialist exploiters of either *M. rubra* or *M. ruginodis*, and also two more generalist populations that survived with – but were less well adapted to – both ants on the same site (as elsewhere in northern Europe, *M. scabrinodis* remained unsuitable on them). Early genetic analyses of Danish *M. alcon* showed considerable variation between populations but no clear-cut pattern that matched host specificity (Gadeberg and Boomsma, 1997). It is too early to determine whether the two generalist populations are recent hybrids between *rubra*- and *ruginodis*-exploiting populations, surviving briefly in evolutionary time in optimal habitat (which results in less-stressed colonies) for both host ants, or whether they evolved from a 'pure' *rubra*- or *ruginodis*-exploiting population. We suspect the former due to the absence of other examples of generalist cuckoo populations, apart from on the boundary of this comparatively minor host shift. On the other hand, Nash *et al.* (2002) report patterns of chemical mimicry in *M. alcon* profiles that suggest local adaptation to individual host populations in Denmark, as well as variation in host ant profiles compatible with coevolutionary arms-races at the population scale.

Further (especially molecular genetic) research is clearly needed to elucidate the evolution of host specificity in *M. alcon* and *M. rebeli*,[1] but on current evidence we conclude that their specializations constrain regional populations of each butterfly to regional populations of a host.

5.3 *Microdon mutabilis* and *Microdon myrmicae* (Diptera, Syrphidae)

High-infestation, low-transmission predators in ant brood chambers, which enter host colonies as eggs

The approximately 350 described species of *Microdon* syrphid resemble most social parasites in having no larval resource other than social insects. *M. mutabilis* adults live a few days above ground, ovipositing in the outer edges of ant nests. The larvae migrate to inner brood chambers, where they live as predators for 2 years before pupating in the outer chambers (Fig. 17.1; Donisthorpe, 1927; Schönrogge *et al.*, 2000).

We studied the ecology of *M. mutabilis*, seeking to understand why this *Red Data Book* species is so rare – inhabiting the same very localized, typically < 0.1 ha sites for many generations – when its listed hosts include several of the most ubiquitous and abundant ants from three subfamilies in Europe (e.g. *Formica fusca*, *Lasius niger*, *Myrmica ruginodis*).

We found in the UK that *M. mutabilis* exploits one of two species, *Formica lemani* or *Myrmica scabrinodis* (another ubiquitous ant not previously listed as a host), but exclusively one host species per population. Knowledge of *Maculinea* suggested that a larval parasite of such different subfamilies would require specific adaptations that were unlikely to coexist in one species, and we duly found morphological differences in the young stages that unequivocally established the existence of two sibling species: *Microdon mutabilis* parasitizing *F. lemani* and the new *Microdon myrmicae* with *Myrmica scabrinodis* (Schönrogge *et al.*, 2002). This, however, failed to explain the extreme local distribution of either parasite in comparison to its host.

A chance remark by *Microdon* specialist Boyd Barr – that he successfully reared *M. mutabilis* only when eggs were introduced to the same colony from which their mother had emerged – led us to test the hypothesis that this social parasite had evolved adaptations to local genotypes of its host. Experiments supported this (Fig. 17.8), confirming 100% survival when eggs were introduced to the natal colony or an adjoining one, but usually 100% mortality in *F. lemani* colonies from only > 1–2 km away, even though all these 'distant' nests were themselves infested with *M. mutabilis* whose eggs, in turn, were killed in the reciprocal tests (Elmes *et al.*, 1999). This extreme host specificity functioned only through the maternal line. To eliminate the remote possibility of an artefact, a redesigned experiment was repeated by Karsten Schönrogge twice, including another locality: the result was the same except that mortality in most 'distant' nests was 65–80%.

We only partly understand how this remarkable system functions (and our detailed results have yet to be ratified by peer review). We know, nevertheless, that the same individual *F. lemani* colonies (about one-third of the total) within a site are infested for generation after generation by *M. mutabilis*. We know also that these successive infestations are by the same family lines: although females fly quite often, their dispersal is so low in the field that an

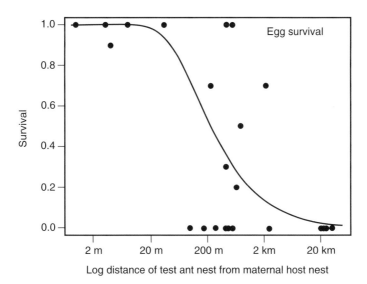

Fig. 17.8. Extreme host specificity in egg survival by *Microdon mutabilis* in colonies of its host ant, *Formica lemani*, at different distances from the natal nest from which the mother social parasite emerged (from Elmes *et al.*, 1999).

individual seldom departs more than 1–2 m from her natal nest, to which she returns to oviposit, while marked males patrol up to 10–30 m away (K. Schönrogge *et al.*, in preparation). This is consistent with a templet for selection for mimicry (or other adaptation) in *M. mutabilis* at the population or supercolony scale of its host. Less easily explained is why, given the substantial predation of host larvae by *M. mutabilis*, should any vulnerable genotype not be replaced by more resistant colonies of *F. lemani* long before local adaptation can evolve. We envisage – and have some evidence of – two processes resulting from the social manipulation of host colonies by *M. mutabilis*: one mechanical, one genetic.

As with *M. arion*, an inevitable consequence of 'cropping' one colony of a polydomous ant species in high-density, optimum habitat is that adjoining colonies bud to occupy the depleted nest site. With *F. lemani*, this might result in mixed 'colonies' within the same nest, in which case these heterogeneous colonies should be more tolerant of aliens especially if, due to the depletion of ant brood, the workers can easily obtain food for the survivors (Fig. 17.5). A further adaptation is required if the social parasite is to (co-)evolve very locally with one subset, or with a particular subspecific mixture, of its host genotype. Unlike *M. arion* caterpillars, which select the largest available brood, *M. mutabilis* (K. Schönrogge *et al.*, in preparation) and at least one congener (Duffield, 1981) prey exclusively on small ant larvae. In theory, this is the least efficient way for a predacious social parasite to exploit its host (Thomas and Wardlaw, 1992), and perhaps explains why *M. mutabilis* and *M. myrmicae* larvae take 2 years to pupation (Schönrogge *et al.*, 2000). On the other hand, the selective cropping of small larvae from an ant colony results in a high probability that the individuals that escape will switch in development to become gyne (queen-ant) females, due to the surplus food available from a worker population geared to feed many mouths (Brian, 1968, 1988). And indeed, we found twice as many queen pupae (but many fewer worker pupae) in colonies of *F. lemani* infested by *M. mutabilis* compared with similar-sized uninfested ones (K. Schönrogge *et al.*, in preparation). In other words, however complex the local genotype to which it is adapted may be, *M. mutabilis* has evolved a mechanism for maintaining – even exporting – rather than eliminating it. This, we believe, is a *sine qua non* for any social parasite that evolves specialization at a submetapopulation scale of its host.

6. Testing and Calibrating the Two Host-specificity Hypotheses

Our six case-histories, representing three systems for exploiting ants found commonly among social parasites, show variations in host specificity that contributed to the two hypotheses described in Section 4.2. Before testing each idea against the wider literature, we recognize that published host lists for other social parasites have almost certainly accumulated similar confusions and errors to the many that existed for *Maculinea* and *Microdon mutabilis* before detailed ecological studies began (see Table 17.1 for summary of types of error, and Thomas *et al.*, 1989, for a fuller critique). If so, host specificity will often be underestimated in other species, especially those that exhibit extreme (subspecific) host specificity at the regional scale but also host shifts across their global ranges: as Schönrogge *et al.* (2002) found with *Microdon*, this can result in a list of multiple hosts being published for certain social parasites, giving the false impression that the morpho-species is a generalist.

With that proviso, in Fig. 17.9 we summarize the published specialization of > 150 comparatively well-studied species of myrmecophile on a graph whose axes represent the two variables in our hypotheses. Current data are too crude to quantify the strength of a species' interaction with ants beyond placing it in one of the 12 classes (H_{sp} scores: minimum 1; maximum 12) that result from each possible combination of our four categories of colony penetration (x axis) and three categories of life cycle (y axis). For clarity, we have shifted the location of certain species within each 'box' to generate smoother patterns of host specificity, which we group into five classes ranging from myrmecophiles that interact with > 1 subfamilies of ant to those that specialize on a local population or supercolony.

We obtained patterns consistent with both hypotheses across a spectrum of myrmecophiles, from commensals to inquilines. At one extreme are the many free-living commensals and mutualists that typically penetrate ant colonies (if at all) only to pupate in the outer edges: these typically interact with > 1 subfamilies of ant. The mutualist *Plebejus argus*, on the other hand, inhabits the outer cells of nests from its first larval instar onwards, but does not exploit colony resources: its modest specializations restrict it to interactions with a single ant genus (black species of *Lasius*) (Jordano and Thomas 1992; Jordano *et al.*, 1992). At the other extreme, we have the microgyne queens of *M. rubra* and *Microdon mutabilis*, whose regional or individual populations appear to be adapted to very local populations, possibly individual supercolonies of a single host species. In between we find intermediate levels of host specificity among species of social parasite whose interactions are less intense. For the present, we have located the 59 ant–ant inquilines listed by Hölldobler and Wilson (1990) as being specific to a single species of ant towards the most specific corner of their 'box'; we predict that most of these will fall into higher categories of host specialization once their ecology is better known.

Atemeles pubicollis, the first and best-studied of all social parasites, presented an obstacle to our hypotheses. The classic work of W.M. Wheeler, Horace Donisthorpe, K. Hölldobler, Bert Hölldobler and others demonstrates that this staphylinid beetle sequentially exploits two subfamilies, *Formica polystena* and various species of *Myrmica*, during a life that is spent almost entirely with ants and includes trophallactic feeding of larvae inside brood chambers (Hölldobler and Wilson, 1990). We suspect that it is the exception that may help prove the rule. We suggest that *A. pubicollis* scarcely qualifies as a social parasite while it coexists with *Myrmica*, for this occurs only in the well-sclerotized adult stage which uses *Myrmica* nests merely as overwintering quarters. *Atemeles* undoubtedly possesses several attributes that enable it to live with *Myrmica*: stealthy movement, a gland that secretes generalist appeasement allomones, and the ability to absorb a colony's gestalt odour and to indulge in trophalaxic bonding with workers. However, it poses

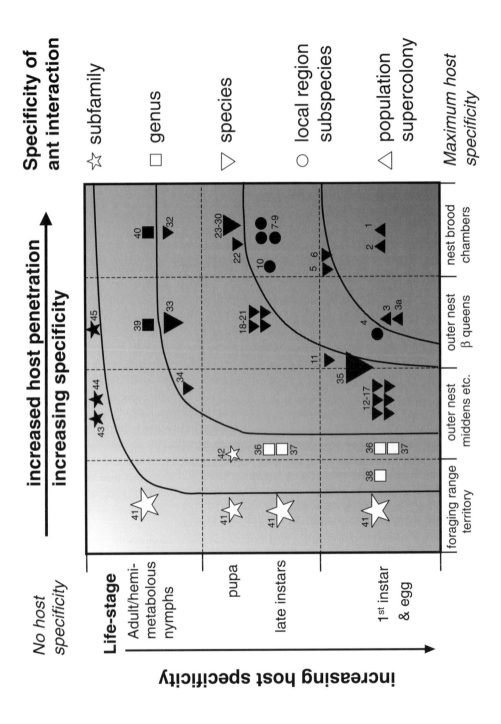

no significant threat to *Myrmica* resources and scores only 2–3 in this period in our rough H_{sp} guide, comparable to other adult myrmecophiles that coexist with ants from different subfamilies (Fig. 17.9). Its 'score' rises steeply to 12 for the vulnerable immature period, when adult females leave their hibernation nests to oviposit in *Formica* colonies, and when the *Atemeles* larvae, like cuckoo *Maculinea*, outcompete the ant larvae for worker attention within ant brood chambers. During this stage, *A. pubicollis* shows closer, albeit imperfect, agreement with our hypotheses in being restricted to one host species, *F. polyctena*, but not, on current knowledge, to local genotypes of its host. We may be guilty of data-fitting in our treatment of *Atemeles* in that we plot it twice, once representing its phase as a roving adult generalist, exploiting ants for protection and minor amounts of food, and again as a specialist immature social parasite adapted to grow from egg to pupa by tapping the food supply of a single host species (Fig. 17.9).

We tentatively conclude that both hypotheses apply independently across a range of myrmecophiles. While each hypothesis is unremarkable in itself, in certain combinations they may entail such extreme specialization for a social parasite to function successfully that it must evolve adaptations to its host at subspecific, population, or even supercolony scales. This inevitably places severe geographical constraints on the deepest-penetrating inquilines, explaining why most are exceedingly localized or rare. These extreme life forms may also evolve frequently but be subject to frequent extinctions.

Fig. 17.9 opposite. A provisional attempt to classify the host specificity of > 150 myrmecophiles with comparatively well-described ant associations according to the twin hypotheses that increased host specificity results from increased penetration of ant societies and by interacting with ants in early stages of the life cycle (homometabolous orders only). Species that interact with ants in different ways at different life stages are entered twice under each of the relevant life-history stages. Specialist myrmecophiles that join ant trails or attack ant-defended domestic herds (aphids, coccids etc.) outside the nest are omitted. Black symbols = social parasites (including ant–ant inquilines); open symbols = mutualists, commensals.

Sources: Wheeler, 1910; Donisthorpe, 1927; Malicky, 1969, 1970; Claassens, 1976; Cottrell, 1984; Pierce *et al.*, 1987, 2002; Fiedler and Maschwitz, 1988; Yamaguchi, 1988; Hölldobler and Wilson, 1990; Thomas and Lewington, 1991.

Key to species: [1]*Telectomyrex schneideri*, [2]*Microdon mutabilis*, [3]*Myrmica microrubra*, [3a]*Myrmica hirsuta*, [4]*M. karawevi*, [5]*Atemeles pubicollis* (juvenile), [6]*Lomechusa strumosa*, [7]*Maculinea rebeli*, [8]*M. alcon*, [9]*Microdon myrmicae*, [10]*Maculinea nausithous*, [11]*Spindasis takanonis*, [12]*Dentarda dontata*, [13]*D. markeli*, [14]*D. hagens*, [15]*D. pygmaea*, [16]*Thestor dicksoni*, [17]*Cremastocheilus* spp. (juveniles), [18]*Maculinea arion*, [19]*M. teleius*, [20]*M. arionides*, [21]*Allotinus apries*, [22]*Niphandra fusca*, [23–30]eight *Lepidochrysops* spp., [32]*Cremastocheilus* spp. (adults), [33]ten temporary ant–ant social parasite spp. (Hölldobler and Wilson, 1990, pp. 438–444), [34]*Hetaerius ferrugineus*, [35]59 spp. workerless ant inquiline spp. (Hölldobler and Wilson, 1990, pp. 438–444), [36]*Plebejus argus*, [37]*P. pylaon*, [38]*Jalmenus evagoras*, [39]*Claviger testaceous*, [40]*Myrmecaphobius excavaticollis*, [41]> 100 species lycaenid butterfly mutualists/commensals that interact with > 1 subfamilies ant (Thomas, 1992a; Thomas and Lewington, 1991; Pierce *et al.*, 2002). [42]*Thecla betulae*, *Quercusia quecus*, *Callophrys rubi*, *Polyommatus icarus*, *Lysandra coridon*, *L. bellargus*, [43]*Atelura formicarra*, [44]*Atemeles pubicollis* (adult), [45]*Myrmecophila acervorum*.

Table 17.1. Nine types of error or causes of confusion encountered in determining host specificity at the scale of individuals or populations during studies of *Maculinea* or *Microdon* species (from Thomas *et al.*, 1989, and subsequent data). Prior to our studies, species in both groups were thought to exploit ants in different subfamilies or genera.

Type or error	Example	Species affected
Speculation	*Lasius flavus* is secondary host of *M. arion* (Ford, 1945; Hinton, 1951)	All social parasites
Misidentification of ant by specialists in taxon of social parasite	On *Maculinea* hosts: '*La fourmi que j'ai dénominée Tetramorium caespitum est en réalité la Tapinoma erraticum et celle que je supposais être une Myrmica se trouve être la Tetramorium caespitum*' (Powell, 1918)	All non-ant social parasites
Revision of ant taxonomy after host name published, especially sibling ant spp.	*Formica fusca* listed for *Microdon mutabilis* before *F. lemani* was recognized (Donisthorpe, 1927). *Myrmica sabuleti/M. scabrinodis* or *M. rubra/M. ruginodis/M. laevinodis* used as synonyms	All hosts, especially *Myrmica*, *Lasius*, *Formica* in Europe
Cryptic speciation of social parasite 'species' each with different host	*Microdon mutabilis/myrmicae*; *Maculinea alcon/rebeli* (Schönrogge *et al.*, 2002)	Especially well-integrated social parasites
'Adoption' by non-host ants	Young *Maculinea* larvae retrieved into the nests by any *Myrmica* species but survive only with one species: sampling nests in late autumn reveals *Maculinea* larvae with all *Myrmica* species. Also, social parasites taken as prey sometimes mistaken for adoption	*Maculinea*
Occasional survival in benign colonies	Occasional individuals survive in 'non-host' nests but too rarely to support a population	All social parasites
Laboratory artefacts	Pampered laboratory ant colonies are typically in such a benign social state that any intruder is tolerated	All social parasites (e.g. *M. rebeli*, see Fig. 17.6)
Host species later replaced by other ants	The damage to host colonies can be so great, especially prior to pupation, that original host ant deserts, leaving the pupa in an abandoned nest that is sometimes colonized by a non-host ant species	Especially predatory social parasites
Host shifts across species' ranges	*M. rebeli* and *M. alcon* each parasitize a total of three different host species, each in a different part of Europe. Predatory *Maculinea* also parasitize different *Myrmica* species in Japan than in Europe	*Maculinea*; *Microdon* possibly all social parasites

7. Questions for the Future

We envisage three areas of research to test or extend the following ideas.

7.1 Case studies of functional ecology and host specificity in additional systems of social parasite

To date, the functional and population ecology of just six out of perhaps 10,000–20,000 species of social parasite has been studied in sufficient detail to test our hypotheses with confidence. Similar research is urgently required for different types of social parasite, especially those that apparently possess low infestation–high transmission and low infestation–low transmission population dynamics. Owing to powerful convergent evolution, we suspect that few radically different strategies for social parasitism exist beyond β queens that cheat, slave-making systems, specialist plunderers of ant middens or food-sources (domestic aphids, coccids, fungi, seeds etc.) or the three systems described here. It is also important that the ecological studies of lycaenid (Lepidoptera) social parasites are balanced by more of other orders, focusing especially on the many species whose basic life-history traits are already described, but for which no detailed account exists of host specificity in the field at the regional and population level.

7.2 Do hot-spots exist for social parasites?

Hölldobler and Wilson (1990) predicted that ant–ant social parasites were likely to be commonest in cool temperate climates due to geographical variation in the vulnerability of host ants; we predicted that non-ant social parasites would be commonest near the centre of their host's range, where the carrying capacity of colonies was highest (Thomas *et al.*, 1998a). At more local scales, it has been noted that an exceptionally high diversity of social parasites exists on a few particular sites or landscapes, for example in the Saas valley, Switzerland (Hölldobler and Wilson, 1990). We doubt that these hot-spots arise by chance, and suggest three factors that might amplify the evolution, accumulation and abundance of social parasites both in individual ant colonies and in competing ant species on certain sites.

1. Social parasites thrive (and probably evolve) in localities where the ecological conditions encourage ants to exist in polydomous-polygynous populations over very long periods of time.
2. Once a polydomous ant species' colony is infested by one social parasite, its social structure and that of neighbouring colonies is inevitably altered, in some systems making it more homogeneous (Foitzik and Herbers, 2001), in others more heterogeneous and in theory more benign (see *M. mutabilis*), and hence more vulnerable to invasion by other social parasites, especially species that inhabit different niches in the ant nest and are unlikely to compete with the original invader.

3. One ecological impact of predatory and cuckoo social parasitism is to reduce the realized niche of its host, to the benefit of competing ant species whose fundamental niches partly overlap with that of the parasitized species (Thomas *et al.*, 1997, 1998a). We have yet to model systems containing two competing ants each infested by a different social parasite, but we are sufficiently familiar with the spatial dynamics of 1-social parasite, and > 1-competing ant models (Hochberg *et al.*, 1994; Clarke *et al.*, 1997; Thomas *et al.*, 1998a) to expect that populations of social parasites and host ant species are likely to show increased stability and persistence on sites in which two competing ants are infested by different parasite species (e.g. *M. rubra* and *M. scabrinodis* by *Maculinea nausithous* and by *M. teleius*, *M. alcon* or *M. myrmicae*, respectively).

7.3 Does local coevolution between social parasites and ants generate speciation in ants?

To our knowledge, only Nash *et al.*'s (2002) studies of whether ants experience local evolutionary arms-races with their social parasites are sufficiently advanced to confirm that local patterns exist in the hydrocarbon profiles (= colony recognition signals), not only of *Maculinea alcon* but also of its host *Myrmica ruginodis*, that are consistent with Thompson's (1994; Thompson and Calsbeek, Chapter 14, this volume) Mosaic theory of coevolution. In addition, the intense population dynamic interactions between *Microdon mutabilis* and *Formica lemani*, and the very local intraspecific gradients in the extreme aggression of *F. lemani* to *Microdon* eggs, which is mirrored in gradients of aggression between the ant colonies themselves (Elmes *et al.*, 1999; K. Schönrogge *et al.*, in preparation), suggest that powerful coevolutionary forces are in play. Logic and some data (e.g. Fig. 17.9) suggest that these interactions generate co-adapted genotypes of social parasite and ant host that approach or achieve incompatibility between community modules of the same species in neighbouring landscapes, perhaps resulting in Hawaiian scales of local diversity in both ants and their social parasites, but extending over continents. It will be a task for future researchers to establish whether this is true. For now, we suspect that it is no coincidence that certain of those ant genera, such as *Myrmica*, which are particularly susceptible to social parasitism, are also believed to comprise many cryptic, sibling species and parasitic forms (Radchenko *et al.*, 2003).

Note

Between sumbission and publication of this paper, molecular studies failed to detect major genetic differentiation within *M. alcon*, or indeed between *M. alcon* and *M. rebeli*. Although it is possible that local selection on (host) specific genes was too rapid to be reflected in other markers, the disparity between molecular and ecological studies is a conundrum. On the other hand, the same study found such major differentiation within predatory *Maculinea* across global ranges that, from a total of just 21 populations sampled, what were hitherto listed as four (morpho-)species should now be regarded as at least nine cryptic species (Als *et al.*, 2004; Thomas and Settele, 2004).

Acknowledgements

We gratefully acknowledge funding from NERC grant GR3/12662 and the European Commission for RTD research grant 'MacMan' (EVK2-CT-2001-00126). We also thank Ralph Clarke, Mike Gardner, Judith Wardlaw, David Simcox, John Pickett, Emma Napper, Toshiharu Akino, Boyd Barr, Alex Radchenko, Michael Hochberg, Koos Boomsma, David Nash, Josef Settele, Konrad Fiedler, John Thompson, Rudi Mattoni, Phil De Vries and many others for discussions and insights on many aspects of social parasitism. Our frequent citations of 'The Ants' are testimony to our huge debt to Bert Hölldobler and E.O. Wilson, both for their original experiments and for the synthesis of ideas which provided a foundation for much in this chapter. Finally, we are indebted to Graham Holloway, Mark Fellowes and Jens Rolff for inviting us to write this chapter.

References

Akino, T., Mochizuki, R., Morimoto, M. and Yamaoka, R. (1996) Chemical camouflage of myrmecophilous cricket *Myrmecophilus* sp. to be integrated with several ant species. *Japanese Journal of Applied Entomology and Zoology* 40, 39–46.

Akino, T., Knapp, J.J., Thomas, J.A. and Elmes, G.W. (1999) Chemical mimicry and host specificity in the butterfly *Maculinea rebeli*, a social parasite of *Myrmica* ant colonies. *Proceedings of the Royal Society of London, Series B* 266, 1419–1426.

Alloway, T.M., Buschinger, M., Stuart, R. and Thomas, C. (1982) Polygyny and polodomy in North American species of the three ant genus *Leptothorax* Mayr (Hymenoptera: Formicidae). *Psyche* 89, 249–274.

Als, T.D., Nash, D.R. and Boomsma, J.J. (2001) Adoption of parasitic *Maculinea alcon* caterpillars (Lepidoptera: Lycaenidae) by three *Myrmica* ant species. *Animal Behaviour* 62, 99–106.

Als, T.D., Nash, D.R. and Boomsma, J.J. (2002) Geographical variation in host-ant specificity of the parasitic butterfly *Maculinea alcon* in Denmark. *Ecological Entomology* 27, 403–414.

Als, T.D., Vila, R., Kandul, N.P., Nash, D.R., Yen, S., Hsu, Y., Mignault, A.A., Boomsma, J.J. and Pierce, N.E. (2004) The evolution of alternative parasitic life histories in large blue butterflies. *Nature* 432, 386–390.

Bourke, A.F.G. and Franks, N.R. (1991) Alternative adaptations, sympatric speciation and the evolution of parasitic, inquiline ants. *Biological Journal of the Linnean Society* 43, 157–178.

Bourke, A.F.G. and Franks, N.R. (1995) *Social Evolution in Ants*. Princeton University Press, Princeton, New Jersey.

Brian, M.V. (1968) Regulation of sexual production in an ant society. *Colloques Internationaux de Centre National de la Recherche Scientifique* 173, 61–76.

Brian, M.V. (1986) Bonding between workers and queens in the ant genus *Myrmica*. *Animal Behaviour* 34, 1135–1145.

Brian, M.V. (1988) Queen selection by worker groups of the ant *Myrmica rubra* L. *Animal Behaviour* 36, 914–925.

Brian, M.V. and Brian, A.D. (1951) Insolation and ant populations in the west of Scotland. *Transactions of the Royal Entomological Society of London* 102, 303–330.

Brian, M.V. and Brian, A.D. (1955) On the two forms macrogyna and microgyna of the ant *Myrmica rubra* L. *Evolution* 9, 280–290.

Buschinger, A. (1970) Neue Vorstellungen zur Evolution des Sozialparasitismus und der Dulosis bei Ameisen (Hym, Formicidae). *Biologisches Zentralblatt* 88, 273–299.

Buschinger, A. (1986) Evolution of social parasitism in ants. *Trends in Ecology and Evolution* 1, 155–160.

Buschinger, A. (1990) Sympatric speciation and radiative evolution of socially parasitic ants: heretic hypotheses and their factual background. *Zeitschrift Fur Zoologische Systematik und Evolutionsforschung* 28, 241–260.

Chapman, T.E. (1920) Contributions to the life history of *Lycaena euphemus* Hb. *Transactions of the Entomological Society of London* 1919, 450–465.

Claassens, V.E. (1976). Observations on the myrmecophilous relationships and the parasites of *Lepidochrysops methymma* (Trimen) and *L. trimeni* (Bethune-Baker). *Journal of the Entomological Society of South Africa* 39, 279–289.

Clarke, R.T., Thomas, J.A., Elmes, G.W. and Hochberg, M.E. (1997) The effects of spatial patterns in habitat quality on community dynamics within a site. *Proceedings of the Royal Society of London, Series B* 264, 247–253.

Clarke, R.T., Thomas, J.A., Elmes, G.W., Wardlaw, J.C., Munguira, M.L. and Hochberg, M.E. (1998) Population modelling of the spatial interactions between *Maculinea*, their initial foodplant and *Myrmica* ants within a site. *Journal of Insect Conservation* 2, 29–38.

Cottrell, C.B. (1984) Aphytophagy in butterflies: its relationship to myrmecophily. *Zoological Journal of the Linnean Society* 79, 1–57.

Dettner, K. and Liepert, C. (1994) Chemical mimicry and camouflage. *Annual Review of Entomology* 39, 129–154.

De Vries, P.J. (1990) Enhancement of symbioses between butterfly caterpillars and ants by vibrational communication. *Science* 248, 1104–1106.

De Vries, P.J. (1991a) Call production by myrmecophilous riodinid and lycaenid butterfly caterpillars (Lepidoptera): morphological, acoustical, functional and evolutionary patterns. *American Museum Novitates* 3025, 1–23.

De Vries, P.J. (1991b) Evolutionary and ecological patterns in myrmecophilous riodinid butterflies. In: Huxley, C.R. and Cutler, D.F. (eds) *Ant–Plant Interactions*. Oxford University Press, Oxford, UK, pp. 143–156.

De Vries, P.J. and Baker, I. (1989) Butterfly exploitation of a plant–ant mutualism: adding insult to herbivory. *Journal of the New York Entomological Society* 97, 332–340.

De Vries, P.J., Cocroft, R.B. and Thomas, J.A. (1993) Comparison of acoustical signals in *Maculinea* butterfly caterpillars and their obligate host *Myrmica* ants. *Biological Journal of the Linnean Society* 49, 229–238.

Donisthorpe, H.St.J.K. (1927) *The Guests of British Ants*. Routledge, London.

Duffield, R.M. (1981) Biology of *Microdon fuscipennis* (Diptera: Syrphidae) with interpretations of the reproductive strategies of *Microdon* found north of Mexico. *Proceedings of the Entomological Society of Washington* 83, 716–724.

Elmes, G.W. (1973) Observations on the density of queens in natural colonies of *Myrmica rubra* (Hymenoptera: Formicidae). *Journal of Animal Ecology* 42, 761–771.

Elmes, G.W. (1974) The effect of colony population on caste size in three species of *Myrmica* (Hymenoptera Formicidae). *Insectes Sociaux* 21, 213–230.

Elmes, G.W. (1976) Some observations on the microgyne form of *Myrmica rubra* L. (Hymenoptera, Formicidae). *Insectes Sociaux* 23, 3–22.

Elmes, G.W. (1978) A morphometric comparison of three closely related species of *Myrmica* (Formicidae), including a new species from England. *Systematic Entomology* 3, 131–145.

Elmes, G.W. (1987) Temporal variation in colony populations of the ant *Myrmica sulcinodis*. I. Changes in queen number, worker number and spring production. *Journal of Animal Ecology* 56, 559–571.

Elmes, G.W. (1991) The social biology of *Myrmica* ants. *Actes Colloques Insectes Sociaux* 7, 17–34.

Elmes, G.W. (1994) A population of the social parasite *Myrmica hirsuta* Elmes (Hymenoptera, Formicidae) recorded from Jutland, Denmark, with a first description of the worker caste. *Insectes Sociaux* 41, 437–442.

Elmes, G.W. (1996) Biological diversity of ants and their role in ecosystem function. In: Lee, B.H., Kim, T.H. and Sun, B.Y. (eds) *Biodiversity Research and its Perspectives in East Asia*. Proceedings of Inaugural Seminar of KIBIO, Chonbuk National University, Korea, pp. 33–48.

Elmes, G.W. and Thomas, J.A. (1992) The complexity of species conservation: interactions between *Maculinea* butterflies and their ant hosts. *Biodiversity and Conservation* 1, 155–169.

Elmes, G.W., Thomas, J.A. and Wardlaw, J.C. (1991a) Larvae of *Maculinea rebeli*, a large-blue butterfly and their *Myrmica* host ants: wild adoption and behaviour in ant-nests. *Journal of Zoology* 223, 447–460.

Elmes, G.W., Wardlaw, J.C. and Thomas, J.A. (1991b) Larvae of *Maculinea rebeli*, a large-blue butterfly and their *Myrmica* host ants: patterns of caterpillar growth and survival. *Journal of Zoology* 224, 79–92.

Elmes G.W., Thomas, J.A., Hammarstedt, O., Munguira, M.L., Martin, J. and van der Made, J.G. (1994) Differences in host-ant specificity between Spanish, Dutch and Swedish populations of the endangered butterfly, *Maculinea alcon* (Denis et Schiff.) (Lepidoptera). *Memorabilia Zoologica* 48, 55–68.

Elmes, G.W., Clarke, R.T., Thomas, J.A. and Hochberg, M.E. (1996) Empirical tests of specific predictions made from a spatial model of the population dynamics of *Maculinea rebeli*, a parasitic butterfly of red ant colonies. *Acta Oecologica* 17, 61–80.

Elmes, G.W., Thomas, J.A., Wardlaw, J.C., Hochberg, M.E., Clarke, R.T. and Simcox, D.J. (1998) The ecology of *Myrmica* ants in relation to the conservation of *Maculinea* butterflies. *Journal of Insect Conservation* 2, 67–78.

Elmes, G.W., Barr, B., Thomas, J.A. and Clarke, R.T. (1999) Extreme host specificity by *Microdon mutabilis* (Diptera, Syrphidae), a social parasite of ants. *Proceedings of the Royal Society of London, Series B* 266, 447–453.

Elmes, G.W., Thomas, J.A., Munguira, M.L. and Fiedler, K. (2001) Larvae of lycaenid butterflies that parasitise ant colonies provide exceptions to normal insect growth rules. *Biological Journal of the Linnean Society* 73, 259–278.

Elmes, G.W., Akino, T., Thomas, J.A., Clarke, R.T. and Knapp, J.J. (2002) Interspecific differences in cuticular hydrocarbon profiles of *Myrmica* ant species are sufficiently consistent to explain host specificity in *Maculinea* (Large blue) butterflies. *Oecologia* 130, 525–535.

Elmes, G.W., Wardlaw, J.C., Schönrogge, K. and Thomas, J.A. (2004) Food stress causes differential survival of socially parasitic larvae of *Maculinea rebeli* (Lepidoptera, Lycaenidae) integrated in colonies of host and non-host *Myrmica* species (Hymenoptera, Formicidae). *Entomologia Experimentalis et Applicata* 110, 53–63.

Emery, C. (1909) Über den ursprung der dulotischen, parasitischen und myrmekophilen Ameisen. *Biologisches Zentralblatt* 29, 352–362.

Evesham, E.J.M. (1984) Queen distribution movements and interactions in a semi-natural nest of the ant *Myrmica rubra* L. *Insectes Sociaux* 31, 5–19.

Fiedler, K. (1990) New information on the biology of *Maculinea nausithous* and *M. teleius* (Lepidoptera: Lycaenidae). *Nota Lepidopterologica* 12, 246–256.

Fiedler, K. (1991) *Systematic, Evolutionary, and Ecological Implications of Myrmecophily within Lycinidae (Insecta: Lepidoptera: Papilionidae).* Zoologisches Forschungsinstitut and Museum Alexander König, Bonn, Germany.

Fiedler, K. (1994) Lycaenid butterflies and ants: is myrmecophily associated with amplified hostplant diversity? *Ecological Entomology* 19, 79–82.

Fiedler, K. (1996) Host-plant relationships of lycaenid butterflies: large-scale patterns, interactions with plant chemistry, and mutualism with ants. *Entomologia Experimentalis et Applicata* 80, 259–267.

Fiedler, K. (1998) Lycaenid-ant interactions of the *Maculinea* type: tracing their historical roots in a comparative framework. *Journal of Insect Conservation* 2, 3–14.

Fiedler, K. and Hölldobler, B. (1992) Ants and *Polyommatus icarus* immatures (Lycaenidae): sex-related development benefits and costs of ant attendance. *Oecologia* 91, 468–473.

Fiedler, K. and Maschwitz, U. (1987) Functional analysis of the myrmecophilous relationships between ants (Hymenoptera: Formicidae) and lycaenids (Lepidoptera: Lycaenidae). III. New aspects of the function of the retractile tentacular organs of lycaenid larvae. *Zoologische Beiträge* 31, 409–416.

Fiedler, K. and Maschwitz, U. (1988) Functional analysis of the myrmecophilous relationships between ants (Hymenoptera: Formicidae) and lycaenids (Lepidoptera: Lycaenidae). II. Lycaenid larvae as trophobiotic partners of ants – a quantitative approach. *Oecologia* 75, 204–206.

Fiedler, K. and Maschwitz, U. (1989) Functional analysis of the myrmecophilous relationships between ants (Hymenoptera: Formicidae) and lycaenids (Lepidoptera: Lycaenidae). I. Release of food recruitment in ants by lycaenid larvae and pupae. *Ethology* 80, 71–80.

Ford, E.B. (1945) *Butterflies.* Collins, London.

Foitzik, S. and Herbers, J.M. (2001) Colony structure of a slavemaking ant. II. Frequency of slave raids and impact on the host population. *Evolution* 55, 316–323.

Frohawk, F.W. (1924) *Natural History of British Butterflies.* Hutchinson, London.

Gadeberg, R.M.E. and Boomsma, J.J. (1997) Genetic population structure of the large blue butterfly *Maculinea alcon* in Denmark. *Journal of Insect Conservation* 1, 99–111.

Gerrish, A.R. (1994) The influence of relatedness and resource investment on the behaviour of worker ants towards brood. PhD thesis, University of Exeter, UK.

Goodisman, M.A.D. and Ross, K.G. (1998) A test of queen recruitment models using nuclear and mitochondrial workers in the fire ant *Solenopsis invicta. Evolution* 52, 1416–1422.

Heath, J. (1981) *Threatened Rhopalocera (Butterflies) in Europe.* Council of Europe, Strasbourg, France.

Herbers, J.M. (1985) Seasonal structuring of a north temperate ant community. *Insectes Sociaux* 32, 224–240.

Hinton, H.E. (1951) Myrmecophilous Lycaendiae and other Lepidoptera – a summary. *Proceedings and Transactions of the South London Entomological and Natural History Society* 1949–1950, 111–175.

Hochberg, M.E., Thomas, J.A. and Elmes, G.W. (1992) A modelling study of the population dynamics of a large blue butterfly, *M. rebeli*, a parasite of red ant nests. *Journal of Animal Ecology* 61, 397–409.

Hochberg, M.E., Clarke, R.T., Elmes, G.W. and Thomas, J.A. (1994) Population dynamic consequences of direct and indirect interactions involving a large blue butterfly and its plant and red ant hosts. *Journal of Animal Ecology* 63, 375–391.

Hölldobler, B. (1967) Zur Physiologie der Gast-Wirt-Beziehungen (Myrmecophilie) bei Ameisen. I. Das Gastverhältnis der *Atemeles-* und *Lomechusa*. Larven (Co. Staphylinidae) zu Formic (Hym. Formicidae). *Zeitschrift für Vergleichenden Physiologie* 56, 1–121.

Hölldobler, B. and Wilson, E.O. (1990) *The Ants*. Springer, Berlin.

Hölldobler, K. (1947) Studien über die Ameisengrille (*Myrmecophila acervorum* Panzer) im mittleren Maingebeit. *Mitteilungen der Schweizerischen Entomologischen Gesellschaft* 20, 607–648.

Huxley, C.R. and Cutler, D.F. (1991) *Ant–Plant Interactions*. Oxford University Press, Oxford, UK.

Jordano, D. and Thomas, C.D. (1992) Specificity of an ant–lycaenid interaction. *Oecologia* 91, 431–438.

Jordano, D., Rodriguez, J., Thomas, C.D. and Haeger, J.F. (1992) The distribution and density of a lycaenid butterfly in relation to *Lasius* ants. *Oecologia* 91, 439–446.

Lenoir, A., D'Ettorre, P., Errard, C. and Hefetz, A. (2001) Chemical ecology and social parasitism in ants. *Annual Review of Entomology* 46, 573–599.

Malicky, H. (1969) Versuch einer Analyse der ökolgiischen Bezeihungen zwischen Lycaeniden (Lepidoptera) und Formiciden (Hymenoptera). *Tijdschrift voor Entomologie* 112, 213–298.

Malicky, H. (1970) New aspects on the association between lycaenid larvae (Lycaenidae) and ants (Formicidae, Hymenoptera). *Journal of the Lepidopterists Society* 24, 190–202.

Mayr, E. (1963) *Animal Species and Evolution*. Harvard University Press, Cambridge, Massachusetts.

Meyer-Hozak, C. (2000) Population biology of *Maculinea rebeli* (Lepidoptera: Lycaendidae) on the chalk grasslands of Eastern Westphalia (Germany) and implications for conservation. *Journal of Insect Conservation* 4, 63–72.

Nash, D.R., Als, T.D., Tentschert, J., Maile, R., Jungnickel, H. and Boomsma, J.J. (2002) Local adaptation and coevolution of chemical mimicry in the butterfly *Maculinea alcon*, a social parasite of *Myrmica* ants. In: *Proceedings of the XIV International Meeting of the IUSSI, Sapporo, Japan*, p. 138.

Pearson, B. (1982) Relatedness of normal queens (macrogynes) in nests of the polygynous ant *Myrmica rubra* L. *Evolution* 36, 107–112.

Pedersen, J.S. and Boomsma, J.J. (1999) Effect of habitat saturation on the number and turnover of queens in the polygynous ant *Myrmica sulcinodis*. *Journal of Evolutionary Biology* 12, 903–917.

Pierce, N.E. (1989) Butterfly–ant mutualisms. In: Grubb, P.J. and Whittaker, J.B. (eds) *Towards a More Exact Ecology*. Blackwell, Oxford, UK, pp. 299–324.

Pierce, N.E. and Elgar, M.A. (1985) The influence of ants on host plant selection by *Jalmenus evagoras*, a myrmecophilous lycaenid butterfly. *Behavioural Ecology and Sociobiology* 16, 209–222.

Pierce, N.E., Kitching, R.L., Buckley, R.C., Taylor, M.F.J. and Benbow, K.F. (1987) The costs and benefits of cooperation between the Australian lycaenid butterfly, *Jalmenus evagoras*, and its attendant ants. *Behavioral Ecology and Sociobiology* 21, 237–248.

Pierce, N.E., Braby, M.F., Heath, A., Lohman, D.J., Mathew, J., Rand, D.B. and Travassos, M.A. (2002) The ecology and evolution of ant association in the Lycaenidae (Lepidoptera). *Annual Review of Entomology* 47, 733–771.

Plateaux, L. (1970) Sur le polymorhisme social de la fourmis *Leptothorax nylanderi* (Förster). I. Morphologie et biologie comparées des castes. *Insectes Sociaux* 12, 373–478.

Powell, H. (1918) Compte rendu de la recherché des chenilles de *Lycaena alcon* à la fin d'aoûet pendant les premier jours de Septembre 1918, à Laillé et dans la lande des Grêles, près Monterfil. *Et Lép Comp* 17, 25–37.

Purefoy, E.B. (1953) An unpublished account of experiments carried out at East Farleigh, Kent, in 1915 and subsequent years on the life history of *Maculinea arion,* the large blue butterfly. *Proceedings of the Royal Entomological Society of London, Series A* 28, 160–162.

Radchenko, A. and Elmes, G.W. (2003) A taxonomic revision of the socially parasitic *Myrmica* ants (Hymenoptera, Formicidae) of the Palaearctic region. *Annales Zoologici* 53, 217–243.

Radchenko, A., Elmes, G.W., Czechowska, W., Stankiewicz, A., Czechowski, W. and Sieleznew, M. (2003) First records of *Myrmica vandeli* BONDROIT and *M. tulinae* Elmes, Radchenko and Aktaç (*Hymenoptera: Formicidae*) for Poland, with a key for the *scabrinodis-* and *sabuleti*-complexes. *Fragmenta Faunistica* 46, 47– 57.

Ross, K.G. (1989) Reproductive and social structure in polygynous fire ant colonies. In: Breed, M.D. and Page, R.E., Jr (eds) *The Genetics of Social Evolution.* Westview Press, Boulder, Colorado, pp. 149–162.

Savolainen, R. and Vepsäläinen, K. (2003) Sympatric speciation through intraspecific social parasitism. *Proceedings of the National Academy of Sciences USA* 100, 7169–7174.

Schönrogge, K., Wardlaw, J.C., Thomas, J.A. and Elmes, G.W. (2000) Polymorphic growth rates in myrmecophilous insects. *Proceedings of the Royal Society of London, Series B* 267, 771–777.

Schönrogge, K., Barr, B., Napper, E., Breen, J., Gardner, M.G., Elmes, G.W. and Thomas, J.A. (2002) When rare species become endangered: cryptic speciation in myrmecophilous hoverflies. *Biological Journal of the Linnean Society* 75, 291–300.

Schönrogge, K., Wardlaw, J.C., Peters, A.J., Everett, S., Thomas, J.A. and Elmes, G.W. (2004) Changes in chemical signature and host specificity from larval retrieval to full social integration in the myrmecophilous butterfly *Maculinea rebeli. Journal of Chemical Ecology* 30, 91–107.

Schroth, M. and Maschwitz, U. (1984) Zur Larvalbiologie und Wirtsfindung van *Maculinea teleius* (Lepidoptera: Lycaenidae), eines Parasiten von *Myrmica laevinodis. Entomologia Generalis* 9, 225–230.

Seifert, B. (1993) Taxonomic description of *Myrmica microrubra* n. sp.: a social parasitic ant so far known as the microgyne of *Myrmica rubra* (L.). *Abhandlungen und Berichte des Naturkundemuseums Görlitz* 67(5), 9–12.

Steiner, F.M., Sieleznew, M., Schlick-Steiner, B.C., Höttinger, H., Stankiewicz, A. and Gornicki, A. (2003) Host specificity revisited: new data on *Myrmica* host ants of the vulnerable lycaenid butterfly *Maculinea rebeli. Journal of Insect Conservation* 7, 1–6.

Thomas, J.A. (1977) *Ecology and Conservation of the Large Blue Butterfly: 2nd Report.* ITE, Huntingdon, UK.

Thomas, J.A. (1980) Why did the large blue become extinct in Britain? *Oryx* 15, 243–247.

Thomas, J.A. (1983) The ecology and conservation of *Lysandra bellargus* (Lepidoptera: Lycaenidae) in Britain. *Journal of Applied Ecology* 20, 59–83.

Thomas, J.A. (1984a) The conservation of butterflies in temperate countries: past efforts and lessons for the future. In: Vane-Wright, R. and Ackery, P. (eds) *Biology of Butterflies. Symposia of the Royal Entomological Society* 11, 333–353.

Thomas, J.A. (1984b) The behaviour and habitat requirements of *Maculinea nausithous* (the dusky large blue) and *M. teleius* (the scarce large blue) in France. *Biological Conservation* 28, 325–347.

Thomas, J.A. (1991) Rare species conservation: case studies of European butterflies. In: Spellerberg, I.F., Goldsmith, F.B. and Morris, M.G. (eds) *The Scientific Management of Temperate Communities for Conservation.* Blackwell Scientific, Oxford, UK, pp. 149–197.

Thomas, J.A. (1992a) Relationships between butterflies and ants. In: Dennis, R.L.H. (ed.) *The Ecology of British Butterflies.* Oxford University Press, Oxford, UK, pp. 149–154.

Thomas, J.A. (1992b) Adaptations to living near ants. In: Dennis, R.L.H. (ed.) *The Ecology of British Butterflies.* Oxford University Press, Oxford, UK, pp. 109–115.

Thomas, J.A. (1995) The ecology and conservation of *Maculinea arion* and other European species of large blue butterfly. In: Pullin, A.S. (ed.) *Ecology and Conservation of Butterflies.* Chapman and Hall, London, pp. 180–119.

Thomas, J.A. (1999) The large blue butterfly – a decade of progress. *British Wildlife* 11, 22–27.

Thomas, J.A. (2002) Larval niche selection and evening exposure enhance adoption of a predacious social parasite, *Maculinea arion* (large blue butterfly), by *Myrmica* ants. *Oecologia* 122, 531–537.

Thomas, J.A. and Elmes, G.W. (1993) Specialised searching and the hostile use of allomones by a parasitoid whose host, the butterfly *Maculinea rebeli*, inhabits ant nests. *Animal Behaviour* 45, 593–602.

Thomas, J.A. and Elmes, G.W. (1998) Higher productivity at the cost of increased host-specificity when *Maculinea* butterfly larvae exploit ant colonies through trophallaxis rather than by predation. *Ecological Entomology* 23, 457–464.

Thomas, J.A. and Elmes, G.W. (2001) Foodplant niche selection rather than the presence of ant nests explains oviposition patterns in the myrmecophilous butterfly genus *Maculinea. Proceedings of the Royal Society of London, Series B* 268, 471–477.

Thomas, J.A. and Lewington, R. (1991) *The Butterflies of Britain and Ireland.* Dorling Kindersley, London.

Thomas, J.A. and Morris, M.G. (1994) Patterns, mechanisms and rates of decline among UK invertebrates. *Philosophical Transactions of the Royal Society of London, Series B* 344, 47–54.

Thomas, J.A. and Wardlaw, J.C. (1990) The effect of queen ants on the survival of *Maculinea arion* larvae in *Myrmica* ant nests. *Oecologia* 85, 87–91.

Thomas, J.A. and Wardlaw, J.C. (1992) The capacity of a *Myrmica* ant nest to support a predacious species of *Maculinea* butterfly. *Oecologia* 91, 101–109.

Thomas, J.A., Elmes, G.W., Wardlaw, J.C. and Woyciechowski, M. (1989) Host specificity among *Maculinea* butterflies in *Myrmica* ant nests. *Oecologia* 79, 452–457.

Thomas, J.A., Munguira, M.L., Martin, J. and Elmes, G.W. (1991) Basal hatching by *Maculinea* butterfly eggs: a consequence of advanced myrmecophily? *Biological Journal of the Linnean Society* 44, 175–184.

Thomas, J.A., Elmes, G.W. and Wardlaw, J.C. (1993) Contest competition among *Maculinea rebeli* butterfly larvae in ant nests. *Ecological Entomology* 18, 73–76.

Thomas, J.A., Elmes, G.W., Clarke, R.T., Kim, K.G., Munguira, M.L. and Hochberg, M.E. (1997) Field evidence and model predictions of butterfly-mediated apparent competition between gentian plants and red ants. *Acta Oecologica* 18, 671–684.

Thomas, J.A., Clarke, R.T., Elmes, G.W. and Hochberg, M.E. (1998a) Population dynamics in the genus *Maculinea* (Lepidoptera: Lycaenidae). In: Dempster, I.P. and McLean, I.F.G. (eds) *Insect Population Dynamics: In Theory and Practice. Symposia of the Royal Entomological Society* 19, 261–290.

Thomas, J.A., Elmes, G.W. and Wardlaw, J.C. (1998b) Polymorphic growth in larvae of the butterfly *Maculinea rebeli*, a social parasite of *Myrmica* ant colonies. *Proceedings of the Royal Society of London, Series B* 265, 1895–1901.

Thomas, J.A., Knapp, J.J., Akino, T., Gerty, S., Wakamura, S., Simcox, D.Y., Wardlaw, J.C. and Elmes, G.W. (2002) Parasitoid secretions provoke ant warfare. *Nature* 417, 505–506.

Thomas, J.A. and Settele, J. (2004) Butterfly mimics of ants. *Nature* 432, 283–284.

Van Dyck, H., Oostermeijer, J.G.B., Talloen, W., Feenstra, V., van der Hidde, A. and Wynhoff, I. (2000) Does the presence of ant nests matter for oviposition to a specialized myrmecophilous *Maculinea* butterfly? *Proceedings of the Royal Society of London, Series B* 267, 861–866.

Vander Meer, R.K. and Morel, L. (1998) Nestmate recognition in ants. In: Vander Meer, R.K., Breed, M.D., Winston, M.L. and Espelie, K.E. (eds) *Pheromone Communication in Social Insects*. Westview Press, Boulder, Colorado, pp. 79–103.

Wardlaw, J.C., Thomas, J.A. and Elmes, G.W. (2000) Do *Maculinea rebeli* caterpillars provide vestigial mutualistic benefits to ants when living as social parasites in *Myrmica* ant nests? *Entomologia Experimentalis et Applicata* 95, 97–103.

Wasmann, E. (1894) *Kritisches Verzeichniss der Myrmecphilen und Termitophilen Arthropoden*. Felix Dames, Berlin.

Wasmann, E. (1901) Zur Lebensweise der Ameisengrillen (*Myrmecophila*). *Natur und Offenbarung* 47, 24.

Wasmann, E. (1912) The ants and their guests. *Smithsonian Report* 1912, 455–474.

Way, M.J. (1953) The relationship between certain ant species with particular reference to biological control of the coreid *Theraptus* sp. *Bulletin of Entomological Research* 45, 93–112.

Way, M.E. (1963) Mutualism between ants and honeydew-producing Homoptera. *Annual Review of Entomology* 8, 307–344.

West-Eberhard, M.J. (1981) Intragroup selection and the evolution of insect societies. In: Alexander, R.D. and Tinkle, D.W. (eds) *Natural Selection and Social Behaviour*. Chiron Press, Concord, Massachusetts, pp. 3–17.

Wheeler, W.M. (1910) *Ants: Their Structure, Development and Behaviour*. Columbia University Press, New York.

Wilson, E.O. (1990) *Success and Dominance in Ecosystems: The Case of the Social Insects*. Ecology Institute, Oldendorf/Luhe, Federal Republic of Germany.

Winterbottom, S. (1981) The chemical basis for species and colony recognition in three species of myrmicine ants. PhD thesis, University of Southampton, UK.

Wynhoff, I. (2001) At home on foreign meadows: the reintroduction of two *Maculinea* butterfly species. PhD thesis, Wageningen University, The Netherlands.

Yamaguchi, S. (1988) *The life histories of five myrmecophilous lycaenid butterflies of Japan*. Kodansha, Tokyo.

18 Evolutionary Changes in Expanding Butterfly Populations

JANE K. HILL, CALVIN DYTHAM AND CLARE L. HUGHES

Department of Biology, University of York, York, UK

1. Introduction

Temperatures warmed by approximately 0.6°C during the 20th century, and are predicted to continue to rise by up to approximately 6°C by the end of this century (IPCC, 2001). Due to their cold-blooded nature, high dispersal ability and rapid reproductive rates, insects would be expected to be sensitive to climate changes, and evidence is accumulating that many insects, including butterflies, are responding to recent warming (see reviews in Hughes, 2000; McCarty, 2001; Walther *et al.*, 2002; Parmesan and Yohe, 2003; Root *et al.*, 2003). Ecological impacts of climate warming on insects include shifts in species' breeding distributions, with species moving to higher latitudes (Parmesan *et al.*, 1999; Warren *et al.*, 2001) and altitudes (Parmesan, 1996; Hill *et al.*, 2002), and changes in species' phenology, with many species flying earlier in the spring (Roy and Sparks, 2000).

Responding to changing climates is not a new phenomenon for insects; most native species currently occur in the UK only because they were able to respond to climate changes and colonize the British Isles after the last ice age. Given that many British butterfly species have higher growth and survival rates at warmer temperatures (within the UK range), it could be argued that many butterfly species will benefit from future warming. However, it is possible that 21st-century climate warming will occur at a rate much higher than any previously observed, making it difficult for insects to track changes. None the less, there is evidence that climates changed rapidly during the Quaternary period (past 2 million years) when increases of approximately 7°C are estimated to have occurred in less than 100 years (Houghton, 1997). Such rates of change are comparable with current predictions for climate changes during the 21st century (IPCC, 2001), and thus butterflies may have experienced rapid rates of climate change in the past. What is new, however, is that butterflies must now respond to changing climates and shift their distributions across landscapes that have been greatly modified by humans. For example, 40–97% declines in important butterfly breeding habitats have been reported in the UK since the 1940s (Asher et al., 2001), and thus butterflies have to cope not only with changing climates but also with the loss and fragmentation of their breeding habitats.

In this chapter we review existing evidence for evolutionary changes in expanding butterfly populations, focusing on the satyrine butterfly *Pararge aegeria* (speckled wood). We study changes in dispersal at expanding range margins, and investigate whether these changes are associated with reduced fecundity. We develop a spatial-explicit model to simulate evolution of dispersal during range expansion and to determine the spatial scales over which evolutionary changes might be evident. Finally, we discuss the conservation implications of these findings for predicting the distribution of species under future climate change.

2. Recent Range Changes in UK Butterflies

Over the past 250 years, many British butterfly species have undergone marked changes in their distributions (Asher et al., 2001). Many species had more extensive distributions during the 19th century, but underwent contraction of their ranges towards the end of the 19th century, at a time when the climate appears to have been cooler (Hulme and Barrow, 1997). Since the 1940s, however, climates have been warming and some butterfly species have expanded their northern range boundaries (Parmesan et al., 1999). Approximately 12 out of 57 species of non-migratory British butterflies are currently expanding their breeding ranges (Asher et al., 2001) at rates of about 1–2 km/year, although these rates vary greatly among species. The most impressive example is the recent range expansion by *Polygonia c-album* (comma), whose range margin has expanded northwards by > 200 km in Britain in the past 60 years. However, the majority of British species have failed to respond to recent climate change, or have declined, and range expansions have generally been confined to mobile, generalist species. Most butterflies are sedentary habitat specialists, and these species have declined during recent climate warming because suitable breeding habitats have continued to disappear. For example, the distribution of the highly specialist butterfly *Agynnis adippe* (high brown fritillary) has declined by 50% (in terms of 10-km grid squares occupied) over the past 30 years (Warren et al., 2001). For these habitat specialists, newly available, climatically suitable habitats are isolated and beyond the reach of potential colonists, making it impossible for species to shift their distributions; the widespread loss of breeding habitats has far outweighed any potential benefits of climate warming. Some northern and/or montane species have also undergone distribution changes and have disappeared from low-elevation sites at their southern range boundaries, which have probably become too warm (Hill et al., 2002). Together, these data indicate that communities in the future are likely to become increasingly dominated by widespread, generalist species, with the continued decline of habitat specialists and montane species.

3. Study Species: Speckled Wood Butterfly *Pararge aegeria*

Although most butterflies are declining, there are some species, including the speckled wood butterfly, which are currently expanding their distributions in the UK. The speckled wood is essentially confined to woodland, at least at the margins of its range, and larvae feed on a wide variety of grasses. *Pararge aegeria* can develop through between 1.5 and 2 generations per year in the UK, and flight periods of different generations overlap, with adults on the wing from April until September (Asher *et al.*, 2001). As with several other British species, *P. aegeria* had a more extensive distribution in the UK during the 19th century and probably occurred throughout most of the British Isles as far north as central Scotland. It declined towards the end of the 19th century and essentially became restricted to southwest England and Wales, with a small refuge population in west Scotland. From the 1940s onwards, however, it has been expanding its range (Hill *et al.*, 1999a; Fig. 18.1). *Pararge aegeria* has a Palaearctic distribution and occurs as far north as the northern UK and central Scandinavia, as far south as North Africa, and reaches an eastern range limit in the Urals. It was first recorded on Madeira in 1976 and has subsequently colonized many areas of the island (Jones and Lace, 1992). There are two subspecies; *aegeria* in the southern part of the range and *tircis* in the north (Tolman, 1997). Previous research shows that the European distribution of *P. aegeria* is limited by climate and that its distribution is predicted to shift northwards during the 21st century such that the species will have the potential to occupy nearly all parts of the UK in the future (Hill *et al.*, 1999a, 2002). Our recent studies indicate, however, that these predictions may be unrealistic and that the distribution of *P. aegeria* is lagging behind current climates. The availability of suitable woodland has declined throughout most of its range in the UK during the 20th century and *P. aegeria* has not yet re-colonized all previously occupied areas in northern England and southern Scotland (Fig. 18.1). The availability of suitable habitat affects patterns of recent range expansion (Hill *et al.*, 2001); comparison of rates of range expansion between England and Scotland have shown that range expansion over the past 30 years is approximately twice as fast in Scotland (Fig. 18.2) where there is approximately 25% more woodland (Hill *et al.*, 2001). Thus, even for those species, such as the speckled wood, that are currently expanding their ranges during a period of climate warming, the availability of breeding habitat is crucial in affecting rates of expansion.

4. Responses to Climate Warming

Evidence from the Quaternary period for beetles indicates that past climate changes were not associated with local adaptation of species; morphologies of extant species which occur in the subfossil record do not appear to have changed over time (Coope, 1995). Nonetheless, there is evidence that evolutionary changes are occurring in populations which are currently expanding during recent climate warming. These include evolutionary changes in dispersal, fecundity and

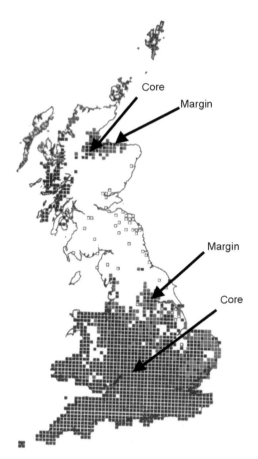

Fig. 18.1. Current distribution of *Pararge aegeria* in Britain. Black squares = recorded (10-km grid resolution) distribution 1940–1989, grey squares = first recent record during the 1990s, white squares = previously occupied (19th century) sites which have not (yet) been re-colonized. Arrows show location of English and Scottish sampling sites (see text). Map reproduced from Hughes *et al.* (2003).

larval host-plant use, which may affect species' responses to current climate change. For example, increased dispersal ability or evolutionary shifts on to new host-plant species which result in broader niches could result in greater-than-expected range expansions. Such effects could be important for making reliable predictions of species' responses to future climate warming. We review the evidence for these changes and discuss their potential impacts below.

4.1 Evolutionary changes in dispersal ability at expanding range margins

Several studies have demonstrated evolutionary increases in dispersal ability in insects in newly established populations (e.g. Niemela and Spence, 1991; Thomas *et al.*, 2001; Hanski *et al.*, 2002). These changes arise as a consequence

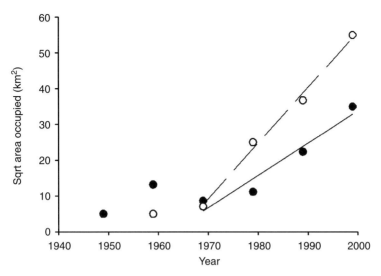

Fig. 18.2. Range expansion of *Pararge aegeria* in England (solid line and solid circles) and Scotland (dashed line and hollow circles) study areas plotted as square root of the area occupied (5-km grid squares with records) against year. Graph reproduced from Hill *et al.* (2001).

of altered costs and benefits of dispersal versus reproduction during colonization. Individuals colonizing empty habitats are not a random selection of the source population and usually share a suite of life-history traits associated with dispersal. In insects, dispersal ability has been related to adult flight morphology (Simberloff, 1981; Dingle, 1986; Leslie, 1990), and in butterflies, the thorax comprises predominantly flight muscle, and individuals which fly faster have relatively larger, broader thoraxes (Dempster *et al.*, 1976; Chai and Srygley, 1990; Srygley and Chai, 1990) and larger wing-spans (Dudley, 1990). Because of these observed relationships, we have used measures of adult flight morphology to quantify dispersal in *Pararge aegeria*. We have focused particularly on measures of thorax mass and shape, and we assume that individuals with larger and/or broader thoraxes have greater dispersal ability. We have quantified evolutionary changes in dispersal by rearing individuals from different populations under common environmental conditions. To take account of allometric effects, all our analyses are carried out on relative values. Here we present results from two studies of *Pararge aegeria*:

- A comparison of evolutionary changes in dispersal in populations from two expanding range margins in Europe;
- A finer spatial-scale comparison of populations from two range margins in the UK.

4.1.1 Dispersal ability in populations at expanding range margins in Europe

We investigated evolutionary changes in dispersal in populations of two subspecies of *P. aegeria* in the southern (subspecies *aegeria*) and northern (subspecies *tircis*) parts of *P. aegeria*'s European range (Hill *et al.*, 1999b). We compared dispersal ability in populations from sites in the UK and Madeira, colonized approximately 20 years previously, with those from sites in the UK and Spain that have been continuously occupied in recent times (Hill *et al.*, 1999b). In general, individuals from the recently colonized site were larger, and had relatively larger thoraxes, indicating increased dispersal ability in recently colonized sites (Fig. 18.3). These effects of colonization were more evident in females than males. This difference between the sexes may be because flight has many functions in addition to long-distance dispersal, for example foraging, mate location and predator avoidance. It is possible that differences between the sexes in colonization effects reflect differences in selection pressures on these various flight functions between the sexes, with males under stronger selection pressure for successful mate location.

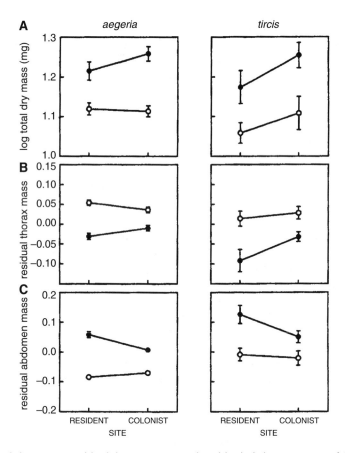

Fig. 18.3. Total dry mass, residual thorax mass and residual abdomen mass of *Pararge aegeria aegeria* and *P. aegeria tircis* from colonized and continuously occupied resident sites. Means and SES are shown (open circles = males, closed circles = females). Graph reproduced from Hill *et al.* (1999a).

4.1.2 Dispersal ability in populations at expanding northern range margins in the UK

We investigated dispersal ability in populations from two range margins in the UK (England and Scotland; Fig. 18.1; Hughes *et al.*, 2003). We compared populations from two sites that had been colonized in the 1990s with two 'core' sites that either had apparently been continuously occupied (England) or had the first recent record for that area (Scotland; Barbour, 1986). As above, the effects of colonization were more evident in females than in males; females from range margin sites had relatively larger and broader thoraxes, indicating increased dispersal ability in expanding populations (Fig. 18.4). There were, however, interactions between country and site, indicating that colonization effects were stronger in England than in Scotland. This difference between England and Scotland may be due to several factors. First, the Scotland core site has been established for only approximately 50 years, and so may still retain some characteristics of more recently colonized sites and thus be more similar to range margin sites. Second, the English and Scottish sites differ in their availability of woodland habitat, with approximately 25% less woodland in England (Hill *et al.*, 2001), and fragmentation of breeding habitat has been related to dispersal ability in butterflies (Thomas *et al.*, 1998a; Hill *et al.*, 1999c), including *P. aegeria* (Berwaerts *et al.*, 1998), which may also have contributed to differences seen here. Third, the Scotland core and margin sites were closer together, potentially resulting in increased gene flow between populations and so reducing any differences between the sites compared with sites in England.

4.2 Evolutionary trade-offs between flight and reproduction at expanding range margins

Trade-offs between dispersal and reproduction have been shown in many insect species, particularly wing-dimorphic species (see review in Zera and Denno, 1997). We investigated whether trade-offs between dispersal and reproduction are also evident in populations of *P. aegeria* that are expanding as a consequence of climate warming (Hughes *et al.*, 2003). Previous studies have used measures of abdomen mass to investigate fecundity in females, assuming that larger abdomens equate to higher fecundity (e.g. Hill *et al.*, 1999b,c). However, results based on such data are often inconclusive, raising doubts as to whether abdomen mass is a good surrogate for more direct measures of fecundity. We measured fecundity in *P. aegeria* females derived from insects collected from core and range margin sites in England and Scotland (the same sites as used in the dispersal study above; Fig. 18.1). On emergence, adult females were paired with males and each pair was kept separately on potted larval host plants (*Poa pratensis*) and allowed to lay eggs.

Overall, females from English sites were more fecund than those from Scotland, and females from core sites laid significantly more eggs than females from range margin sites (Fig. 18.4), indicating reduced investment in repro-

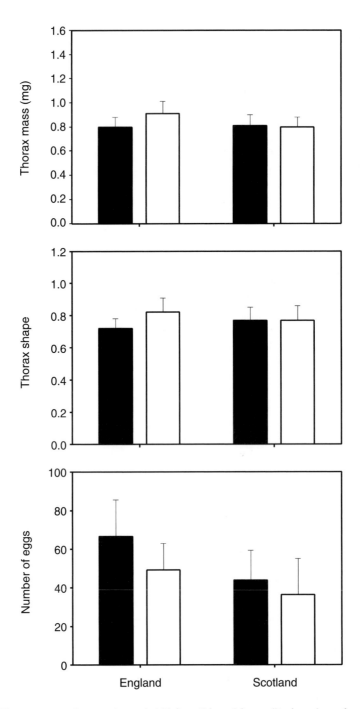

Fig. 18.4. Thorax mass, thorax shape (width/length) and fecundity (number of eggs laid in first 7 days after mating) of female *Pararge aegeria* from populations at core (solid bars) and range margin (hollow bars) sites in England and Scotland. Means and SDs are shown. Graph reproduced from Hughes *et al.* (2003).

duction at the range margin. Again, this effect was greater in England than in Scotland. None of these effects, however, were evident from adult morphology data and there were no differences in relative abdomen mass of females between sites or countries. These data suggest that measures of abdomen mass may not accurately reflect differences in fecundity; a lack of any difference between populations in abdomen mass may not necessarily mean no differences in fecundity.

As with the dispersal study above, differences between range and core sites were more pronounced in England than in Scotland. As discussed above, this may reflect differences in effective time-since-colonization between the two countries; the Scotland core site, which has been established for only approximately 50 years, may still retain some characteristics of more recently colonized sites. The lower fecundity in Scotland versus England populations overall may be due to differences in sizes of refuge populations during early-20th-century range contractions. Current populations in Scotland derive from a small refuge population around Oban (western Scotland). Although the size of this refuge population is not known, it is likely to have undergone a more severe genetic bottleneck than did populations in England, which are derived from more extensive refuge populations in Wales and southwest England. These potential differences in genetic diversity may have affected fecundity through inbreeding depression (Charlesworth and Charlesworth, 1987). It is also possible that rearing conditions may have been more favourable for individuals from the England rather than the Scotland populations, resulting in higher fecundity in England. The results presented above provide evidence of the expected trade-off between flight and reproduction (Zera and Denno, 1997). These results also indicate that any benefits of increased dispersal ability of individuals at range margins, and thus increased potential to colonize new habitats, will be offset by reduced fecundity in these colonizing individuals and reduced growth rates in newly established populations.

4.3 Theoretical studies of evolution of dispersal

The empirical data presented above indicate that evolutionary increases in dispersal occur at expanding range margins. However, the spatial and temporal scales over which these effects may be evident is not clear. The results discussed above also indicate that habitat fragmentation may affect evolution of dispersal. In order to study these factors, we developed an individual-based spatially explicit model to simulate evolution of dispersal in an expanding population (C.L. Hughes, C. Dytham and J.K. Hill, unpublished data). The model has been described and tested in detail elsewhere (Travis and Dytham, 1998, 1999, 2002; Travis et al., 1999) and so we will only briefly describe it here. The landscape is represented by a square grid measuring 300×300 cells. Each cell potentially contains a habitat patch and all patches have the same density of individuals at equilibrium. The model is highly stylized for simplicity and simulates expansion in an asexual haploid species with discrete populations. Individuals are born, undergo local 'contest' competition, disperse, reproduce and die. Each individual has a probability of dispersing into an adjacent cell from 0 (no offspring disperse) to 1 (all offspring disperse). If an individual disperses, the adjacent cell it moves into is chosen at random. Offspring inherit their dispersal genotype from their parent with a small (3%) probability of error. We assume a cost to dispersal of 0.3 (i.e. 30% of dispersers die without reproducing), a growth rate of ten offspring per parent, and an equilibrium patch density of ten individuals. Individuals die if they disperse into cells with no habitat.

We used the model to investigate impacts of habitat availability on evolution of dispersal. We investigated the effect of: (i) altering habitat availability (seven treatments, ranging from 100% availability to 40% availability); and (ii) altering the distribution of habitat (two treatments; random distribution of habitat across the grid, or declining gradient of habitat from seed location to margin; C.L. Hughes, C. Dytham and J.K. Hill, unpublished data). Each realization of the model was seeded with ten individuals placed in a single cell near the bottom of the grid. The model was then run for 200 generations and data on dispersal ability were collected along a straight-line transect from the seed location to the range margin. After 200 generations, individuals had not reached cells at the top of the grid, and there were still empty cells available for colonization. Models were also run for 1500 generations in order to investigate evolution of dispersal ability in a fully occupied landscape.

During range expansion, output from the simulations show that, for all levels of habitat availability and distribution, there was an increase in dispersal ability from the seed location to the range margin (Fig. 18.5A,C). However, the distribution of habitat affected the pattern of this relationship; in a random landscape, dispersal increased more or less linearly with distance from seed location, whereas in a gradient landscape increased dispersal was most evident close to the range margin. This difference between the two landscapes was more pronounced at increasing levels of habitat loss. Figure 18.5 shows changes in dispersal ability along a transect from seed location to range margin in random (Fig. 18.5A) and gradient landscapes (Fig. 18.5C) with 70% availability of habitat. These patterns were not observed after 1500 generations of the model in fully occupied landscapes, where dispersal rates generally were similar across the grid (Fig. 18.5B,D). At the highest level of habitat loss (40%) in the random landscape, the probability of dispersal was greatly decreased, approaching zero, presumably because most migrants failed to arrive in cells with any habitat and so removing any benefits of dispersal in such landscapes.

These results from the model simulations indicate that evolutionary increases in dispersal at range margins would be expected to be a general phenomenon in species that are expanding their ranges. However, evolution of dispersal in the model was affected by interactive effects between habitat fragmentation and distribution which affected the spatial pattern over which these effects were manifest. At the highest levels of habitat loss, dispersal declined greatly; thus those specialist species which are already facing widespread loss of their breeding habitats may also experience evolutionary decreases in dispersal ability, which will result in isolated patches being even less likely to be colonized by migrants.

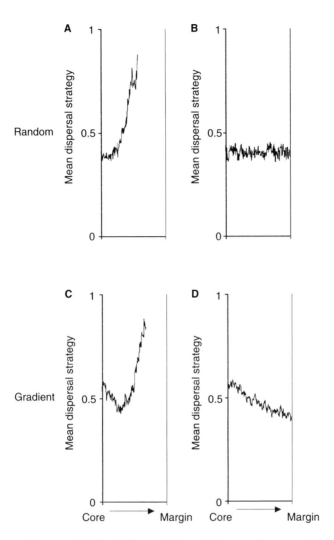

Fig. 18.5. Output from model simulating evolution of dispersal in expanding (**A**, **C**) and equilibrium (**B**, **D**) populations. Results show dispersal potential when habitat is distributed at random (**A**, **B**) or distributed in a decreasing gradient (**C**, **D**) across the grid. Mean values for ten replicate model runs are shown. Models were run with seven treatment values for habitat availability (100%–40% availability, see text), but we only show results for 70% habitat availability.

4.4 Evolutionary changes in larval host-plant use in expanding populations

Many species have more restricted niches towards the edges of their ranges (Hengeveld and Haeck, 1982; Brown, 1984), resulting in more patchy distributions and lower abundance at range margins (Thomas *et al.*, 1998b). Butterflies, for example, are often restricted to particularly warm microhabitats at their cool, northern range margins, but are able to occupy a wider variety of habitats towards the core of their range. For instance, *P. aegeria* occurs only in sheltered woodlands at its range margins but is found in an increasingly wide variety of habitats, in addition to woodlands, towards the core of its range (Hill *et al.*, 1999a). Thus, we might expect climate warming to benefit some species through increasing niche breadth and abundance at range margins, thus assisting species in tracking climate warming. The fact that many species have failed to do so (Warren *et al.*, 2001) indicates that for many species these changes do not compensate for the large-scale losses of natural and semi-natural habitats (Asher *et al.*, 2001). In addition, relationships between range expansion and availability of habitat may not be linear (Hill *et al.*, 2001), with the possibility that relatively small changes in habitat availability around a critical threshold might be crucial to whether or not species are able to expand their distributions. Thus, increases in the range of host plants used, or incorporation of more widespread host-plants into the larval diet, could have non-linear effects on niche breadth, resulting in greater-than-expected range expansions.

For example, *Polygonia c-album* (comma) has undergone marked changes in its UK distribution over the past 150 years, similar to those of *Pararge aegeria*. *Polygonia c-album* occurred throughout most of the British Isles during the early 19th century, but it subsequently declined and by the early 20th century was restricted to a small area of southwest England and Wales (Pratt, 1986/1987). During this period, larvae of *P. c-album* were commonly found on hops (*Humulus lupulus*) and its decline was associated both with a decline in the planting of hops as well as cooler temperatures. However, from the early 20th century onwards, the distribution of *P. c-album* expanded in the UK, and its range margin has shifted northwards by approximately 200 km in the past 60 years (Asher *et al.*, 2001); patterns of current range expansion are consistent with recent climate warming (Warren *et al.*, 2001). Nonetheless, the larval host-plant species on which *P. c-album* larvae develop are reported to have changed; most populations now exploit common nettle (*Urtica dioica*) (Asher *et al.*, 2001), an ancestral host-plant commonly used by other closely related Nymphalid species (Janz *et al.*, 2001). *Urtica dioica* is ubiquitous in the UK, and thus the ability of *P. c-album* to keep track of recent climate warming may have been facilitated by its exploitation of alternative host-plant species. Changes in larval host-plant use have also been associated with greater-than-expected current range expansions in other butterfly species such as *Aricia agestis* (brown argus; Thomas *et al.*, 2001). These results are in contrast with most other butterfly species, such as *Pararge aegeria*, for which there is no evidence of host shifts, and where distributions are currently lagging behind climate warming (Hill *et al.*, 1999a). Results do, nonetheless, raise the possibility that some species may show unexpected responses to future climate warming that could not have been predicted from past distribution changes.

5. Conservation Implications

Understanding how species will respond to future climate warming, and making realistic predictions of distribution of biodiversity, are crucial tasks for conservationists and ecologists. Current evidence indicates that many insect species are failing to shift their distributions because of the loss and fragmentation of breeding habitats (Warren *et al.*, 2001). Theoretical and empirical evidence presented in this chapter suggest that evolutionary increases in dispersal are expected at expanding range margins, which may help counteract these effects and make it more likely that isolated habitat patches will be colonized by migrants with increased dispersal ability. However, trade-offs between flight and reproduction show that these migrants would have reduced fecundity; and any benefits of increased dispersal ability would be balanced by reduced population growth in newly established populations. Most species that are currently expanding their ranges during current climate warming are widespread, generalist species. This indicates that communities will become increasingly dominated by generalists in the future, as specialists continue to decline (Hill *et al.*, 2002). Availability of habitat appears to be the most important factor determining whether or not species can respond to current climate changes, but evolutionary changes in expanding populations may affect patterns and rates of range change and may produce unexpected range expansions in some species.

Acknowledgements

Much of the research reported here has been carried out in collaboration with Chris Thomas (University of Leeds), Brian Huntley (University of Durham), Martin Warren and Richard Fox (Butterfly Conservation) and funded by NERC grants GR9/3016 and GR3/12542. C.L.H. was funded by a NERC studentship.

References

Asher, J., Warren, M., Fox, R., Harding, P., Jeffcoate, G. and Jeffcoate, S. (2001) The *Millennium Atlas of Butterflies in Britain and Ireland*. Oxford University Press, Oxford, UK.

Barbour, D.A. (1986) Expansion in range of the speckled wood butterfly, *Pararge aegeria* L., in northeast Scotland. *Entomologists Record* 98, 98–105.

Berwaerts, K., Van Dyck, H., Dongen, S. and Matthysen, E. (1998) Morphological and genetic variation in the speckled wood butterfly (*Pararge aegeria*) among differently fragmented landscapes. *Netherlands Journal of Zoology* 48, 241–253.

Brown, J.H. (1984) On the relationship between abundance and distribution. *American Naturalist* 124, 255–279.

Chai, P. and Srygley, R.B. (1990) Predation and the flight, morphology, and temperature of neotropical rain-forest butterflies. *American Naturalist* 135, 748–765.

Charlesworth, D. and Charlesworth, B. (1987) Inbreeding depression and its evolutionary consequences. *Annual Review of Ecology and Systematics* 18, 237–268.

Coope, R. (1995) The effects of Quaternary climatic change on insect populations: lessons from the past. In: Harrington, R. and Stork, N.E. (eds) *Insects in a Changing Environment: 17th Symposium of the Royal Entomological Society*. Academic Press, London, pp. 29–48.

Dempster, J.P., King, M.L. and Lakhani, K.H. (1976) The status of the swallowtail butterfly in Britain. *Ecological Entomology* 1, 71–84.

Dingle, H. (1986) Evolution and genetics of insect migration. In: Danthanarayana, W. (ed.) *Insect Flight*. Springer, Berlin, pp. 11–26.

Dudley, R. (1990) Biomechanics of flight in neotropical butterflies: morphometrics and kinematics. *Journal of Experimental Biology* 150, 37–53.

Hanski, I., Breuker, C.J., Schops, K., Setchfield, R. and Nieminen, M. (2002) Population history and life history influence the migration rate of female Glanville fritillary butterflies. *Oikos* 98, 87–97.

Hengeveld, R. and Haeck, J. (1982) The distribution of abundance. I. Measurements. *Journal of Biogeography* 9, 303–316.

Hill, J.K., Thomas, C.D. and Blakeley, D.S. (1999a) Evolution of flight morphology in a butterfly that has recently expanded its geographic range. *Oecologia* 121, 165–170.

Hill, J.K., Thomas, C.D. and Huntley, B. (1999b) Climate and habitat availability determine 20th century changes in a butterfly's range margins. *Proceedings of the Royal Society of London, Series B* 266, 1197–1206.

Hill, J.K., Thomas, C.D. and Lewis, O.T. (1999c) Flight morphology in fragmented populations of a rare British butterfly, *Hesperia comma*. *Biological Conservation* 87, 277–283.

Hill, J.K., Collingham, Y.C., Thomas, C.D., Blakeley, D.S., Fox, R., Moss, D. and Huntley, B. (2001) Impacts of landscape structure on butterfly range expansion. *Ecology Letters* 4, 313–321.

Hill, J.K., Thomas, C.D., Fox, R., Telfer, M.G., Willis, S.G., Asher, J. and Huntley, B. (2002) Responses of butterflies to twentieth century climate warming: implications for future ranges. *Proceedings of the Royal Society of London, Series B* 269, 2163–2171.

Houghton, J. (1997) *Global Warming: The Complete Briefing*, 2nd edn. Cambridge University Press, Cambridge, UK.

Hughes, C.L., Hill, J.K. and Dytham, C. (2003) Evolutionary trade-offs between reproduction and dispersal in populations at expanding range boundaries. *Proceedings of the Royal Society of London, Series B* 270, 147–150.

Hughes, L. (2000) Biological consequences of global warming: is the signal already apparent? *Trends in Ecology and Evolution* 15, 57–61.

Hulme, M. and Barrow, E. (1997) *Climates of the British Isles; Present, Past and Future*. Routledge, London, UK.

IPCC (2001) *Climate Change 2001: The Scientific Basis*. Cambridge University Press, Cambridge, UK.

Janz, N., Nyblom, K. and Nylin, S. (2001) Evolutionary dynamics of host-plant specialization: a case study of the tribe Nymphalini. *Evolution* 55, 783–796.

Jones, M.J. and Lace, L.A. (1992) The speckled wood butterflies *Pararge xiphia* and *P. aegeria* (Satyridae) on Madeira: distribution, territorial behaviour and possible competition. *Biological Journal of the Linnean Society* 46, 77–89.

Leslie, J.F. (1990) Geographic and genetic structure of the life-history variation in milkweed bugs (Hemiptera: Lygaeidae, *Oncopeltus*). *Evolution* 44, 295–304.

McCarty, J.P. (2001) Ecological consequences of recent climate change. *Conservation Biology* 15, 320–331.

Niemela, J. and Spence, J.R. (1991) Distribution and abundance of an exotic ground-beetle (Carabidae): a test of community impact. *Oikos* 62, 351–359.

Parmesan, C. (1996) Climate and species' range. *Nature* 382, 765–766.

Parmesan, C. and Yohe, G. (2003) A globally coherent fingerprint of climate change impacts across natural systems. *Nature* 421, 37–42.

Parmesan, C., Ryrholm, N., Stefanescu, C., Hill, J.K., Thomas, C.D., Descimon, H.,

Huntley, B., Kaila, L., Kullberg, J., Tammaru, T., Tennant, J., Thomas, J.A. and Warren, M. (1999) Polewards shifts in geographical ranges of butterfly species associated with regional warming. *Nature* 399, 579–583.

Pratt, C. (1986/1987) A history and investigation into the fluctuations of *Polygonia c-album* L. *Entomologists Record and Journal of Variation* 98, 197–203, 244–250; 99, 21–27.

Root, T.L., Price, J.T., Hall, K.R., Schneider, S.H., Rosenzweig, C. and Pounds, J.A. (2003) Fingerprints of global warming on wild animals and plants. *Nature* 421, 57–60.

Roy, D.B. and Sparks, T.H. (2000) Phenology of British butterflies and climate change. *Global Change Biology* 6, 407–416.

Simberloff, D. (1981) What makes a good island colonist? In: Denno, R.F. and Dingle, H. (eds) *Insect Life History Patterns: Habitat and Geographic Variation*. Springer, Berlin, pp. 195–205.

Srygley, R.B. and Chai, P. (1990) Flight morphology of Neotropical butterflies: palatability and distribution of mass to the thorax and abdomen. *Oecologia* 84, 492–499.

Thomas, C.D., Hill, J.K. and Lewis, O.T. (1998a) Evolutionary consequences of habitat fragmentation in a localised butterfly. *Journal of Animal Ecology* 67, 485–497.

Thomas, C.D., Jordano, D., Lewis, O.T., Hill, J.K., Sutcliffe, O.L. and Thomas, J.A. (1998b) Butterfly distributional patterns, processes and conservation. In: Mace, G.M., Balmford, A. and Ginsberg, J.R. (eds) *Conservation in a Changing World: Conservation Biology Series 1*. Cambridge University Press, Cambridge, UK, pp. 107–138.

Thomas, C.D., Bodsworth, E.J., Wilson, R.J., Simmons, A.D., Davies, Z.G., Musche, M. and Conradt, L. (2001) Ecological and evolutionary processes at expanding range margins. *Nature* 411, 577–581.

Tolman, T. (1997) *Butterflies of Britain and Europe*. HarperCollins, London.

Travis, J.M.J. and Dytham, C. (1998) The evolution of dispersal in a metapopulation: a spatially explicit, individual-based model. *Proceedings of the Royal Society of London, Series B* 265, 17–23.

Travis, J.M.J. and Dytham, C. (1999) Habitat persistence, habitat availability and the evolution of dispersal. *Proceedings of the Royal Society of London, Series B* 266, 723–728.

Travis, J.M.J. and Dytham, C. (2002) Dispersal evolution during invasions. *Evolutionary Ecology Research* 4, 1119–1129.

Travis, J.M.J., Murrell, D.J. and Dytham, C. (1999) The evolution of density-dependent dispersal. *Proceedings of the Royal Society of London, Series B* 266, 1837–1842.

Walther, G.R., Post, E., Convey, P., Menzel, A., Parmesan, C., Beebee, T.J., Fromentin, J.-M., Hoegh-Guldberg, O. and Bairleen, F. (2002) Ecological responses to recent climate change. *Nature* 416, 389–395.

Warren, M.S. (and 14 others) (2001) Climate versus habitat change: opposing forces underlie rapid changes to the distribution and abundances of British butterflies. *Nature* 414, 65–69.

Zera, A.J. and Denno, R.F. (1997) Physiology and ecology of dispersal polymorphism in insects. *Annual Review of Entomology* 42, 207–231.

Index